Forging New Frontiers: Fuzzy Pioneers II

Studies in Fuzziness and Soft Computing, Volume 218

Editor-in-chief
Prof. Janusz Kacprzyk
Systems Research Institute
Polish Academy of Sciences
ul. Newelska 6
01-447 Warsaw
Poland
E-mail: kacprzyk@ibspan.waw.pl

Further volumes of this series can be found on our homepage: springer.com

Vol. 203. Gloria Bordogna, Giuseppe Psaila (Eds.)
Flexible Databases Supporting Imprecision and Uncertainty, 2006
ISBN 978-3-540-33288-6

Vol. 204. Zongmin Ma (Ed.)
Soft Computing in Ontologies and Semantic Web, 2006
ISBN 978-3-540-33472-9 Vol. 205. Mika Sato-Ilic, Lakhmi C. Jain
Innovations in Fuzzy Clustering, 2006
ISBN 978-3-540-34356-1

Vol. 206. A. Sengupta (Ed.)
Chaos, Nonlinearity, Complexity, 2006
ISBN 978-3-540-31756-2

Vol. 207. Isabelle Guyon, Steve Gunn, Masoud Nikravesh, Lotfi A. Zadeh (Eds.)
Feature Extraction, 2006
ISBN 978-3-540-35487-1

Vol. 208. Oscar Castillo, Patricia Melin, Janusz Kacprzyk, Witold Pedrycz (Eds.)
Hybrid Intelligent Systems, 2007
ISBN 978-3-540-37419-0

Vol. 209. Alexander Mehler, Reinhard Köhler
Aspects of Automatic Text Analysis, 2007
ISBN 978-3-540-37520-3

Vol. 210. Mike Nachtegael, Dietrich Van der Weken, Etienne E. Kerre, Wilfried Philips (Eds.)
Soft Computing in Image Processing, 2007
ISBN 978-3-540-38232-4

Vol. 211. Alexander Gegov
Complexity Management in Fuzzy Systems, 2007
ISBN 978-3-540-38883-8

Vol. 212. Elisabeth Rakus-Andersson
Fuzzy and Rough Techniques in Medical Diagnosis and Medication, 2007
ISBN 978-3-540-49707-3

Vol. 213. Peter Lucas, José A. Gámez, Antonio Salmerón (Eds.)
Advances in Probabilistic Graphical Models, 2007
ISBN 978-3-540-68994-2

Vol. 214. Irina Georgescu
Fuzzy Choice Functions, 2007
ISBN 978-3-540-68997-3

Vol. 215. Paul P. Wang, Da Ruan, Etienne E. Kerre (Eds.)
Fuzzy Logic, 2007
ISBN 978-3-540-71257-2

Vol. 216. Rudolf Seising
The Fuzzification of Systems, 2007
ISBN 978-3-540-71794-2

Vol. 217. Masoud Nikravesh, Janusz Kacprzyk, Lofti A. Zadeh (Eds.)
Forging the New Frontiers: Fuzzy Pioneers I, 2007
ISBN 978-3-540-73181-8

Vol. 218. Masoud Nikravesh, Janusz Kacprzyk, Lofti A. Zadeh (Eds.)
Forging the New Frontiers: Fuzzy Pioneers II, 2008
ISBN 978-3-540-73184-9

Masoud Nikravesh · Janusz Kacprzyk ·
Lofti A. Zadeh

Editors

Forging New Frontiers:
Fuzzy Pioneers II

With 177 Figures and 57 Tables

 Springer

Masoud Nikravesh
University of California Berkeley
Department of Electrical Engineering
and Computer Science - EECS
94720 Berkeley
USA
nikravesh@cs.berkeley.edu

Janusz Kacprzyk
PAN Warszawa
Systems Research Institute
Newelska 6
01-447 Warszawa
Poland
kacprzyk@ibspan.waw.pl

Lofti A. Zadeh
University of California Berkeley
Department of Electrical Engineering
and Computer Science (EECS)
Soda Hall 729
94720-1776 Berkeley
USA
zadeh@cs.berkeley.edu

Library of Congress Control Number: 2007933148

ISSN print edition: 1434-9922
ISSN electronic edition: 1860-0808
ISBN   978-3-540-73184-9 Springer Berlin Heidelberg New York

Springer is a part of Springer Science+Business Media
springer.com
© Springer-Verlag Berlin Heidelberg 2008

Typesetting: Integra Software Services Pvt. Ltd., India
Cover design: WMX Design, Heidelberg

Printed on acid-free paper   SPIN: 11865315   89/3180/Integra   5 4 3 2 1 0

Fuzzy Community

Preface

The 2005 BISC International Special Event-BISCSE'05 "FORGING THE FRONTIERS" was held in the University of California, Berkeley, "WHERE FUZZY LOGIC BEGAN, from November 3–6, 2005. The successful applications of fuzzy logic and it's rapid growth suggest that the impact of fuzzy logic will be felt increasingly in coming years. Fuzzy logic is likely to play an especially important role in science and engineering, but eventually its influence may extend much farther. In many ways, fuzzy logic represents a significant paradigm shift in the aims of computing – a shift which reflects the fact that the human mind, unlike present day computers, possesses a remarkable ability to store and process information which is pervasively imprecise, uncertain and lacking in categoricity.

The BISC Program invited pioneers, the most prominent contributors, researchers, and executives from around the world who are interested in forging the frontiers by the use of fuzzy logic and soft computing methods to create the information backbone for global enterprises. This special event provided a unique and excellent opportunity for the academic, industry leaders and corporate communities to address new challenges, share solutions, and discuss research directions for the future with enormous potential for the future of the Global Economy.

During this event, over 300 hundred researchers and executives were attended. They were presented their most recent works, their original contribution, and historical trends of the field. In addition, four high level panels organized to discuss the issues related to the field, trends, past, present and future of the field.

Panel one was organized and moderated by Janusz Kacprzyk and Vesa A. Niskanen, "SOFT COMPUTING: PAST, PRESENT, FUTURE (GLOBAL IS-SUE)". The Panelists were Janusz Kacprzyk, Vesa A. Niskanen, Didier Dubois, Masoud Nikravesh, Hannu Nurmi, Rudi Seising, Richard Tong, Enric Trillas, and Junzo Watada. The panel discussed general issues such as role of SC in social and behavioral sciences, medicin, economics and philosophy (esp. philosophy of science, methodology, and ethics). Panel also considered aspects of decision making, democracy and voting as well as manufacturing, industrial and business aspects.

Second Panel was organized and moderated by Elie Sanchez and Masoud Nikravesh, "FLINT AND SEMANTIC WEB". The Panelists were C. Carlsson, N. Kasabov, T. Martin, M. Nikravesh, E. Sanchez, A. Sheth, R. R. Yager and L.A. Zadeh. The Organizers think that, it is exciting time in the fields of Fuzzy Logic and Internet (FLINT) and the Semantic Web. The panel discussion added to the excitement, as it did focus on the growing connections between these two fields. Most of today's Web content is suitable for human consumption. The Semantic Web is presented as an extension of the current web in which information is given well-defined meaning, better enabling computers and people to work in cooperation. But while the vision of the Semantic Web and associated research attracts attention, as long as bivalent-based logical methods will be used, little progress can be expected in handling ill-structured, uncertain or imprecise information encountered in real world knowledge. During recent years, important initiatives have led to reports of connections between Fuzzy Logic and the Internet (FLINT). FLINT meetings ("Fuzzy Logic and the Internet") have been organized by BISC ("Berkeley Initiative in Soft Computing"). Meanwhile, scattered papers were published on Fuzzy Logic and the Semantic Web. Special sessions and workshops were organized, showing the positive role Fuzzy Logic could play in the development of the Internet and Semantic Web, filling a gap and facing a new challenge. Fuzzy Logic field has been maturing for forty years. These years have witnessed a tremendous growth in the number and variety of applications, with a real-world impact across a wide variety of domains with humanlike behavior and reasoning. And we believe that in the coming years, the FLINT and Semantic Web will be major fields of applications of Fuzzy Logic. This panel session discussed concepts, models, techniques and examples exhibiting the usefulness, and the necessity, of using Fuzzy Logic and Internet and Semantic Web. In fact, the question is not really a matter of necessity, but to recognize where, and how, it is necessary.

Third Panel was organized and moderated by Patrick Bosc and Masoud Nikravesh. The Panelist were P. Bosc, R. de Caluwe, D. Kraft, M. Nikravesh, F. Petry, G. de Tré, Raghu Krishnapuram, and Gabriella Pasi, FUZZY SETS IN INFORMATION SYSTEMS. Fuzzy sets approaches have been applied in the database and information retrieval areas for more than thirty years. A certain degree of maturity has been reached in terms of use of fuzzy sets techniques in relational databases, object-oriented databases, information retrieval systems, geographic information systems and the systems dealing with the huge quantity of information available through the Web. Panel discussed issues related to research works including database design, flexible querying, imprecise database management, digital libraries or web retrieval have been undertaken and their major outputs will be emphasized. On the other hand, aspects that seem to be promising for further research has been identified.

The forth panel was organized and moderated by L. A. Zadeh and Masoud Nikravesh, A GLIMPSE INTO THE FUTURE. Panelists were Janusz Kacprzyk, K. Hirota, Masoud Nikravesh, Henri Prade, Enric Trillas, Burhan Turksen, and Lotfi A. Zadeh. Predicting the future of fuzzy logic is difficult if fuzzy logic is interpreted in its wide sense, that is, a theory in which everything is or is allowed to be a matter of degree. But what is certain is that as we move further into the age of machine intelligence and mechanized decision-making, both theoretical and applied aspects of fuzzy logic will gain in visibility and importance. What will stand out is the unique capability of fuzzy logic to serve as a basis for reasoning and computation with information described in natural language. No other theory has this capability.

The chapters of the book are evolved from presentations made by selected participant at the meeting and organized in two books. The papers include report from the different front of soft computing in various industries and address the problems of different fields of research in fuzzy logic, fuzzy set and soft computing. The book provides a collection of forty two (42) articles in two volumes.

We would like to take this opportunity to thanks all the contributors and reviewers of the articles. We also wish to acknowledge our colleagues who have contributed to the area directly and indirectly to the content of this book. Finally, we gratefully acknowledge BT, OMRON, Chevron, ONR, and EECS Department, CITRIS program, BISC associate members, for the financial and technical support and specially, Prof. Shankar Sastry—CITRIS Director and former EECS Chair, for his special support for this event, which made the meeting and publication and preparation of the book possible.

November 29, 2006   Masoud Nikravesh, Lotfi A. Zadeh,
Berkeley, California   and Janusz Kacprzyk
USA   Berkeley Initiative in Soft Computing (BISC)
University of California, Berkeley

Contents

Morphic Computing: Quantum and Fields
Germano Resconi and Masoud Nikravesh 1

Decision-Based Query and Questionnaire Systems
Masoud Nikravesh 21

Qualitative Possibility Theory in Information Processing
Didier Dubois and Henri Prade 53

Fuzzy Modeling of Nonlinear Stochastic Systems by Learning from Examples
Mohammad-R. Akbarzadeh-T and Amir-H. Meghdadi 85

Conceptual Soft-Computing based Web search: FIS-CRM, FISS Metasearcher and GUMSe Architecture
José A. Olivas, Pablo J. Garcés, Javier de la Mata, Francisco P. Romero and Jesús Serrano-Guerrero 107

Mediative Fuzzy Logic: A New Approach for Contradictory Knowledge Management
Oscar Montiel, Oscar Castillo, Patricia Melin and Roberto Sepulveda 135

Fuzzy Cognitive Maps Structure for Medical Decision Support Systems
Chrysostomos D. Stylios and Voula C. Georgopoulos 151

Fuzzy-Neuro Systems for Local and Personalized Modelling
Tian Min Ma, Qun Song and Nikola Kasabov 175

Dynamic Simulation of a Supply Chain with and without Flexibility in the Lead Time Agreements
Kaj-Mikael Björk, Anton Sialiauka and Christer Carlsson 199

Robust Statistics and Fuzzy Industrial Clustering
Barbara Diaz and Antonio Morillas 219

Fuzzy Label Semantics for Data Mining
Zengchang Qin and Jonathan Lawry 237

Aggregating Subjective and Objective Measures of Web Search Quality using Modified Shimura Technique
Rashid Ali and M. M. Sufyan Beg 269

Interpolative Realization of Boolean Algebra as a Consistent Frame for Gradation and/or Fuzziness
Dragan Radojević 295

Systematic Design of a Stable Type-2 Fuzzy Logic Controller
Oscar Castillo, Luís Aguilar, Nohé Cázarez and Patricia Melin 319

Soft Computing for Intelligent Reservoir Characterization and Decision Analysis
Masoud Nikravesh 333

Fuzzy-Based Nosocomial Infection Control
Klaus-Peter Adlassnig, Alexander Blacky and Walter Koller 343

Fuzzy Association Rules for Query Refinement in Web Retrieval
M. Delgado, M.J. Martín-Bautista, D.Sánchez, J.M. Serrano and M.A. Vila .. 351

Fuzzy Logic in a Postmodern Era
Mory M. Ghomshei, John A. Meech and Reza Naderi 363

Implementation of Fuzzy Set Theory in the Field of Energy Processes Applied in Nuclear Power Technology
Traichel Anke, Kästner Wolfgang, Hampel and Rainer 377

Methods of Possibilistic Optimization with Applications
A.V. Yazenin 385

Using Nth Dimensional Fuzzy Logic for Complexity Analysis, Decision Analysis and Autonomous Control Systems
Tony Nolan OAM JP and Emily Baker 391

Pattern Trees: An Effective Machine Learning Approach
Zhiheng Huang, Masoud Nikravesh, Tamás D. Gedeon and Ben Azvine 399

List of Contributors

Antonio Morillas
Dept. of Statistics and Econometrics.
University of Malaga Plaza El
Ejido s/n 29017 Malaga, Spain
morillas@uma.es.

Anton Sialiauka
Rheinisch Westfälische Tech-
nische Hochschule, Aachen /
DE-252074 Aachen / Germany /
Anton.Sialiauka@rwth-aachen.de

Alexander Blacky
Division of Hospital Hygiene, Clinical
Institute for Hygiene and Medical
Microbiology, Medical University
of Vienna, Währinger Gürtel 18–20,
A-1090 Vienna, Austria

Amir-H. Meghdadi
Department of Electrical Engineering
Ferdowsi University of Mashhad, Iran
akbarzadeh@ieee.org,
ahmehgdadi@ee.umanitoba.ca

Barbara Diaz
BISC Program, Computer Sciences
Division, EECS Department University
of California, Berkeley, CA 94720,
USA
bdiaz@eecs.berkeley.edu

Ben Azvine
Computational Intelligence Re-
search Group, Intelligent Systems
Research Center BT Group Chief
Technology Office, British Telecom
ben.azvine@bt.com

Christer Carlsson
IAMSR / Åbo Akademi University /
Lemminkäinengatan 14, FIN-20520 /
ÅBO / Finland /
kbjork@abo.fi

Chrysostomos D. Stylios
Lab. of Knowledge and Intelligent
Computing, Dept. of Informatics and
Telecommunications Technology, TEI
of Epirus,47100 Artas, Epirus, Greece
stylios@teleinfom.teiep.gr

M. Delgado
Department of Computer Science and
Artificial Intelligence, University of
Granada, 18071 Granada, Spain

Didier Dubois
IRIT - 118, route de Narbonne 31062
Toulouse Cedex 4, France
{dubois, prade}@irit.fr

Dragan Radojević
Mihajlo Pupin Institute, Volgina 15,
11000 Belgrade, Serbia & Montenegro
Dragan.Radojevic@automatika.
imp.bg.ac.yu

Emily Baker
Decision Intelligence
Group Sydney Australia
Emily.Baker@optusnet.com.au

Francisco P. Romero
Department of Computer Science,
Universidad de Castilla La
Mancha, Paseo de la Universi-
dad 4, 13071-Ciudad Real, Spain,
`JoseAngel.Olivas@uclm.es`
Soluziona, Centro Mixto de
Investigación y Desarrollo
Soluciona-UCLM, Ronda de Toledo
s/n, 13003-Ciudad Real, Spain,
`fpromero@soluziona.com`

Germano Resconi
Catholic University, Brescia , Italy,
`resconi@numerica.it`

Hampel
University of Applied Sciences
Zittau/Goerlitz, Institute of Process
Technology, Process Automation, and
Measuring Technology Zittau, Germany
`a.traichel@hs-zigr.de`

Henri Prade
IRIT - 118, route de Narbonne 31062
Toulouse Cedex 4, France {dubois,
`prade}@irit.fr`

Javier de la Mata
Department of Computer Science,
Universidad de Castilla La
Mancha, Paseo de la Universidad
4, 13071-Ciudad Real, Spain,
`JoseAngel.Olivas@uclm.es`

Jesús Serrano-Guerrero
Department of Computer Science,
Universidad de Castilla La
Mancha, Paseo de la Universidad
4, 13071-Ciudad Real, Spain,
`JoseAngel.Olivas@uclm.es`

John A. Meech
The University of British Columbia,
The Centre for Environmental Research
in Minerals, Metals, and Materials,
Vancouver, British Columbia, V6T 1Z4,
Canada
`mory<@>interchange.ubc.ca`,
`cerm3dir<@>mining.ubc.ca`

Jonathan Lawry
Artificial Intelligence Group, Depart-
ment of Engineering Mathematics,
University of Bristol, BS8 1TR, UK.
`j.lawry@bris.ac.uk`

José A. Olivas
Department of Computer Science,
Universidad de Castilla La
Mancha, Paseo de la Universidad
4, 13071-Ciudad Real, Spain,
`JoseAngel.Olivas@uclm.es`

Kaj-Mikael Björk
IAMSR / Åbo Akademi University /
Lemminkäinengatan 14, FIN-20520 /
ÅBO / Finland /
`kbjork@abo.fi`

Kästner Wolfgang
University of Applied Sciences
Zittau/Goerlitz, Institute of Process
Technology, Process Automation, and
Measuring Technology Zittau, Germany
`a.traichel@hs-zigr.de`

Klaus-Peter Adlassnig
Section on Medical Expert and
Knowledge-Based Systems, Core Unit
for Medical Statistics and Informatics,
Medical University of Vienna, A-1090
Vienna, Austria

Luís Aguilar
Tijuana Institute of Technology,
Calzada del Tecnológico S/N, Tijuana,
BC, México

M. J. Martín-Bautista
Department of Computer Science
and Artificial Intelligence, University
of Granada, 18071 Granada, Spain
`mbautis@decsai.ugr.es`

Masoud Nikravesh,
BISC Program, Computer Sciences
Division, EECS Department University
of California, Berkeley, and Life
Sciences-Lawrence Berkeley
National Lab., CA 94720, USA
Nikravesh@eecs.berkeley.edu

Mohammad-R. Akbarzadeh-T.
Department of Electrical Engineering
Ferdowsi University of Mashhad, Iran
akbarzadeh@ieee.org,
ahmehgdadi@ee.umanitoba.ca

Mory M. Ghomshei
The University of British Columbia,
The Centre for Environmental
Research in Minerals, Metals,
and Materials, Vancouver, British
Columbia, V6T 1Z4, Canada
mory<@>interchange.ubc.ca,
cerm3dir<@>mining.ubc.ca

Nikola Kasabov
Knowledge Engineering & Discov-
ery Research Institute Auckland
University of Technology Private
Bag 92006, Auckland 1020,
New Zealand mmaa@aut.ac.nz,
qsong@aut.ac.nz
nkasabov@aut.ac.nz

Nohé Cázarez
Tijuana Institute of Technology,
Calzada del Tecnológico S/N, Tijuana,
BC, México

Oscar Castillo
Tijuana Institute of Technology,
Calzada del Tecnológico S/N, Tijuana,
BC, México

Oscar Castillo
Department of Computer Science,
Tijuana Institute of Technology, P.O.
Box 4207, Chula Vista CA 91909,
USA.

Oscar Montiel
CITEDI-IPN, Av. del Parque #1310,
Mesa de Otay Tijuana, B. C., Méxi-co.

Pablo J. Garcés
Department of Computer Science and
Artificial Intelligence. Universidad de
Alicante, Carretera San Vicente del
Raspeig s/n, 03080 – Alicante, Spain,
pgarces@dccia.ua.es

Patricia Melin
Tijuana Institute of Technology,
Calzada del Tecnológico S/N, Tijuana,
BC, México

Patricia Melin
Department of Computer Science,
Tijuana Institute of Technology, P.O.
Box 4207, Chula Vista CA 91909,
USA.

Qun Song
Knowledge Engineering & Discovery
Research Institute Auckland University
of Technology Private Bag 92006,
Auckland 1020, New Zealand
mmaa@aut.ac.nz,
qsong@aut.ac.nz
nkasabov@aut.ac.nz

Rainer
University of Applied Sciences
Zittau/Goerlitz, Institute of Process
Technology, Process Automation, and
Measuring Technology Zittau, Germany
a.traichel@hs-zigr.de

Rashid Ali
Department of Computer Engineering,
A. M. U. Aligarh, 202002, India,
rashidaliamu@rediffmail.com,
pet04msb@amu.ac.in

Reza Naderi
Department of Philosophy, University
of Toronto, Toronto, ON, Canada
naderi<@>acumentechnologies.com

Roberto Sepulveda
CITEDI-IPN, Av. del Parque #1310,
Mesa de Otay Tijuana, B. C., Méxi-co.

D. Sánchez
Department of Computer Science and
Artificial Intelligence, University of
Granada, 18071 Granada, Spain

J. M. Serrano
Department of Computer Science,
University of Jaén, 20371 Jaén, Spain

M. M. Sufyan Beg
Department of Computer Engineering,
A. M. U. Aligarh, 202002, India,
rashidaliamu@rediffmail.com,
pet04msb@amu.ac.in

Tamás D. Gedeon
Department of Computer Science,
The Australian National University,
Canberra, ACT 0200, Australia
tom@cs.anu.edu.au

Tian Min Ma
Knowledge Engineering & Discovery
Research Institute Auckland University
of Technology Private Bag 92006,
Auckland 1020, New Zealand
mmaa@aut.ac.nz, qsong@aut.ac.nz
nkasabov@aut.ac.nz

Tony Nolan OAM JP
Australian Taxation Office 100 Market
Street Sydney, NSW Australia 2000
t.nolan@uts.edu.au

Traichel, Anke
University of Applied Sciences
Zittau/Goerlitz, Institute of Process
Technology, Process Automation, and
Measuring Technology Zittau, Germany
a.traichel@hs-zigr.de

M. A. Vila
Department of Computer Science and
Artificial Intelligence, University of
Granada, 18071 Granada, Spain

Voula C. Georgopoulos
Department of Speech and Language
Therapy, TEI of Patras, 26334 Patras,
Greece voula@teipat.gr

Walter Koller
Division of Hospital Hygiene, Clinical
Institute for Hygiene and Medical
Microbiology, Medical University
of Vienna, Währinger Gürtel 18–20,
A-1090 Vienna, Austria

A. V. Yazenin
Computer Science Department, Tver
State University, Ul. Zhelyabova 33,
170000 Tver,
Russia.
Alexander.Yazenin@tversu.ru

Zengchang Qin
Berkeley Initiative in Soft Computing
(BISC), Computer Science Division,
EECS Department, University of
California, Berkeley, CA 94720, US.
zqin@eecs.berkeley.edu

Zhiheng Huang
Electrical Engineering and Computer
Science, University of California
at Berkeley, CA 94720, USA
{zhiheng,nikravesh}@cs.
berkeley.edu

Morphic Computing: Quantum and Fields

Germano Resconi and Masoud Nikravesh

Abstract *Morphic Computing* was first introduced by Resconi and Nikravesh [2006, 2007a, and 2007b] which makes use of Morphic and Morphogenetic Fields. *Morphic Fields* and its subset *Morphogenetic Fields* have been at the center of controversy for many years in mainstream science and the hypothesis is not accepted by some scientists who consider it a pseudoscience. Resconi and Nikravesh [2006, 2007a, and 2007b] claim that *Morphic Computing is* a natural extension of Holographic Computation, Quantum Computation, Soft Computing, and DNA Computing. In this paper, we introduce extensions of the Morphic Computing such as *Quantum Logic and Entanglement in Morphic Computing, Morphic Systems and Morphic System of Systems (M-SOS)*. Its applications to the field of computation by words as an example of the *Morphic Computing, Morphogenetic Fields in neural network and Morphic Computing, Morphic Fields - concepts and Web search, and agents and fuzzy in Morphic Computing will also be discussed.*

Key words: Morphic Computing · Morphogenetic Computing · Morphic Fields · Morphogenetic Fields · Quantum Computing · DNA Computing · Soft Computing · Computing with Words · Morphic Systems · Morphic Neural Network · Morphic System of Systems · Optical Computation by Holograms · Holistic Systems

1 Introduction

Morphic Fields and it's subset *Morphogenetic Fields* have been at the center of controversy for many years in mainstream science and the hypothesis is not accepted by some scientists who consider it a pseudoscience. *Morphogenetic Fields* is a hypothetical biological fields and it has been used by environmental biologists since 1920's which deals with living things. However, *Morphic Fields* are more general than *Morphogenetic Fields* and are defined as universal information for both organic (living things) and abstract forms.

We claim that Morphic Fields reshape multidimensional space to generate local contexts. For example, the gravitational fields in general relativity are the *Morphic Fields* that reshape space-time space to generate local context where particles move.

M. Nikravesh et al. (eds.), *Forging the New Frontiers: Fuzzy Pioneers II.*

Morphic Computing reverses the ordinary N input basis fields of possible data to one field in the output system. The set of input fields form a N dimension space or context. The N dimensional space can be obtained by a deformation of an Euclidean space and we argue that the *Morphic Fields* is the cause of the deformation. In line with Rupert Sheldrake [1981] our *Morphic Fields* is the formative causation of the context.

Resconi and Nikravesh [2006, 2007a, and 2007b] claimed that *Morphic Computing* is a natural extension of Optical Computation by holograms [Gabor 1972] and holistic systems [Smuts 1926] however in form of complex System of Systems [Kotov 1997], Quantum Computation [Deutsch 1985, Feynman, 1982, Omnès 1994, Nielsen and Chuang 2000], Soft Computing [Zadeh 1991], and DNA Computing [Adleman 1994]. All natural computation bonded by the Turing Machine [Turing 1936–7 and 1950] such as classical logic and AI [McCarthy et al. 1955] can be formalized and extended by our new type of computation model – *Morphic Computing*. In holistic systems (biological, chemical, social, economic, mental, linguistic, etc.) one cannot determined or explained the properties of the system by the sum of its component parts alone; *"The whole is more than the sum of its parts"* (Aristotle).

In this paper, we introduce extensions of Morphic Computing such as Quantum Logic and Entanglement in Morphic Computing, Morphic Systems and Morphic System of Systems (M-SOS). Then Morphic Computing's applications to the field of computation by words [Zadeh and Kacprzyk 1999a and 1999b, Zadeh and Nikravesh 2002] will be given. Finally, we present *Morphogenetic Fields* in the neural network and *Morphic Computing, Morphic Fields* - concepts and Web search, and Agents [Resconi and Jain 2004] and fuzzy [Zadeh, 1965] in *Morphic Computing*.

2 The Basis for Morphic Computing

The basis for *Morphic Computing* is *Field Theory* and more specifically *Morphic Fields*. *Morphic Fields* were first introduced by Rupert Sheldrake [1981] from his hypothesis of formative causation [Sheldrake, 1981 and 1988] that made use of the older notion of *Morphogenetic Fields*. Rupert Sheldrake developed his famous theory, Morphic Resonance [Sheldrake, 1981 and 1988], on the basis of the work by French philosopher Henri Bergson [1896, 1911]. *Morphic Fields* and its subset *Morphogenetic Fields* have been at the center of controversy for many years in mainstream science. The hypothesis is not accepted by some scientists who consider it a pseudoscience.

Gabor [1972] and H. Fatmi and Resconi [1988] discovered the possibility of computing images made by a huge number of points as output from objects as a set of huge number of points as input by reference beams or laser (holography). It is also known that a set of particles can have a huge number of possible states that in classical physics, separate one from the other. However, only one state (position and velocity of the particles) would be possible at a time. With respect to quantum mechanics, one can have a superposition of all states with all states presented in the superposition at the same time. It is also very important to note that at the same time one cannot separate the states as individual entities but consider them as one entity. For this very peculiar property of quantum mechanics, one can change all the superpose states at the same time. This type of global computation is the conceptual principle by which we think one can build quantum computers. Similar phenomena can be used to develop DNA computation where a huge number of DNA as a field of DNA elements are transformed (replication) at the same time and filtered (selection) to solve non polynomial problems. In addition, soft-computing or computation by words extend the classical local definition of true and false value for logic predicate to a field of degree of true and false inside the space of all possible values of the predicates. In this way, the computational power of soft computing is extended similar to that which one can find in quantum computing, DNA computing, and Holographic computing. In conclusion, one can expect that all the previous approaches and models of computing are examples of a more general computation model called *"Morphic Computing"* where "Morphic" means "form" and is associated with the idea of holism, geometry, field, superposition, globality and so on.

3 Quantum Logic and Entanglement in Morphic Computing

In the Morphic Computing, we can make computation on the context H as we make the computation on the Hilbert Space. Now we have the algebras among the context or spaces H. in fact we have

$$H = H_1 \oplus H_2 \text{ where } \oplus \text{ is the direct sum. For example given}$$
$$H_1 = (h_{1,1}, h_{1,2}, \ldots\ldots, h_{1,p}) \text{ where } h_{1,k} \text{ are the basis fields in } H_1$$
$$H_2 = (h_{2,1}, h_{2,2}, \ldots\ldots, h_{2,q}) \text{ where } h_{2,k} \text{ are the basis fields in } H_2$$

So we have

$$H = H_1 \oplus H_2 = (h_{1,1}, h_{1,2}, \ldots\ldots, h_{1,p}, h_{2,1}, h_{2,2}, \ldots\ldots, h_{2,q})$$

The intersection among the context is

$$H = H_1 \cap H_2$$

The space H is the subspace in common to H_1 , H_2 . In fact for the set V_1 and V_2 of the vectors

$$V_1 = S_{1,1} h_{1,1} + S_{1,2} h_{1,2} + \ldots\ldots + S_{1,p} h_{1,p}$$
$$V_2 = S_{2,1} h_{2,1} + S_{2,2} h_{2,2} + \ldots\ldots + S_{2,q} h_{2,q}$$

The space or context $H = H_1 \cap H_2$ include all the vectors in $V_1 \cap V_2$.

Given a space H, we can also built the orthogonal space $H^\perp$ of the vectors that are all orthogonal to any vectors in H.

No we have this logic structure

$$Q\,(H_1 \oplus H_2) = Q_1 \vee Q_2 = Q_1 \text{ OR } Q_2$$

where Q_1 is the projection operator on the context H_1 and Q_2 is the projection operator on the context H_2.

$$Q\,(H_1 \cap H_2) = Q_1 \wedge Q_2 = Q_1 \text{ AND } Q_2$$

$$Q\,(H^{\perp)} = Q - ID = \neg Q = \text{NOT } Q$$

In fact we know that $Q\,X - X = (Q - ID)\,X$ is orthogonal to Y and so orthogonal to H. In this case the operator $(Q - ID)$ is the not operator.

Now it easy to show [9] that the logic of the projection operator is isomorphic to the quantum logic and form the operator lattice for which the distributive low (interference) is not true. In Fig. 1, we show an expression in the projection lattice for the Morphic Computing

Now we give an example of the projection logic and lattice in this way :

Given the elementary field references

$$H_1 = \begin{bmatrix} 1 \\ 0 \end{bmatrix},\; H_2 = \begin{bmatrix} 0 \\ 1 \end{bmatrix},\; H_3 = \begin{bmatrix} \dfrac{1}{\sqrt{2}} \\ \dfrac{1}{\sqrt{2}} \end{bmatrix}$$

For which we have the projection operators

$$Q_1 = H_1(H_1{}^T\,H_1)^{-1}H_1{}^T = \begin{bmatrix} 1 & 0 \\ 0 & 0 \end{bmatrix}$$

$$Q_2 = H_2(H_2{}^T\,H_2)^{-1}H_2{}^T = \begin{bmatrix} 0 & 0 \\ 0 & 1 \end{bmatrix}$$

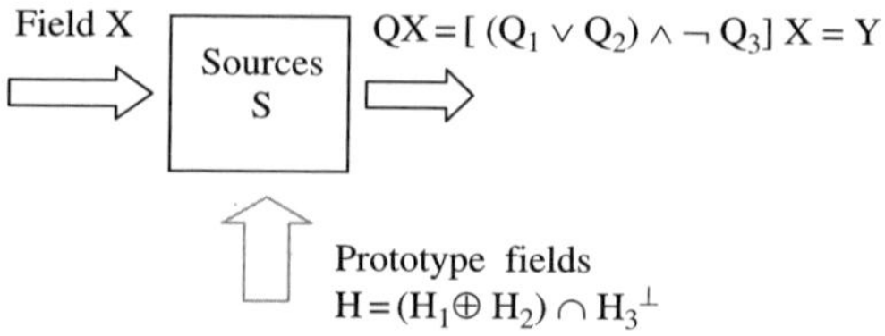

Fig. 1 Expressions for projection operator that from the Morphic Computing Entity

$$Q_3 = H_3(H_3{}^T H_3)^{-1} H_3{}^T = \begin{bmatrix} \dfrac{1}{2} & \dfrac{1}{2} \\ \dfrac{1}{2} & \dfrac{1}{2} \end{bmatrix}$$

With the lattice logic we have

$$H_{1,2} = H_1 \oplus H_2 = \begin{bmatrix} 1 & 0 \\ 0 & 1 \end{bmatrix}, \; Q_1 \vee Q_2 = H_{1,2}(H_{1,2}{}^T H_{1,2})^{-1} H_{1,2} = \begin{bmatrix} 1 & 0 \\ 0 & 1 \end{bmatrix}$$

$$H_{1,3} = H_1 \oplus H_3 = \begin{bmatrix} 1 & \dfrac{1}{\sqrt{2}} \\ 0 & \dfrac{1}{\sqrt{2}} \end{bmatrix}, \; Q_1 \vee Q_3 = H_{1,3}(H_{1,3}{}^T H_{1,3})^{-1} H_{1,3} = \begin{bmatrix} 1 & 0 \\ 0 & 1 \end{bmatrix}$$

$$H_{2,3} = H_2 \oplus H_3 = \begin{bmatrix} 0 & \dfrac{1}{\sqrt{2}} \\ 1 & \dfrac{1}{\sqrt{2}} \end{bmatrix}, \; Q_2 \vee Q_3 = H_{2,3}(H_{2,3}{}^T H_{2,3})^{-1} H_{2,} = \begin{bmatrix} 1 & 0 \\ 0 & 1 \end{bmatrix}$$

And

$$H = H_1 \cap H_2 = H_1 \cap H_3 = H_2 \cap H_3 = \begin{bmatrix} 0 \\ 0 \end{bmatrix}$$

$$and \; Q_1 \wedge Q_2 = Q_1 \wedge Q_3 = Q_2 \wedge Q_3 = \begin{bmatrix} 0 & 0 \\ 0 & 0 \end{bmatrix}$$

And in conclusion we have the lattice

We remark that

$$(Q_1 \vee Q_2) \wedge Q_3 = Q_3 \text{ but } (Q_1 \wedge Q_3) \vee (Q_2 \wedge Q_3) = 0 \vee 0 = 0$$

When we try to separate Q_1 from Q_2 in the second expression the result change. Between Q_1 and Q_2 we have a connection or relation (Q_1 and Q_2 generate the two dimensional space) that we destroy when we separate one from the other. In fact $Q_1 \wedge Q_3$ project in the zero point. A union of the zero point cannot create the two dimensional space. the non distributive property assume that among the projection operators there is an entanglement or relation that we destroy when we separate the operators one from the other.

Given two references or contexts H_1, H_2 the tensor product $H = H_1 \otimes H_2$ is the composition of the two independent contexts in one.

We can prove that the projection operator of the tensor product H is the tensor product of Q_1 , Q_2 . So we have

$$Q = H(H^{\mathrm{T}}H)^{-1}H^{\mathrm{T}} = Q_1 \otimes Q_2$$

The sources are $S^{\alpha\beta} = S^{\alpha}{}_1 S^{\beta}{}_2$
So we have

$$Y = Y_1 \otimes Y_2 = (H_1 \otimes H_2)S^{\alpha\beta}$$

The two Morphic System are independent from one another. The output is the product of the two outputs for the any Morphic System. Now we give some examples

$$H_1 = \begin{bmatrix} 1 & \dfrac{1}{2} \\ \dfrac{1}{2} & 1 \\ 1 & \dfrac{1}{2} \end{bmatrix}, H_2 = \begin{bmatrix} \dfrac{1}{2} & 1 \\ 0 & \dfrac{1}{2} \\ \dfrac{1}{2} & \dfrac{1}{2} \end{bmatrix}$$

$$H_1 \oplus H_2 = H_1{}^{\alpha,\beta} H_2{}^{\gamma,\delta} = \begin{bmatrix} 1 H_2 & \dfrac{1}{2} H_2 \\[2ex] \dfrac{1}{2} H_2 & 0 H_2 \\[2ex] 1 H_2 & \dfrac{1}{2} H_2 \end{bmatrix} = \begin{bmatrix} 1 \begin{bmatrix} \dfrac{1}{2} & 1 \\[1ex] 0 & \dfrac{1}{2} \\[1ex] \dfrac{1}{2} & \dfrac{1}{2} \end{bmatrix} & \dfrac{1}{2} \begin{bmatrix} \dfrac{1}{2} & 1 \\[1ex] 0 & \dfrac{1}{2} \\[1ex] \dfrac{1}{2} & \dfrac{1}{2} \end{bmatrix} \\[6ex] \dfrac{1}{2} \begin{bmatrix} \dfrac{1}{2} & 1 \\[1ex] 0 & \dfrac{1}{2} \\[1ex] \dfrac{1}{2} & \dfrac{1}{2} \end{bmatrix} & 1 \begin{bmatrix} \dfrac{1}{2} & 1 \\[1ex] 0 & \dfrac{1}{2} \\[1ex] \dfrac{1}{2} & \dfrac{1}{2} \end{bmatrix} \\[6ex] 1 \begin{bmatrix} \dfrac{1}{2} & 1 \\[1ex] 0 & \dfrac{1}{2} \\[1ex] \dfrac{1}{2} & \dfrac{1}{2} \end{bmatrix} & \dfrac{1}{2} \begin{bmatrix} \dfrac{1}{2} & 1 \\[1ex] 0 & \dfrac{1}{2} \\[1ex] \dfrac{1}{2} & \dfrac{1}{2} \end{bmatrix} \end{bmatrix}$$

$$= H^{\alpha,\beta,\gamma,\delta}$$

At every value of the basis fields we associate the basis field H_2 multiply for the value of the basis fields in H_1. Now because we have

$$Q_1 = H_1 (H_1^T H_1)^{-1} H_1^T = \begin{bmatrix} \dfrac{1}{2} & 0 & \dfrac{1}{2} \\[2ex] 0 & 1 & 0 \\[2ex] \dfrac{1}{2} & 0 & \dfrac{1}{2} \end{bmatrix}$$

$$Q_2 = H_2 (H_2^T H_2)^{-1} H_2^T = \begin{bmatrix} \dfrac{2}{3} & \dfrac{1}{3} & \dfrac{1}{3} \\[2ex] \dfrac{1}{3} & \dfrac{2}{3} & -\dfrac{1}{3} \\[2ex] \dfrac{1}{3} & \dfrac{1}{3} & \dfrac{2}{3} \end{bmatrix}$$

$$Q = Q_1 \oplus Q_2 = \begin{bmatrix} \frac{1}{2}Q_2 & 0Q_2 & \frac{1}{2}Q_2 \\ 0Q_2 & 1Q_2 & 0Q_2 \\ \frac{1}{2}Q_2 & 0Q_2 & \frac{1}{2}Q_2 \end{bmatrix}$$

$$X_1 = \begin{bmatrix} \frac{1}{3} \\ 0 \\ \frac{1}{3} \end{bmatrix}, X_2 = \begin{bmatrix} \frac{1}{4} \\ 0 \\ \frac{1}{4} \end{bmatrix}, X = X_1 \oplus X_2 = \begin{bmatrix} \frac{1}{3}X_2 \\ 0X_2 \\ \frac{1}{3}X_2 \end{bmatrix}$$

$$S_1 = (H_1^T H_1)^{-1} H_1^T X_1 = \begin{bmatrix} \frac{4}{9} \\ -\frac{2}{9} \end{bmatrix}, S_2 = (H_2^T H_2)^{-1} H_2^T X_1 = \begin{bmatrix} -\frac{3}{2} \\ \frac{4}{3} \end{bmatrix}$$

And

$$S^{\alpha\beta} = S_1 \otimes S_2 = \begin{bmatrix} \frac{4}{9}S_2 \\ -\frac{2}{9}S_2 \end{bmatrix}$$

For

$$Y_1 = H_1 S_1 = \begin{bmatrix} \frac{1}{3} \\ 0 \\ \frac{1}{3} \end{bmatrix}, Y_2 = H_2 S_2 = \begin{bmatrix} \frac{7}{12} \\ \frac{2}{3} \\ -\frac{1}{12} \end{bmatrix}$$

$$Y = Y_1 \otimes Y_2 = \begin{bmatrix} \frac{1}{3}Y_2 \\ 0Y_2 \\ \frac{1}{3}Y_2 \end{bmatrix}$$

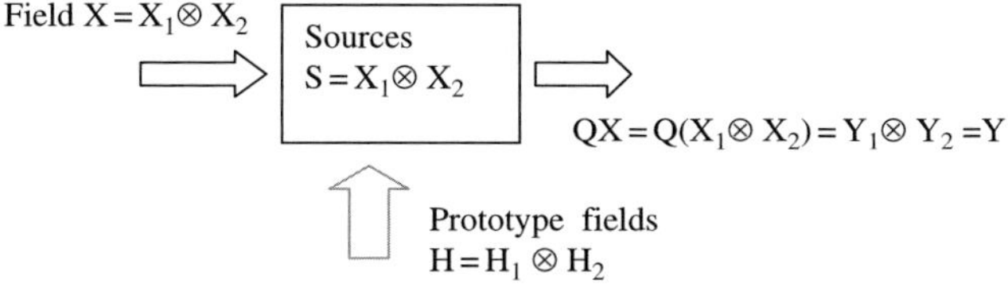

Fig. 2 Tensor product for independent projection operators or measures

In conclusion, we can say that the computation of Q, S and Y by H and X can be obtained only with the results of Q_1, S_1, Y_1 and Q_2, S_2, Y_2 independently. Now when H and X cannot be write as a scalar product of H_1, X_1 and H_2, X_2, the context H and the input X are not separable in the more simple entities so are in entanglement.

In conclusion with the tensor product, we can know if two measures or projection operators are dependent or independent from one another.

So we have for the tensor product of the context that all the other elements of the entity are obtained by the tensor product (Fig. 2).

4 Field Theory, Concepts and Web Search

To search in the web, we use the term-document matrix to obtain the information retrieved. In Table 1, we show the data useful in obtaining the desired information in the Web.

Where $K_{i,j}$ is the value of the word$_j$ in the document$_i$. The word in Table 1 is a source of a field which values are the values in the position space of the documents. Any document is one of the possible positions of the word. In a Web search, it is useful to denote the words and complex text in a symbolic form as *queries*, the*answers* are the fields generated by the words or text as sources. Any ontology map is a conceptual graph in RDF language where we structure the query as a structured variable. The conceptual map in Fig. 3 is the input X of a complex field obtained by the superposition of the individual words in the map.

With the table two we have that the library H is the main reference for which we have

Table 1 Term (word), document and complex text G

	Word$_1$	Word$_2$	...	Word$_N$	Concept X
Document$_1$	$K_{1,1}$	$K_{1,2}$	...	$K_{1,N}$	X_1
Document$_2$	$K_{2,1}$	$K_{2,2}$	...	$K_{2,N}$	X_2
...	...	...	...	...	...
Document$_M$	$K_{M,1}$	$K_{M,2}$	...	$K_{M,N}$	X_M

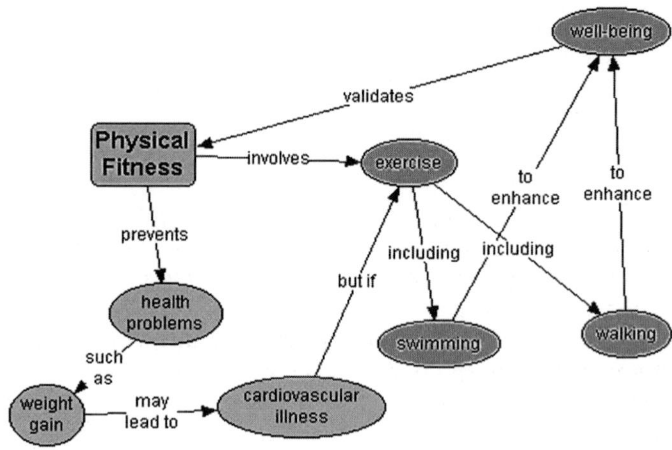

Fig. 3 Conceptual map as structured query. The map is a structured variable whose answer or meaning is the field G in the documents space located in the Web

$$
H = \begin{bmatrix}
K_{1,1} & K_{1,2} & \dots & K_{1,N} \\
K_{2,1} & K_{2,2} & \dots & K_{2,N} \\
\dots & \dots & \dots & \dots \\
K_{M,1} & K_{M,2} & \dots & K_{M,N}
\end{bmatrix}
$$

Where M > N.

For any new concept in input $X = \begin{bmatrix} X_1 \\ X_2 \\ \dots \\ X_M \end{bmatrix}$ we can found the suitable sources S by which we can project X into the knowledge inside the library H.

The library can be represented in a geometric way in this form (Fig. 4)

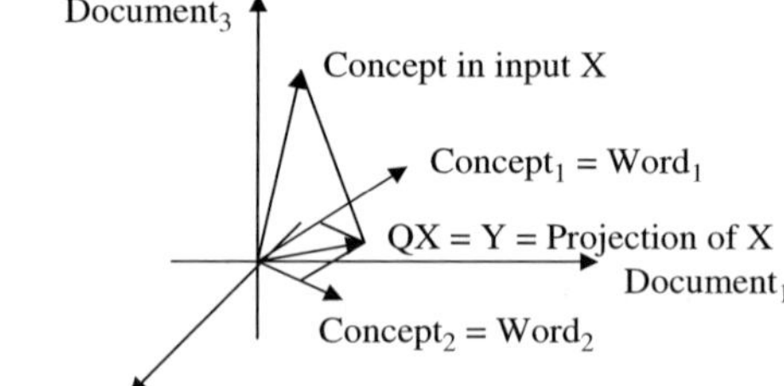

Fig. 4 Geometric representation of the Library (Documents) and concepts (words)

Recently much work has been done to realise conceptual maps oflexicon for natural language. This work has a foundation in Description Logic. With a conceptual map of natural language and Description Logic, we can create a semantic web. The problem with a semantic web is the integration of different concepts in a geometric frame. We are in the same situation as the quantum computer where the Geometric Hilber space is the geometric tool to show the global properties of the classical logic and quantum logic. We propose in this paper to give a geometric image of knowledge. The meaning of the words and concepts are not located in the axiomatic description logic, in the conceptual map or in the lexical structure. The meaning is inside the space of the fields of instances of the words or documents. A prototype set of concepts, inside the space of the documents as object, generate an infinite number of other concepts in a space. Any concept inside the space of the concepts is obtained by an integration process of the prototype concepts. Now any new concept can be compared with the concepts that belong to the space of the concepts. The MS can reshape, with the minimum change the new concept X into another concept Y. Concept Y belongs to the set of concepts obtained by the integration of the prototype concepts. Y is inside the space of the prototype fields that represent prototype concepts.

MS can compute the part of X that cannot be integrated in the space of the concepts. Transformation of the space of the concepts generate a set of new prototype concepts by which we can represent the dynamical process of the concepts and the possible conceptual invariance. In this way a huge set of concepts can be reduced to a more simple set of prototype concepts and invariance. In the semantic web based on description logic and the conceptual map, we cannot reduce the concepts to a simple set of basic concepts which. With the description of invariance conceptual frame and the type of geometry (metric, Euclidean and non -Eucliden) we believe we can improve the primitive semantic structure base on the lexical description of the concepts.

5 Morphogenetic Field in Neural Network and Morphic Computing: Morphic Systems and Morphic System of Systems (M-SOS)

We know that in traditional neural network one of the problems was the realisation of the Boolean function XOR. The solution of this problem was realised by the well known back propagation. Now the physiological realisation in the brain of the Boolean functions are obtained by the genetic change on the synaptic form. The morphogenetic field is the biochemical system that change the synaptic structure. The solution of the XOR problem and other similar problems is relate to the hidden layer of neurons which activity is not used directly to the output computation but are an intermediary computation useful to obtain the wanted output. In Fig. 5, we show the neural network that realise the function

$$Z = X \text{ XOR } Y = X \oplus Y$$

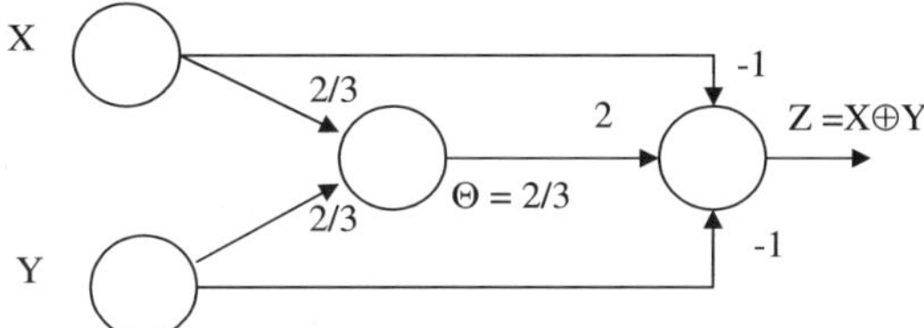

Fig. 5 Neural network for the Boolean function XOR

For the previous neural network we have

$$Z = X \oplus Y = \Phi\left(\frac{2}{3}X + \frac{2}{3}Y - \frac{2}{3}\right) - (X + Y)$$

The weights for the neural network can be obtained by different methods as back propagation or others.

Now we study the same Boolean function by the Morphic Computing network or System of System. In fact at the begin we consider the simple Morphic System with

$$H = X \cup Y = \begin{bmatrix} 0 \\ 0 \\ 1 \\ 1 \end{bmatrix} \cup \begin{bmatrix} 0 \\ 1 \\ 0 \\ 1 \end{bmatrix} = \begin{bmatrix} 0 & 0 \\ 0 & 1 \\ 1 & 0 \\ 1 & 1 \end{bmatrix}, Input = Z = X \oplus Y = \begin{bmatrix} 0 \\ 1 \\ 1 \\ 0 \end{bmatrix}$$

where the input of the Morphic System is the field of all possible values for Z. The field reference H or the context is made by the fields one are all possible values for the input X and the other are all possible values for Y.

With the context H (field references) and the input Z we compute the sources and the output. So we have

$$S = (H^T H)^{-1} H^T Input = \begin{bmatrix} \dfrac{1}{3} \\ \dfrac{1}{3} \end{bmatrix}, Output = HS = \frac{1}{3}\begin{bmatrix} 0 \\ 0 \\ 1 \\ 1 \end{bmatrix} + \frac{1}{3}\begin{bmatrix} 0 \\ 1 \\ 0 \\ 1 \end{bmatrix} = \begin{bmatrix} 0 \\ \dfrac{1}{3} \\ \dfrac{1}{3} \\ \dfrac{2}{3} \end{bmatrix}$$

we remark that the space generate by the fields X and Y in H is given by the superposition

$$Output = S_1 \begin{bmatrix} 0 \\ 0 \\ 1 \\ 1 \end{bmatrix} + S_2 \begin{bmatrix} 0 \\ 1 \\ 0 \\ 1 \end{bmatrix} = \begin{bmatrix} 0 \\ S_2 \\ S_1 \\ S_1 + S_2 \end{bmatrix}$$

Now we know that the Q projection operator for

$$H = X \cup Y = \begin{bmatrix} 0 \\ 0 \\ 1 \\ 1 \end{bmatrix} \cup \begin{bmatrix} 0 \\ 1 \\ 0 \\ 1 \end{bmatrix} = \begin{bmatrix} 0 & 0 \\ 0 & 1 \\ 1 & 0 \\ 1 & 1 \end{bmatrix} \tag{1}$$

is

$$Q = H(H^T H)^{-1} H^T = \begin{bmatrix} 0 & 0 & 0 & 0 \\ 0 & \dfrac{2}{3} & -\dfrac{1}{3} & \dfrac{1}{3} \\ 0 & -\dfrac{1}{3} & \dfrac{2}{3} & \dfrac{1}{3} \\ 0 & \dfrac{1}{3} & \dfrac{1}{3} & \dfrac{2}{3} \end{bmatrix}$$

Because the determinant of Q is equal to zero, the colon vectors are not independent so for

$$H = \begin{bmatrix} \dfrac{2}{3} & -\dfrac{1}{3} \\ -\dfrac{1}{3} & \dfrac{2}{3} \\ \dfrac{1}{3} & \dfrac{1}{3} \end{bmatrix}, \ X = \begin{bmatrix} \dfrac{1}{3} \\ \dfrac{1}{3} \\ \dfrac{2}{3} \end{bmatrix}$$

we have

$$S = (H^T H)^{-1} H^T X = \begin{bmatrix} 1 \\ 1 \end{bmatrix} \text{ and } y_4 = S_1 y_2 + S_2 y_3 = y_2 + y_3 \qquad (2)$$

So we have the constrain by the projection operator that any Y must have this form.

$$Y = \begin{bmatrix} 0 \\ y_2 \\ y_3 \\ y_2 + y_3 \end{bmatrix} \qquad (3)$$

Now because the output that we want must have this form

$$y_2 > 0, \ y_3 > 0, \ y_2 + y_3 < min(y_2, y_3)$$

but this is impossible so the projection operator and reference H that we use put a limitation or a constraint that the XOR Boolean function cannot satisfy.

Now for the reference or context $H = \begin{bmatrix} 0 & 0 & 0 \\ 0 & 1 & 1 \\ 1 & 0 & 1 \\ 1 & 1 & 1 \end{bmatrix}$

That we can reduce to $H = \begin{bmatrix} 0 & 1 & 1 \\ 1 & 0 & 1 \\ 1 & 1 & 1 \end{bmatrix}$ Now because H is a quadratic matrix, in the projection we have no limitation or constrain so when in input we have

$$Input = Z = X \oplus Y = \begin{bmatrix} 0 \\ 1 \\ 1 \\ 0 \end{bmatrix}$$

by the projection operator Q we have that the output = input. In this case we can compute the sources S in a traditional way for the context

$$H = \begin{bmatrix} 0 & 1 & 1 \\ 1 & 0 & 1 \\ 1 & 1 & 1 \end{bmatrix}$$

The sources are

$$S = (H^T H)^{-1} H^T X = \begin{bmatrix} -1 \\ -1 \\ 2 \end{bmatrix}$$

Now given the reference (4) we have the constrain (5). Now for the input

$$Input = \begin{bmatrix} 0 \\ 1 \\ 1 \\ 1 \end{bmatrix} \tag{4}$$

The output is (6) that is coherent or similar to the input. In fact we have that for

$$Y = \begin{bmatrix} 0 \\ y_2 \\ y_3 \\ y_2 + y_3 \end{bmatrix}$$

because in input we have the property

$x_1 = 0, x_2 > 0, x_3 > 0, x_4 > 0$ and in output we have the same property $y_1 = 0,$ $y_2 > 0, y_3 > 0, y_4 = y_2 + y_3 > 0$

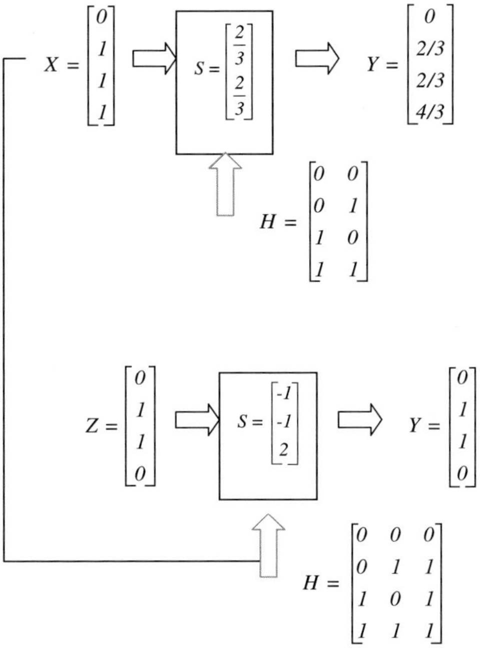

$$X = \begin{bmatrix} 0 \\ 1 \\ 1 \\ 1 \end{bmatrix} \Rightarrow S = \begin{bmatrix} \dfrac{2}{3} \\ \dfrac{2}{3} \end{bmatrix} \Rightarrow Y = \begin{bmatrix} 0 \\ 2/3 \\ 2/3 \\ 4/3 \end{bmatrix}$$

$$H = \begin{bmatrix} 0 & 0 \\ 0 & 1 \\ 1 & 0 \\ 1 & 1 \end{bmatrix}$$

$$Z = \begin{bmatrix} 0 \\ 1 \\ 1 \\ 0 \end{bmatrix} \Rightarrow S = \begin{bmatrix} -1 \\ -1 \\ 2 \end{bmatrix} \Rightarrow Y = \begin{bmatrix} 0 \\ 1 \\ 1 \\ 0 \end{bmatrix}$$

$$H = \begin{bmatrix} 0 & 0 & 0 \\ 0 & 1 & 1 \\ 1 & 0 & 1 \\ 1 & 1 & 1 \end{bmatrix}$$

Fig. 6 Network of morphic systems or system of systems

For the context (4) and the input (1) the sources are $S = (H^T H)^{-1} H^T X = \begin{bmatrix} \dfrac{2}{3} \\ \dfrac{2}{3} \end{bmatrix}$ In conclusion we found the same weights as in the Fig. 6 with the Morphic System.

We can conclude that in the ordinary back propagation we compute the weights in neuron network by a recursive correction of the errors in output. In Morphic Computing we are interested in the similarity between the input and output. So we are interested more in the morphos (form or structure) that in the individual values in input and output.

6 Agents and Fuzzy in Morphic Computing

Given the word "warm" and three agents that are named Carlo, Anna and Antonio (Fig. 7), we assume that any agent is a source for a Boolean fields in Fig. 8.

The superposition of the three Boolean fields in Fig. 7 is the fuzzy set

$$\mu(A, B) = S_1 F_1(A, B) + S_2 F_2(A, B) + S_3 F_3(A, B) \tag{5}$$

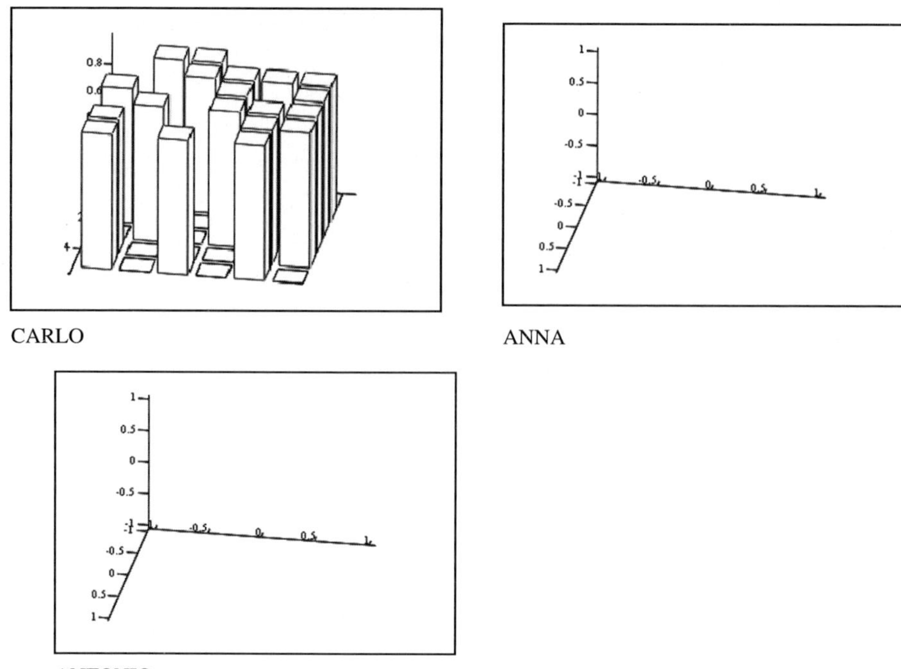

Fig. 7 Three Boolean fields $F_1(A, B)$, $F_2(A, B)$, $F_3(A, B)$ for three agents

At the word "warm" that is the query we associate the answer $\mu(A, B)$ that is a field. With the superposition process, we can introduce the logic of the fuzzy fields in this way.

When the logic variables values for p agents assume the usual 1 and 0 (boolean fields), we have the operations

$$(X_1, X_2, \ldots, X_p)(Y_1, Y_2, \ldots, Y_p) = (Y_1 X_1, Y_2 X_2, \ldots, Y_p X_p)$$
$$\overline{(X_1, X_2, \ldots, X_p) + (Y_1, Y_2, \ldots, Y_p) = (Y_1 + X_1, Y_2 + X_2, \ldots, Y_p + X_p)} \quad (6)$$
$$\overline{(X_1, X_2, \ldots, X_p)} = (\overline{X_1}, \overline{X_2}, \ldots, \overline{X_p})$$

That are the vector product, the vector sum and the vector negation.

For any vector we can compute the scalar product

Fig. 8 Because in the projection operation we loss information the original boolean field for the input agent X is transformed in the fuzzy field $QX = Y = \mu(X) = S_1 X_1 + S_2 X_2 + \ldots + S_p X_p$

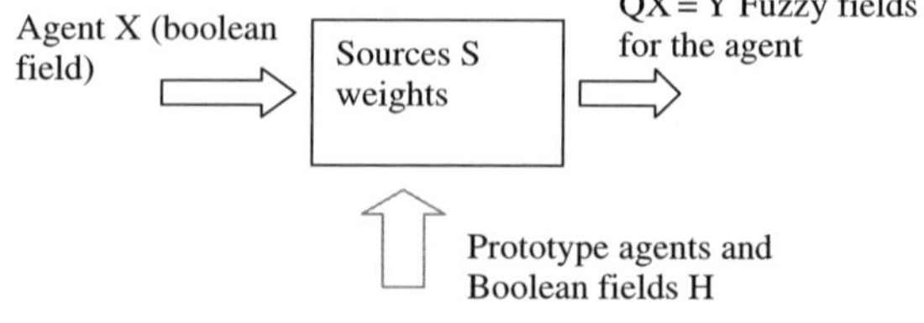

$$\mu(X_1, X_2, \ldots, X_p) = \mu(X) = \frac{m_1 X_1 + m_2 X_2 + \ldots + m_p X_p}{m_1 + m_2 + \ldots + m_p} =$$

$$S_1 X_1 + S_2 X_2 + \ldots + S_p X_p$$

where μ is the membership function and m_k where $k = 1, \ldots, p$ are the weights of any component in the Boolean vector. So the membership function is the weighted average of the 1 or 0 value of the variables X_k

We can show that

$$\text{when } \mu(Y_1, \ldots, Y_p) \leq \mu(X_1, \ldots, X_p) \text{ we have}$$

$$\mu(Y_1 X_1, \ldots, Y_p X_p) = \min[\mu(X), \mu(Y)] - \mu(Y\overline{X}) \tag{7}$$

$$\mu(Y_1 + X_1, \ldots, Y_p + X_p) = \max[\mu(X), \mu(Y_1)] + \mu(Y\overline{X}) \tag{8}$$

In fact we have $X_k Y_k + \overline{X_k} Y_k = Y_k$ and $X_k Y_k = Y_k - \overline{X_k} Y_k$ in the same way we have

$$X_k + Y_k = X_k + \overline{X_k} Y_k + X_k Y_k = X_k(1 + Y_k) + \overline{X_k} Y_k = X_k + \overline{X_k} Y_k$$

because in the fuzzy set for the Zadeh rule we have

$$\mu[(X_1, \ldots, X_p)\overline{(X_1, \ldots, X_p)}] = \min[\mu(X_1, \ldots, X_p), \mu\overline{(X_1, \ldots, X_p)}] > 0$$
$$\mu[(X_1, \ldots, X_p)\overline{(X_1, \ldots, X_p)}] = \max[\mu(X_1, \ldots, X_p), \mu\overline{(X_1, \ldots, X_p)}] < 1$$

the negation in (5) is not compatible with the other operations. So in the decomposition of the vector

$$\overline{(X_1, X_2, \ldots, X_p)} = (F(X_1), F(X_2), \ldots, F(X_p)) \tag{9}$$

at the negation $\overline{X_k}$ we substitute the negation

$$F(X_k) = \overline{X_k} \oplus \Gamma_k \text{ and } \oplus \text{ is the XOR operation}$$

We remark that when $\Gamma_k = 1$ we have $F(X_k) = X_k$ when $\Gamma_k = 0$ we have $F(X_k) = \overline{X_k}$ For the negation F we have

$$F(F(X_k) = X_k..$$

In fact

$$F(F(X)) = F(\overline{X} \oplus \Gamma(X)) = \overline{(\overline{X} \oplus \Gamma(X))} \oplus \Gamma(X) = (\overline{X}\Gamma(X) + X\overline{\Gamma(X)}) \oplus \Gamma(X) =$$
$$[X + \overline{\Gamma(X)}][\overline{X} + \Gamma(X)]\Gamma(X) + (\overline{X}\Gamma(X) + X\overline{\Gamma(X)})\Gamma(X) = X\Gamma(X) + X\overline{\Gamma(X)} = X$$

So the function F is an extension of the negation operation in the classical logic.

We remark that in (9) we introduce a new logic variable S_k. The vector $\Gamma = (\Gamma_1, \Gamma_2, \ldots, \Gamma_p)$ is the inconsistent vector for which the Tautology is not always true and the contradiction is not always false as in the Zadeh rule. In fact we have for the contradiction expression

$$C = F(X_k)X_k = (\overline{X_k} \oplus \Gamma_k)X_k = X_k\Gamma_k$$

For the previous case the fuzzy contradiction C is not always equal to zero. We remember that the classical contradiction is $\overline{X_k}X_k = 0$ always. When $\Gamma = (\Gamma_1, \Gamma_2, \ldots, \Gamma_p) = (0, 0, .., 0)$ we came back to the classical negation and $C_k = 0$. For the tautology we have

$$T = F(X) + X = (\overline{X_k} \oplus \Gamma_k) + X_k = \overline{X_k\Gamma_k} + X_k$$

The reference space is the space of all possible propositions and any agent is a Boolean field or prototype field. Given a Boolean field X in input, external agent, by the projection operator we generate the fuzzy field Y in output by the sources S.

References

1. L. M. Adleman (1994) "Molecular Computation Of Solutions To Combinatorial Problems". Science (journal) 266 (11): 1021–1024 (1994).
2. H. Bergson (1896), Matter and Memory 1896. (Matière et mémoire) Zone Books 1990: ISBN 0-942299-05-1 (published in English in 1911).
3. D. Deutsch (1985). "Quantum Theory, the Church-Turing Principle, and the Universal Quantum Computer". Proc. Roy. Soc. Lond. A400, 97–117.
4. R. P. Feynman (1982). "Simulating Physics with Computers". International Journal of Theoretical Physics 21: 467–488.
5. H. A. Fatmi and G.Resconi (1988), A New Computing Principle, Il Nuovo Cimento Vol.101 B,N.2 - Febbraio 1988 - pp. 239–242
6. D. Gabor (1972), Holography 1948–1971, Proc.IEEE Vol60 pp 655–668 June 1972.
7. V. Koto (1997)v, "Systems-of-Systems as Communicating Structures," Hewlett Packard Computer Systems Laboratory Paper HPL-97–124, (1997), pp. 1–15.
8. J. McCarthy, M. L. Minsky, N. Rochester, and C.E. Shannon (1955), Proposal for Dartmouth Summer Research Project on Artificial Intelligence, August 31, 1955.
9. M. Nielsen and I. Chuang (2000). Quantum Computation and Quantum Information. Cambridge: Cambridge University Press. ISBN 0-521-63503-9.
10. M. Nikravesh, Intelligent Computing Techniques for Complex systems, in *Soft Computing and Intelligent Data Analysis in Oil Exploration2003*, pp. 651–672, Elsevier, 2003.
11. R. Omnès (1994), The Interpretation of Quantum Mechanics, Princeton Series in Physics, 1994.
13. G. Resconi and M. Nikravesh, Morphic Computing, to be published in Forging New Frontiers, 2007.
13. G. Resconi and M. Nikravesh, Morphic Computing, Book in Preparation, 2007.
14. G. Resconi and M. Nikravesh, Field Theory and Computing with Words, FLINS 2006, 7th International FLINS Conference on Applied Artificial Intelligence, August 29–31, 2006 Italy
15. G. Resconi and L. C.Jain (2004), Intelligent Agents, Springer, 2004

16. R. Sheldrake (1981), A New Science of Life: The Hypothesis of Morphic Resonance (1981, second edition 1985), Park Street Press; Reprint edition (March 1, 1995).
17. R. Sheldrake (1988), The Presence of the Past: Morphic Resonance and the Habits of Nature (1988), Park Street Press; Reprint edition (March 1, 1995).
18. J. C. Smuts (1926),Holism and Evolution, 1926 MacMillan, Compass/Viking Press 1961 reprint: ISBN 0-598-63750-8, Greenwood Press 1973 reprint: ISBN 0-8371-6556-3, Sierra Sunrise 1999 (mildly edited): ISBN 1-887263-14-4.
19. A. M. Turing (1936–7), 'On computable numbers, with an application to the Entscheidungsproblem', Proc. London Maths. Soc., ser. 2, 42: 230–265; also in (Davis 1965) and (Gandy and Yates 2001)
20. A. M. Turing (1950), 'Computing machinery and intelligence', Mind 50: 433–460; also in (Boden 1990), (Ince 1992)
21. L. A. Zadeh, Fuzzy Logic, Neural Networks and Soft Computing, Communications of the ACM, 37(3):77–84, 1994.
22. L. A. Zadeh and M. Nikravesh (2002), Perception-Based Intelligent Decision Systems; Office of Naval Research, Summer 2002 Program Review, Covel Commons, University of California, Los Angeles, July 30th-August 1st, 2002.
23. L. A. Zadeh, and J. Kacprzyk, (eds.) (1999a), Computing With Words in Information/Intelligent Systems 1: Foundations, Physica-Verlag, Germany, 1999a.
24. L. A. Zadeh, L. and J. Kacprzyk (eds.) (1999b), Computing With Words in Information/Intelligent Systems 2: Applications, Physica-Verlag, Germany, 1999b.
25. L. A. Zadeh, Fuzzy sets, Inf. Control 8, 338–353, 1965.

Decision-Based Query
and Questionnaire Systems

Masoud Nikravesh

Abstract Searching database records, ranking the results and making decision based on multi-criteria queries is central for many database applications and information systems used within organizations in finance, business, industrial and other fields. Most of the available systems are modeled using crisp logic and queries, which result in rigid systems with imprecise results. In this paper, we introduce an intelligent decision and information support based on evolutionary-based optimization as a flexible tool allowing approximation where the selected objects do not need to match exactly the decision criteria resembling natural human behavior.

1 Introduction

Searching database records and ranking the results based on multi-criteria queries is central for many database applications used within organizations in finance, business, industrial and other fields. Most of the available systems' 'software' are modeled using crisp logic and queries, which result in rigid systems with imprecise and subjective processes and results. In this chapter we introduce fuzzy querying and ranking as a flexible tool allowing approximation where the selected objects do not need to match exactly the decision criteria resembling natural human behavior. The model consists of five major modules: the Fuzzy Search Engine (FSE), the Application Templates (AT), the User Interface (UI), the Database (DB) and the Evolutionary Computing (EC). We developed the software with many essential features. It is built as a web-based software system that users can access and use over the Internet. The system is designed to be generic so that it can run different application domains. To this end, the Application Template module provides information of a specific application as attributes and properties, and serves as a guideline structure for building a new application. The Fuzzy Search Engine (FSE) is the core module of the system. It has been developed to be generic so that it would fit any application. The main FSE component is the query structure, which utilizes membership functions, similarity functions and aggregators. Through the user interface a user can enter and save his profile, input criteria for a new query, run different queries and display results. The user can manually eliminate the results he disapproves of or change the ranking according to his preferences. The Evolutionary Computing (EC) module monitors ranking preferences of the user's queries. It learns to adjust to the intended meaning of the users' preferences.

We present our approach with three important applications: ranking (scoring), which has been used to make financing decisions concerning credit cards, cars and mortgage loans; the process of college admissions where hundreds of thousands of applications are processed yearly by U.S. universities; and date matching as one of the most popular internet programs. Even though we implemented three applications, the system is designed in a generic form to accommodate more diverse applications and to be delivered as stand-alone software to academia and businesses.

2 BISC-DSS

The BISC-DSS Components include:

- Data Management: database(s) which contains relevant data for the decision process
- User Interface: users and DSS communication
- Model Management and Data Mining: includes software with quantitative and fuzzy models including aggregation process, query, ranking, and fitness evaluation
- Knowledge Management and Expert System

  o model representation including linguistic formulation

- Evolutionary Kernel and Learning Process
- Data Visualization and Visual Interactive Decision Making

3 Model Framework

The DSS system starts by loading the application template, which consists of various configuration files for a specific application (see Sect. 4) and initializing the database for the application (see Sect. 6), before handling a user's requests, see Fig. 1. Once the DSS system is initialized, users can enter their own profiles in the user interface or make a search with their preferences. These requests are handled by the control unit of the system. The control unit converts user input into data objects that are recognized by the DSS system. Based on the request types, it forwards them to the appropriate modules. If the user wants to create a profile, the control unit will send the profile data directly to the database module which stores the data in the database for the application. If the user wants to query the system, the control unit will direct the user's preferences to the Fuzzy Search Engine which queries the database. The query results will be sent back to the control unit and displayed to the users.

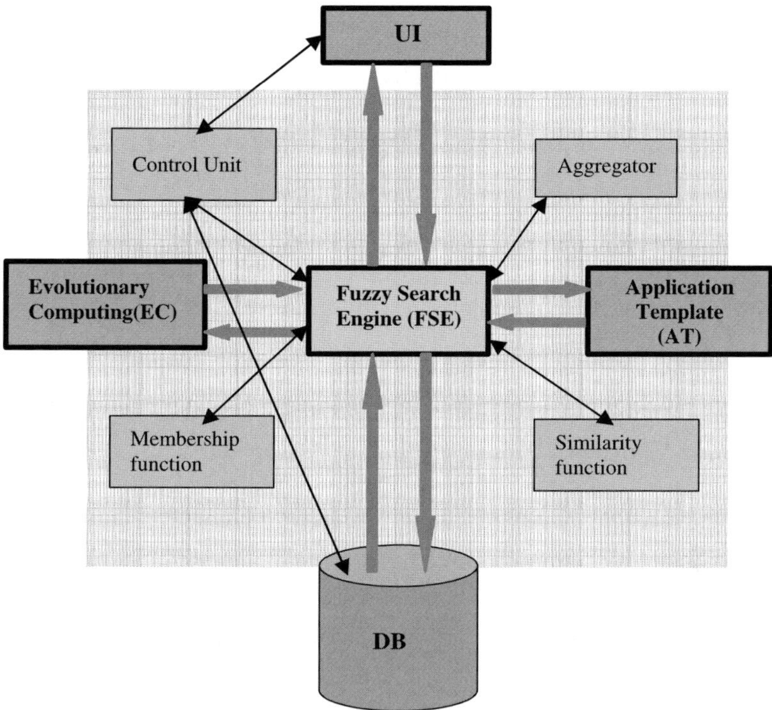

Fig. 1 The BISC-DSS general framework

4 Fuzzy Engine

Fuzzy Query, search and Ranking: To support generic queries, the fuzzy engine has been designed to have a tree structure. There are two types of nodes in the tree, category nodes and attribute nodes, as depicted in Fig. 2. While multiple category levels are not necessary, they are designed to allow various refinements of the query through the use of the type of aggregation of the children. Categories only act to aggregate the lower levels. The attribute nodes contain all the important information about a query. They contain the membership functions for the fuzzy comparison as well as the use of the various aggregation methods to compare two values. The attribute nodes handle the compare method slightly differently than the category nodes. There are two different ways attributes may be compared. The attribute nodes contain a list of membership functions comprising the fuzzy set. The degrees of membership for this set are passed to the similarity comparator object, which currently has a variety of different methods to calculate the similarity between the two membership vectors. In the other method, the membership vector created by having full membership to a single membership function specified in the fuzzy data object, but no membership value for the other functions.

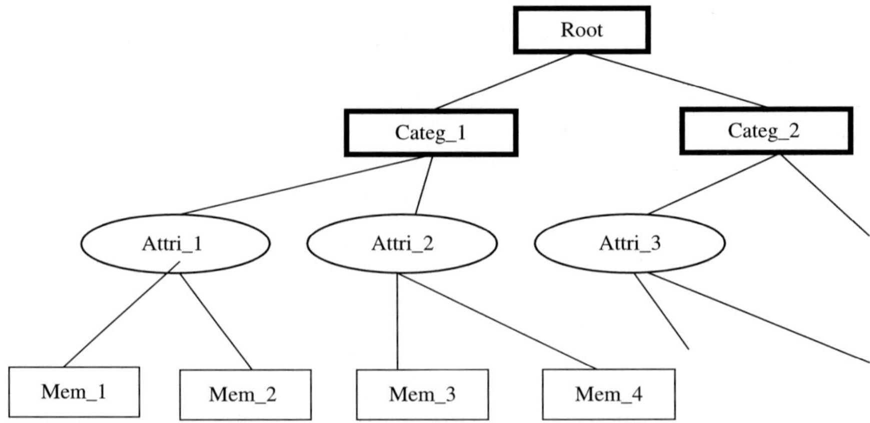

Fig. 2 The Fuzzy search engine tree structure

Membership function: Currently there are three membership functions implemented for the Fuzzy Engine. A generic interface has been created to allow several different types of membership functions to be added to the system. The three types of membership functions in the system are: Gaussian, Triangular and Trapezoidal. These functions have three main points, for the lower bound, upper bound and the point of maximum membership. For other functions, optional extra points may be used to define the shape (an extra point is required for the trapezoidal form).

5 Application Template

The DSS system is designed to work with different application domains. The application template is a format for any new application we build; it contains data of different categories, attributes and membership functions of that application. The application template module consists of two parts: 1) the application template data file specifies all the membership functions, attributes and categories of an application. We can consider it as a configuration data file for an application. It contains the definition of membership functions, attributes and the relationship between them; 2) The application template logic parses and caches data from the data file so that other modules in the system can have faster access to definitions of membership functions, attributes and categories. It also creates a tree data structure for the fuzzy search engine to transverse.

6 User Interface

It is difficult to design a generic user interface that suits different kind of applications for all the fields. For example, we may want to have different layouts for user interfaces for different applications. To make the DSS system generic while preserving

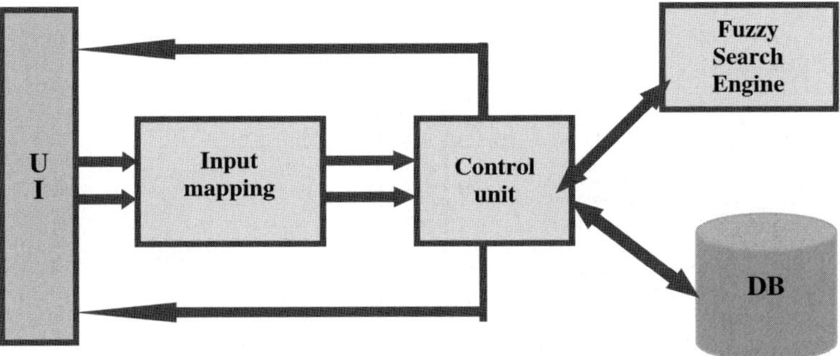

Fig. 3 User interface data flow

the user friendliness of the interfaces for different applications, we developed the
user interfaces into two parts (see Fig. 3).

First, we designed a specific HTML interface for each application we developed. Users can input their own profiles; make queries by specifying preferences
for different attributes. Details for the DSS system are encapsulated from the HTML
interface so that the HTML interface design would not be constrained by the DSS
system. The second part of our user interface module is a mapping between the
parameters in the HTML files and the attributes in the application template module
for the application. The input mapping specifies the attribute names to which each
parameter in the HTML interface corresponds.. With this input mapping, a user
interface designer can use input methods and parameter names freely.

7 Database (DB)

The database module is responsible for all the transactions between the DSS system
and the database. This module handles all queries or user profile creations from
the Fuzzy Engine and the Control Unit respectively. For queries from the Fuzzy
Search Engine, it retrieves data from the database and returns it in a data object
form. Usually queries are sets of attribute values and their associated weights. The
database module returns the matching records in a format that can be manipulated
by the user, such as eliminating one or more record or changing their order. To
create a user profile, it takes data objects from the Control Unit and stores it in the
database. There are three components in the DB module (see Fig. 4).

The DB Manager is accountable for two things: 1) setting up database connections and allocating database connections to DB Accessor objects when needed. It
also supplies information to the database for authentication purposes (e.g. username,
password, path to the database etc); 2) The DB Accessor Factory creates DB Accessor objects for a specific application. For example, if the system is running the date
matching application, DB Accessor Factory will create DB Accessor objects for the
date matching application. The existence of this class serves the purpose of using
a generic Fuzzy Search Engine; 3) the DB Accessor is responsible for storing and

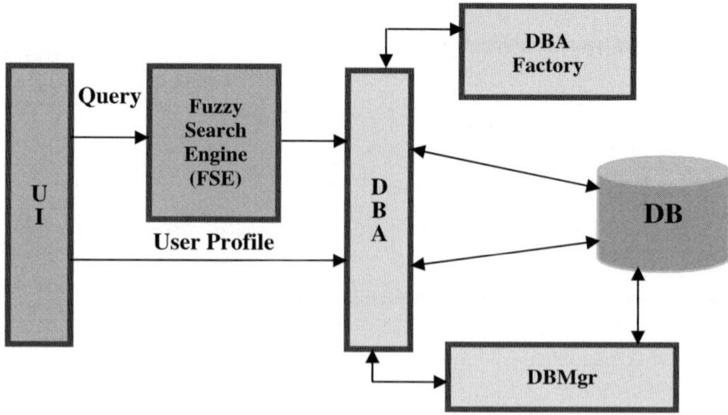

Fig. 4 Database module components

getting user profiles to and from the database. It is the component that queries the database and wrap result from the database into data objects that are recognized by our application framework.

8 Measure of Association and Fuzzy Similarity

As in crisp query and ranking, an important concept in fuzzy query and ranking applications is the measure of association or similarity between two objects in consideration. For example, in a fuzzy query application, a measure of similarity between two queries and a document, or between two documents, provides a basis for determining the optimal response from the system. In fuzzy ranking applications, a measure of similarity between a new object and a known preferred (or non-preferred) object can be used to define the relative goodness of the new object. Most of the measures of fuzzy association and similarity are simply extensions from their crisp counterparts. However, because of the use of perception based and fuzzy information, the computation in the fuzzy domain can be more powerful and more complex.

Various definitions of similarity exist in the classical, crisp domain, and many of them can be easily extended to the fuzzy domain. However, unlike in the crisp case, in the fuzzy case the similarity is defined on two fuzzy sets. Suppose we have two fuzzy sets A and B with membership functions $\mu_A(x)$ and $\mu_B(x)$, respectively. The arithmetic operators involved in the fuzzy similarity measures can be treated using their usual definitions while the union and the intersection operators need to be treated specially. It is important for these operator pairs to have the following properties: (1) conservation, (2) monotonicity, (3) commutativity, and (4) associativity. It can be verified that the triangular norm (T-norm) and triangular co-norm (T-conorm) (Nikravesh, 2001a; Mizumoto, 1989; Fagin, 1998 and 1999) conform to these properties and can be applied here. A detailed survey of some commonly used T-norm and T-conorm pairs along with other aggregation operators can be find at Nikravesh and Azvine (2002).

In many situations, the controlling parameters, including the similarity metric, the type of T-norm/conorm, the type of aggregation operator and associated weights, can all be specified based on the domain knowledge of a particular application. However, in some other cases, it may be difficult to specify a priori an optimal set of parameters. In those cases, various machine learning methods can be employed to automatically "discover" a suitable set of parameters using a supervised or un-supervised approach. For example, the Genetic Algorithm (GA) and DNA-based computing, as described in later sections, can be quite effective.

9 Evolutionary Computing

In the Evolutionary Computing (EC) module of the BISC Decision Support System, our purpose is to use an evolutionary method to allow automatic adjusting of the user's preferences. These preferences can be seen as parameters of the fuzzy logic model in form of weighting of the used variables. These preferences are then represented by a weight vector and genetic algorithms will be used to fix them.

In the Evolutionary Computation approach, Genetic Programming, which is an extension of Genetic Algorithms, is the closest technique to our purpose. It allows us to learn a tree structure, which represents the combination of aggregators. The selection of these aggregators is included to the learning process using the Genetic Programming.

In this section, we describe the GA (Holland, 1992) and GP (Koza, 1992) application to our problem. Our aim is learning fuzzy-DSS parameters which are the weight vectors representing the user preferences associated to the variables that have to be aggregated on the one hand, and the adequate decision tree representing the combination of the aggregation operators that have to be used on the other hand.

Weight vector being a linear structure, can be represented by a binary string in which weight values are converted to binary numbers. This binary string corresponds to the individual's DNA in the GA learning process. The goal is to find the optimal weighting of the variables. A general GA module can be used by defining a specific fitness function for each application as shown in Fig. 5.

Aggregators can be combined in the form of a tree structure which can be built using a Genetic Programming learning module (Figs. 6 and 7). It consists in evolving a population of individuals represented by tree structures. The evolution principle remains the same as in a conventional GP module but the DNA encoding needs to be defined according to the considered problem. We propose to define an encoding for aggregation trees which is more complex than for classical trees and which is common to all considered applications. As shown in Fig. 7, we need to define a specific encoding, in addition to the fitness function specification.

Tree structures are generated randomly as in the conventional GP. But, since these trees are augmented according the properties defined above, the generation process has to be updated. So, we decided to randomly generate the number of arguments

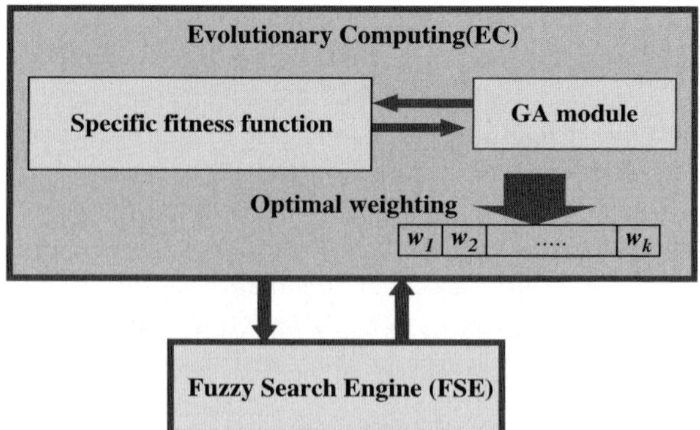

Fig. 5 Evolutionary Computing Module: preferences learning

when choosing an aggregator as a node in the tree structure. And for the weights, we chose to generate them randomly for each node during its creation.

Concerning the fitness function, it is based on performing the aggregation operation and the root node of the tree that has to be evaluated. For the university admissions application, the result of the root execution corresponds to the score that has to be computed for each value vector in the training data set. The fitness function, as in the GA learning of the user preferences, consists in simple or combined similarity measures. In addition, we can include to the fitness function a complementary measure that represents the individual's size which has to be minimized in order to avoid over-sized trees.

We have described the use of evolutionary computation methods for optimization problems in the BISC decision support system. It is an original idea in combining fuzzy logic, machine learning and evolutionary computation. We gave some implementation precisions for the university admissions application. We also plan to apply our system to many other applications.

Another important and unique component of our system is compactification algorithm or Z-Compact (Nikravesh 2005, Zadeh 1976, and Zadeh and Nikravesh 2002). The fuzzified model of Z-Compact algorithm developed by Nikravesh based on original idea by Lotfi A. Zadeh (Zadeh 1976) and it has been implemented for

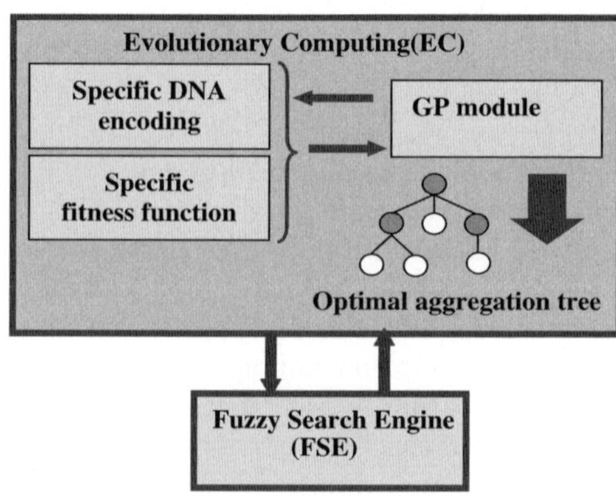

Fig. 6 Evolutionary Computing Module: aggregation tree learning

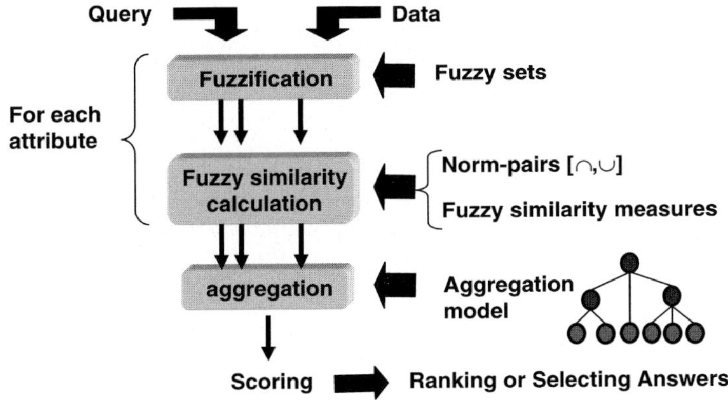

Fig. 7 Multi-Criteria Decision Model – aggregation tree learning

the first time as part of BISC-DSS for automatons multi-agents modeling as part of ONR project (Zadeh and Nikravesh 2002). The algorithm has been extended to handle fuzzy and linguistic variables and currently is part of the BISC-DSS software and it has been applied in several applications (Nikravesh 2005 and Zadeh and Nikravesh 2002). Details of step by step of this algorithm can be find (Zadeh and Nikravesh 2002 and Nikravesh 2005).

10 Compactification Algorithm

Let's consider a set of attributes Atrb1, Atrb2, . . . , AtrbN, each Atrbj defined over a discrete domain. We consider a representation space as the cartesian product of the attribute domains over these attributes, and its mapping into a set of K classes represented by class labels C1, C2, . . . , CK. The goal is to build a rule-based classifier in which each class is represented by a set of rules, and these rules are the result of the compactification of the original training data for each class. One can use Z(n)-compact algorithm to represent the Data-Attribute-Class (DAC) matrix with rule-base model. Table 2 shows the intermediate results based on Z(n)-compact algorithm for concept 1. Table 3 shows how the original DAC (Table 1) matrix is represented in final pass with a maximally compact representation. Once the DAC matrix represented by a maximally compact representation (Table 3), one can translate this compact representation with rules as presented in Tables 4 and 5. Table 6 shows the Z(n)-compact algorithm.

As it has been proposed, the DAC entries could not be crisp numbers. The following cases would be possible:

A – The basis for the k_{ii}^{jj}s are [0 and 1]. This is the simplest case and Z(n)-compact will work as presented

B – The basis for the k_{ii}^{jj}s are frequency/probability or any similar statistical based

Table 1 Data-Attribute-Class (DAC) Matrix

Data	$Atrb_1$	$Atrb_2$	$Atrb_3$	C_{class}
D_1	$a_1{}^1$	$a_1{}^2$	$a_1{}^3$	c_1
D_2	$a_1{}^1$	$a_2{}^2$	$a_1{}^3$	c_1
D_3	$a_2{}^1$	$a_2{}^2$	$a_1{}^3$	c_1
D_4	$a_3{}^1$	$a_2{}^2$	$a_1{}^3$	c_1
D_5	$a_3{}^1$	$a_1{}^2$	$a_2{}^3$	c_1
D_6	$a_1{}^1$	$a_2{}^2$	$a_2{}^3$	c_1
D_7	$a_2{}^1$	$a_2{}^2$	$a_2{}^3$	c_1
D_8	$a_3{}^1$	$a_2{}^2$	$a_2{}^3$	c_1
D_9	$a_3{}^1$	$a_1{}^2$	$a_1{}^3$	c_2
D_{10}	$a_1{}^1$	$a_1{}^2$	$a_2{}^3$	c_2
D_{11}	$a_2{}^1$	$a_1{}^2$	$a_1{}^3$	c_2
D_{12}	$a_2{}^1$	$a_1{}^2$	$a_2{}^3$	c_2

values. In this case, we use fuzzy granulation to granulate into series of granular, two ([0 or 1] or [high and low]), three (i.e. low, medium, and high), etc. Then the Z(n)-compact will work as it is presented.

C – The basis for the $k_{ii}{}^{jj}$s are set value created based on set theory which can be created based on traditional statistical based methods, human made, or fuzzy-set. In this case, the first step is to find the similarities between set values using statistical or fuzzy similarly measures. BISC-DSS software has a set of similarity measures, T-norm and T-conorm, and aggregator operators for this purpose. The second step is to use fuzzy granulation to granulate the similarities values into series of granular, two ([0 or 1] or [high and low]), three (i.e. low, medium, and high), etc. Then the Z(n)-compact will work as it is presented.

Table 2 Intermediate results for Z(n)-compact algorithm

Data	$Atrb_1$	$Atrb_2$	$Atrb_3$	C_1
D_1	$a_1{}^1$	$a_1{}^2$	$a_1{}^3$	c_1
D_2	$a_1{}^1$	$a_2{}^2$	$a_1{}^3$	c_1
D_3	$a_2{}^1$	$a_2{}^2$	$a_1{}^3$	c_1
D_4	$a_3{}^1$	$a_2{}^2$	$a_1{}^3$	c_1
D_5	$a_3{}^1$	$a_1{}^2$	$a_2{}^3$	c_1
D_6	$a_1{}^1$	$a_2{}^2$	$a_2{}^3$	c_1
D_7	$a_2{}^1$	$a_2{}^2$	$a_2{}^3$	c_1
D_8	$a_3{}^1$	$a_2{}^2$	$a_2{}^3$	c_1
D_2, D_3, D_4	*	$a_2{}^2$	$a_1{}^3$	c_1
D_6, D_7, D_8	*	$a_2{}^2$	$a_2{}^3$	c_1
D_5, D_8	$a_3{}^1$	*	$a_2{}^3$	c_1
D_2, D_6	$a_1{}^1$	$a_2{}^2$	*	c_1
D_3, D_7	$a_2{}^1$	$a_2{}^2$	*	c_1
D_4, D_8	$a_3{}^1$	$a_2{}^2$	*	c_1
D_2, D_6	*	$a_2{}^2$	*	c_1
$D_2, D_3, D_4, D_6, D_7, D_8$				

Table 3 Maximally Z(n)-compact representation of DAC matrix

Data	$Atrb_1$	$Atrb_2$	$Atrb_n$	Class
D_1	$a_1{}^1$	$a_1{}^2$	$a_1{}^3$	c_1
D_5, D_8	$a_3{}^1$	$*$	$a_2{}^3$	c_1
$D_2, D_3, D_4, D_6, D_7, D_8$	$*$	$a_2{}^2$	$*$	c_1
D_9	$a_2{}^1$	$a_1{}^2$	$a_1{}^3$	c_2
$D_1 0$	$a_1{}^1$	$a_1{}^2$	$a_2{}^3$	c_2
D_9, D_{11}	$a_2{}^1$	$a_1{}^2$	$*$	c_2

D – It is also important to note that classes may also not be crisp. Therefore, steps B and C could also be used to granulate concepts as it is used to granulate the keyword entries values ($k_{ii}{}^{jj}$s).

E – Another important case is how to select the right attributes in first place. One can use traditional statistical or probabilistic techniques, non traditional techniques such as GA-GP, fuzzy-logic or clustering techniques. These techniques will be used as first pass to select the first set of initial attributes. The second step will be based on feature selection technique based on to maximally separating the classes. This techniques are currently part of the BISC-DSS toolbox, which includes the Z(n)-Compact-Feature-Selection technique (Z(n)-CFS)).

F – Another very important case is when the cases are not possibilistic and are in general form "IF Atrb$_i$ isr$_i$ A$_i$ and... Then Class isr$_c$ C$_j$; where isr can be presented as:

r: =	*equality constraint: X=R is abbreviation of X is=R*
r: ≤	*inequality constraint: X ≤ R*
r: ⊂	*subsethood constraint: X ⊂ R*
r: blank	*possibilistic constraint; X is R; R is the possibility distribution of X*
r: v	*veristic constraint; X isv R; R is the verity distribution of X*
r: p	*probabilistic constraint; X isp R; R is the probability distribution of X*
r: rs	*random set constraint; X isrs R; R is the set-valued probability distribution of X*
r: fg	*fuzzy graph constraint; X isfg R; X is a function and R is its fuzzy graph*
r: u	*usuality constraint; X isu R means usually (X is R)*
r: g	*group constraint; X isg R means that R constrains the attribute-values of the group*

- *Primary constraints: possibilistic, probabilisitic and veristic*
- *Standard constraints: bivalent possibilistic, probabilistic and bivalent veristic*

Once the DAC matrix has been transformed into maximally compact concept based on graph representation and rules based on possibilistic relational universal

Table 4 Rule-based representation of Z(n)-compact of DAC matrix

Data	Rules
D_1	If $Atrb_1$ is $a_1{}^1$ and $Atrb_2$ is $a_1{}^2$ and $Atrb_3$ is $a_1{}^3$ THEN Class is c_1
D_5, D_8	If $Atrb_1$ is $a_3{}^1$ and $Atrb_3$ is $a_2{}^3$ THEN Class is c_1
$D_2, D_3, D_4, D_6, D_7, D_8$	If $Atrb_2$ is $a_2{}^2$ THEN Class is c_1
D_9	If $Atrb_1$ is $a_2{}^1$ and $Atrb_2$ is $a_1{}^2$ and $Atrb_3$ is $a_1{}^3$ THEN Class is c_2
D_{10}	If $Atrb_1$ is $a_1{}^1$ and $Atrb_2$ is $a_1{}^2$ and $Atrb_3$ is $a_2{}^3$ THEN Class is c_2
D_9, D_{11}	If $Atrb_1$ is $a_2{}^1$ and $Atrb_2$ is $a_1{}^2$ THEN Class is c_2

fuzzy–type I, II, III, and IV (pertaining to modification, composition, quantification, and qualification), one can use Z(n)-compact algorithm and transform the DAC into a decision-tree and hierarchical graph that will represents a Q&A and/or questionnaire model (Figs. 8a and 8b). The decision-tree and hierarchical graph will be a more general form of traditional tree since at each node the operation will be more general and tree could be constructed based on "isr" rather than "is" which in this

Table 5 Rule-based representation of Maximally Z(n)-compact of DAC matrix (Alternative representation for Table 4)

Data	Rules
D_1	If $Atrb_1$ is $a_1{}^1$ and $Atrb_2$ is $a_1{}^2$ and $Atrb_3$ is $a_1{}^3$ OR
D_5, D_8	If $Atrb_1$ is $a_3{}^1$ and $Atrb_3$ is $a_2{}^3$ OR If $Atrb_2$ is $a_2{}^2$ THEN Clas is c_1
$D_2, D_3, D_4, D_6, D_7, D_8$	
D_9	If $Atrb_1$ is $a_2{}^1$ and $Atrb_2$ is $a_1{}^2$ and $Atrb_3$ is $a_1{}^3$ OR
D_{10}	If $Atrb_1$ is $a_1{}^1$ and $Atrb_2$ is $a_1{}^2$ and $Atrb_3$ is $a_2{}^3$ OR Concept is c_2
D_9, D_{11}	If $Atrb_1$ is $a_2{}^1$ and $Atrb_2$ is $a_1{}^2$
	THEN Concept is c_2

Table 6 $Z(n)$-Compactification Algorithm

Z(n)-Compact Algorithm:
The following steps are performed successively for each column j; j=1 ... n

1. Starting with k_{ii}^{jj} (ii=1, jj=1) check if for any k_{ii}^{1} (ii=1, ...,3 in this case) all the columns are the same, then k_{ii}^{1} can be replaced by *

    - For example, we can replace k_{ii}^{1} (ii=1, ...,3) with * in rows 2,3, and 4. One can also replace k_{ii}^{1} with * in rows 6, 7, and 8. (Table 1, first pass).

2. Starting with k_{ii}^{jj} (ii=1, jj=2) check if for nay k_{ii}^{2} (i=1, ...,3 in this case) all the columns are the same, then k_{ii}^{2} can be replaced by *

    - For example, we can replace k_{ii}^{2} (ii=1, ...,3) with * in rows 5 and 8. (Table 1, first pass).

3. Repeat steps one and 2 for all jj.
4. Repeat steps 1 through 3 on new rows created Row* (Pass 1 to Pass nn, in this case, Pass 1 to Pass 3).

    - For example, on Rows* 2,3,4 (Pass 1), check if any of the rows given columns jj can be replaced by *. In this case, k_{ii}^{3} can be replaced by *. This will gives: * k_{2}^{2} *.
    - For example, on Pass 3, check if any of the rows given columns jj can be replaced by *. In this case, k_{ii}^{1} can be replaced by *. This will gives: * k_{2}^{2} *

5. Repeat steps 1 through 4 until no compactification would be possible

---

case at each connection we may also have qualifiers such as "most likely", "probably", etc. Finally, the concept of semantic equivalence and semantic entailment based on possibilistic relational universal fuzzy will be used as a basis for question-answering (Q&A), questionnaire and inference from fuzzy premises (Figs. 8a and 8b). This will provide a foundation for approximate reasoning, language for representation of imprecise knowledge, a meaning representation language for natural

(a)

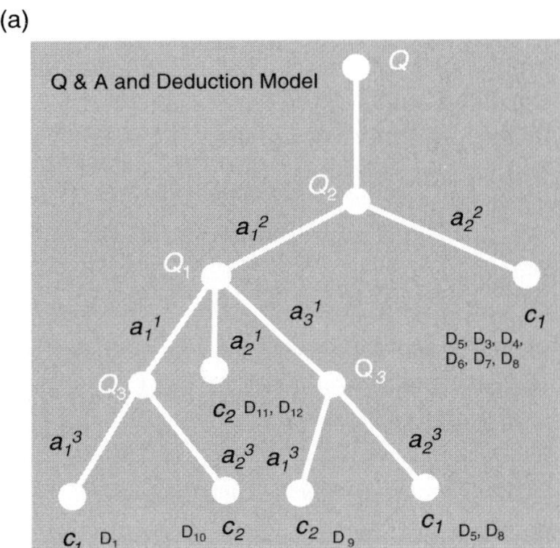

Fig. 8a Q & A model of DAC matrix and maximally Z(n)-compact rules and concept-based query-based clustering

Fig. 8b Tree model of DAC   (b)
matrix for Classes 1 and 2

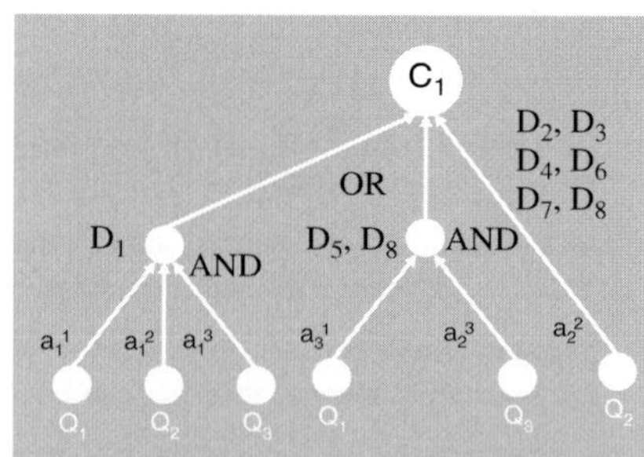

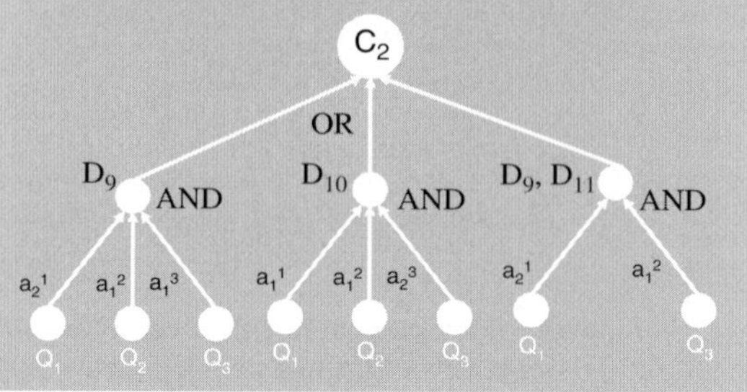

languages, precisiation of fuzzy propositions expressed in a natural language, and as a tool for Precisiated Natural Language (PNL) and precisation of meaning. The maximally compact dataset based on Z(n)-compact algorithm and possibilistic relational universal fuzzy–type II will be used to cluster the data based on class-based query-based

search criteria. Tables 7 shows the technique based on possibilistic relational universal fuzzy–type I, II, III, and IV (pertaining to modification, composition, quantifica-

Table 7 Four Types of PRUF, Type I, Type II, Type III and Type IV

PRUF Type I: pertaining to modification	*PRUF – Type III: pertaining to quantification*
X is very small	*Most Swedes are tall*
X is much larger than Y	*Many men are much taller than most men*
Eleanor was very upset	*Most tall men are very intelligent*
The Man with the blond hair is very tall	
PRUF Type II: pertaining to composition	*PRUF – Type IV: pertaining to qualification*
X is small and Y is large (conjunctive composition)	• *Abe is young is not very true (truth qualification)*
X is small then Y is large (disjunctive composition)	• *Abe is young is quite probable (probability qualification)*
If X is small then Y is large (conditional and conjunctive composition)	• *Abe is young is almost impossible (possibility qualification)*
If X is small then Y is large else Y is very large (conditional and conjunctiv composition)	

Table 8 Possibilistic Relational Universal Fuzzy–Type I; Pertaining to modification, rules of type I, basis is the modifier

Proposition p
p : N is F

Modified proposition p^+
$p^+ = N$ is mF
m: not, very, more or less, quite, extremely, etc.

$\mu_{\text{YOUNG}} = 1 - S(25, 35, 45)$
Lisa is very young $\rightarrow \Pi_{\text{Age(Lisa)}} = YOUNG^2$
$\mu_{\text{YOUNG}^2} = (1 - S(25, 35, 45))^2$

p: Vera and Pat are close friends
Approximate p $\rightarrow$ p*: Vera and Pat are friends[2]
$\pi(FRIENDS) = \mu FRIENDS^2(Name1 = Vera; Name2 = Pat)$

Table 9 Possibilistic Relational Universal Fuzzy–Type II – operation of composition – to be used to compact rules presented in Table 5.

R	X_1	X_2	$\cdots$	X_n
	F_{11}	F_{12}	$\cdots$	F_{1n}
	$\cdots$	$\cdots$	$\cdots$	$\cdots$
	F_{m1}	$\cdots$	$\cdots$	F_{mn}

$R = X_1$ is F_{11} and X_2 is F_{12} and $\ldots X_n$ is F_{1n} OR
$ X_1$ is F_{21} and X_2 is F_{22} and $\ldots X_n$ is F_{2n} OR

$\cdots \quad \cdots \quad \cdots$

$ X_1$ is F_{m1} and X_2 is F_{m2} and $\ldots X_n$ is F_{mn}

$R \rightarrow (F_{11} \times \cdots \times F_{1n}) + \cdots + (F_{m1} \times \cdots \times F_{mn})$

Table 10 Possibilistic Relational Universal Fuzzy–Type II – pertaining to composition – main operators

$p = q^*r$
$q:M$ is F
$r:N$ is G

M is F and N is G : $\overline{F} \cap \overline{G} = F \times G$ If M is F then N is G else N is H

M is F or N is G $= \overline{F} + \overline{G}$ $\Pi_{(X,\ldots,X_m,Y_1,\ldots,Y_n)} = (\overline{F'} \oplus \overline{G}) \cap (\overline{F} \oplus \overline{H})$

if M is F then N is G $= \overline{F} \oplus \overline{G}$ *or*

or

If M is F then N is G $= F \times G + F' \times V$ If M is F then N is G else N is H $\leftrightarrow$

$\overline{F'} = F' \times V$ $\left(\begin{array}{l} \text{(If M is F then N is G) and} \\ \text{(If M is not F then N is H)} \end{array} \right)$

$\overline{G} = U \times G$

$\mu_{F \times G}(u, v) = \mu_F(u) \wedge \mu_G(V)$

$\mu_{\overline{F'} \oplus \overline{G}}(u, v) = 1 \wedge (1 - \mu_F(u) + \mu_G(V))$

$\wedge$: min, $+$ arithmetic sum,

$-$ arithmetic difference

Table 11 Possibilistic Relational Universal Fuzzy–Type II – Examples

Example:

$U = V = 1 + 2 + 3,$

M: X,N:Y
F: SMALL : 1/1+0.6/2+0.1/3

G: LARGE : 0.1/1+0.6/2+1/3.

X is small and Y is large:

$$\Pi_{(x,y)} = \begin{bmatrix} 0.1 & 0.6 & 1 \\ 0.1 & 0.6 & 0.6 \\ 0.1 & 0.1 & 0.1 \end{bmatrix}$$

X is small or Y is large:

$$\Pi_{(x,y)} = \begin{bmatrix} 1 & 1 & 1 \\ 0.6 & 0.6 & 1 \\ 0.1 & 0.6 & 1 \end{bmatrix}$$

if X is small then Y is large:

$$\Pi_{(x,y)} = \begin{bmatrix} 0.1 & 0.6 & 1 \\ 0.5 & 1 & 1 \\ 1 & 1 & 1 \end{bmatrix}$$

if X is small then Y is large:

$$\Pi_{(x,y)} = \begin{bmatrix} 0.1 & 0.6 & 1 \\ 0.4 & 0.6 & 0.6 \\ 0.9 & 0.9 & 0.9 \end{bmatrix}$$

Table 12 Possibilistic Relational Universal Fuzzy–Type III, Pertaining to quantification, Rules of type III; p: Q N are F

p = Q N are F
Q : fuzzy quantifier
quantifier : most, many, few, some, almost

Most Swedes are tall

N is F $\rightarrow$ $\Pi_X = F$
$QN\ are$ F $\rightarrow$ $\Pi_{count(F)} = Q$

$mp \leftrightarrow q$
$q \rightarrow mp$

$(\Pi^p)'$ if m : not
$(\Pi^p)^2$ if m : very
$(\Pi^p)^{0.5}$ if m : more or less

Table 13 Possibilistic Relational Universal Fuzzy–Type III, Pertaining to quantification, Rules of type III; p: Q N are F, Examples

m (N is F) $\leftrightarrow$ N is mF
 not (N is F) $\leftrightarrow$ N is not F
 very (N is F) $\leftrightarrow$ N is very F
 more of less (N is F) $\leftrightarrow$ N is more of less F

m(M is F and N is G) $\leftrightarrow$ (X, Y) is m (F $\times$ G)
 not (M is F and N is G) $\leftrightarrow$ (X, Y) is (F $\times$ G)$'$
         $\leftrightarrow$ (M is not F)or
          (N is not G)
very (M is F and N is G) $\leftrightarrow$ (M is very F)and
         (N is very G)
more of less (M is F and N is G) $\leftrightarrow$ (M is more or less F) and
         (N is more or less G)

m (QN are F) $\leftrightarrow$ (m Q) N are F
 not (Q N are F) $\leftrightarrow$ (not Q) N are F

Table 14 Possibilistic Relational Universal Fuzzy–Type IV, Pertaining to qualification, Rules of type IV; q: p is ?, Example 1

Y: truth value, a probability -value, a possibility value, . . .

for p: N is F and q be a truth - qualified value for p
q: N is F is T
   where T is a liguistic truth - value

N is F is T ↔ N is G	q: N is F is T
$T = \mu_F(G)$	q: N is small is very true
$\mu_G(u) = \mu_T(\mu_F(u))$	$\mu_{SMALL} = 1 - S(5, 10, 15)$
N is F → $\Pi_X = F$	$\mu_{TRUE} = 1 - S(0.6, 0.8, 1.0)$
Then N is F is $T \to \Pi_X = F^+$	
$\mu_{F^+}(u) = \mu_T(\mu_F(u))$	$q \to \Pi_X(u) = S^2(1 - S(u; 5, 10, 15); 0.6, 0.8, 1.0))$
if $T = $ u - true	
$\mu_{u\text{-true}}(v) = v \quad v \in [0, 1]$	
N is F is u - true → N is F	

Table 15 Possibilistic Relational Universal Fuzzy–Type IV, Pertaining to qualification, Rules of type IV; q: p is ? Example 2

Y : truth value, a probability - value, a possibility value,. . .

for p: N is F and q be a truth - qualified value for p
q: N is F is T
where T is a liguistic truth - value

N is F is T ↔ N is G	q: N is F is T
$T = \mu_F(G)$	q: N is small is very true
$\mu_G(u) = \mu_T(\mu_F(u))$	$\mu_{SMALL} = 1\text{-}S(5, 10, 15)$
N is F → $\Pi_X = F$	$\mu_{TRUE} = 1\text{-}S(0.6, 0.8, 1.0)$
Then N is F is $T \to \Pi_X = F^+$	
$\mu_{F^+}(u) = \mu_T(\mu_F(u))$	$q \to \Pi_X(u) = S^2(1 - S(u; 5, 10, 15); 0.6, 0.8, 1.0))$
If $T = $ u- true	
$\mu_{u\text{-true}}(v) = v \quad v \in [0, 1]$	
N is F is u-true → N is F	

Table 16 Possibilistic Relational Universal Fuzzy–Type IV, Pertaining to qualification, Rules of type IV; q: p is ?, Example 3

m (N is F is T) ↔ N is F is m T
  not (N is F is T) ↔ N is F is not T
  very (N is F is T) ↔ N is F is very T
  more or less (N is F is T) ↔ N is F is more or less T

  N is not F is T ↔ N is F is ant T
    ant: antonym
    false:ant true

  N is very F is T ↔ N is F is $^{0.5}T$
    $\mu_{0.5T}(V) = \mu_T(V^2)$
    $\mu_{a_T}(V) = \mu_T(V^{1/a})$

Table 17 Possibilistic Relational Universal Fuzzy–Type IV, Pertaining to qualification, Rules of type IV; q: p is ?, Example 4

"*Barbara* is not very rich"
Semantically equivalent propositions:
   Barbara is not very rich
   Barbara is not very rich is u - true
   Barbara is very rich is ant u - true
   Barbara is rich is $^{0.5}(ant$ u - true)
      where $\mu_{0,5(ant\ \text{u-true})}(V) = 1 - V^2$
      and true approximately semantically $= u - true$
   the proposition will be approximated to:
   "Barbara is rich is not very true"
   "Is Barbara rich? T" is "not very true".

---

tion, and qualification), [3–4] and series of examples to clarify the techniques. Types I through Types IV of possibilistic relational universal fuzzy in connection with Z(n)-Compact can be used as a basis for precisation of meaning. Tables 8 through 17 show the technique based on possibilistic relational universal fuzzy–type I, II, III, and IV (pertaining to modification, composition, quantification, and qualification) [11–13] and series of examples to clarify definitions.

11 Implementation - Fuzzy Query and Ranking

In this section, we introduce fuzzy query and fuzzy aggregation for credit scoring, university admissions and date matching.

Credit Scoring: Credit scoring was first developed in the 1950's and has been used extensively in the last two decades. In the early 1980's, the three major credit bureaus, Equitax, Experian, and TransUnion worked with the Fair Isaac Company to develop generic scoring models that allow each bureau to offer an individual score based on the contents of the credit bureau's data. When you apply for financing, whether it's a new credit card, car or student loan, or a mortgage, about 40 pieces of information from your credit card report are fed into a model (Nikravesh and Azvine 2002 and Nikravesh et al. 2003). This information is categorized into the following five categories with different levels of importance (% of the score):

   Past payment history (35%)
   Amount of credit owed (30%)
   Length of time credit established (15%)
   Search for and acquisition of new credit (10%)
   Types of credit established (10%)

   Given the factors presented earlier, a simulated model has been developed. A series of excellent, very good, good, not good, not bad, bad, and very bad credit scores have been recognized (without including history). Then, fuzzy similarity and ranking have been used to rank the new user and define his/her credit score.. In the inference engine, the rules based on factual knowledge (data) and knowledge drawn from human experts (inference) are combined, ranked, and clustered based on the confidence level of human and factual support. This information is then used to build the fuzzy query model with associated weights. In the query level, an intelligent knowledge-based search engine provides a means for specific queries. Initially we blend traditional computation with fuzzy reasoning. This effectively provides validation of an interpretation, model, hypothesis, or alternatively, indicates the need to reject or reevaluate. Information must be clustered, ranked, and translated to a format amenable to user interpretation.

University Admissions: Hundreds of millions of applications were processed by U.S. universities resulting in more than 15 million enrollments in the year 2000 for a total revenue of over \$250 billion. College admissions are expected to reach over 17 million by the year 2010, for total revenue of over \$280 billion.

The UC Berkeley campus admits its freshman class on the basis of an assessment of the applicants' high school academic performance (approximately 50%) and through a comprehensive review of the application including personal achievements of the applicant (approximately 50%) (University of California-Berkeley).

At Stanford University, in addition to the academic transcript, close attention is paid to other factors such as student's written application, teacher references, the short responses and one-page essay (carefully read for quality, content, and creativity), and personal qualities.

Given the factors and general admission *criteria*, a simulated-hypothetical model (a Virtual Model) was developed. A series of excellent, very good, good, not good, not bad, bad, and very bad student given the criteria for admission has been recognized. These criteria over time can be modified based on the success rate of students admitted to the university and their performances during the first, second, third and fourth years of their education with different weights and degrees of importance given for each year. Then, fuzzy similarity and ranking can evaluate a new student rating and find it's similarity to a given set of criteria.

This will provide the ability to answer "what if" questions in order to decrease uncertainty and provide a better risk analysis to improve the chance for "increased success" on student selection or it can be used to select students on the basis of "diversity" criteria. The model can be used as for decision support and for a more uniform, consistent and less subjective and biased way. Finally, the model could learn and provide the mean to include the feedback into the system through time and will be adapted to the new situation for defining better criteria for student selection.

In this study, it has been found that ranking and scoring is a very subjective problem and depends on user perception and preferences in addition to the techniques used for the aggregation process which will effect the process of the data mining in reduced domain. Therefore, user feedback and an interactive model are recommended tools to fine-tune the preferences based on user constraints. This will allow the representation of a multi-objective optimization with a large number of constraints for complex problems such as credit scoring or admissions. To solve such subjective and multi-criteria optimization problems, GA-fuzzy logic and DNA-fuzzy logic models are good candidates. To solve such subjective and multi-criteria optimization problems with a large number of constraints for complex problems such as university admissions, the BISC Decision Support System is an excellent candidate.

Date Matching: The main objective is to find the best possible match in the huge space of possible outputs in the databases using the imprecise matching such as fuzzy logic concepts, by storing the query attributes and continuously refining the query to update the user's preferences. We have also built a Fuzzy Query system, which is a Java application that sits on top of a database.

With traditional SQL queries (relational DBMS), one can select records that match the selection criteria from a database. However, a record will not be selected if any one of the conditions fails. This makes searching for a range of potential candidates difficult.

In this program, one can basically retrieve all the records from the database, compare them with the desired record, aggregate the data, compute the ranking, and then output the records in the order of their rankings. Retrieving all the records from the database is a naïve approach because with some preprocessing, some very different records are not needed from the database. However, the main task is to compute the fuzzy rankings of the records so efficiency is not the main concern here.

The major difference between this application and other date matching system is that a user can input his hobbies in a fuzzy sense using a slider instead of choosing crisp terms like "kind of" or "love it". These values are stored in the database according to the slider value.

12 BISC-DSS: Applications

The followings are the potential applications of the BISC Decision Support System:

I. *Physical Stores or E-Store:* A computer system that could instantly track sales and inventory at all of its stores and recognize the customer buying trends and provide suggestion regarding any item that may interest the customer

- to arrange the products
- on pricing, promotions, coupons, etc
- for advertising strategy

II. *Internet-Based Advising:* A computer system that uses the expert knowledge and the customer data (Internet brokers and full-service investment firms) to recognize the good and bad traders and provide intelligent recommendation to which stocks buy or sell

- reduce the expert needs at service centers
- increase customer confidence

- ease-of-use
- Intelligent coaching on investing through the Internet
- allow customers access to information more intelligently

III. *Profitable Customers:* A computer system that uses customer data (Fig. 9) that allows the company to recognize good and bad customer by the cost of doing business with them and the profits they return

- keep the good customers
- improve the bad customers or decide to drop them
- identify customers who spend money
- identify customers who are profitable
- compare the complex mix of marketing and servicing costs to access to new customers

IV. *Managing Global Business:* A computer system responding to new customers and markets through integrated decision support activities globally using global enterprise data warehouse

- information delivery in minutes
- lower inventories
- intelligent and faster inventory decisions in remote locations

V. *Resource Allocator:* A computer system that intelligently allocate resources given the degree of match between objectives and resources available

- resource allocation in factories floor
- for human resource management
- find resumes of applicants posted on the Web and sort them to match needed skill and can facilitate training and to manage fringe benefits programs
- evaluate candidates predict employee performance

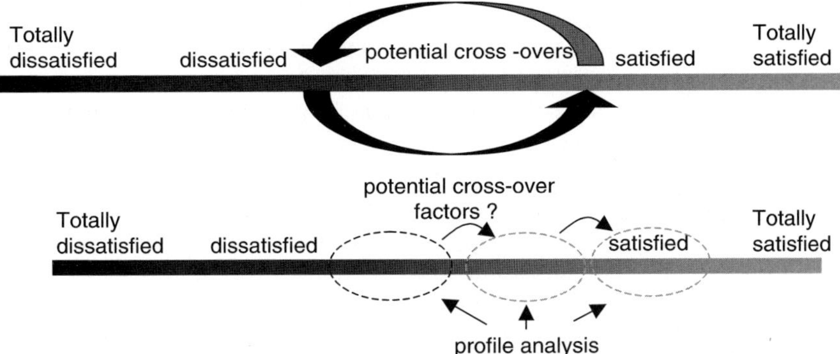

Fig. 9 Successful efforts to proactively convert marginally dissatisfied customers to satisfied ones by even a few percentage points will benefit most companies. Preventing "decay" in the other way is equally important and beneficial

VI. *Intelligent Systems to Support Sales:* A computer system that matching products and services to customers needs and interest based on case-based reasoning and decision support system to improve

- sale
- advertising

VII. *Enterprise Decision Support:* An interactive computer-based system that facilitates the solution of complex problems by a group of decision makers either by speeding up the process of the decision-making process and improving the quality of the resulting decisions through expert and user (company-customer) collaboration and sharing the information, goals, and objectives.

VIII. *Fraud Detection:* An Intelligent Computer that can learn the user's behavior through in mining customer databases and predicting customer behaviours (normal and irregularities) to be used to uncover, reduce or prevent fraud.

- in credit cards
- stocks
- financial markets
- telecommunication
- insurance

IX. *Supply-Chain Management (SCM):* Global optimization of design, manufacturing, supplier , distribution, planning decisions in a distributed environment.

X. *BISC-DSS and Autonomous Multi-Agent System:* A key component of any autonomous multi-agent system –especially in an adversarial setting – is decision module, which should be capable of functioning in an environment of imprecision, uncertainty and imperfect reliability. BISC-DSS will be focused on the development of such system and can be used as a decision-support system for ranking of decision alternatives. BISC-DSS can be used :

- As global optimizer for planning decisions in a distributed environment
- To facilitates the solution of complex problems by a group of autonomous agents by speeding up the process of decision-making, collaboration and sharing the information, goals, and objectives
- To intelligently allocate resources given the degree of match between objectives and resources available
- Assisting autonomous multi-agent system in assessing the consequences of decision made in an environment of imprecision, uncertainty, and partial truth and providing a systematic risk analysis
- Assisting multi-agent system answer "What if Questions", examine numerous alternatives very quickly, ranking of decision alternatives, and find the value of the inputs to achieve a desired level of output

XI. *Knowledge Mobilization and Intelligent Augmentation (KnowMInA); Application to Homeland Security:* KnowMInA (Fig. 10–12) is an intelligent system that recognizes terrorism activities. *KnowMInA* includes behavior/profile modeling, reasoning engines, decision-risk analysis, and visual data mining-analytic software. KnowMInA is used to find suspicious pattern in data using

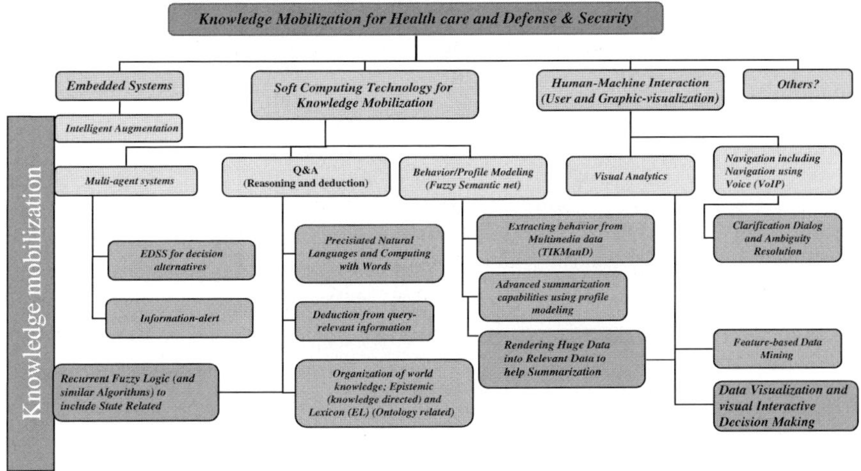

Fig. 10 Knowledge Mobilization and Intelligent Augmentation (KnowMInA)

geographical maps and recognition technology. It uses multimedia/Multimodal information or huge number of "tips" received prior or immediately after the attack. The system filters out the irrelevant part and acts on the useful subset of such data in a speedy manner. KnowMInA shares this relevant and summarized information with authorities using knowledge mobilization

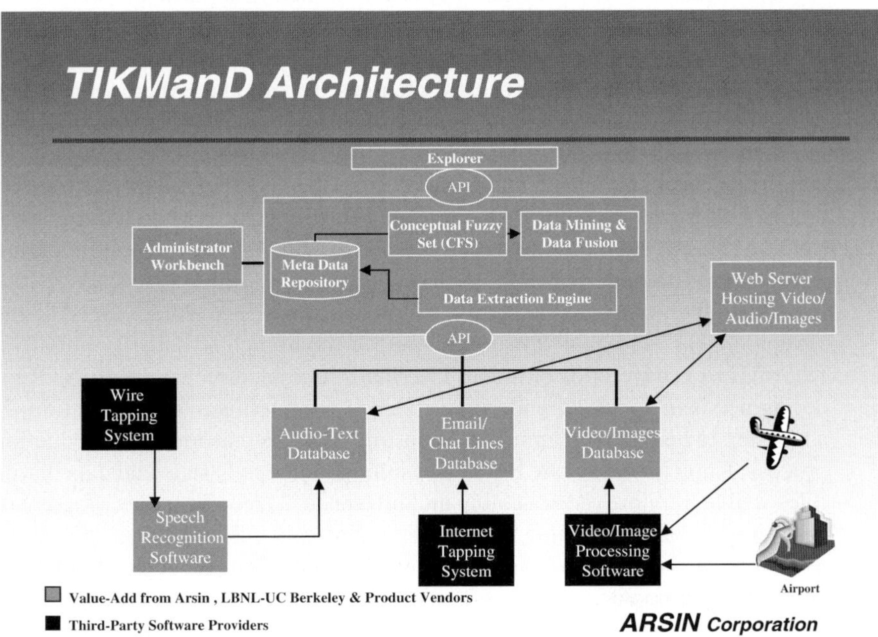

Fig. 11 TIKMand Architecture

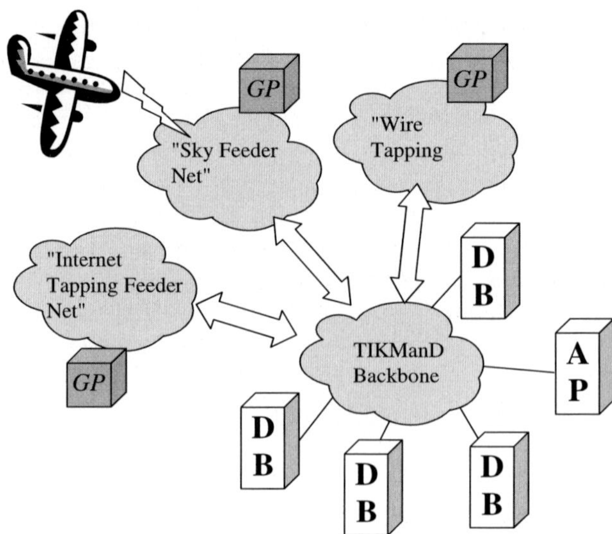

Fig. 12 High Level Network Design; the TIKManD Network Architecture high-level view defines two major functional layers: feeder networks and a backbone

infrastructure. KnowMInA can detect unusual patterns; raise alarms based on classification of activities and offer explanations based on automatic learning techniques for why a certain activity is placed in a particular class such as "Safe", "Suspicious", "Dangerous" etc. The underlying techniques can combine expert knowledge and data driven rules to continually improve its classification and adapt to dynamic changes in data and expert knowledge. This will also provide the ability to answer "What if?" questions in order to decrease uncertainty and provide a better risk analysis. For Example, US authorities have announced that 5 out of 19 suspected September 11 hijackers met in Las Vegas before the attack. Some of the terrorist attended the same training school, had communication tracks, and flew at the same time period to the same location in the same flight. KnowMInA intends to capture and recognize such activities and raise alarms. KnowMInA would apply to the "Preparedness" and also "Response" part of the homeland security initiative. KnowMInA is developed based on TIKManD architecture. KnowMInA - Rendering needs includes 1) rendering is needed for intelligent data reduction, 2) a key step to help better Summarization, and 3) helps reasoning and deduction process. KnowMInA - State Based Reasoning includes 1) state and data driven rules will help improved reasoning and inference, 2) recurrent fuzzy logic will help simplify understanding and implementation of rules, 3) recurrent neural net-based learning for ongoing adaptation and improvement. KnowMInA - Web Interface includes 1) for efficient communication, data may be posted on a web by various sources all over the country, 2) KnowMInA may also use other web based information for reasoning, behavioral profile modeling and man-machine interaction, 3) web based "rendering" will be needed to get right data from the web. KnowMInA-visual analytic and visual data mining and decision analysis will address the following issues 1)how to go from massive amount of data into something meaningful?, 2) how to represent events from spatio-temporal data? 3) how to couple sensitivity/uncertainty analysis with information retrieval? 4) how to construct a workbench for quantitative comparison of multidimensional spatio-temporal and multimedia data that are only correlated conceptually? 5) browsing the database against the schema, and 6) navigation through annotated documents and corresponding images. We intend to design a secure and distributed IT infrastructure for TIKManD to provide a means for secure communication between resources and to build textual, voice and image databases online. Given the distributed nature of the information sources, a federated database framework is required to distribute storage and information access across multiple locations. A high-level summary of this architecture and possible implementation options follows. TIK-ManD Architecture consists of following components:

- Data Mining & Data Fusion Component
- Conceptual Fuzzy Search (CFS) Model
- Administrator Workbench
- Explorer
- Meta Data Repository
- Data Extraction Engine
- Audio-Text Database
- Speech Recognition Software
- Wire Tapping System
- Email/Chatline Database
- Internet Tapping System
- Video/Image analysis software and database, e.g., face detection and matching

The relationships between these major components is represented in the following diagram. The TIKManD Network Architecture high-level view defines two major functional layers: feeder networks and a backbone. The system deals with a large size of completely different type of information, gathered from a wide range of sources. It is conceivable, that the transport of this information will be via diverse networks and media. These feeder networks will have high security and bandwidth requirements. Archival will be distributed and a naming context, e.g. LDAP or another database, will provide geographical location for a particular database over the wide area network. These technologies (security, resource allocation and brokering, and data localization) have been developed under DOE National Collaboratories program, which will be leveraged for integration. Finally, we will use our stat-of-the-art technology as a tool for effective data fusion and data mining of textual and voice information. This would apply to the "Preparedness" part of the initiative. Data fusion part would attempt to collect seemingly related data to recognize:

- Increased activities in a given geographic area,
- Unusual movements and travel patterns,

- Intelligence gathered through authorized wire tapping of suspected subgroups
- Unexplainable increase or decrease in routine communication
- Integration of reports and voice (text/audio) from different sources

Data mining would use the above data to pin point and extract useful Information from the rest. This would also apply to the "Response" part of the initiative on how to fuse huge number of "tips" received immediately after the attack, filter out the irrelevant part and act on the useful subset of such data in a speedy manner.

XII. *Soft Computing-based Recognition Technology for Monitoring and Automation of Fault Detection and Diagnosis:* The project specifically will focus on the applications of soft computing to the next-generation sensor information processing, analysis and interpretation, perception-based information processing and hybrid rule-based/case-based reasoning and its application for medical application, oil industry applications, industrial inspection and diagnosis systems, and speech recognition. Specifically, we intend to develop the next-generation of fault diagnosis, interpretation and prediction system with anticipatory capabilities (Fig. 13 through 20). The technologies can be grouped according to the functionality of different components of the system as follows:

- Signal processing and feature extraction:

  - Determine relevant acoustic wave attributes associated with specific properties of the products and integrate various attributes and features to expand the diagnostic and predictive capability;
  - Pseudo signal generation for characterization and calibration: As an example, pseudo signals can be generated with constraints to expand the size of

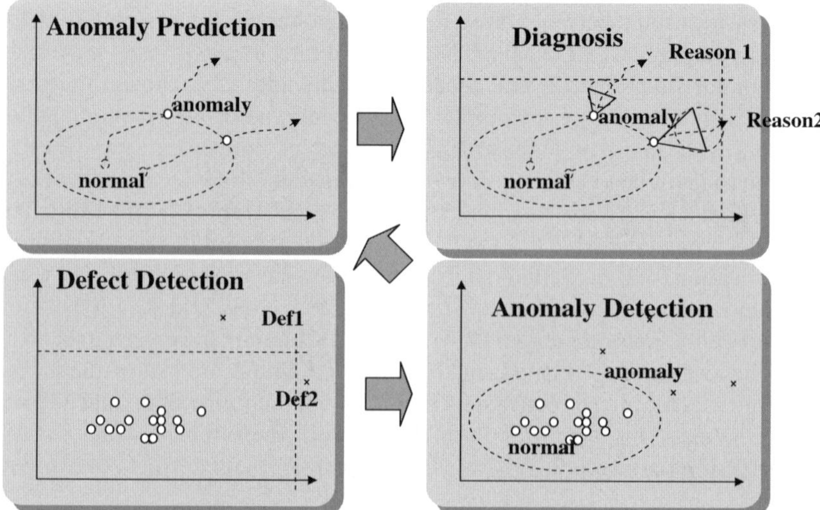

Fig. 13 Four Phases of Problem (Predictive and Monitoring Tools)

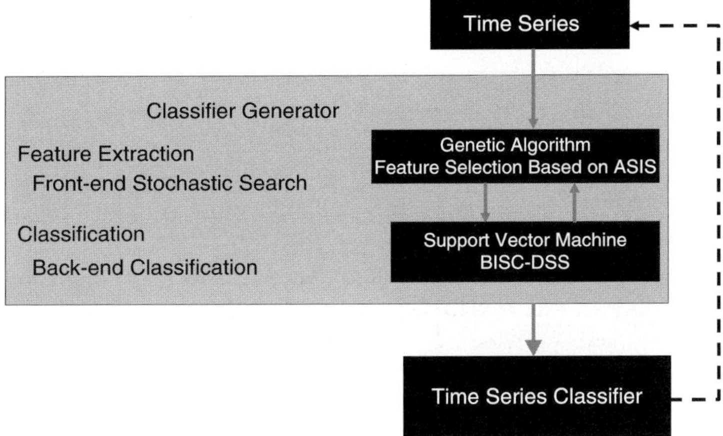

Fig. 14 Front/Back-end Architecture (BISC-DSS/-Zeus)

the datasets to ensure statistical significance, addressing a crucial problem of data scarcity in the current project;

- Signal attributes/characteristics analysis such as conventional Hilbert attributes, FFT-based frequency-domain attributes, modulation spectral features, and other techniques such as wavelet analysis and empirical correlation of attributes, as well as linear and non-linear aggregation operators for attribute fusion.

- Feature pre-processing and dimension reduction:

  - Various data compression techniques for data mining, to simplify similarity evaluation between signals and cases;
  - Evaluation of the relative importance, synergy and redundancy of various attributes for preliminary feature selection, and reduction of data dimension, using linear and non-linear techniques including conventional

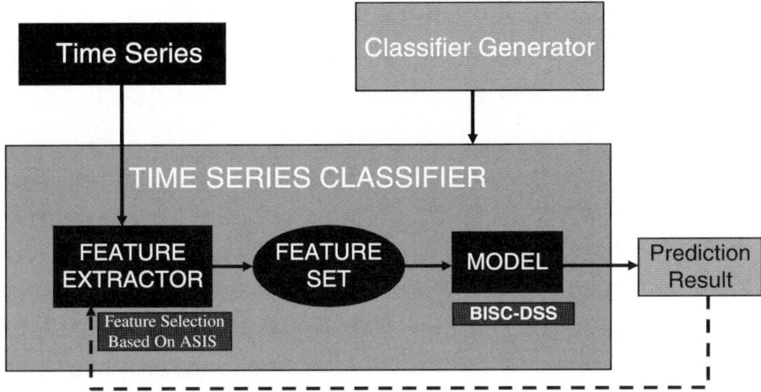

Fig. 15 Classifier Architecture (BISC-DSS/Zeus)

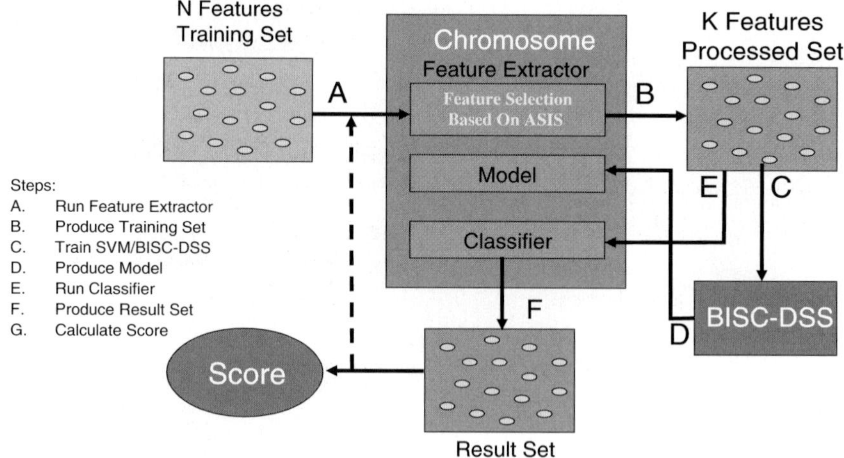

Fig. 16 Fitness: In-sample Rate; Feature selection

statistical methods such as PCA and SVD, as well as fuzzy measures and fuzzy integrals.

- Conventional classification methods:

  - Radial-basis function (RBF) networks and artificial neural networks with supervised training;
  - Support-vector machines (SVM) for classification and novelty detection, as well as its improvement from hard decision to soft decision.

- Hybrid case-based and rule-based classification and decision making:

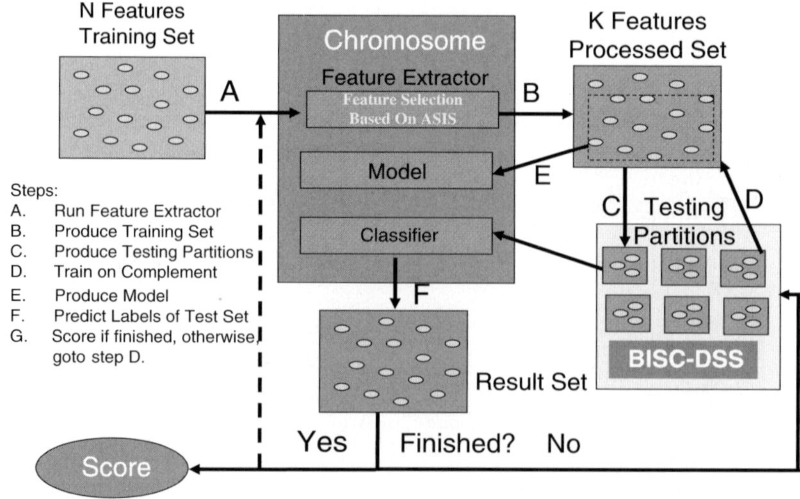

Fig. 17 Fitness: N-Fold Cross Valid; Feature selection

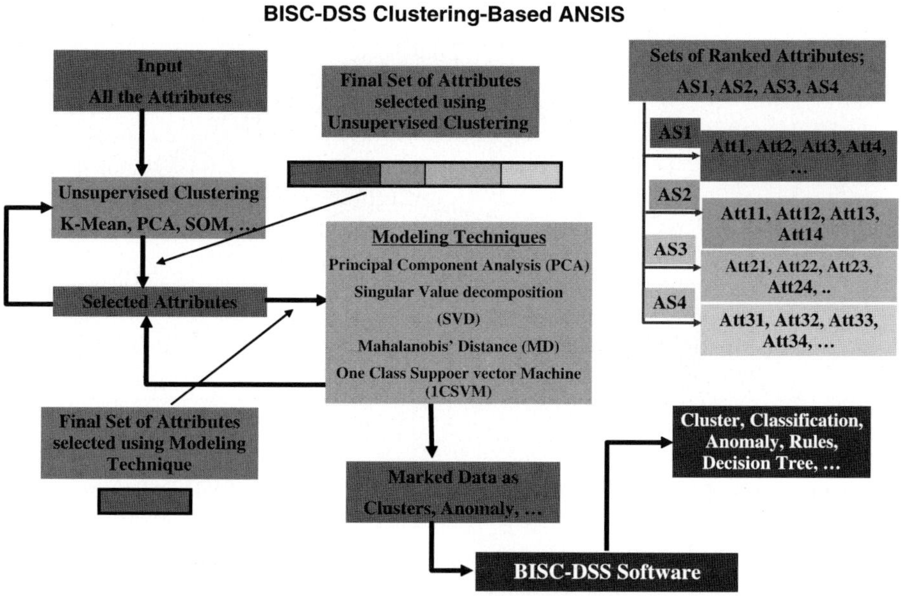

BISC-DSS Clustering-Based ANSIS

Fig. 18 BISC-DSS Clustering/Time Series-Based ANSIS

- Neuro-fuzzy systems that integrate rule-based and case-based reasoning and provide a smooth transition between the two depending on the availability of data and expert knowledge. This may be constructed in a hierarchical, decision-tree like structure and also include unsupervised clustering.
- Related methodologies and techniques for data fusion and mining, machine learning, and prediction and interpretation will be developed based on adaptation and integration of existing tools.

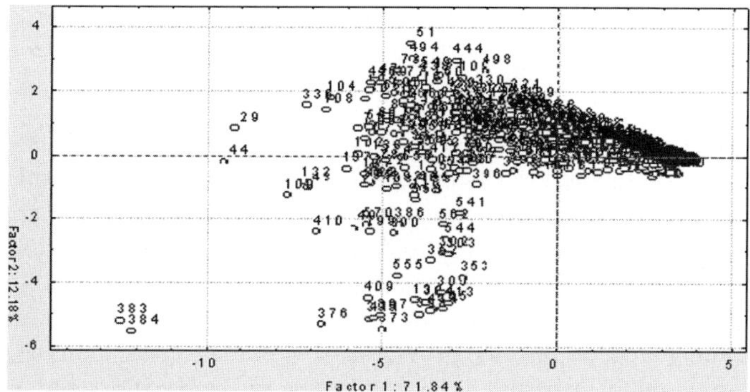

Fig. 19 PCA representation of the Data; all the features have been used; the cluster and the dynamics of the system and trend are not detected

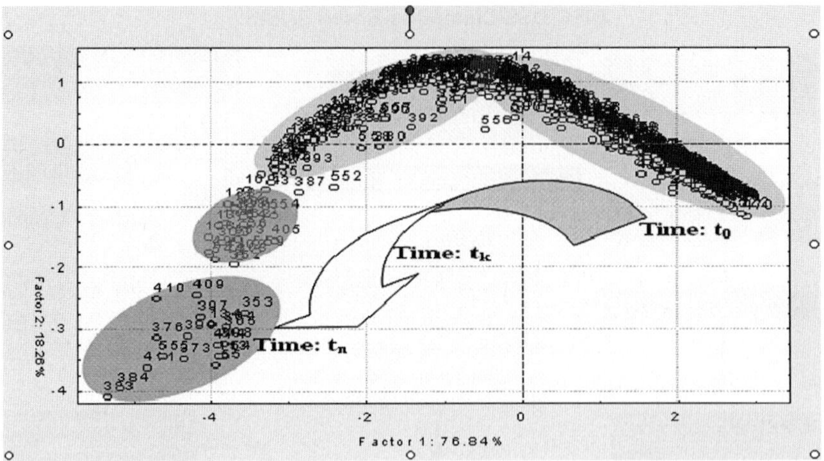

Fig. 20 PCA representation of the Data; Selected recognized features by BISC-DSS (Figs. 14–18) have been used; the cluster and the dynamics of the system and trend are well detected

- Further recognition and decision making technologies, and feature selection methods:

  - Fuzzy-integration-based aggregation techniques and hybrid fuzzy-logic/genetic-algorithm for pattern recognition, decision analysis, multi-criteria decision-making and multi-attribute optimization;
  - Self organization map (SOM) for building communities and clusters from datasets;
  - Genetic algorithms (GA) and reinforcement learning for feature set optimization and preference learning, within a filter or wrapper model of feature selection;
  - Fuzzy reinforcement learning techniques for online learning, maintenance, and adaptation of developed systems.

13 Conclusion

In this study, we introduced fuzzy query, fuzzy aggregation, evolutionary computing and the BISC decision support system as an alternative for ranking and predicting the risk for credit scoring, university admissions, and several other applications, which currently utilize an imprecise and subjective process. The BISC decision support system key features are 1) intelligent tools to assist decision-makers in assessing the consequences of decision made in an environment of imprecision, uncertainty, and partial truth and providing a systematic risk analysis, 2) intelligent tools to be used to assist decision-makers answer "what if",questions examine numerous alternatives very quickly and find the value of the inputs to achieve a desired level of output, and 3) intelligent tools to be used with human interaction and feedback to achieve a capability to learn and adapt through time In addition, the following important points have been found in this study: 1) no single ranking function works well for all contexts, 2) most similarity measures work about the same regardless of the model, 3) there is little overlap between successful ranking functions, and 4) the same model can be used for other applications such as the design of a more intelligent search engine which includes the user's preferences and profile (Nikravesh, 2001a and 2001b).

Acknowledgments Funding for this research was provided by the British Telecommunication (BT), OMRON and the BISC Program of UC Berkeley.

References

1. Fagin R. (1998) Fuzzy Queries in Multimedia Database Systems, Proc. ACM Symposium on Principles of Database Systems, pp. 1–10.
2. Fagin R. (1999) Combining fuzzy information from multiple systems. J. Computer and System Sciences 58, pp. 83–99.
3. Koza J. R., Genetic Programming : On the Programming of Computers by Means of Natural Selection, Cambridge, Mass. : MIT Press, USA 1992, 819 pages.
4. Holland J. H., Adaptation in Natural and Artificial Systems: An Introductory Analysis with Applications to Biology, Control and Artificial Intelligence. MIT Press, 1992. First Published by University of Michigan Press 1975.
5. Mizumoto M. (1989) Pictorial Representations of Fuzzy Connectives, Part I: Cases of T-norms, T-conorms and Averaging Operators, Fuzzy Sets and Systems 31, pp. 217–242.
6. Nikravesh M. (2001a) Perception-based information processing and re-trieval: application to user profiling, 2001 research summary, EECS, ERL, University of California, Berkeley, BT-BISC Project. (http://zadeh.cs.berkeley.edu/ & http://www.cs.berkeley.edu/~nikraves/ & http://www-bisc.cs.berkeley.edu/).
7. Nikravesh M. (2001b) Credit Scoring for Billions of Financing Deci-sions, Joint 9th IFSA World Congress and 20th NAFIPS International Conference. IFSA/NAFIPS 2001 "Fuzziness and Soft Computing in the New Millenium", Vancouver, Canada, July 25-28, 2001.
8. Nikravesh M. and Azvine B. (2002), Fuzzy Queries, Search, and Decision Support System, Journal of Soft Computing, Volum 6, # 5, August 2002.
9. Nikravesh M., Azvine B.,Yagar R., and Zadeh L.A. (2003) "New Directions in Enhancing the power of the Internet", Editors:, to be published in the Series Studies in Fuzziness and Soft Computing, Physica-Verlag, Springer (August 2003)
10. Nikravesh M. (2005), Evolutionary-Based Intelligent Information and Decision Systems, FuzzIEEE, May 22-25, Reno-Nevada, 2005 (Invited Talk)
11. Zadeh L. A. , (1976), A fuzzy-algorithmic approach to the definition of complex or imprecise concepts, Int. Jour. Man-Machine Studies 8, 249–291, 1976.
12. Zadeh L. A. and Nikravesh M., Perception-Based Intelligent Decision Systems; Office of Naval Research, Summer 2002 Program Review, Covel Commons, University of California, Los Angeles, July 30th-August 1st, 2002.
13. Zadeh, L. A., Toward a Perception-based Theory of Probabilistic Reasoning with Imprecise Probabilities, *Journal of Statistical Planning and Inference*, 105 233–264, 2002.

Qualitative Possibility Theory in Information Processing

Didier Dubois and Henri Prade

Abstract This chapter provides a general overview of advances made in qualitative possibility theory and its application to information processing in the last decade. These methodological results reflect a series of works done in few research groups, including the one of the authors. More precisely, the ability of possibility theory to handle positive and negative information (knowledge as well as preferences) in a bipolar way is emphasized including an original revision process, together with its use for representing information in different formats, for formalizing various types of reasoning pervaded with uncertainty, for providing a qualitative framework for decision making, and for learning from imperfect data.

1 Introductive Summary

Possibility theory offers a general framework, either qualitative or quantitative (according as possibility degrees belong to a discrete linearly ordered scale or to the real unit interval), for modeling uncertainty as well as preferences. In terms of uncertainty, possibility degrees are understood as assessing the level of plausibility of possible states of the world, while in preference modeling, they estimate the level of feasibility of alternative choices.

The representation framework provided by possibility theory is dual and bipolar. It is dual since necessity measures, which correspond to the impossibility of the contrary, enables us to express certainty or priority levels. In the quantitative setting, possibility and necessity degrees provide a simple, but non-trivial, system of upper and lower probabilities. It is bipolar because negative pieces of information restricting the possible states of the world, or the set of possible choices, can be represented, as well as positive pieces of information. Positive statements correspond to reports of actually observed states or to sets of choices that are guaranteed to be feasible or satisfactory at some minimal level (estimated through a so-called guaranteed possibility measure that is non-increasing w. r. t. set inclusion).

In possibility theory, different representation formats have been developed. Indeed the possibilistic logic setting (where classical logic formulas are associated with lower bounds of necessity degrees) has a Bayesian net-like graphical counterpart, and also enables the representation of default conditional pieces of information.

This framework is suitable for modeling reasoning from knowledge and factual information pervaded with uncertainty, for merging information, either when fusing multiple-source information or when looking for compromises between conflicting goals. Recent applications to temporal reasoning, to case-based reasoning, or to description logic are examples of the variety of reasoning problems where the representation capabilities of possibility theory appear to be useful. Besides, an axiomatic setting for qualitative decision, based on possibility theory, has been developed and further refined in the last ten years. Flexible querying in face of incomplete and fuzzy information is an example of such a decision process.

The paper is organized in three main parts. First, the framework of possibility theory is recalled in Sect. 2. Then its capability for representing knowledge or preferences in different representation formats is emphasized and illustrated in Sect. 3. Lastly, Sect. 4 points out its key role in the formalization of different forms of reasoning, and stresses its interest in query evaluation, decision making and learning. Each section only presents the main ideas and their formalization, without discussing technical results; however, it refers to the appropriate references for details. This paper provides a kind of guided tour of recent advances, made in the last decade and maybe viewed as an update of a previous paper written in the same spirit by the authors (Dubois and Prade, 1996).

With the exception of general references that are explicitly referred to in the text as usual, the more specific references focusing on particular points are only listed at the end in a structured bibliography that closely follows the structure of the paper. Then a specific reference can be easily identified from its title and the content description provided in the text. References to related works can be found in the papers of this structured bibliography.

2 The Possibility Theory Framework

This section provides a brief refresher on possibility theory, emphasizing its basic notions in five short subsections, each of them focusing on a key aspect, including the bipolar view that has been more recently introduced. The reader is referred to (Dubois and Prade, 1997; 1988) for a more detailed introduction and a treatise.

2.1 A Possibility Distribution is Associated with a Pair of Set Functions

Zadeh (1978) introduced possibility theory as a new framework for representing information, especially when expressed in linguistic terms. Thus, a piece of information of the form « X is F », understood as « the possible values of the (single-valued) variable X (supposed to range on a universe U) are restricted by a subset F of U (which may be fuzzy) ». In other words, given the granule of information « X is F », we know that the value of X must/should be in F. Since F may be fuzzy, this is understood as a graded possibility assignment, namely

$$\forall u, \pi(u) = \mu_F(u)$$

where π is a *possibility distribution*, defined as a function from U to $[0, 1]$. The following conventions are assumed : i) $\pi(u) = 0$ means that $X = u$ is impossible, totally excluded ; ii) $\pi(u) = 1$ means that $X = u$ is fully possible, but several distinct u and u' are allowed to be simultaneously such that $\pi(u) = 1$ and $\pi(u') = 1$; iii) the larger $\pi(u)$, the more possible, i.e., plausible or feasible u is. Thus, a possibility distribution encodes a *flexible* restriction (constraint). The above equality $\pi(u) = \mu_F(u)$ states that $X = u$ is possible, inasmuch as u is compatible with F, given the piece of information "X is F". The compatibility of u with F is estimated as the degree of membership function $\mu_F(u)$ of u to F. Two more remarks are of importance here. First, the range $[0, 1]$ of π may be replaced by any linearly ordered, possibly finite, scale. Second, although in the examples originally used by Zadeh, and many authors after him, U is a continuum (U is a subset of the real line), U may be finite and is not necessarily ordered. Thus, U may be a set of possible states of the world that are encoded by the interpretations associated with a classical propositional logic language, and then π rank-orders the possible states by their level of possibility according to the available information.

Two set functions, also called " measures ", are associated with a possibility distribution π, namely
– a possibility measure Π defined by

$$\Pi(A) = \sup\{\pi(u) : u \in A\},$$

which estimates to what extent the classical event A is *not inconsistent with* the information "X is F" represented by π ;
– a dual measure of necessity N, expressing that an event is all the more necessarily (certainly) true as the opposite event is more impossible. N is usually defined by

$$N(A) = 1 - \Pi(A^c) = 1 - \sup\{\pi(u) : u \notin A\},$$

where $A^c = U - A$, is the complement of A. N(A) estimates to what extent event A is *implied* by the information "X is F" represented by π (inasmuch as this information entails that any realization of A^c is more or less impossible).

As a consequence, the possibility and necessity measures Π and N satisfy the characteristic properties respectively (for classical or fuzzy events)

$$\Pi(A \cup B) = \max(\Pi(A), \Pi(B))$$

and

$$N(A \cap B) = \min(N(A), N(B)).$$

When A is a classical event, one can thus distinguish between the following epistemic states of knowledge, assuming that the possibility distribution π is normalized

($\exists u, \pi(u) = 1$), which expresses that there is no underlying contradiction pervading the available information (at least one value of U is completely possible) :

- $N(A) = 1 = \Pi(A)$: Given that X is F, *A is certain* ("X is A" is true in all non-impossible situations, knowing that "X is F");
- $N(A) > 0$ and $\Pi(A) = 1$: Given that X is F, A is normally true ("X is A" is true in all the most plausible situations)
- $N(A) = 0$ and $\Pi(A) = 1$: this is a state of *complete ignorance* about A, since both A and A^c are completely possible ($\Pi(A) = 1 = \Pi(A^c)$). Note that we should have here $A \neq U$, since $N(U) = 1$ and $\Pi(\emptyset) = 0$ are assumed.
- $N(A) = 0$ and $\Pi(A) < 1$: Given that X is F, A is normally false (all situations u where "X is A" is true are somewhat impossible, i.e. $\pi(u) < 1$)
- $N(A) = 0$ and $\Pi(A) = 0$: Given that X is F, *A is certainly false* ($\forall u \in A$, $\pi(u) = 0$, and then "X is A" is false and equivalently, "X is A^c" is true in all non-impossible situations u, that is in all situations somewhat compatible with "X is F")

Indeed the normalization of π ensures, that $\forall A$, $\Pi(A) \geq N(A)$, i.e. any event A should be possible before being certain. When A is a classical event, the above inequality strengthens into $N(A) > 0 \Rightarrow \Pi(A) = 1$. Besides, complete ignorance about any non-trivial event $A \neq \emptyset$ is represented by $\forall u \in U, \pi(u) = 1$.

These two measures extend to fuzzy events. Then the degree of consistency of A and F, is usually defined by

$$\Pi(A) = \sup_u \min(\mu_A(u), \pi(u)),$$

where μ_A is the membership function of A.

2.2 Possibilistic Information Combination Agrees with Classical Logic

Observe that any state of knowledge, represented by a possibility distribution π, compatible with a piece of information of the form "X is F", when F is a classical subset, should be such that $\forall u$ s. t. $\pi(u) > 0$, $u \in F$, i.e. there is no situation u that is non completely impossible which makes F false.

Note that there may be some $u \in F$ s. t. $\pi(u) = 0$ due to other piece(s) of information already taken into account. More generally, when F may be fuzzy, the granule of knowledge "X is F", viewed as a constraint on the epistemic state, translates into the inequality

$$\forall u, \pi(u) \leq \mu_F(u).$$

Thus given several pieces of knowledge of the form "X is F_i", for i = 1, ..., n, each of them translates into the constraint

$$\pi \leq \mu_{Fi} \text{ for all } i = 1, \ldots, n$$

and by consequence, several constraints "X is F_i" are equivalent to

$$\pi \leq \min_i \mu_{Fi}$$

By after taking into account *all* the available pieces of information, it is reasonable to apply a principle of minimal commitment that stipulates that *anything not declared impossible should remain possible* (in other words, let us not more restrictive about the possible situations than what is imposed by the available pieces of information. This principle is often called the "minimal specificity principle". Thus, given that all the available information is constituted by the set of pieces of knowledge "X is F_i", for $i = 1, \ldots, n$, this available information should be represented by the possibility distribution

$$\pi = \min_i \mu_{Fi}$$

Clearly, the above conjunctive combination of the constraints followed by the application of the minimal specificity principle would also make sense for pieces of knowledge of the form "$(X_1, \ldots X_m)$ is R" relating together tuples of variables, replacing X by a vector of variables. Note that a constraint encoding "**X** is F" (where **X** may be a vector of variables) can be always extended to any tuple (X, Y), where Y stands for another sub-vector of distinct variables, by rewriting the constraint $\forall u, \pi(u) < \mu_F(u)$ under the form

$$\forall u, \forall v_1, \ldots, \forall v_k, \pi(u, v_1, \ldots, v_k) < \mu_F(u, v_1, \ldots, v_k)$$

where $\forall u, \forall v_1, \ldots, \forall v_k, \pi(u, v_1, \ldots, v_k) = \pi(u)$ and $\mu_F(u, v_1, \ldots, v_k) = \mu_F(u)$. Indeed, we are just repeating the same inequality.

It is noticeable that this way of combining pieces of information fully agrees with the combination of information in logic, where a classical logic base is generally viewed as equivalent to the logical conjunction of the logical formulas that belong to the base. Indeed, in the case of propositional logic, asserting p, where p is a proposition, amounts to saying that any interpretation (situation) that makes p false is impossible, because it would not be compatible with the state of knowledge.

2.3 Possibility Theory May be Qualitative or Quantitative

We have already pointed out that a possibility distribution may range on scales, other than the unit interval [0, 1]. This gives birth to the *qualitative* or *quantitative* possibility theory settings according to whether possibility degrees respectively belong to

- a discrete linearly ordered scale;
- the unit real interval [0, 1].

At the formal level, the two settings only differ on the way conditioning is defined. In the qualitative setting, we have

$$\Pi(A \cap B) = \min(\Pi(B|A), \Pi(A))$$

in order to express in a qualitative way that having " A and B true " is possible as much as having " A true " is possible and having " B true " is possible in the context where A is true. Then $\Pi(B|A)$ is defined from the above equation by looking for its greatest solution and by applying the minimal specificity principle. Namely, since $\Pi(A \cap B) \leq \Pi(A)$ always hold by monotony of the set function Π, we have

$$\Pi(B|A) = 1 \text{ if } \Pi(A \cap B) = \Pi(A)$$
$$\Pi(B|A) = \Pi(A \cap B) \text{ if } \Pi(A \cap B) < \Pi(A)$$

Then, the conditional necessity is defined by duality

$$N(B|A) = 1 - \Pi(B^c|A)$$

Then it can be checked that

$$N(B|A) > 0 \Leftrightarrow \Pi(A \cap B) > \Pi(A \cap B^c)$$

which expresses that B is an accepted belief to some extent when knowing that A is true if and only if having $A \cap B$ true is strictly more possible than having $A \cap B^c$ true. This is the basis for representing default rules of the form " if A is true then generally B is true " by the above inequality.

In the quantitative setting, conditioning is defined by means of the product for continuity reasons, in a way that parallels the Dempster rule of conditioning in Shafer belief functions theory, namely,

$$\Pi(A \cap B) = \Pi(B|A) \cdot \Pi(A)$$

Quantitative possibility theory can be directly compared and related to probabilities. Indeed possibility and necessity degrees then provide a simple, but non-trivial, system of upper and lower probabilities where an ill-known probability P is described by means of constraints of the form
$\Pi(A) \geq P(A) \geq N(A)$. We do not develop further this point here, since the paper focuses on qualitative possibility theory.

2.4 Possibility Theory can be Used for Modeling both Uncertainty and Preferences

A practical feature of possibility theory that is worth emphasizing is its interest for modeling uncertainty as well as preferences. Originally, possibility theory has been introduced as a framework for dealing with the uncertainty induced by states of partial knowledge that are linguistically expressed. Then, possibility degrees have to be understood as *levels of plausibility* attached to possible states of the world, and the measure of necessity of event A, N(A), estimates how *certain* we are to be in A, given the available information represented by the possibility distribution underlying N. Dually, $\Pi(A)$ estimates to what extent being in A is *not impossible*, not surprising given the available information.

When possibility is used for representing preferences, possibility degrees reflect the *satisfaction levels* of alternative choices. In other words, the possibility distribution restricts the set of possible alternatives that are more or less acceptable for the decision maker. Then $\Pi(A)$ estimates how much it is *feasible to choose* an alternative in A. N(A) is then an estimate of how *imperative* it is to choose the solution in A, i.e., how dissatisfactory it is to make the choice outside A. N(A) is the imperativeness of taking "the choice should be in A" as a *goal*.

2.5 Bipolar Possibility Theory Processes Positive and Negative Information Differently

When representing knowledge, or preferences, it may be fruitful to distinguish between negative and positive information. Possibility theory provides a dual, *bipolar* representation framework for such a representation. Negative knowledge is usually generic, like in integrity constraints, laws, necessary conditions, which state what is impossible, forbidden. Observed cases, examples of solutions, sufficient conditions are positive pieces of information. The bipolar view applies to preference representation as well when telling apart positive desires from what is not more or less rejected. More precisely, negative preferences express solutions that are rejected or unacceptable to some extent, while positive preferences express solutions that are desired, pursued, among feasible ones.

When modeling*negative statements,* the values of a necessity measure for different events correspond to estimates of the *impossibility* of the opposite events, due to the duality between necessity and possibility, namely $N(A) = 1 - \Pi(\neg A)$. In particular, $\Pi(\neg A) = 0 \Leftrightarrow N(A) = 1$, i. e. A is completely certain or imperative is and only if "not A" (A^c) is fully impossible or forbidden. As already noted, knowing that N(A) = 1 provides much *more information* than just knowing that $\Pi(A) = 1$. Indeed, by *duality* a necessity measure N expresses *certainty* or *priority* levels. A necessity measure N is the expression of a more or less strong impossibility, through duality, and corresponds to negative information.

Positive statements are reports on *actually observed* states when dealing with pieces of knowledge, or on sets of choices that are *guaranteed to be* feasible or satisfactory (at some minimal level). Thus, positive information refers to cases. Then, a granule of the form " X is F ", no longer means " X *should be* in F ", but now means " X *can be* in F ". In other words, any value in F can be taken for X if we are modeling preferences, or any value in F may take place if we are dealing with uncertainty. Naturally, this may become a matter of degree: choosing any value in F may ensure some minimal guaranteed satisfaction level, while in case of uncertainty, any value in F is guaranteed to have a minimal level of "happen-ability". A granule of the type " X *can be* in F " will be represented by the assignment equation

$$\delta = \mu_F$$

where δ is a *guaranteed* possibility distribution that evaluates evidential support, where F may be fuzzy. δ is a mapping from U to a linearly ordered scale (in practice, a finite scale, or [0, 1] in a quantitative setting). At first glance, it looks very similar to the assignment $\pi = \mu_F$ used to represent " X *should be* in F ", but the conventions for interpreting δ are very different. Indeed,

- $\delta(u) = 1$ means that $X = u$ is *guaranteed to be possible* (u has been actually observed, or found satisfactory if u is a potential solution);
- $\delta(u) = 0$ means that $X = u$ has *not* been *observed* (yet), or that it is not known how much satisfactory is u as a potential solution (because it has been never tried).

Thus, $\delta(u) = \alpha$ is all the more a useful piece of information as α is high. This contrasts with π, where $\pi(u) = \alpha$ is all the more a useful piece of information as α is small and expresses impossibility. In contrast with negative information (represented by π), complete ignorance about positive information is represented by $\forall u \in U, \delta(u) = 0$.

A measure of *guaranteed possibility* is associated with δ, namely,

$$\Delta(A) = \inf\{\delta(u) : u \in A\}.$$

It estimates to what extent *all* situations where A is true are indeed possible for sure, as induced by " X *can be* in F ". Measures of *guaranteed possibility* are characterized by the property

$$\Delta(A \cup B) = \min(\Delta(A), \Delta(B)).$$

This expresses that guaranteeing all the values in A∪B as being possible amounts as guaranteeing as possible all the values in A on the one hand, and all the values in B on the other hand.

Combining pieces of positive information turns to be a disjunctive process, while combining pieces of negative information is a conjunctive process, as explained in Sect. 2.2. Indeed, an observation « X *can be* in F » should be understood as the inequality

$$\delta \geq \mu_F$$

expressing that any representation δ compatible with this observation should be point-wisely at least equal to the membership function of F (understood as a minimal guaranteed possibility degree). Then several pieces of positive information "X *can be* in F_j" for $j = 1, \ldots, m$ are equivalent to a collection of inequalities $\delta \geq \mu_{F_j}$, and thus to

$$\delta \geq \max_i \mu_{F_i}.$$

Then a principle of minimal commitment, which here corresponds to a closed world assumption, i. e., anything not observed as actually possible, or asserted as actually satisfactory is ruled out. This leads to

$$\delta = \max_i \mu_{F_i}.$$

This corresponds to a maximal specificity principle. Combining positive information is an accumulative (hence disjunctive) aggregation process.

Although positive and negative information are represented in separate and different ways via δ and π respectively, there is a coherence condition that should hold between positive and negative information. Indeed observed information should not be impossible, or more generally observed information should not be highly impossible inasmuch the observation is regarded as certain, and in terms of preferences solutions that are guaranteed to be satisfactory to some extent should not be rejected (at least strongly). This leads to enforce the coherence constraint

$$\delta \leq \pi$$

between the two representations. This should be maintained when new information arrives and is combined with the previous one. This does not go for free since δ tends to increase while π tends to decrease due to the disjunctive and conjunctive processes that respectively govern the information combination. This coherence maintenance requires a revision process that works as follows. If the current information state is represented by the pair (δ, π), receiving a new positive (resp. negative) piece of information represented by δ^{new} (resp. π^{new}) that we want to enforce, yields to revise (δ, π) into $(\max(\delta, \delta^{new}), \pi^{revised})$ (resp. into $(\delta^{revised}, \min(\pi, \pi^{new}))$), using respectively

$$\pi^{revised} = \max(\pi, \delta^{new});$$
$$\delta^{revised} = \min(\pi^{new}, \delta).$$

It is important to notice that when both positive and negative pieces of information are collected, there are two options. Either priority is given to positive information (i. e. to δ-information) over negative information (i. e. to π-information) or vice-versa. The first option means that (past) positive information cannot be ruled out by (future) negative information. This may be found natural when observations (represented by δ) contradict beliefs (represented by N through π), at least provided that strong observations do not contradict fully entrenched beliefs (such a contradiction would take place if $\exists u_0$ s. t. $\delta(u_0) = 1$ and $\pi(u_0) = 0$). Up to this reservation, revising (δ, π) by $(\delta^{new}, \pi^{new})$ yields the new pair

$$(\delta^{\text{revised}}, \pi^{\text{revised}}) = (\max(\delta, \delta^{\text{new}}), \max(\min(\pi, \pi^{\text{new}}), \max(\delta, \delta^{\text{new}})))$$

The other option is chosen, i. e., priority is given to negative information over positive information, makes sense when handling preferences. Indeed, then, positive information may be viewed as stating what is pleasant or satisfactory, independently of what is unacceptable due to feasibility or integrity constraints stated by the negative part of the information. Then, revising (δ, π) by $(\delta^{\text{new}}, \pi^{\text{new}})$ would yield the new pair

$$(\delta^{\text{revised}}, \pi^{\text{revised}}) = (\min(\min(\pi, \pi^{\text{new}}), \max(\delta, \delta^{\text{new}})), \min(\pi, \pi^{\text{new}}))$$

It can be checked that the two latter formulas generalize the two previous expressions. With both revision options, it can be checked that if $\delta \leq \pi$ and $\delta^{\text{new}} \leq \pi^{\text{new}}$ hold, revising (δ, π) by $(\delta^{\text{new}}, \pi^{\text{new}})$ yields a new coherent pair. This revision process should not be confused with another one pertaining only to the negative part of the information, namely computing $\min(\pi, \pi^{\text{new}})$ may yield a possibility distribution that is not normalized, which means that some inconsistency exists inside the negative information. If such an inconsistency takes place, it should be resolved (by some appropriate renormalization) before one of the two above bipolar revision mechanisms be applied.

3 Possibility Theory-based Representation of Knowledge and Preferences

A noticeable feature of the possibilistic framework for representing information is its compatibility with different types of representation format. Apart from the semantic format expressed in terms of distributions, there exist three other types of (more compact) format: logical ones, graphical ones, and conditional ones, which correspond to different "natural" modes for expressing information.

The *possibilistic logic* format manipulates weighted formulas of the form (p, α) where p is a classical logic formula, and α is a weight belonging to a linearly ordered scale with top element 1. It is supposed to encode a constraint of the form $N([[p]]) \geq \alpha$ ($[[p]]$ denotes the set of interpretations that make p true). A N-possibilistic logic base is a set of such formulas, $\Sigma = \{(a_i, \gamma_i) : i = 1, \ldots, n\}$, which corresponds to a series of constraints $N(A_i) \geq \gamma_i$ (A_i is the set of interpretations that make a_i true) from which a possibility distribution π_Σ over the set of interpretations induced by the a_i's can be rebuilt. Namely, an interpretation u is all the less possible as it falsifies only formulas with higher weights, i. e.

$$\pi_\Sigma(u) = 1 \text{ if } \forall \, (a_i, \gamma_i) \in \Sigma, u \models a_i$$
$$\pi_\Sigma(u) = 1 - \max\{\gamma_i \, (a_i, \gamma_i) \in \Sigma, u \not\models a_i\} \text{ otherwise.}$$

where $1 - ()$ denotes the order-reversing map of the scale. This distribution is obtained by applying the minimal specificity principle, since it is the largest one that satisfy the constraints $N(A_i) \geq \gamma_i$. If the classical logic base $\{a_i, : i = 1, \ldots, n\}$ is inconsistent, π_Σ is not normalized, a level of inconsistency equal to inc $(\Sigma) = 1 - \max_u \pi_\Sigma(u)$ can be attached to the base Σ. The set of formulas $\{a_i: (a_i, \gamma_i) \in \Sigma$ and $\gamma_i > inc(\Sigma)\}$ is consistent. Deduction is governed for propositional possibilistic logic by the cut rule

$$(a \vee b, \alpha), (\neg a \vee c, \beta) \mid - (b \vee c, \min(\alpha, \beta)),$$

generalizing resolution. Constraints of the form $\Pi([[p]]) \geq \alpha$ can also be handled, but are less interesting since they convey very poor information. Besides, constraints of the form $\Delta([[p]]) \geq \alpha$, can also be processed, and are denoted by $[p, \alpha]$. Similarly, a Δ-possibilistic logic base is a set of such formulas, $\Xi = \{[b_j, \delta_j] : j = 1, \ldots, m\}$, where $[b_j, \delta_j]$ encodes $\Delta(B_j) \geq \delta_j$ (B_j is the set of models of b_j). Associated with this base is a guaranteed possibility distribution defined by

$$\delta_\Xi(u) = 0 \text{ if}/\exists j, u| = b_j \text{ and } [b_j, \delta_j] \in \Xi\}$$
$$\delta_\Xi(u) = \max\{\delta_j : u| = b_j, [b_j, \delta_j] \in \Xi\} \text{ otherwise}$$

that obeys the maximal specificity principle. $\delta_\Xi(u)$ is all the greater as u satisfies formulas b_j associated with a high degree of guaranteed possibility δ_j. Inference is governed by the cut rule

$$[a \wedge b, \alpha], [\neg a \wedge c, \beta] \mid - [b \wedge c, \min(\alpha, \beta)].$$

The graphical format is based on a *possibilistic counterpart of Bayesian nets*. This format relies on the definition of conditional possibility (which may be based on minimum or product, depending if we are in a qualitative or a quantitative setting, as already mentioned in Sect. 2.3). Bayesian nets are directed acyclic graphs, where each node corresponds to a variable. Normalized conditional measures relate a variable to the variables of the father nodes. A qualitative possibilistic belief net reflects the decomposition of a joint possibility distribution associated with variables $X_1, \ldots, X_n$, by the chain rule

$$\pi(x_1, \ldots, x_n) = \min(\pi(x_n|x_1, \ldots, x_{n-1}), \ldots, \pi(x_2|x_1), \pi(x_1))$$

Such a decomposition can be simplified by assuming conditional independence relations between variables, which are reflected by the structure of the graph. The form of independence between variables at work here is conditional *non-interactivity*: Two variables X and Y are independent in the context Z, if for each instance (x, y, z) of (X, Y, Z) we have:

$$\pi(x, y | z) = \min(\pi(x|z), \pi(y|z)).$$

Translation procedures without loss of information between a possibilistic logic base and a minimum (or product)-based possibilistic Bayesian net have been developed. Nets based on a conditional measure $\Delta(q|p)$ can be also defined, using the relation $\Delta(p \wedge q) = \max(\Delta(q|p), \Delta(p))$.

A possibilistic conditional default is a constraint of the form $\Pi(p \wedge q) > \Pi(p \wedge \neg q)$ (which is equivalent to $N(q|p) > 0$). A possibilistic conditional default base is a set of such constraints. At the semantic level such a base is equivalent to a qualitative possibility distribution, which is a partition of the set of interpretations that is stratified. Each element in the partition forms a class of interpretations having the same level of possibility, each class being associated with a different level. Constraints of the form $\Pi(p \wedge q) > \Pi(p \wedge \neg q)$ express that in the context where p, it is strictly more possible to have q true than q false. They were proved to provide a proper representation framework for default rules of the form "if p then generally q", and their processing in non-monotonic reasoning. Translation procedures without loss of information exist between a possibilistic conditional default base and a possibilistic logic base. Similar results have been obtained for constraints of the form $\Delta(p \wedge q) > \Delta(p \wedge \neg q) \Leftrightarrow \Delta(q|p) > 0$.

This is a distinctive feature of the possibilistic setting to have these different representation formats and to have procedures for going from one format to another, being insured to obtain a new representation that is semantically equivalent to the initial one in terms of the underling possibility distribution. It applies both to *knowledge* and *preference* representation. The following toy example illustrates the use of the different representation formats for expressing preferences.

Example 1. Assume we want to represent the following preferences: "I prefer to take tea (t)' and 'If there is no tea then I will take coffee (c)". This directly translates into the conditional format. Namely, we have the two constraints

$$\Pi(t) > \Pi(\neg t)$$
$$\Pi(\neg t \wedge c) > \Pi(\neg t \wedge \neg c)$$

There exists a unique, minimally specific, possibility distribution satisfying a set of consistent constraints encoding possibilistic conditional defaults. In the example this qualitative possibility distribution is equal to

$$\pi(ct) = 1; \pi(\neg ct) = 1; \pi(c\neg t) = \alpha; \pi(\neg c\neg t) = \beta, \text{ where } \alpha > \beta.$$

This expresses that having tea with or without coffee available are the most satisfactory situations, then having coffee when there is no tea is less satisfactory, and finally having neither tea nor coffee is still less satisfactory.

The above possibility distribution can be associated to a N-possibilistic base

$$\Sigma = \{(c \vee t, 1 - \beta), (t, 1 - \alpha)\}.$$

It expresses that it is quite imperative to have tea or coffee, and if the former is satisfied, it is more imperative to have tea rather than coffee ($1 - \alpha < 1 - \beta$),

It can also be modeled by a Δ-type one (note that here we are not working in a bipolar perspective since one deals with only one distribution that is "read" in two different ways):

$$\Xi = \{[t, 1], [c \wedge \neg t, \alpha], [\neg c \wedge \neg t, \beta]\},$$

which expresses that it is fully satisfactory to have tea, less satisfactory to have coffee but no tea, and still less satisfactory to have no coffee and no tea. Generally speaking, a N-possibilistic base can be equivalently put in a conjunctive normal form, and a Δ-possibilistic base in a disjunctive normal form.

Lastly, the graphical encoding of the preferences is given by the possibilistic Bayesian network

$$\pi(t) = 1 \; ; \; \pi(\neg t) = \alpha.$$
$$\pi(c|\neg t) = 1; \; \pi(\neg c|\neg t) = \beta \; ; \; \pi(c|t) = 1 \; ; \; \pi(\neg c|t) = 1.$$

The interest of being able to jointly handle the different formats is that a format may provide a more concise, and thus natural, way for expressing preferences than another format in a given situation, as illustrated by the following example.

Example 2. Our preferences for a house to let for a week is a location 'near the sea' and an 'affordable' price. This can be expressed by the N-possibilistic base

$$\Sigma \quad = \{(d \leq 15, 1), (d \leq 10, .8), (d \leq 5, .3), (p \leq 400, 1), (p \leq 200, .5)\},$$

which states that being at a distance d less than 15 km is imperative, being at less than 10 km is slightly less imperative and so on (p denotes the price). Interestingly enough, the Δ-possibilistic base representation is much more complex here:

$$\Xi = \{[(d \leq 5) \wedge (p \leq 200), 1], [(5 < d \leq 10) \wedge (p \leq 200), .7], [(d \leq 10)$$
$$\wedge (200 < p \leq 400), .5], [(10 < d \leq 15) \wedge (p \leq 400), .2]\}$$

This would be the converse with the *disjunctive* preference 'near the sea' *or* 'affordable price'. Indeed the two corresponding representations would then be:

$$\Xi' = \{[d \leq 5, 1], [d \leq 10, .7], [d \leq 15, .2], [p \leq 400, .5], [p \leq 200, 1]\}$$
$$\Sigma' = \{(d \leq 15 \vee p \leq 400, 1), (d \leq 10 \vee p \leq 400, .8), (d \leq 10 \vee p \leq 200, .5),$$
$$(d \leq 5 \vee p \leq 200, .3)\}$$

Comparing alternatives described by multi-factorial ratings is a matter of defining a pre-ordering between these ratings. Rather than using an aggregation function for that purpose as generally done, the preferences of a user between evaluation vectors may be expressed by means of generic rules and prototypical examples. Indeed let $u = (x_1, \ldots, x_n)$ and $v = (y_1, \ldots, y_n)$ be two evaluation vectors

(x_i denotes the value of attribute i for the considered alternative). A preference between two vectors u and v can be encoded in terms of a possibility distribution, stating that u is more satisfactory than v as an inequality of the form $\pi(u) > \pi(v)$. This enables us to specify *generic* requirements, expressing for instance agreement with Pareto ordering, or the greater importance of a criterion with respect to another another, by means of inequalities between partially specified vectors encoding ceteris paribus preferences (e. g., criterion i is more important than j : $\pi(\ldots x_i, \ldots, y_j, \ldots) > \pi(\ldots y_i, \ldots, x_j, \ldots)$) holds for all values of x and y such that $x > y$, where the values of criteria i and j are swapped in the vector). Ceteris paribus ("everything else being equal") preferences express preferences between partially specified situations that apply in any context described by a particular instantiation of the other variables. Besides, preferences between fully *specificied instances* $\pi(a_1, \ldots, a_n) > \pi(b_1, \ldots, b_n)$ can be given as examples of preferences that should hold between typical situations. Then, the application of the minimal specificity principle to these generic and these instantiated constraints yields a complete pre-ordering encoded by a qualitative possibility distribution that stratifies the set of alternatives according their level of satisfaction. The application of the minimal specificity principle means here that any alternative is by default assumed to be satisfactory unless constraints lead to consider it as less satisfactory than others.

Note that the information completion principle at work always presupposes that the non-committal level is chosen by default. On [0, 1] viewed as a possibility scale, which is a negative one, this level is 1, hence the principle of minimal specificity is encoded by maximizing degrees of possibility. On a guaranteed possibility scale, the non-committal level is 0, hence minimizing guaranteed possibility levels is appropriate.

Besides, a comparison of the representation capabilities of the possibilistic setting for handling preferences with other approaches such as conditional preference networks (CP-nets) that are based on a systematic ceteris paribus assumption and that induce partial orders rather than pre-orders, has started to be done. The possibilistic framework offers a large variety of representation formats that may be more or less natural to use depending on what one has to express, and that can be translated into each other. It provides a rich language for representing preferences in an accurate way.

4 Possibilistic Reasoning, Decision and Learning

In this section, we survey recent developments of the use of possibility theory for modeling different types of reasoning. The first three subsections respectively deal with advances in possibilistic logic inference machineries, discuss possibilistic description logics, and show the interest of possibilistic logic for syntactic information fusion. The next three subsections will cover temporal reasoning, bipolar approximate reasoning, and case-based reasoning. The last two subsections briefly surveys possibilistic decision and learning.

4.1 Possibilistic Logic Inference Machineries

As already recalled in Sect. 3, a possiblistic logic formula is a pair (p, α) of a classical logic formula p and a weight α belonging to a linearly ordered scale. It can be used for handling qualitative uncertainty or preferences, as well as exceptions (via the computation of a level of inconsistency).

A partially ordered extension of possibilistic logic has been proposed, whose semantic counterpart consists of partially ordered models. Another approach for handling partial orderings between weights is to encode formulas with partially constrained weights in a possibilistic-like many-sorted propositional logic. Namely, a formula (p, α) is rewritten as a classical two-sorted clause p $\vee$ A, where A means "the situation is A-abnormal", and thus the clause expresses that "p is true or the situation is abnormal", while more generally (p, $\min(\alpha, \beta)$) is rewritten as the clause p $\vee$ A $\vee$ B. Then a known constraint between weights such as $\alpha \geq \beta$ is translated into a clause $\neg A \vee B$ with the understanding that "if the situation is at least very abnormal (A), it is at least abnormal (B)" (the larger α in (p, α), the more certain p, and the more abnormal a situation where p is false). In this way, a possibilistic logic base, where only partial information about the relative ordering between the weights is available under the form of constraints, can be handled as a set of classical logic formulas that involve symbolic weights. Then an efficient inference process has been proposed using the notion of forgetting variables. This approach is also of interest for ordinary possibilistic logic bases, where the weights are completely ordered, and provides a way for compiling a possibilistic knowledge base in order to process inference from it in polynomial time. Let us also mention quasi-possibilistic logic, an extension of possibilistic logic based on quasi-classical logic, a paraconsistent logic whose inference mechanism is close to classical inference (except that if p is obtained as a conclusion, it is not allowed to infer p $\vee$ q from it). This enables us to cope with inconsistency between formulas having the same weight.

There is a major difference between possibilistic logic and weighted many-valued logics. Namely, in the latter, a weight τ attached to a (many-valued, thus non-classical) formula p acts as a truth-value threshold, and (p, τ) in a fuzzy knowledge base expresses the requirement that the truth-value of p should be at least equal to τ for (p, τ) to be valid. So in such fuzzy logics, while truth is many-valued, the validity of a weighted formula is two-valued. On the contrary, in possibilistic logic, truth is two-valued (since p is Boolean), but the validity of a possibilistic formula (p, α) is many-valued (indeed the possibility distribution that is its semantic counterpart is $\pi_{\{(p,\alpha)\}}(u) = 1 - \alpha$ if $u \not\models p$ and $\pi_{\{(p,\alpha)\}}(u) = 1$ if $u \models p$). In particular, it is possible to cast possibilistic logic inside a (regular) many-valued logic. The idea is to consider many-valued atomic sentences ϕ of the form (p, α) where p is a formula in classical logic. Then, one can define well-formed formulas such as $\phi \vee \psi$, $\phi \wedge \psi$ or $\phi \rightarrow \psi$, where the " external " connectives linking ϕ and ψ are those of the chosen many-valued logic. From this point of view, possibilistic logic can be viewed as a fragment of a many-valued logic that uses only one external connective: conjunction $\wedge$ interpreted as minimum. This approach involving a Boolean algebra embedded in a non-classical one has been proposed by Boldrin and Sossai with a view to augment possibilistic logic with fusion modes cast at the object level. In a different way, Lehmke has cast fuzzy logics and possibilistic logic inside the same framework, considering weighted many-valued formulas of the form (p, θ), where p is a many-valued formula with truth set T, and θ is a "label" defined as a monotone mapping from the truth-set T to a validity set L. T and L are supposed to be complete lattices, and the set of labels has properties that make it a fuzzy extension of a filter in L^T. Labels encompass "fuzzy truth-values" in the sense of Zadeh, such as "very true", "more or less true" that express uncertainty about (many-valued) truth in a graded way.

Rather than expressing, for instance, that "it is half-true that John is tall", which presupposes a state of complete knowledge about John's height, one may be interested in handling states of incomplete knowledge, such as "all we know is that John is tall". One way to do it is to introduce fuzzy constants in a possibilistic first ordered logic. Dubois, Prade, and Sandri have noticed that an imprecise restriction on the scope of an existential quantifier can be handled in the following way. From the two premises $\forall x \in A \neg p(x, y) \vee q(x, y)$, and $\exists x \in B\ p(x, a)$, where a is a constant, we can conclude that $\exists x \in B\ q(x, a)$ *provided that* $B \subseteq A$. Thus, letting $p(B, a)$ stand for that $\exists x \in B\ p(x, a)$, one can write $\forall x \in A \neg p(x, y) \vee q(x, y), p(B, a) \mid - q(B, a)$ if $B \subseteq A$, B being an imprecise constant. Letting A and B be fuzzy sets, the following pattern can be established

$$(p(x, y) \vee q(x, y), \min(\mu_A(x), \alpha)), (p(B, a), \beta) \mid - (q(B, a), \min(N_B(A), \alpha, \beta)$$

where $N_B(A) = \inf_t \max(\mu_A(t), 1 - \mu_B(t))$ is the necessity measure of the fuzzy event A based on fuzzy information B. Note that A, which appears in the weight slot of the first possibilistic formula plays the role of a fuzzy predicate, since the formula expresses that "the more x is A, the more certain (up to level α) that if p is true for (x, y), q is true for them as well". Alsinet and Godo have further developed similar ideas in a logic programming perspective. In particular the above pattern can be strengthened, replacing B by the cut B_β in $N_B(A)$ and extended to a sound resolution rule. These authors have cast possibilistic logic in the framework of Gödel many-valued logic, and have developed programming environments based on possibility theory and that allows for fuzziness and more recently argumentation.

Lastly, in order to improve the knowledge representation power of the answer set programming paradigm, the stable model semantics has been extended by taking into account a certainty level, expressed in terms of necessity measure, on each rule of a normal logic program. It leads to the definition of a possibilistic stable model for a normal logic program, with default negation.

4.2 Description Logics

Description logics (initially named 'terminological logics') are tractable fragments of first-order logic representation languages that handle the notions of concepts, roles and instances, thus directly relying at the semantic level on the respective notions of set, binary relations, membership, and cardinality. They are useful for describing ontologies that consist in hierarchies of concepts in a particular domain, for the semantic web.

Two ideas that respectively come from fuzzy sets and possibility theory, and that may be combined, may be used for extending the expressive power of description logics. On the one hand, vague concepts can be approximated in practice by pairs of nested sets corresponding to the cores and the supports of fuzzy sets, thus sorting out the typical elements, in a way that agrees with fuzzy set operations and inclusions. On the other hand, a possibilistic treatment of uncertainty and exceptions can be performed on top of a description logic in a possibilistic logic style. In both cases, the underlying principle is to remain as close as possible to classical logic for preserving computational efficiency as much as possible. Thus, formal expressions such as $(P \sqsubseteq^X_\alpha Q, \beta)$ intend to mean that "it is *certain* at least at level β, that the degree of subsumption of concept P in Q is at least α in the sense of some X-implication" (e.g., Gödel, or Dienes implication). In particular, it can be expressed that "typical P's are Q's", or that "typical P's are typical Q's", or that an instance is typical of a concept.

4.3 Syntactic Information Fusion

Information fusion can be performed at the semantic level on possibility distributions, or syntactically on more compact representations such as possibilistic logic bases (the merging of possibilistic networks has been also recently considered). The latter type of fusion may be of interest both from a computational and from representational points of view. Still it is important to make sure that the syntactic operations are counterparts of semantic ones. A similar problem exists in belief revision where an epistemic state, represented either by a possibility distribution or by a possibilistic logic base, is revised by an input information p. Then different operations can be considered depending if in the revised epistemic state one wants to enforce $N(p) = 1$, or $N(p) > 0$, or if we are dealing with an uncertain input (p, α). Then, the uncertain input may be understood as enforcing $N(p) > \alpha$ in any case, or as taking it into account only if it is sufficiently certain w. r. t. the current epistemic state. All these operations can be performed both at the semantic and at the syntactic levels equivalently.

Fuzzy set theory offers a large panoply of aggregation operations for combining possibility distributions that represent epistemic states in a conjunctive, disjunctive, or adaptive manner. Since possibilistic logic bases are associated semantically with possibility distributions, syntactic counterparts of these operations can be developed. Namely, given a semantic combination rule C, one can define a syntactic combination C where $\Sigma^C = C(\Sigma^1, \ldots, \Sigma^n)$ is the result of merging $\Sigma^1, \ldots, \Sigma^n$ that semantically agrees with C, i.e. such that $C(\pi_{\Sigma 1}, \ldots, \pi_{\Sigma n}) = \pi_{\Sigma C}$, where π_Σ denotes the possibility distribution associated with a base Σ.

For instance, given two bases $\Sigma_1 = \{(p_i, \alpha_i), i = 1, \ldots, n\}$ and $\Sigma_2 = \{(q_j, \beta_j), j = 1, \ldots, m\}$, and a monotonic combination operation $\otimes$ such that $1 \otimes 1 = 1$, then

$$\Sigma_C = \{(p_i \vee q_j, 1 - (1 - \alpha_i) \otimes (1 - \beta_j)) \mid (p_i, \alpha_i) \in \Sigma_1 \text{ and } (q_j, \beta_j) \in \Sigma_2\} \cup$$
$$\{(p_i, 1 - (1 - \alpha_i) \otimes 1)) \mid (p_i, \alpha_i) \in \Sigma_1\} \cup \{(q_j, 1 - 1 \otimes (1 - \beta_j))$$
$$\mid (q_j, \beta_j) \in \Sigma_2\}.$$

where $1 -$ is the reversing map of the scale. For instance, if $\otimes$ is the arithmetic mean, we have

$1 - (1 - \alpha_i) \otimes (1 - \beta_j) = (\alpha_i + \beta_j)/2$ and $1 - (1 - \alpha_i) \otimes 1 = \alpha_i/2$. If $\otimes$ is the minimum, Σ_C is simply the union of Σ_1 and Σ_2.

The distance-based approach that applies to the fusion of classical logic bases can be embedded in this setting as well. To this end it is necessary to encode the distance $d(u, K)$ between an interpretation u and each classical base K (usually defined as $d(u, K) = \min\{H(u, u^*) \mid u^* \in [[K]]\}$ where $H(u, u^*)$ is the Hamming distance that evaluates the number of literals with different signs in u and u^*) into a possibilistic knowledge base. Given a classical base K, and a $\in (0, 1)$, the associated possibility distribution is defined by $\pi(u) = a^{d(u, K)}$. The result of the fusion is a possibilistic knowledge base of which the layer with the highest weight is the classical database that is searched for.

Besides, the fusion of bipolar information raises specific issues. Assume that n sources or agents provide both positive and negative information under the form of Δ- or N-possibilistic logic formulas respectively. As already explained in Sects. 2.2 and 2.5 respectively, a conjunctive combination of the negative parts $\Sigma_k = \{(a^k_i, \gamma^k_i) \mid i = 1, \ldots, n\}$ and a disjunctive combination of the positive parts $\Xi_k = \{[b^k_j, \delta^k] \mid i = 1, \ldots, n\}$ have to be performed. Then, if Σ (resp. Ξ) represents the result of the fusion of the negative (resp. positive) parts, then one has also to maintain the consistency condition semantically expressed by $\delta_\Xi \leq \pi_\Sigma$. For enforcing consistency, one may give priority either to positive information (in case of observations and beliefs), or to negative information (when combining both strong preferences modeled by constraints and weak preferences expressing "optional" wishes represented by positive information). These different operations can be performed directly at the syntactic level.

4.4 Temporal Reasoning

Temporal reasoning may refer to time intervals or to time points. When handling time intervals, the basic building block is the one provided by Allen relations between time intervals. There are thirteen relations that describe the possible relative locations of two intervals. For instance, given two intervals A = [a, a'] and B = [b, b'], A is *before* (resp. *after*) B iff $a' < b$ (resp. $b' < a$), A *meets* (resp. is *met by*) B iff a'= b (resp. b'= a), A *overlaps* (resp. is *overlapped by*) B iff *b > a and a' > b and b' > a'* (resp. *a > b and b' > a and a' > b'*). The introduction of fuzzy features in temporal reasoning can be related to two different issues.

First, it can be motivated by the need of a gradual, linguistic-like description of temporal relations even in the face of complete information. Then an extension of

Allen relational calculus has been proposed, which is based on fuzzy comparators expressing linguistic tolerance, which are used in place of the exact relations '>', '=', and '<'. Fuzzy Allen relations are thus defined from three fuzzy relations between dates *approximately equal (L)*, *clearly greater (K)*, and *clearly smaller* (K^{ant}), where, e.g., the extent to which x is approximately equal to y is the degree of membership of x − y to L. Note that L, K, K^{ant} make a fuzzy partition, which provides a way of looking to the relative position of two intervals in a new manner, since the use of '>' does enable us to make a difference between "much before" or "slightly before" for instance; see Fig. 1.

Second, the possibilistic handling of fuzzy or incomplete information leads to pervade *classical* Allen relations, and more generally fuzzy Allen relations, with uncertainty. Then patterns for propagating uncertainty and composing the different (fuzzy) Allen relations in a possibilistic way have been laid bare.

Besides, the handling of temporal reasoning in terms of relations between time points can be also extended in case of uncertain information. Uncertain relations between temporal points are represented by means of possibility distributions over the three basic relations '>', '=', and '<'. Operations for computing inverse relations, for composing relations, for combining relations coming from different sources and pertaining to the same temporal points, or for handling negation, have been defined. This shows that possibilistic temporal uncertainty can be handled in the setting of point algebra. The possibilistic approach can then be favorably compared with a probabilistic approach previously proposed (first, the approach can be purely qualitative, thus avoiding the necessity of quantifying uncertainty if information is poor, and second, it is capable of modeling ignorance in a non-biased way).

4.5 Bipolar Approximate Reasoning

Observe that a rule 'if A then B' creates a 3-partition of the situations that can be encountered: i) there are the examples of the rule that satisfy both A and B; ii) the counter-examples of the rule that satisfy A but not B ; iii) the situations that are irrelevant w. r. t. the rule since they do not satisfy A. Thus, the set of situations that are not counter-examples of the rule are those that satisfy not A or B. This leads

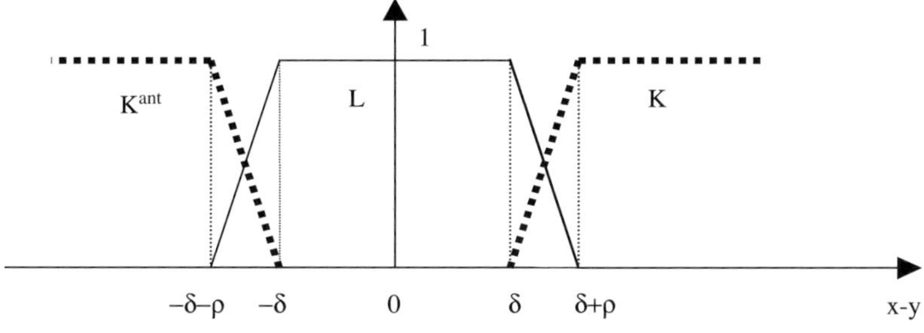

Fig. 1 A fuzzy partition of the possible relative positions of two points x and y

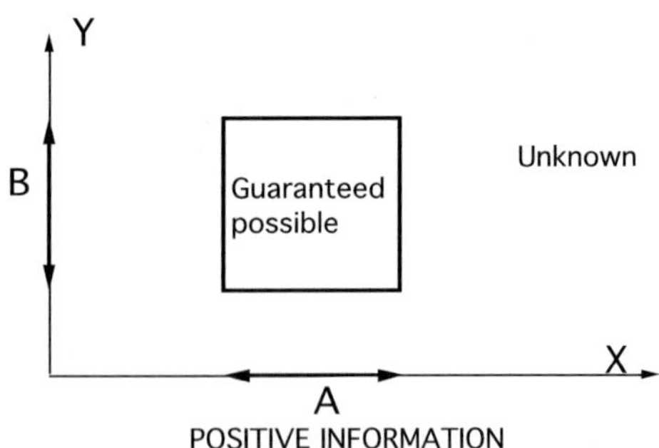

Fig. 2a The bipolar view of a rule 'if A then B'

to a bipolar view of the rule 'if X is A then Y is B', supposed to be represented by a relation R from the range U of X to the range V of Y. In the negative view, the rule is a *constraint* that expresses that "if X is A then Y *must be* B" and that translates into $(x, y) \in A \times V$ and $(x, y) \notin U \times B$ implies $(x, y) \notin R$. Thus, this leads to the *implicative* form of the rule, namely $R = A^c \times V \cup U \times B$ defines the complement of the set of counter-examples. In the positive view, the rule is reporting *cases*, i.e. the rule means "if X is A then Y *can be* B", which translates into $x \in A$ and $y \in B$ implies $(x, y) \in R$. It points out examples of the rule gathered in $A \times V \cap U \times B$. This is the *"conjunctive"* form of the rule. See Figs. 2a and 2b. This bipolar view extends to fuzzy rules (when A and/or B become fuzzy). The bipolar view of "if-then" rules can be exploited for building a typology of fuzzy "if then" rules, based on multiple-valued implications or conjunctions, where each type of fuzzy rules serves a specific purpose. Moreover, it makes sense to use conjointly implicative rules (encoding negative information) and conjunctive rules (encoding positive information) in the same rule-based system.

The bipolar view has important consequences for the proper handling of fuzzy rules in inference. First, considering a set of parallel fuzzy rules 'if X is A_i then Y is B_i', their implicative representations should be combined conjunctively, while their conjunctive representations have to be combined disjunctively (as in Mamdani's approach).

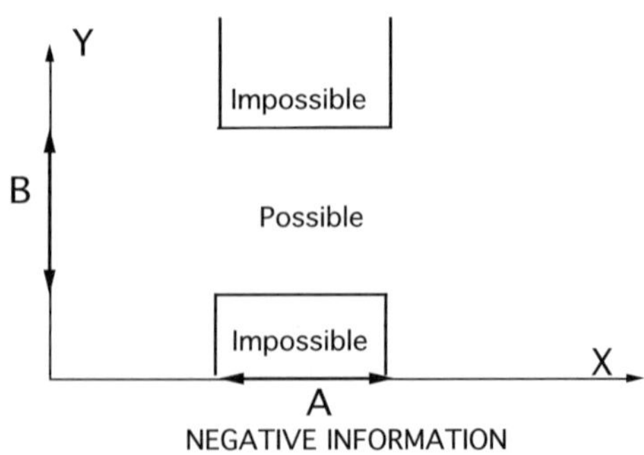

Fig. 2b The bipolar view of a rule 'if A then B'

Moreover, in presence of a fuzzy or imprecise input, the generalized modus ponens should not be defined in the same way for implicative and for conjunctive rules. Let us consider the case of implicative rules. Then, the input "X is $A°$" and the representations $A_i \rightarrow B_i$ of the rules, express fuzzy constraints, and the output "Y is $B°$" is computed as $B° = A° o \cap_i (A_i \rightarrow B_i)$, i.e. in terms of membership functions

$$\mu_{B°}(y) = \sup_x \min(\mu_{A°}(x), \min_i \mu_{Ai}(x) \rightarrow \mu_{Bi}(y)).$$

Note that, letting $B_i° = A° o (A_i \rightarrow B_i)$ then $\cap_i B_i° \supseteq A° o (\cap_i A_i \rightarrow B_i)$, which means that the rule-by-rule approach is not complete.

In case of conjunctive rules and fuzzy inputs, the usual approach is to compute

$$B° = A° o (\cup_i A_i \times B_i)$$
$$= \cup_i (A° o (A_i \times B_i)) \quad \text{(which allows for a rule by rule evaluation)}$$
$$= \cup_i(\text{Cons}(A°, A_i) * B_i)) \text{ (which is a weighted union of the } B_i\text{'s)}$$

where $\text{Cons}(A°, A_i) = \sup_x \min(\mu_{A°}(x), \mu_{Ai}(x))$ and $*$ stands for a conjunction operator such as minimum or product. This is questionable because it views $\cup_i A_i \times B_i$ as a constraint (i.e. as negative information). Moreover, the more rules are triggered the more imprecise the conclusion, adding a rule leads to more imprecise conclusions, and adding a redundant rule destroys information (for instance: "if anything then anything", because $(\cup_i A_i \times B_i) \cup (U \times V) = U \times V$. It is has also been noticed that with a fuzzy partition, there is no way of achieving $A_j o (\cup_i A_i \times B_i) = B_j$ except if the A_i's are disjoint.

Interpreting the conjunctive rules as *fuzzy cases* leads to restoring a proper approach to inference: Then adding a case leads to more possibilities, adding $U \times V$ means that for any input value, any output value is guaranteed possible, that is, there is no link at all between inputs and outputs. Thus, reasoning with conjunctive rules should not be based on the sup-min inference, since the input $A°$ has to be viewed as a restriction of focus on the input space, and the aim of the inference is to locate a set B_o of values on the output space, each of which is guaranteed attainable from any value compatible with the input. So the inference amounts at finding the output values guaranteed to be possible according to the case base, despite the imprecision of the input. In the crisp case, we have

$$B° = \{y, \forall x \in A°, \exists i, x \in A_i \text{ and } y \in B_i\}$$
$$= \cap_{x \in A°} (\cup_{i:x \in A_i} B_i)$$

$B°$ is the lower image $\cap_{x \in A°} R(x)$ of A through the relation $R = \cup_i A_i \times B_i$. In the fuzzy case, this leads to

$$\mu_{B_o}(y) = \inf_x \mu_{A°}(x) \rightarrow \max_i \min(\mu_{Ai}(x), \mu_{Bi}(y))$$
$$= \Delta_y(A°) \text{ based on } \delta = \mu_R(., y)$$

where $\rightarrow$ is Gödel implication ($a \rightarrow b = 1$ if $a \leq b$ and $a \rightarrow b = b$ otherwise).

This can be denoted $B_\circ = A^\circ \bigoplus (\cup_i A_i \times B_i)$. Then indeed, if the A_i's form a fuzzy partition, then we do have

$$B_j = A_j \bigoplus (\cup_i A_i \times B_i).$$

It is also interesting to see how the bipolar view of if-then rules can be handled in the bipolar possibilistic logic setting. Although $\_$-based formulas and N-based formulas are dealt with separately in the inference machinery, their parallel processing may be of interest, as shown now. Observe that a $\_$-possibilistic logic formula $[p\_q, \beta]$ is semantically equivalent to the formula $[q, \min(v(p), \beta)]$ where $v(p) = 1$ if p is true and $v(p) = 0$ if p is false. For N-possibilistic logic formulas, it is also possible to "move" a part of the formula in the weight slot, but in a different manner, namely $(\neg p\_q, \alpha)$ is equivalent to $(q, \min(v(p), \alpha))$. This remark enables us to deal with the application of a set of parallel uncertain "if then" rules, say "if pl then qi typically, and qi_ for sure", where each rule is represented by a pair of possibilistic formulas of each type, namely $([pi\_qi, \beta i], (\neg pi\_qi\_, \alpha i))$. Then, the rules can be rewritten as $([qi, \min(v(pi), \beta i)], (qi\_, \min(v(pI), \alpha i))$. Note that we should have here $qi| = qi\_$ for ensuring consistency. Now, observe that $[q1, \min(v(p1), \beta 1)]$ semantically entails $[q1\_q2, \min(v(p1), \beta 1)]$, and similarly $[q2, \min(v(p2), \beta 2)]$ entails $[q1\_q2, \min(v(p2), \beta 2)]$ since $[q, \beta]$ entails $[q\_r, \beta]$. Finally, since $[q, \beta]$ and $[q, \beta\_]$ entail $[q, \max(\beta, \beta\_)]$, we obtain $[q1\_q2, \max(\min(v(p1), \beta 1), \min(v(p2), \beta 2))]$. For N-possibilistic logic formulas, we have that (p, α) entails (p_s, α) on the one hand, while (p, β) and $(p, \beta\_)$ entail $(p, \max(\beta, \beta\_))$. So for the two rules, we obtain $(q1\_q2\_, \max(\min(v(p1), \alpha 1), \min(v(p2), \alpha 2)))$. For $\beta 1 = \beta 2 = 1$ and $\alpha 1 = \alpha 2 = 1$, this expresses that if it is known that $p1\_p2$ is true (since $v(p1\_p2) = \max(v(pl), v(p2))$), then $q1\_q2$ is guaranteed to be possible, and interpretations which falsifies $q1\_q2\_$ are impossible. This is now illustrated.

Example 3. Let us consider the two rules R1: "if an employee is in category 1 (pl), his monthly salary (in euros) is necessarily in the interval [1000, 2000] (q1_) and typically in the interval [1500, 1800] (q1)", R2: "if an employee is in category 2 (p2), his salary (in euros) is necessarily in the interval [1500, 2500] (q2_) and typically in [1700, 2000] (q2)". Typical values are here the values known as being guaranteed to be possible. Let us examine the case of a person who is in category 1 or 2 (p1_p2); then we can calculate that his salary is necessarily in the interval [1000, 2500] (q1_q2_), while values in [1700, 1800] (q1_q2) are for sure possible for his salary. Note that one might get an empty conclusion with other numerical values, but this would not have meant a contradiction! This bipolar conclusion may, for instance, be instrumental when proposing a salary to a person whose category is ill-determinate.

4.6 Case-based Reasoning and Flexible Querying

Case-based reasoning (CBR), in general, assumes the following implicit principle: "*similar situations may lead to similar outcomes*". Thus, a similarity relation S between problem descriptions or situations, and a similarity measure T between outcomes are needed. This implicit CBR-principle can be expressed in the framework of fuzzy rules as, *"the more similar (in the sense of S) are the attribute values describing two situations, the more possible the similarity (in the sense of T) of the values of the corresponding outcome attributes"*. Given a situation s_0 associated to an unknown outcome t_0 and a current case (s, t), this principle enables us to conclude on the *possibility* of t_0 being equal to a value similar to t. This acknowledges the fact that, often in practice, a database may contain cases that are rather similar with respect to the problem description attributes, but which may be distinct with respect to outcome attribute(s). This emphasizes that case-based reasoning can only lead to cautious conclusions.

This can be modeled in terms of the *possibility rule* "the more similar s and s_0, the more *possible* t and t_0 are similar", where 'possible' is understood in terms of a guaranteed possibility measure. This leads to enforce the inequality $\Delta_{t0}(T(t, \cdot)) \geq \mu_S(s, s_0)$, which expresses that the guaranteed possibility that t_0 belongs to a high degree to the fuzzy set of values that are T-similar to t, is lower bounded by the S-similarity of s and s_0. Then the fuzzy set F of possible values t' for t_0 with respect to case (s, t) is given by

$$F_{t0}(t') = \min(\mu_T(t, t'), \mu_S(s, s_0)),$$

since the maximally specific distribution such that $\Delta(A) \geq \alpha$ is $\delta = \min(\mu_A, \alpha)$. As it can be seen, what is obtained is the fuzzy set T(t, .) of values t' that are T-similar to t, whose possibility level is truncated at the global degree $\mu_S(s, s_0)$ of similarity of s and s_0. The max-based aggregation of the various contributions obtained from the comparison with each case (s, t) in the memory M of cases, acknowledges the fact that each new comparison may suggest new possible values for t_0 and agrees with the positive nature of the information constituted by the repertory of cases. Thus, we obtain the following fuzzy set E_{s0} of possible values t' for t_0:

$$E_{s0}(t') = \max_{(s,t)\in M} \min(S(s, s_0), T(t, t')).$$

This latter expression can be put in parallel with the evaluation of a flexible query. Indeed here a situation is described in terms of a collection of attributes, and we are looking for cases that are constrained to be similar to s_0 w. r. t. the attributes. But rather than exhibiting the fuzzy set of cases similar to s_0, we produce the fuzzy set of the fuzzy sets of values similar to the t-attribute values of these cases (a level 2 fuzzy set reduced there to an ordinary fuzzy set by performing the union of the fuzzy sets of similar values). This approach has been generalized to situations s_0 imprecisely or fuzzily described, and has been related to other approaches to instance-based prediction.

The evaluation of a flexible query in face of *incomplete* or *fuzzy* information amounts to the possibility and the necessity of the fuzzy event corresponding to the high satisfaction of the query. This evaluation, known as fuzzy pattern matching, corresponds to the extent to which fuzzy sets (representing the query) overlap, or include the possibility distributions (representing the available information). Such an evaluation procedure has been extended to symbolic labels that are no longer represented by possibility distributions, but which belong to possibilistic ontologies where approximate similarity and subsumption between labels are estimated in terms of possibility and necessity degrees respectively. Besides, flexible querying has been extended to collections of XML documents having heterogeneous structures. Then, priorities introduced in the expression of a query may refer to the tags structure as well as to the information content. Requirements of the form 'A and *preferably* B' (to be understood as it is more satisfactory to have A and B than A alone) are then especially useful for expressing such priorities. Lastly in bipolar queries, flexible constraints that are more or less compulsory are distinguished from additional *wishes* that are optional, as for instance in the request "find the apartments that are cheap and *maybe* near the train station". Indeed negative preferences express what is (more or less, or completely) impossible or undesirable, and by complementation state flexible constraints restricting the possible or acceptable values. Positive preferences are not compulsory, but rather express wishes; they state what attribute values would be really satisfactory.

4.7 Qualitative Decision

Possibility theory offers a rich setting for qualitative decision under uncertainty (which can be also transposed in multiple criteria decision). Let π_d denote the possibility distribution restricting the plausible states in which decision d is applied, let d(s) be the consequence of d in state s, and let μ_G be the membership function of the fuzzy set of good results. Two quantitative criteria that evaluate the worth of decision d have been put forward in the literature, provided that a commensurability assumption between plausibility and preference is made:
– a pessimistic criterion

$$QU^-(\pi_d|\mu_G) = \min_{s\in S} \max(n(\pi_d(s)), \mu_G(d(s))),$$

where n is an order-reversing map from the plausibility scale to the preference scale (both scales are only assumed to be linearly ordered scales; in case of a numerical encoding within [0, 1], n(x) = 1 - x). This is an inclusion degree of the set of plausible states into set of states that lead to a good result when d is applied. It is small as soon as there exists a possible consequence of d which is both highly plausible and bad with respect to preferences. This is clearly a risk-averse and thus a pessimistic attitude. It generalizes the max-min Wald criterion in the absence of probabilistic knowledge since if π is the characteristic function of a subset A of states, the criterion is the utility of the worst consequence of states in A, however unlikely they are. But the possibilistic criterion is less pessimistic. It focuses on the idea of usuality and relies on the worst *plausible* consequences induced by the decision. Some unlikely states are neglected by a variable thresholding and the threshold is determined by comparing the values of π_d and μ_G via the mapping n.

– an optimistic criterion

$$QU^+(\pi_d \mid \mu_G) = \max_{s \in S} \min(h(\pi_d(s)), \mu_G(d(s))),$$

where h maps the plausibility scale on the preference scale (in case of a numerical encoding, $h(x) = x$). This is an intersection degree of π with μ_G. It corresponds to an optimistic attitude since it is high as soon as there exists a possible consequence of d that is both highly plausible and highly prized. It generalizes the maximax optimistic criterion. This latter criterion can be used as a secondary criterion, for breaking ties between decisions which are equivalent w.r.t. the pessimistic criterion.

$QU^-(\pi_d \mid \mu_G)$ and $QU^+(\pi_d \mid \mu_G)$ are respectively the necessity measure and the possibility measure of the fuzzy event G given the possibility distribution π_d. As such, they are special cases of Sugeno integrals. These decision criteria, as well as the more general Sugeno integral, have been axiomatized on the basis of postulates that parallel the construct made by Savage for justifying expected utility theory and the use of probabilities for representing uncertainty.

A logical handling of qualitative decision under uncertainty has been developed where π_d and μ_G are the semantic counterparts of two distinct possibilistic logic bases that respectively describe the pieces of available knowledge with their certainty levels on the one hand, and the goals pursued by the user with their priority levels on the other hand. Then the computation of the best decision that maximizes the pessimistic (or the optimistic) criterion can be performed directly at the syntactic level. Intuitively, a good decision (from the pessimistic point of view) is such that the most certain part of the knowledge is sufficient, when the decision is applied, to ensure that all the goals from the ones having the higher priority to less important ones are reached. This logical view of decision making can then be used for exhibiting the arguments in favor or against a decision, with their respective strength.

The two above criteria are based on min-max / max-min combinations and may lead to ties, although a more detailed inspection of the vectors of values to be combined (before the min or the max aggregation) may suggest some preference between the candidate decisions. For instance, in a min aggregation, only the smallest value is retained and the others are lost although they might be useful in case of ties for refining the ordering under the form of a leximin ordering that takes advantage of the smallest discriminating value when comparing two vectors. Then, introducing a proper definition of leximin of leximax, it becomes possible to refine the above criteria in a way that is identical to what could be obtained with classical expected utility provided that we use very particular types of probability distributions (called big-stepped distributions $p_1 \geq \ldots \geq p_n$ such that for all i, $p_i \geq \sum_{k=i,n} p_k$ holds) for encoding the information conveyed by π_d and μ_G.

This approach to qualitative decision finds applications to soft constraint satisfaction problems with non-controllable parameters (as for instance in scheduling some operations have uncertain durations that cannot be adjusted by the decision-maker), in multi-stage decision-making, in possibilistic partially observed Markov decision processes, or in possibilistic influence diagrams. It is also worth mentioning that decision criteria, formally expressed as the possibility or the necessity of a fuzzy event, may be used in classification, or in diagnosis problems. Thus, in diagnosis, we can compute the fuzzy set of malfunctions that are consistent with the observations (possibility measure), and the fuzzy set of malfunctions that alone explains all the observations, i.e. the ones that are such that all the observations are among their possible effects (abductive point of view captured by a necessity measure).

Lastly, the commensurability assumption made above between plausibility and preference is open to discussions. A decision-maker may only supply a relative likelihood ordering on events that describes his knowledge about the state of the world, and a preference ordering on the consequences of feasible acts. Then a natural, intuitively appealing lifting technique that solves the act-ranking problem is as follows: namely, an act d is preferred to another d' if the (relative) likelihood that d outperforms d' is greater than the likelihood that d' outperforms d. This idea has been the starting point for the axiomatization (in a Savage-like manner) of an original framework for a genuinely qualitative decision. When starting with a regular possibilistic ordering on events, lifting it to acts and re-projecting the relation thus obtained on acts back to events, yields an auto-dual relation (defined by, A is more likely than B if and only if $\Pi(A \cap B^c) > \Pi(B \cap A^c)$), which refines the possibilistic ordering we start with.

4.8 Learning

Applications of possibility theory to learning have started to be investigated rather recently in different directions. Thus, taking advantage of the proximity between reinforcement learning and partially observed Markov decision processes, a possibilistic counterpart of reinforcement learning has been proposed after developing the possibilistic version of the latter. Besides, by looking for big-stepped probability distributions (that have been shown to provide a faithful representation of default rules, which is equivalent to the possibilistic one), one can mine data bases for discovering default rules.

The version space approach to learning presents interesting similarities with the *binary* bipolar possibilistic representation setting, thinking of examples as positive information and of counter-examples as negative information. The general bipolar setting, where intermediary degrees of possibility are allowed, provides a basis for extending version space approach in a graded way, where examples and counter-examples can be weighted according to their importance.

The graded version space approach agrees with the possibilistic extension of inductive logic programming, where the background knowledge may be associated with certainty levels, where the examples may be more or less important to cover, and where the set of rules that is learnt may be stratified in order to have a better management of exceptions in multiple-class classification problems, in agreement with the possibilistic approach to non-monotonic reasoning.

5 Conclusion

This paper has surveyed contributions of qualitative possibility theory to information processing that have been developed in the last decade. Their main features of this framework can be summarized as follows:

- it can represent imprecise information with confidence levels.
- its setting is rich enough for expressing bipolarity;
- it is compatible both with logical and Bayesian net settings;
- it provides a way to handle priorities in inference, decision and learning;
- it is useful for representing qualitative uncertainty and preferences;
- it is usually computationally tractable.

However, this paper has not covered achievements that are more specific of *quantitative* possibility theory such has its applications to imprecise statistics or to fuzzy intervals calculations.

Some General References on Possibility Theory

Dubois D., Prade H. Possibility Theory: An Approach to Computerized Processing of Uncertainty, Plenum Press, New York, 1988.

Dubois D., Prade H. Fuzzy sets in approximate reasoning: A personal view. In: Implementations and Applications for Fuzzy Logic, (M. J. Patyra, D.M. Mlynek, eds.), J. Wiley and B. G. Teubner, New York and Stuttgart, 3–35, 1996.

Dubois D., Prade H. Possibility theory: qualitative and quantitative aspects. In: Quantified Representation of Uncertainty and Imprecision, (D. M. Gabbay, Ph. Smets, eds.), Vol. 1 of the Handbook of Defeasible Reasoning and Uncertainty Management Systems, Kluwer Acad. Publ., 169–226, 1998.

Zadeh L. A. (1978) Fuzzy sets as a basis for a theory of possibility. Fuzzy Sets and Systems, 1, 3–28.

Specific References (in the Last Decade)

Possibilistic Logic

Alsinet T., Logic Programming with Fuzzy Unification and Imprecise Constants : Possibilistic Semantics and Automated Deduction. Ph. D. Thesis, Technical University of Catalunya, Barcelona, 2001.

Alsinet T., Godo L. A complete calculus for possibilistic logic programming with fuzzy propositional variables. Proc. of the 16th Conference on Uncertainty in Artificial Intelligence (UAI'00), San Francisco, California, 1–10, 2000.

Alsinet T., Godo L., Sandri S. On the semantics and automated deduction for PLFC, a logic of possibilistic uncertainty and fuzziness. Proc. of the 15th Conference on Uncertainty in Artificial Intelligence, (UAI'99), Stockholm, Sweden, 3–20, 1999.

Alsinet T., Godo L., Sandri S. Two formalisms of extended possibilistic logic programming with context-dependent fuzzy unification: a comparative description. Elec. Notes in Theor. Computer Sci. 66(5), 2002.

Ben Amor N., Benferhat S., Dubois D., Mellouli K., Prade H. A theoretical framework for possibilistic independence in a weakly ordered setting, Inter. J. of Uncertainty, Fuzziness and Knowledge-Based Systems, 10, 117–155, 2002.

Benferhat S., Dubois D., Garcia L., Prade H. On the transformation between possibilistic logic bases and possibilistic causal networks. Inter. J. of Approximate Reasoning, 29, 135–173, 2002.

Benferhat S., Lagrue S., and Papini O. Reasoning with partially ordered information in a possibilistic logic framework. Fuzzy Sets and Systems, 144, 25–41, 2004.

Benferhat S., Prade H. Encoding formulas with partially constrained weights in a possibilistic-like many-sorted propositional logic. Proc. of the 9th Int. Joint Conf. on Artificial Intelligence (IJCAI'05), Edinburgh, Scotland, July 31-Aug. 5, 2005, 1281–1286, 2005.

Benferhat S., Prade H. Compiling possibilistic knowledge bases. Proc. of the 17th Europ. Conf. on Artificial Intelligence, Riva del Garda, Italy, Aug. 28th- Sept. 1st, 2006

Boldrin L., Sossai C. Local possibilistic logic. J. Applied Non-Classical Logics, 7, 309–333, 1997.

Dubois D., Konieczny S., Prade H. Quasi-possibilistic logic and its measures of information and conflict. Fundamenta Informaticae, 57, 101–125, 2003.

Dubois D., Lehmke S., and Prade H. A comparative study of logics of graded uncertainty and logics of graded truth. In Proc. of the 18th Linz Seminar on Fuzzy Set Theory Enriched Lattice Structures for Many-Valued and Fuzzy Logics (S. Gottwald and E. P. Klement, eds.), pp. 10–15, Linz, Austria, Feb. 25 - Mar. 1, 1997.

Dubois D., Prade H. Possibilistic logic: a retrospective and prospective view. Fuzzy Sets and Systems, Elsevier, 144, 3–23, 2004.

Dubois D., Prade H. and Sandri S. A. Possibilistic logic with fuzzy constants and fuzzily restricted quantifiers. In: Logic Programming and Soft Computing (T. P. Martin, F. Arcelli-Fontana, eds.), Research Studies Press Ltd, Baldock, England, 69–90, 1998.

Dupin de Saint-Cyr F., Prade H. Possibilistic handling of uncertain default rules with applications to persistence modeling and fuzzy default reasoning. Proc. 10th Inter. Conf. on Principles of Knowledge Representation and Reasoning (KR 2006), Lake District, UK, June 2-5, 2006, 440–451.

Nicolas P., Garcia L., Stéphan I. A possibilistic inconsistency handling in answer set programming. Proc. Europ. Conf. on Symbolic and Quantitative Approaches to Reasoning with Uncertainty (ECSQARU'05) - LNAI 3571, Barcelone, July 6 -8, 2005. (L. Godo, ed.), Springer-Verlag, 402–414.

Nicolas P., Garcia L., Stéphan I. Possibilistic stable models. Proc. of the 19th Inter. Joint Conf. on Artificial Intelligence (IJCAI '05), Edinburgh, Scotland, July 30- Aug. 5, 2005, 248–253.

Lehmke S. Logics which Allow Degrees of Truth and Degrees of Validity. PhD dissertation, Universität Dortmund, Germany, 2001.

Possibilistic Representations of Knowledge and Preferences

Benferhat S., Dubois D., Kaci S., Prade H. Bridging logical, comparative and graphical possibilistic representation frameworks. Proc. 6th.European Conf. (ESCQARU 2001), Toulouse, France, Sept. 19–21, 2001. Springer-Verlag, LNAI 2143, 422–431.

Benferhat S., Dubois D., Kaci S., Prade H. Bipolar possibilistic representations. Proc. 18th. Conf. on Uncertainty in Artificial Intelligence (UAI 2002), Edmonton, Aug. 1–4, 2002, Morgan Kaufmann Publ., 45–52.

Benferhat S., Dubois D., Prade H. Possibilistic and standard probabilistic semantics of conditional knowledge bases. J. of Logic and Computation, 9, 873–895, 1999.

Benferhat S., Dubois D., Prade H. Towards a possibilistic logic handling of preferences. Applied Intelligence, 14, 303–317, 2001.

Dubois D., Kaci S., Prade H. Expressing preferences from generic rules and examples - A possibilistic approach without aggregation function. Proc. Europ. Conf. on Symbolic and Quantitative Approaches to Reasoning with Uncertainty (ECSQARU'05) - LNAI 3571, Barcelone, July 6 -8, 2005. (L. Godo, ed.), Springer-Verlag, 293–304.

Dubois D., Kaci S., Prade H. Approximation of conditional preferences networks "CP-nets" in possibilistic logic. Proc. 2006 IEEE International Conference on Fuzzy Systems, Vancouver

Dubois D., Prade H. _A bipolar possibilistic representation of knowledge and preferences and its applications. In : Fuzzy Logic and Applications, 6th Inter. Workshop, (WILF 2005), (I. Bloch, A. Petrosino, A. Tettamanzi, eds.) Crema, Italy, Sept. 15–17, 2005, LNCS 3849, Springer, 1–10, 2006.

Possibilistic Description Logic

Dubois D., Mengin J., Prade H. Possibilistic uncertainty and fuzzy features in description logic. A preliminary discussion. In: Fuzzy Logic and the Semantic Web, (E. Sanchez, ed.), Elsevier, 2005.

Possibilistic Information Fusion

Benferhat S. Merging possibilistic networks. Proc. of the 17th Europ. Conf. on Artificial Intelligence, Riva del Garda, Italy, Aug. 28th- Sept. 1st, 2006

Benferhat S., Dubois D., Kaci S., Prade H. Possibilistic merging and distance-based fusion of propositional information. Annals of Mathematics and Artificial Intelligence, 34, 217–252, 2002.

Benferhat S., Dubois D., Kaci S., Prade H. Bipolar representation and fusion of preferences in the possibilistic logic framework. Proc. 8th Int. Conf. on Principles of Knowledge Representation and Reasoning, Toulouse, France, April 22–25, 2002. Morgan Kaufmann Publ., 421–432.

Benferhat S., Dubois D., Kaci S., Prade H. Bipolar possibility theory in preference modeling: Representation, fusion and optimal solutions. Inter. J. on Information Fusion, 7, 135-150, 2006.

Dubois D., Prade H. A synthetic view of belief revision with uncertain inputs in the framework of possibility theory. Int. J. of Approximate Reasoning, 17(2/3), 1997, 295–324.

Dubois D., Prade H., Smets P. Not impossible vs. guaranteed possible in fusion and revision. Proc. 6th.European Conf. (ESCQARU 2001), Toulouse, France, Sept. 19–21, 2001. Springer-Verlag, LNAI 2143, 522–531.

Kaci S., van der Torre L. Algorithms for a nonmonotonic logic of preferences. Proc. 8th European Conference on Symbolic and Quantitative Approaches to Reasoning with Uncertainty (ECSQARU'05), 281–292, 2005.

Possibilistic Temporal Reasoning

Dubois D., Hadj Ali A., Prade H. Fuzziness and uncertainty in temporal reasoning. J. of Universal Computer Science, 9, 1168–1194, 2003.

Hadj Ali A., Dubois D., Prade H. A possibility theory-based approach to the handling of uncertain relations between temporal points. Proc. of the 11th Inter. Symp. on Temporal Representation and Reasoning (TIME'04), Tatihou Island, France, July1-3, 2004, IEEE, 36-43. Revised version in Int. J. of Intelligent Systems, to appear

Possibilistic Approximate Reasoning

Dubois D., Prade H. What are fuzzy rules and how to use them. Fuzzy Sets and Systems, 84, 169–185, 1996.

Dubois D., Prade H. Bipolarity in possibilistic logic and fuzzy rules. Proc. 29th Conf. on Current Trends in Theory and Practice of Informatics (SOFSEM 2002), Milovy, Czech Republic, Nov. 22 -29, 2002. LNCS 2540, Springer, 168–173.

Dubois D., Prade H., Ughetto L. A new perspective on reasoning with fuzzy rules. Inter. J. of Intelligent Systems, 18 , 541–567, 2003.

Ughetto L., Dubois D., Prade H. Implicative and conjunctive fuzzy rules- A tool for reasoning from knowledge and examples. Proc. National Conference in Articial Intelligence (AAAI-99), Orlando, Florida, July 18-22, 1999, AAAI Press/The MIT Press, 214–219.

Possibilistic Case-based Reasoning

Dubois D., Hüllermeier E., Prade H. Fuzzy set-based methods in instance-based reasoning. IEEE Trans. on Fuzzy Systems, 10, 322–332, 2002.

Dubois D., Hüllermeier E., Prade H. Fuzzy methods for case-based recommendation and decision support. J. of Intelligent Information Systems, to appear

Hüllermeier E., Dubois D., Prade H. Model adaptation in possibilistic instance-based reasoning. IEEE Trans. on Fuzzy Systems, 10, 333–339, 2002.

Flexible Querying and Possibilistic Data

De Calmès M., Prade H., Sedes F. Flexible querying of semi-structured data: a fuzzy set-based approach. Inter. J. of Intelligent Systems, to appear, 2006.

De Tré G., De Caluwe R., Prade H. Null values revisited in prospect of data integration. In: Semantics of a Networked World (First International IFIP conference, ICSNW 2004), Springer Verlag, LNCS 3226, Paris, (M. Bouzeghoub, C. Goble, V. Kashyap, S. Spaccapietra, eds.), 79–90.

Dubois D., Prade H. Bipolarity in flexible querying. Proc. 5th Int. Conf. Flexible Query Answering Systems (FQAS'02), Copenhagen, Denmark, Oct. 27–29, 2002, LNAI, 2522, Springer-Verlag, 174–182.

Loiseau Y., Prade H., Boughanem M. Qualitative pattern matching with linguistic terms. AI Communication, 17 (1), 25–34, 2004.

Possibilistic Decision and Related Topics

Amgoud L., Prade H. Towards argumentation-based decision making: A possibilistic logic approach. Proc. IEEE Int. Conf. on Fuzzy Systems, Budapest, July 25–29, 2004, 1531–1536.

Amgoud L., Prade H. Explaining qualitative decision under uncertainty by argumentation. Proc. 21st National Conference on Artificial Intelligence (AAAI-06), Boston, July 16–20, 2006

Benferhat S., Dubois D., Fargier H., Prade H. Sabbadin R. Decision, nonmonotonic reasoning and possibilistic logic. In: Logic-Based Artificial Intelligence, (J. Minker, ed.), Kluwer Academic Publishers, 333–358, 2000.

Da Costa Pereira C., Garcia F., Lang J., Martin-Clouaire R., Planning with graded nondeterministic actions: a possibilistic approach, Inter. J. of Intelligent Systems 12, 935–962, 1997.

de Mouzon O., Guérandel X., Dubois D., Prade H., Boverie S. Online possibilistic diagnosis based on expert knowledge for engine dyno test benches. Proc. of 18th IFIP World Computer Congress - Artificial Intelligence Applications and Innovations (AIAI '04), Toulouse, France, Aug. 22–27, 2004, Kluwer, 435–448.

Dubois D., Fargier H., Fortemps P. Fuzzy scheduling: Modelling flexible constraints vs. coping with incomplete knowledge. European J. of Operational Reasearch, 147, 231–252, 2003.

Dubois D., Fargier H., Perny P. Qualitative decision theory with preference relations and comparative uncertainty: An axiomatic approach. Artificial Intelligence, 148, 219–260, 2003.

Dubois D., Fargier H., Prade H. Possibility theory in constraint satisfaction problems: Handling priority, preference and uncertainty. Applied Intelligence, 6, 287–309, 1996.

Dubois D., Fargier H., Perny P., Prade H. Qualitative decision theory: From Savage's axioms to nonmonotic reasoning. J. of the ACM, 49, 455–495, 2002.

Dubois D., Fargier H., Perny P., Prade H.A characterization of generalized concordance rules in multicriteria decision making. Inter. J. of Intelligent Systems, 18, 751–774, 2003.

Dubois D., Godo L., Prade H., Zapico A. On the possibilistic decision model: from decision under uncertainty to case-based decision. Inter. J. of Uncertainty, Fuzziness and Knowledge-Based Systems, 7, 631–670, 1999.

Dubois D., Godo L., Prade H., Zapico A. Advances in qualitative decision theory: Refined rankings. Proc. of Int. Joint Conf. 7th Ibero-American Conference on AI and 15th Brazilian Symposium on AI (IBERAMIA-SBIA'00), Atibaia, Brazil, Springer Verlag, LNAI n° 1952, 427–436, 2000.

Dubois D., Le Berre D., Prade H., Sabbadin R. Using possibilistic logic for modeling qualitative decision: ATMS-based algorithms. Fundamenta Informaticae, 37, 1–30, 1999.

Dubois D., Prade H. Sabbadin R. Decision-theoretic foundations of qualitative possibility theory. European J. of Operational Research, 128, 459–478, 2001.

Fargier H., Lamothe J. Handling soft constraints in hoist scheduling problems: the fuzzy approach. Engineering Applications of Artificial Intelligence, 14, 387–399, 2001.

Fargier H., Lang J., Sabbadin R. Towards qualitative approaches to multi-stage decision making. Inter. J. of Approximate Reasoning, 19, 441–471, 1998.

Fargier H., Lang J., Schiex T. Mixed constraint satisfaction: A framework for decision problems under incomplete knowledge. Proc. of the 13th National Conf. on Artificial Intelligence (AAAI'96), Portland, Oregon, AAAI Press and The MIT Press, USA, 175–180, Aug. 1996.

Fargier H., Sabbadin R. Qualitative decision under uncertainty: back to expected utility. Artificial Intelligence, 164, 245–280, 2005.

Garcia L., Sabbadin R. Possibilistic influence diagrams. Proc. 17th Europ. Conf. on Artificial Intelligence (ECAI'06),_ Riva del Garda, Italy, Aug. 28 - Sept. 1, 2006.

Godo L., Zapico A. On the possibilistic-based decision model: Characterization of preference relations under partial inconsistency. Appl. Intell.14 319–333 (2001)

Moura Pires J., Prade H. Flexibility as relaxation - the fuzzy set view. In: Soft Constraints: Theory and Practice (workshop associated to CP'00 - 6th Int. Conf. on Principles and Practice of Constraint Programming, Singapore, September 22, 2000.

Sabbadin R. Empirical comparison of probabilistic and possibilistic Markov decision processes algorithms. ECAI 2000: 586–590.

Sabbadin R. A possibilistic model for qualitative sequential decision problems under uncertainty in partially observable environments. UAI 1999: 567–574.

Zapico A. Weakening commensurability hypothesis in possibilistic qualitative decision theory. Proc. IJCAI 2001: 717–722

Possibilistic Learning

Prade H., Serrurier M. Version space learning for possibilistic hypotheses, Proc. 17th Europ. Conf. on Artificial Intelligence (ECAI'06),_ Riva del Garda, Italy, Aug. 28 - Sept. 1, 2006, 801–802.

Sabbadin R. Towards possibilistic reinforcement learning algorithms. FUZZ-IEEE 2001: 404–407.

Serrurier M., Prade H. Possibilistic inductive logic programming. Proc. Europ. Conf. on Symbolic and Quantitative Approaches to Reasoning with Uncertainty (ECSQARU'05), LNAI 3571, Barcelone, July 6–8, 2005. (L. Godo, ed.), Springer-Verlag, 675–686.

Serrurier M., Prade H. Coping with exceptions in multiclass ILP problems using possibilistic logic. Proc. of the 19th Inter. Joint Conf. on Artificial Intelligence (IJCAI '05), Edinburgh, Scotland, July 30-Aug. 5, 2005, 1761–1762.

Fuzzy Modeling of Nonlinear Stochastic Systems by Learning from Examples

Mohammad-R. Akbarzadeh-T and Amir-H. Meghdadi

Abstract A variety of approaches have used fuzzy logic for modeling nonlinear and complex systems due to fuzzy logic's inherent ability in handling complexities and uncertainties. Such techniques, however, generally ignore the statistical nature that many such complex systems may exhibit. When the system's behavior is significantly influenced by stochastic parameters such as in human behavior, it is reasonable to expect that the modeling performance would be improved if the effect of such parameters is taken into consideration. In this chapter, a novel modification to the *Table Look-up Scheme* is proposed and applied to modeling of two different systems that are both highly stochastic in nature. The first problem is a randomized version of a benchmark chaotic Mackey Glass time series problem and is used to establish the significance of the modeling method in terms of lower overall error by considering the system's stochastic nature at various noise levels. The second problem is experimental learning of human behavior in a manually controlled overhead crane problem. A user interface is developed for the simulation environment in which a human operator manually controls an overhead crane for the purpose of generating required training examples as well as testing the learnt behaviors.

1 Introduction

Conventional systems and control methodologies often treat the random nature of processes as undesirable, labeling them as noise. Much of the current literature is hence attributed to rejecting noise and/or building robustness against it. While this may be a reasonable treatment for many systems, there are also many systems in which randomness is an unavoidable, inherent and perhaps even desirable part of their characteristics. Consider, for instance, modeling an individual human's behavior, his interaction with others, or that of his cumulative society. While we may still wish to reach a representative norm in such systems, we may need to exploit their inherent *randomness* instead of avoiding *noise*.

Modeling and control of such complex systems presents both major challenges in terms of computational complexity as well as significant rewards in forecasting and thereby controlling their behavior. Black box modeling by learning from examples is a widely used approach for many such real world applications when the internal dynamics is either unknown or poorly known [1]. In current literature, various soft computing paradigms such as neural networks and fuzzy systems are generally classified as black box and model-free approaches for modeling complex systems [2, 3, 4, 5, 6, 7, 8, 9, 10, 11, 12, 13, 14, 15]. Much of this interest is due to the universal approximation property of these paradigms as shown in [5, 15, 16].

There are different soft computing based techniques which employ neural or fuzzy approaches such as "Table Look-up Scheme" [5], fuzzy clustering [17], fuzzy ARTMAP [12], ANFIS [11], evolutionary fuzzy systems [18, 19], and genetically optimized fuzzy polynomial neural networks [14]. Accuracy of the first two methods is analyzed and approximation bounds are established in [16]. There are also other important requirements such as avoiding complexity, interpretability and robust convergence which are considered for different learning methods in [3, 16, 20, 21]. Fuzzy systems have the advantage of being interpretable and transparent, i.e. they are easily comprehensible by humans [3]. In situations where there is a priori human expert knowledge about the system, this knowledge may be used to further enhance the learning process [5].

However, in high dimensional spaces with large number of inputs and/or finer fuzzy partitioning of the input space, fuzzy systems suffer from the *curse of dimensionality*, i.e. an exponential growth in the number of fuzzy system's parameters. This limitation has been investigated using many approaches based on rule reduction and other means of reducing complexity [3, 8, 20]. Therefore, a trade off exists between accuracy (better fitting sampled data) and simplicity. In this perspective, more complex systems generally require a more complex fuzzy model with larger number of fuzzy rules and fuzzy sets to achieve desirable accuracy. Many of the fuzzy modeling approaches simplify the fuzzy model by eliminating inconsistencies and redundancies which may exist among the most promising rules in training data [3, 9, 20, 21]. Proper handling of the observed inconsistencies and redundancies in training stage is hence an important concern in data-driven fuzzy modeling approaches.

In this paper, we claim that the *source* of these inconsistencies should be considered in their proper handling, i.e. the source of redundancies and inconsistencies may not be merely too few partitioning of the universe or a result of inadequate measurement, but the randomness inherent in many natural processes. With this vision, we consider a common rule generation algorithm, i.e. the Table Look-up Scheme [5], and propose appropriate modifications to treat the inconsistencies and redundancies in order to achieve better results in modeling systems with stochastic nature.

Conventional learning methods generally address the complexity of real systems through deterministic methodologies and disregard the randomness which may inherently exist in such systems [3, 5, 8, 20, 21]. They treat the existing inconsistencies in the training data set as undesirable noise and hence attempt to reject such "inconsistencies" and "redundancies" without paying attention to their statistics. This can be accomplished by defining a certainty degree and selecting the most prominent fuzzy rule with higher certainty degree [21]. This may be an efficient approach in situations where the system under consideration is indeed a deterministic nonlinear system which has been corrupted by noisy data in its worst case [9]. However, in nonlinear stochastic system, the stochastic nature of signals is an inherent characteristic of the system and should not be disregarded.

To demonstrate this approach, the proposed modified Table Look-up Scheme considers the statistical source of system uncertainty to improve learning accuracy by assigning a new reliability factor in addition to its certainty factor. In order to demonstrate its potential capabilities in modeling non-deterministic systems, the method is then applied in two different function approximation problems. The first problem, a randomized Mackey-Glass time series [7] is chosen as a highly nonlinear and also stochastic benchmark time-series prediction problem. Statistical data analysis of the modified learning methodology reveals that consideration of stochastic nature can improve the learning accuracy of the algorithm for highly stochastic systems. In the second problem, we show this different approach to handling of inconsistency leads to successful learning of human control strategy with embedded uncertainties and randomness in human behavior [22, 23, 24].

The remainder of this paper is as follows. In Sect. 2, the Table Look-up Scheme as originally developed by Wang and Mendel [5] is considered. Subsequently, using a new concept of *reliability factor*, a modified version of the method is introduced for modeling stochastic systems. In Sect. 3, the first case study, the modified method is implemented on the time series prediction problem for a stochastic-chaotic Mackey-Glass time series [7]. It is shown that the proposed modified method is considerably superior to the standard method for this *randomized-chaotic* series, unlike the case for deterministic series. The effect of this randomness on learning enhancement is then studied. In Sect. 4, human control strategy is considered as the second case study in modeling a nonlinear stochastic system. Human control strategy is defined here as the complex mapping between sensory inputs and control outputs of a human in manual control of dynamic systems [23]. Human control strategy in controlling an overhead carne is considered as an example of a challenging problem for both manual and automatic control strategies. A computer simulator has been developed [22] which can record sensory inputs and control actions of a human operator as well as simulate this dynamic system. A human operator is then asked to control the simulated crane, and both of the standard and modified *Table Look-up Schemes* are used. Simulation results show that modified method produces a more similar performance to the original manual control. Finally, Sect. 5 concludes the paper with a brief discussion about the effect of both noise power and model complexity on the learning performance.

2 The Proposed Modified Table Look-up Scheme

The standard five-step learning algorithm by Wang [5, 17] is first considered here. Using samples of input-output data pairs, this algorithm generates fuzzy rules in an input space which has been first partitioned by predefined membership functions. Without loss of generality, we use a 2-input 1-output system to explain how the algorithm works. Suppose that we are given N input-output data pairs as follows,

$$\left(x_1^p, x_2^p, y^p\right) \text{ for } p = 1, \dots, N \tag{1}$$

The Standard five steps of the method are as follow:

Step 1- Fuzzy Partitioning of the input and output variables

Having the range of variation of each input or output variable (universe of discourse), we can partition the universe of discourse by a number of fuzzy regions, which are represented by fuzzy sets. The only requirement is that the fuzzy sets must be *complete* in the universe of discourse. A set of fuzzy sets is complete in a universe of discourse U if for every value x in U, there exist at least one fuzzy set A such that $\mu_A(x) \neq 0$.

Figure 1 shows a sample fuzzy partitioning for variable x_1, x_2 which consists of 5 symmetrical fuzzy regions with their associated membership functions A_i^j which represents jth membership functions of the ith variable. Similar partitioning is repeated for output y variable as shown in Fig. 2.

Step 2- Generate one fuzzy rule for each of the given data pairs

In this step a fuzzy rule is generated for each data pair using the membership values of the data pairs in the fuzzy sets. In an attempt to discover the fuzzy rule which can best represent each data pair, the fuzzy set with greatest membership value is selected for each variable and is associated with that variable. Each rule is then represented by the selected fuzzy sets.

For example, consider a given data pair (x_1^p, x_2^p, y^p). x_1^p, x_2^p, and y^p are each evaluated through their corresponding membership functions as in Figs. 1 and 2. In this example, x_1^p has the maximum membership value in A_1^2, x_2^p has maximum membership value in A_2^4 and y^p has maximum membership value in B^1, then the fuzzy rule associated with this data pair will be:

$$If \; x_1 \; is \; A_1^2 \; and \; x_2 \; is \; A_2^4, \; then \; y \; is \; B^1 \tag{2}$$

In this manner, we can obtain one rule for each available data pair. The antecedent parts of these rules are related to each other using *AND* operators because x_1 and x_2 are part of the input vector x and should be evaluated simultaneously.

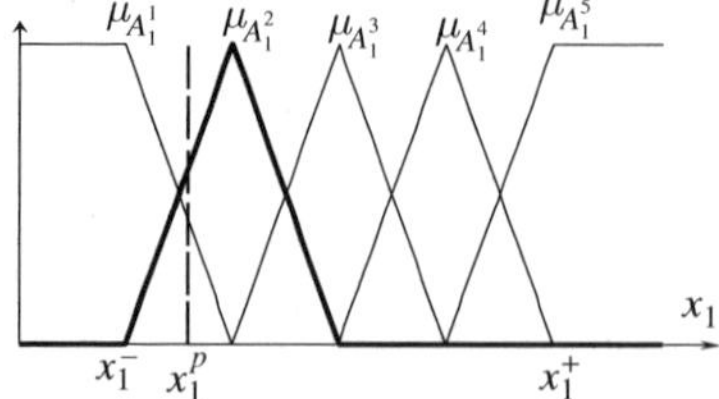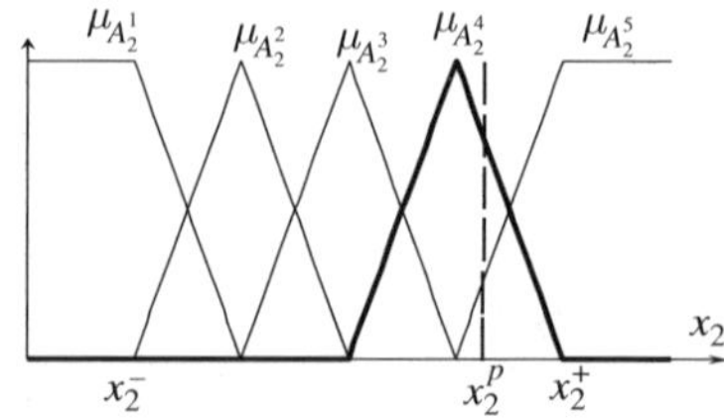

Fig. 1 A sample fuzzy partitioning for variables x_1 and x_2

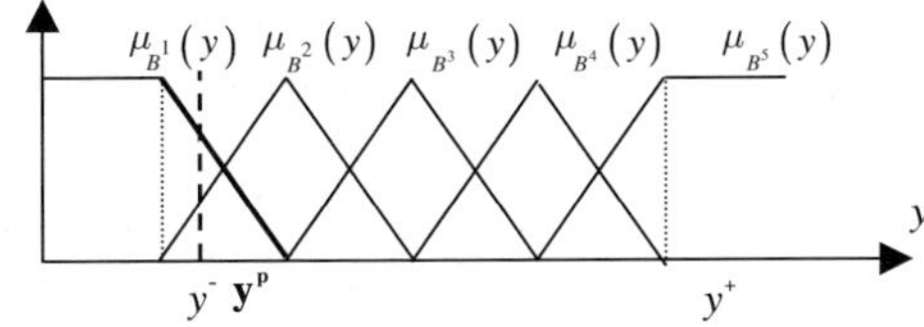

Fig. 2 A sample fuzzy partitioning for the output variable y

Step 3- Assign a certainty degree to each rule

The number of data pairs is usually much larger than the maximum number of possible rules. However, due to the above reasons in Sect. 1, we can expect *inconsistent* and *redundant* rules in the initial rule set obtained from the previous step. *Inconsistent* rules are rules with the same antecedent but different consequent parts while *redundant* rules are identical rules which have been repeated in the rule set and usually do not have any more information about the system.

Handling this *inconsistency* and *redundancy* is the main key difference between the standard method and the modified method which has been first presented in [7]. Standard algorithm [5] assigns a degree of *compliance* to each rule in order to show the degree to which a given rule complies with the data pair. Then from a group of *inconsistent* rules, the rule with maximum degree of compliance is selected and other rules in that group are removed. Normally, this degree is defined as the product of membership values of the data pairs in the associated fuzzy sets. For example, consider a fuzzy rule:

$$If\ x_1\ is\ A_1^i\ and\ x_2\ is\ A_2^j,\ then\ y\ is B^k \tag{3}$$

Here, the *compliance degree* will be:

$$Compliance\ Degree = \mu_{A_1^i}(x_1).\mu_{A_2^j}(x_2).\mu_{B^k}(y) \tag{4}$$

As stated before, Table look-up scheme has the ability of incorporating human expert knowledge into design process. Moreover, an expert may assign a degree to each data pair to emphasis its importance and contribution. This degree can be represented with a real number $d \in [0, 1]$. Here, we call this number as *Human Certainty Factor (HCF)*. In the above example in (3), if we suppose that there is a human certainty factor of d, then the new certainty degree for this rule will be the product of *compliance degree* by d:

$$Certainty\ Degree = \mu_A(x_1).\mu_B(x_2).\mu_C(y).d \tag{5}$$

Step 4- Create final fuzzy rule base

After eliminating *inconsistent* and *redundant* rules from the initial rule set, the final rule set remains. The number of generated rules in this step depends on the complexity of the system. Without loss of generality, suppose that there are n inputs and

1 output variables, and let N_i be the number of membership functions for the ith input variable, then the maximum number of rules will be:

$$R_{\max} = \prod_{i=1}^{n} N_i \qquad (6)$$

Usually the number of generated rules by this algorithm is less than R_{max}.

Step 5- Determine the overall fuzzy system

After creating the above rules, the *domain knowledge* of the fuzzy system is completely defined. It is then necessary to define the parameters of the fuzzy system's *meta knowledge* such as the Defuzzification method. In [5], the product operation was used as the T-norm operator and the Centroid method was used for defuzzification as in the following formula [17]:

$$y = \frac{\sum_{i=1}^{M} \mu^i . \bar{y}^i}{\sum_{i=1}^{M} \mu^i} \qquad (7)$$

Where y is the actual output of the fuzzy system, M is the number of fuzzy rules in the final rule base, $\bar{y}^i$ denotes the center value of the output membership function in the ith rule and μ^i is the rule firing strength for the ith rule.

2.1 Modification of the method

We believe that for modeling of systems such as the stochastic Mackey-Glass time series in Sect. 3, a different strategy is needed to deal with *inconsistency* and *redundancy* in initial fuzzy rule set. Indeed the number of occurrences of each possible rule in the initial rule set is an important factor for selection of that rule in the final rule set.

Thus the modified algorithm modifies step 3 of the standard algorithm by introducing a new *repeatability factor* and including it in the rule certainty degree as below. Suppose that in the simple example of (3), there are k rules in the initial rule set with the same *antecedent* parts.

$$If \ x_1 \ is \ A \ and \ x_2 \ is \ B \qquad (8)$$

These k rules may have different *consequent* parts that result in a set of conflicting (*inconsistent*) rules. Now suppose that there are k_1 rules in the form of:

$$If \ x_1 \ is \ A \ and \ x_2 \ is \ B \ then \ y \ is \ C \qquad (9)$$

Then, *repeatability factor,* η for these k_1 rules is defined as:

$$\eta = \frac{k_1}{k} \tag{10}$$

The modified rule degree is hence defined as the product of previous certainty degree as in (5) by the *repeatability factor.*

$$Degree = \mu_A(x_1).\ \mu_B(x_2).\ \mu_C(y).d.\eta \tag{11}$$

3 Case study 1: Time Series Prediction as a Learning Problem

Prediction of the current and future samples of a time series by examining the previous samples can be formulated as a function approximation problem with past samples as the inputs and current sample as the output of a system to be learned. Prediction of chaotic time series is a challenging problem. Such systems are so complex that they may appear to be random at first glance. However, they are generated by simple nonlinear and *deterministic* functions. A standard deterministic Mackey-Glass chaotic time series is defined by the following equation:

$$\frac{dx(t)}{dt} = \frac{0.2x(t - \tau)}{1 + x^{10}(t - \tau)} - 0.1x(t) \tag{12}$$

where τ is a constant representing the dimension of the chaos and $x(t)$ denotes the value of the series at time t. Figure 3 shows a section of the series where $\tau = 30$. To predict the current sample of the time series $x(t)$, four past samples of the series are considered here as the inputs of the system as is demonstrated in Fig. 4. Also, 7 similar triangular membership functions have been chosen for each input variable. The figure is not shown here due to brevity.

After learning the fuzzy system with first 700 samples of the time series, samples 701 through 1000 are used as the testing data set to test the validity of the learned model. Figure 5 shows the main and predicted values of the time series using the standard method. All the signals have been normalized between 0 and 1.

To determine accuracy of the learned model an error measure is needed. Although no quantitative error measure is used in [5], errors based on mean of absolute or mean of square error are commonly used for calculating the prediction error [12, 17]. Here the mean of absolute values of errors (*MAE*) has been selected to achieve greater values which are more convenient for comparison between different methods. Let $e(t)$ be the prediction error (the difference between the main signal and predicted value) at any time instance t. Then mean of absolute errors for standard and modified *Table Look-up Scheme* is named $\overline{|e|}_{TLS}$ and $\overline{|e|}_{MTLS}$ respectively and is defined as in (13) where n is the number of samples.

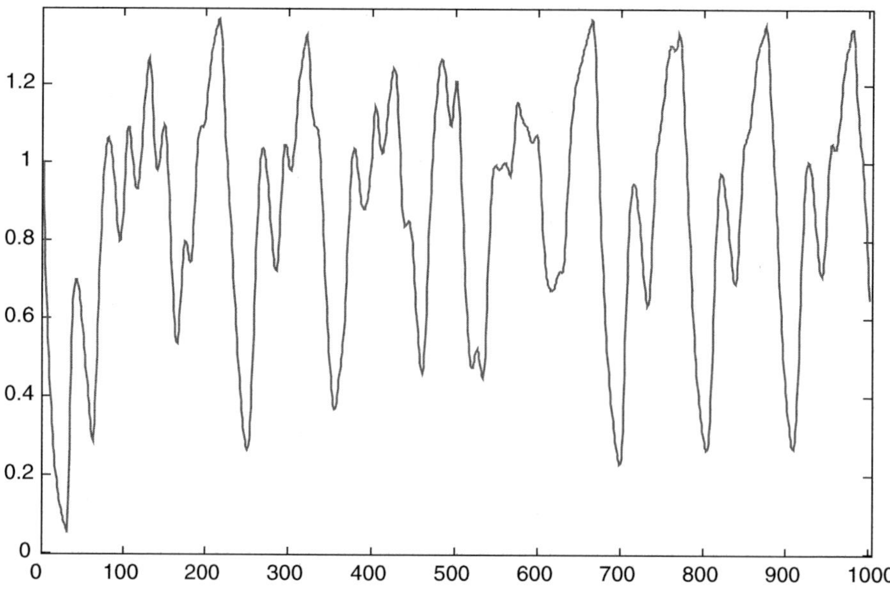

Fig. 3 Standard Mackey-Glass with $x(0) = 1$ and $\tau = 30$

$$\overline{|e|}_{TLS}\,\% = \frac{\sum\limits_{i=1}^{n} |e(i)|}{n} \times 100\% \tag{13}$$

For a simple standard time series and using the standard learning method, predicted values which are shown in Fig. 6, have an error measure of 4.62% and the resulting fuzzy system will have 45 rules where the maximum number of rules is $7^4 = 2401$.

3.1 Stochastic-Chaotic Time Series

To investigate the learning problem for nonlinear systems with stochastic structure, we define stochastic Mackey-Glass time series with the following equation:

$$\frac{dx(t)}{dt} = \frac{0.2x(t-\tau)}{1 + x^{10}(t-\tau)} - 0.1x(t) + n(t) \tag{14}$$

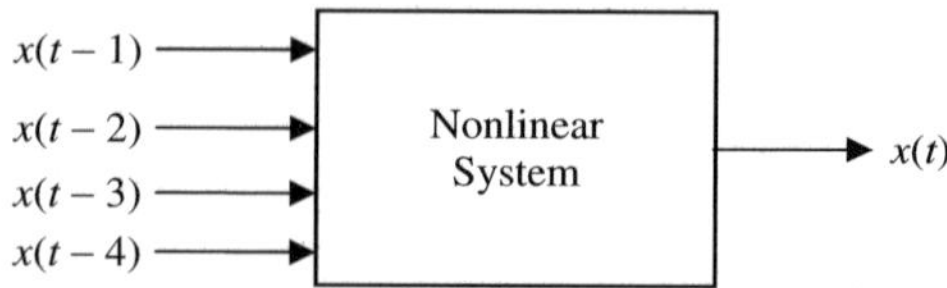

Fig. 4 Input/output definition

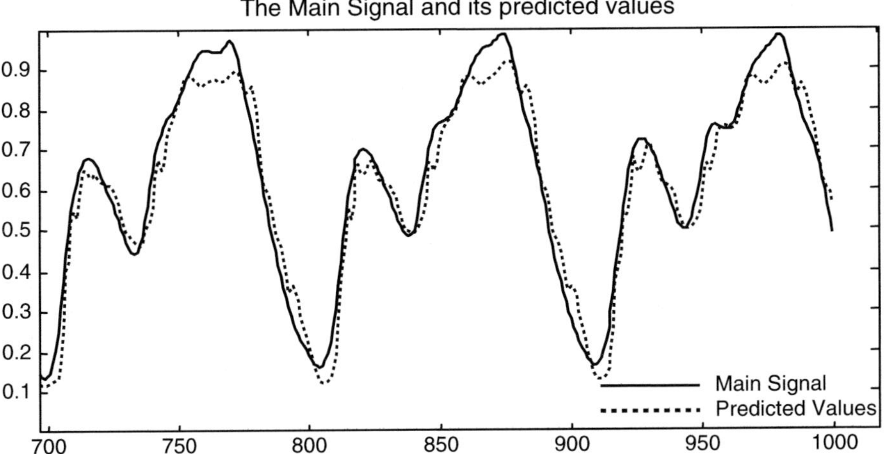

Fig. 5 Prediction of the normalized time series using 4 inputs and 7 membership functions

Where $n(t)$ is white noise with zero mean. Let

$$Noise\ power = Cov(n(t)) \tag{15}$$

The shape of the time series in this case is completely random and depends on $n(t)$. It is important to note that this noise is not simply a measurement noise but a process noise resulting in a randomized Mackey-Glass series. A sample of such stochastic series is shown in Fig. 6 for $1 < t < 1000$ where *Noise power=2.*

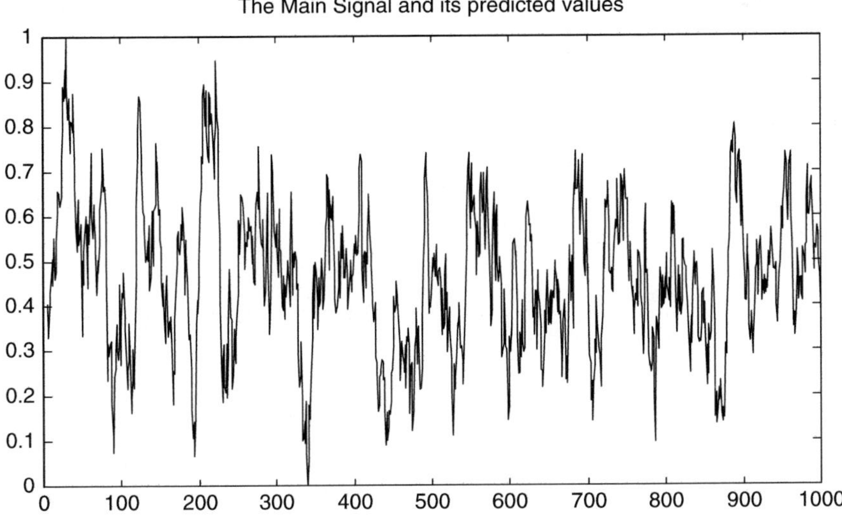

Fig. 6 A sample stochastic Mackey-Glass series

3.2 Simulation Results

The standard and modified *Table Look-up Scheme* is used to predict this randomized time series. In order to adequately compare the modified learning algorithm with the standard algorithm, we consider two different cases as follow. In Figs. 7 and 8, predicted values are shown in comparison to real values for testing data and for complex and simple fuzzy models respectively. The accuracy of the methods in each case has been compared in Table 1.

> *Case 1: A high dimensional complex fuzzy system with 7 membership functions for each input variable.*
> *Case 2: A simple low dimensional fuzzy system with 3 membership functions for each input variable.*

The number of resulting rules is nearly the same for each fuzzy model and does not depend on the learning method.

It is clear that modified learning method performs better than standard method particularly in lower dimensional and more simplified fuzzy systems with fewer rules.

3.3 Discussions and Results

Chaotic nature of the Mackey-Glass time series makes it very sensitive to process noise $n(t)$. To investigate the effect of noise power ($Cov(n(t))$) on the performance of the learning algorithm, *MAE* errors for both the standard and modified algorithms have been compared at different noise powers in a broad range from 10^{-7} to 10^{+3}.

Figure 9 illustrate the error measures versus noise power when the numbers of membership functions are 3 (a) and 7 (b) respectively. The solid line shows error

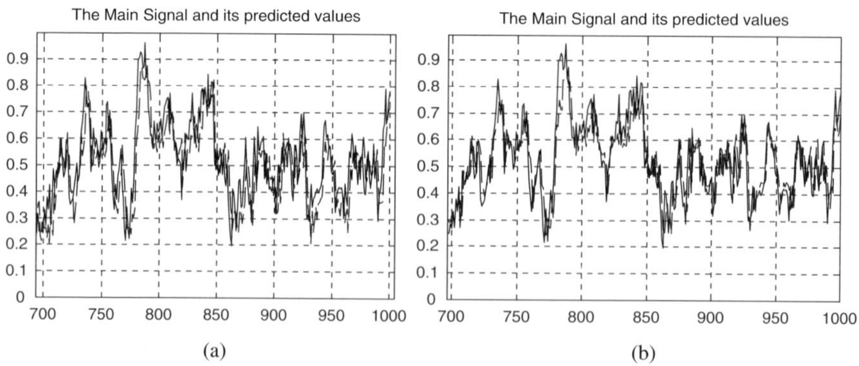

Fig. 7 Time series prediction with (**a**)-standard and (**b**)-modified algorithm $700 < t < 1000$ for complex fuzzy system with 7 membership functions for each variable: $N = 7$

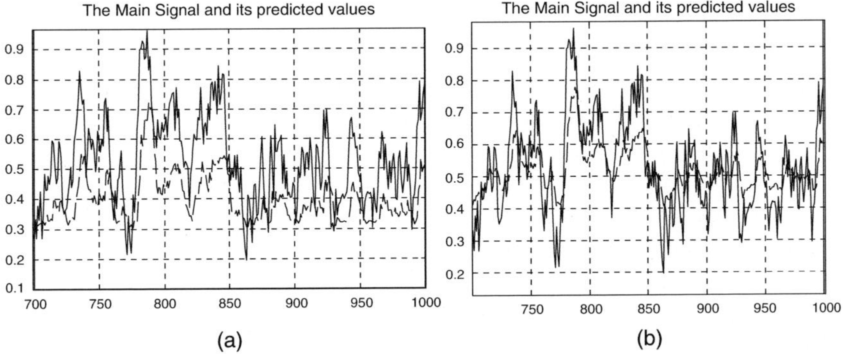

Fig. 8 Time series prediction with (**a**)-standard and (**b**)-modified algorithm $700 < t < 1000$ for complex fuzzy system for each variable: $N = 3$

Table 1 *MAE* errors for standard and modified algorithms

$Cov(n(t))=2$	$N = 7$	$N = 3$
$\overline{|e|}_{TLS}$ (*Standard Algorithm*)	*7.9%*	*14.2%*
$\overline{|e|}_{MTLS}$ (*Modified Algorithm*)	*6.8%*	*8.4%*
Rule numbers	*117*	*31*

measure for standard algorithm and dashed line shows this error for modified algorithm. Note that x-axis is logarithmic to show a wider range of the values.

Moreover, in each of the above cases, the number of resulting fuzzy rules has been shown at different noise powers in Figs. 10(a) and 10(b) this number is nearly the same for both the methods.

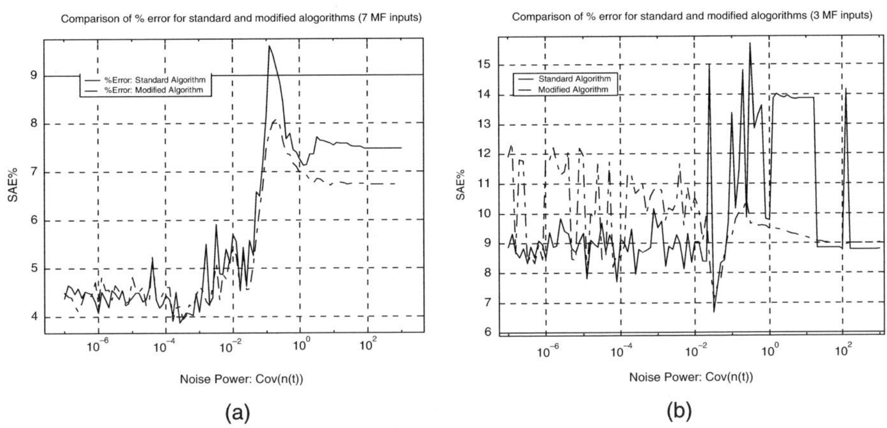

Fig. 9 *MAE* ($\overline{|e|}\%$) Error measures at different noise powers for (**a**) $N = 7$ and (**b**) $N = 3$

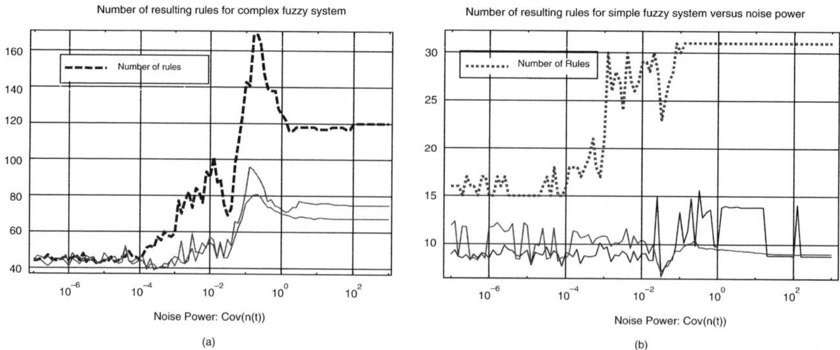

Fig. 10 Number of the rules in resulting fuzzy system for (**a**) $N = 7$ and (**b**) $N = 3$

Some of the important results are:

- When noise level is high, with a sufficient and equal number of (seven) membership functions, the modified algorithm is consistently superior to the standard algorithm as in Fig. 9(a).
- For both algorithms, as expected, increasing noise level degrades the learning performance through increasing both the rule number and error measure as in Figs. 9 and 10.
- When noise level is low and a lower dimensional fuzzy system is considered, the proposed algorithm is not however superior over the standard algorithm. Indeed it may even have lower performance for very few number of fuzzy membership functions. (Figure 9(b)). This is also expected, since at very low noise levels, the system is indeed deterministic.
- In comparison, when noise level is significant, the modified algorithm is considerably superior and has smaller error. In particular, this enhancement of the performance is usually greater when a more simple fuzzy system is considered (as in Table 1).

4 Case Study 2: Human Control Strategy Modeling

Another modeling problem which has been studied here is modeling human control strategy in controlling an overhead crane as an example. *Human Control Strategy* can be generally defined as: *The complex mapping between human sensory inputs and his control actions in performing a specific task such as controlling a system* [23].

This problem has been selected here because human behavior is also believed to be non-deterministic at least partially. Consequently, standard and modified Table Look up Scheme has been tested again for this modeling problem.

4.1 Over Head Crane

Cranes are one of the complex human operated systems in industry. A human operator controls the motion of the heavy payload carried by the crane. Dock mounted container cranes, as shown in Fig. 11, are supposed to move very heavy loads under performance and safety requirements [25]. Safety requirements prohibit fast changes in load and trolley positions and velocities while performance criterion suggests higher velocities, which means smaller required time for loading/unloading the ship.

Although several approaches have been used to design an automatic controller for these systems [25, 26, 27], optimal and robust control of the crane is still a challenging problem for both the conventional and intelligent control methods. However, this system is chosen for modeling human control strategy, in particular, because of its relatively large time constants which make it a suitable plant for control by human. This modeling is carried on using a computer simulator.

4.2 System Dynamics:

Figure 12 shows the geometry, coordinates and torques. There are two inputs, T_1 and T_2 and three outputs y, Φ and x where $y = b_2\theta_2$ and $x = x_1 + x_2 \, sin(\Phi)$ (x_1, x_2 and Φ are defined in the figure).

System variables are as follow:

θ_1: Angle of rotation of trolley drive motor (rad), J_1: Total moment of inertia of the trolley drive motor mechanism including a brake, a drum and a set of reduction gears (kgm^2), b_1: equivalent radius of the drum of the trolley drive motor (m), θ_2: Angle of rotation of the hoist motor (rad), J_2: Total moment of inertia of the hoist motor mechanism including a brake, a drum and a set of reduction gears (kgm^2), b_2: equivalent radius of the drum of the hoist motor (m); Φ: The load swing angle (m), m: Total mass of the trolley (kg), M: Total mass of the container load and

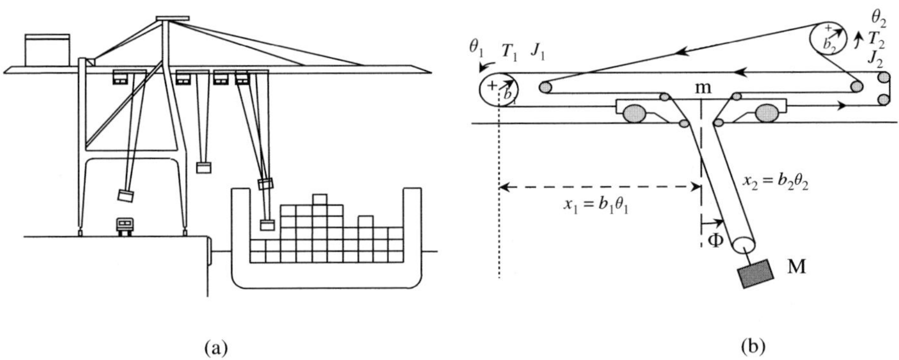

Fig. 11 (a) Dock mounted gantry crane, (b) Crane Coordinates and Torques

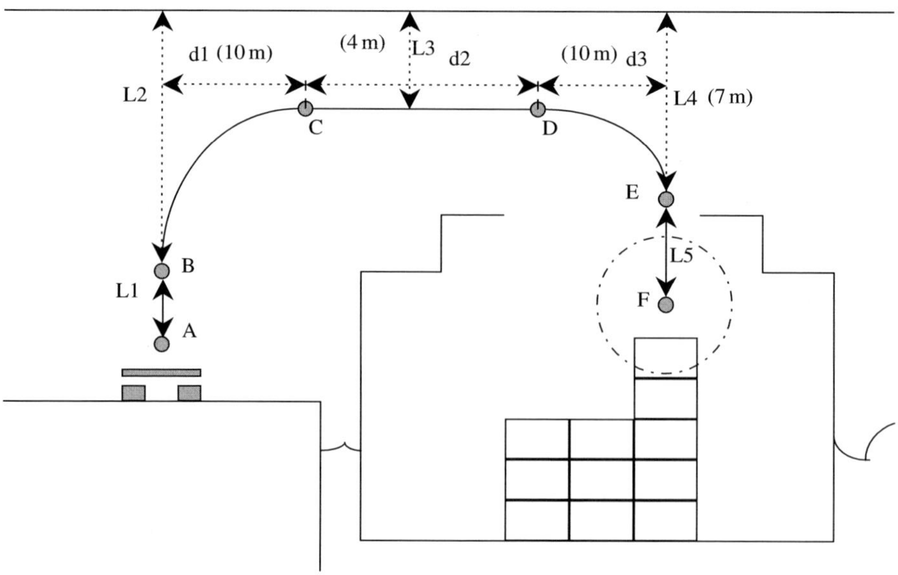

Fig. 12 A desired trajectory

equipment (kg), T_1: Driving torque generated by the trolley drive motor (Nm), T_2: Driving torque generated by the hoist motor (Nm), g: The acceleration of gravity $(9.81\ \mathrm{ms}^{-2})$.

Under the assumptions that the cable does not stretch, there are no damping effects, and the motor dynamics are negligible, differential equations of the system are derived in [23] and are not shown here. System equations can be more simplified into a state space form as in the following equations by using the small angle assumption, which holds under normal operating conditions [25, 26, 27].

$$\begin{bmatrix} \dot{x}_1 \\ \dot{x}_2 \\ \dot{x}_3 \\ \dot{x}_4 \\ \dot{x}_5 \\ \dot{x}_6 \end{bmatrix} = \begin{bmatrix} b_1\theta_1 \\ b_2\theta_2 \\ \Phi \\ x_1 \\ x_2 \\ x_3 \end{bmatrix} = \begin{bmatrix} x_4 \\ x_5 \\ x_6 \\ v_1 - \delta_1 x_3 v_2 + \delta_1 g\, x_3 \\ -\delta_2 x_3 v_1 + v_2 \\ -\dfrac{1}{x_2}[v_1 - \delta_1 x_3 v_2 + (1+\delta_1)\, g\, x_3 + 2x_5 x_6] \end{bmatrix} \tag{16}$$

Where

$$\delta_1 = \frac{Mb_1^2}{J_1 + mb_1^2} \qquad \delta_2 = \frac{Mb_2^2}{J_2 + Mb_2^2} \tag{17}$$

And v_1 and v_2 are normalized torques of the motors defined as the new inputs:

$$v_1 = \frac{b_1 T_1}{J_1 + mb_1^2} \qquad v_2 = \frac{b_2(T_2 + Mb_2 g)}{J_2 + Mb_2^2}. \tag{18}$$

The objective of the control problem is to move the load in a desired trajectory while satisfying performance and safety requirements. A desired trajectory is shown in Fig. 12.

4.3 Computer Simulation

Using SIMULINK®, a computer simulator with a graphical user interface for manual control is designed and used to simulate human control strategy. Figure 13 shows the simulator environment. The position of the mouse pointer (as in Fig. 13 - part 6) on a control-panel (part 1), on the computer screen, controls the values of the horizontal and vertical torques of the corresponding electric motors (T_1 and T_2). The states of the system and the applied control signals by a human operator are recorded during simulation. Although an inexperienced human operator *may* not be able to achieve the high level of performance and optimality as in the automatic control methods developed yet, our objective here is to just learn human control strategy (whether it is optimal or non optimal).

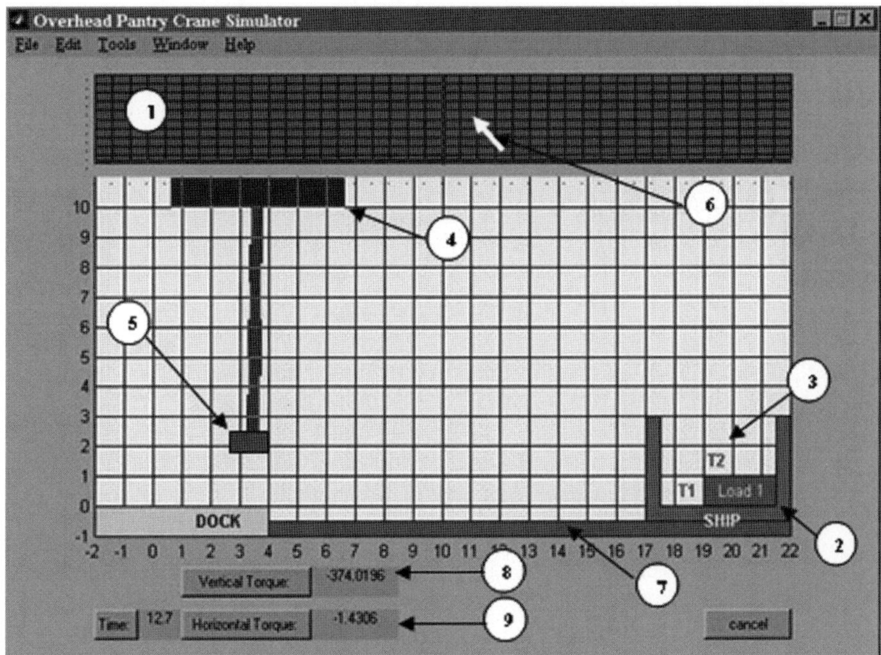

Fig. 13 Developed simulator user interface (1. Control Panel, 2. Ship, 3. Target (T2), 4. Trolley 5. Load, 6. Mouse pointer, 7. Sea, 8. Hoist motor torque indicator, 9. Trolley motor torque indicator)

4.4 Learning Procedures

First, a human operator controls the crane for a number of times where at each time, the following parameters are recorded with a sampling time of 0.2 sec which is considerably smaller than the time constant and delay time of the human as a controller. Later, a fuzzy system with 5-inputs and 2-outputs is learned to model human control strategy using the standard and modified Table Look-up Scheme. Training data for this problem are N sampled input-output data pairs corresponded to 10 different times of successful control of the crane.

Input variables are:

1. x-position of the load: $xpos = x_1 + x_2 sin(\Phi)$
2. y-position of the load: $ypos = x_2 sin(\Phi)$
3. Swing angle of the load: Φ
4. Rate of change of the swing angle: $d\Phi/dt$
5. Rate of change of the rope length: dx_2/dt

Output variables are trolley and hoist motor torques (T_1 and T_2) where five simple symmetric triangular membership functions are used for each input variable and

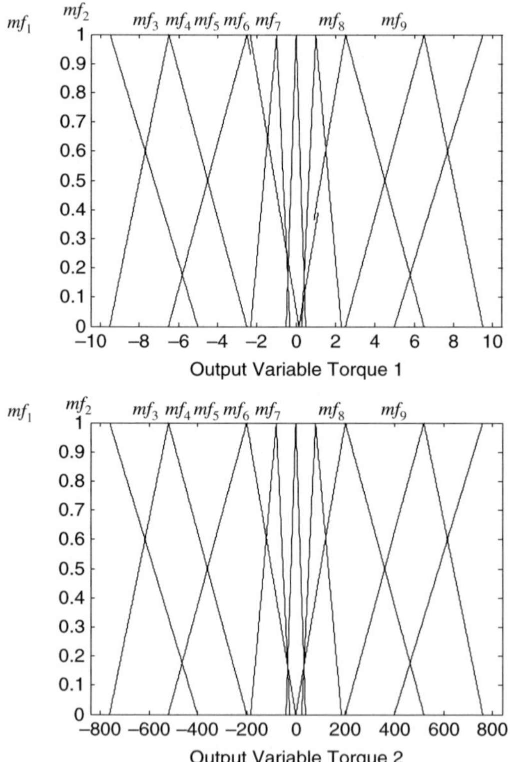

Fig. 14 Output membership functions for T_1 and T_2

are not shown here. This makes up for a total of $5^5 = 625$ potential rules. The corresponding output membership functions are shown in Fig. 14.

Load trajectories for different successful manual control of the crane are sketched in Fig. 15. Significant inconsistencies in load trajectories show that the operator is not experienced enough. However, one of the sub-optimal trajectories is used as the training data and is shown in Fig. 16. We believe the inconsistencies in observed resulting trajectories is mainly because of the inherent randomness in human behavior that cause human not to repeat himself at various times and under similar conditions.

After training the fuzzy system using the standard and modified learning methods, the resulting fuzzy controllers are tested on the system to examine the controller performance. The first fuzzy controller is trained using the standard Table Look-up Scheme and used to control the crane and the resulting trajectory is shown in Fig. 17. Learning performance is not desirable at all and the load falls into the sea 5.5 meters before reaching the target. The error between desired and actual trajectory is very high.

The fuzzy controller which is learned using the modified method, however, has better performance. Its trajectory is more similar to manual control trajectory in Fig. 16 but has 2-meter error in hitting the target (as in Fig. 18).

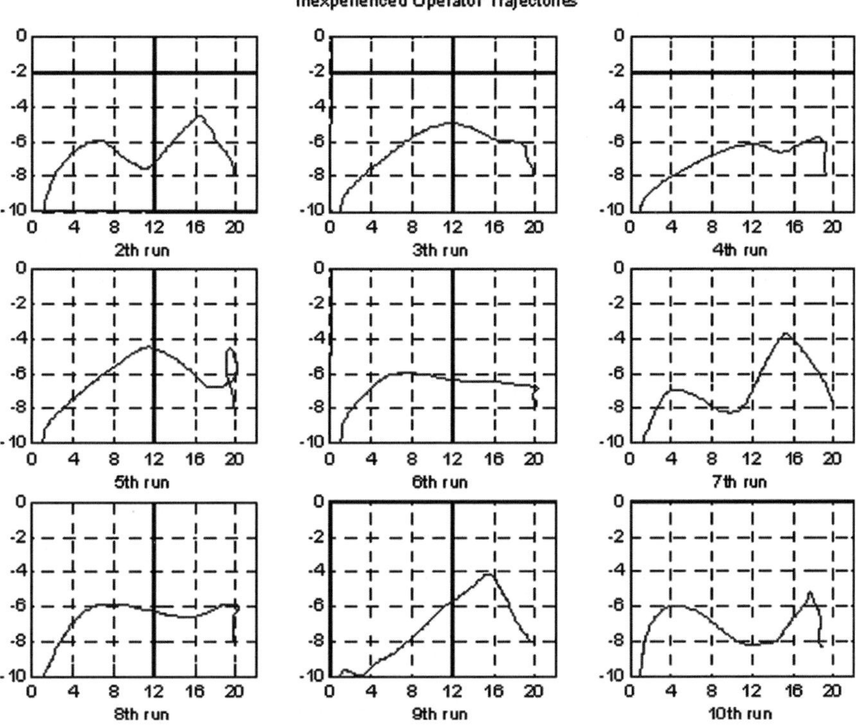

Fig. 15 Several human produced trajectories in the x-y plane

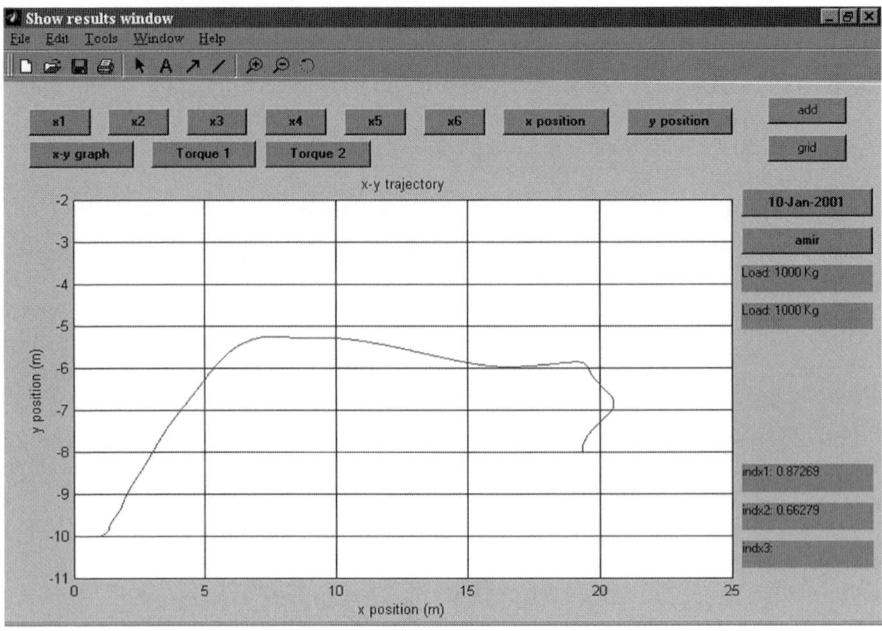

Fig. 16 A sample manual control load trajectory

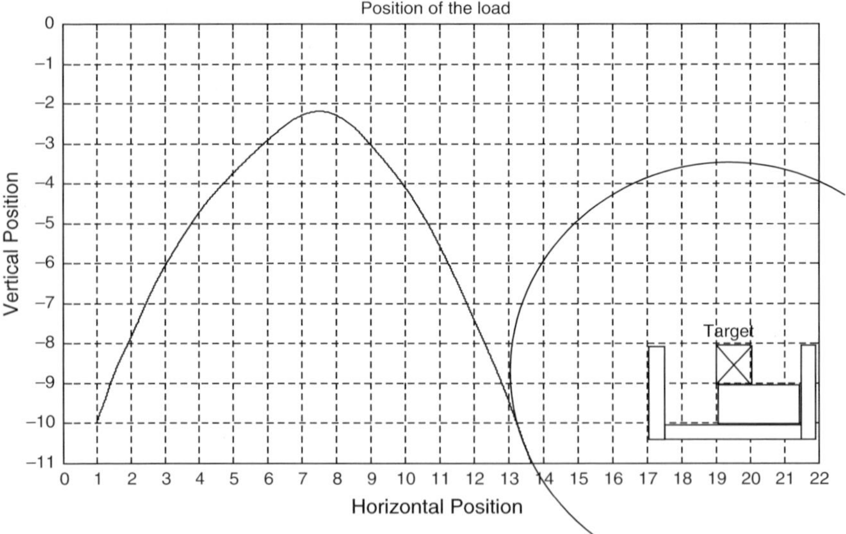

Fig. 17 Fuzzy controller that is trained with standard Table Look-up Scheme

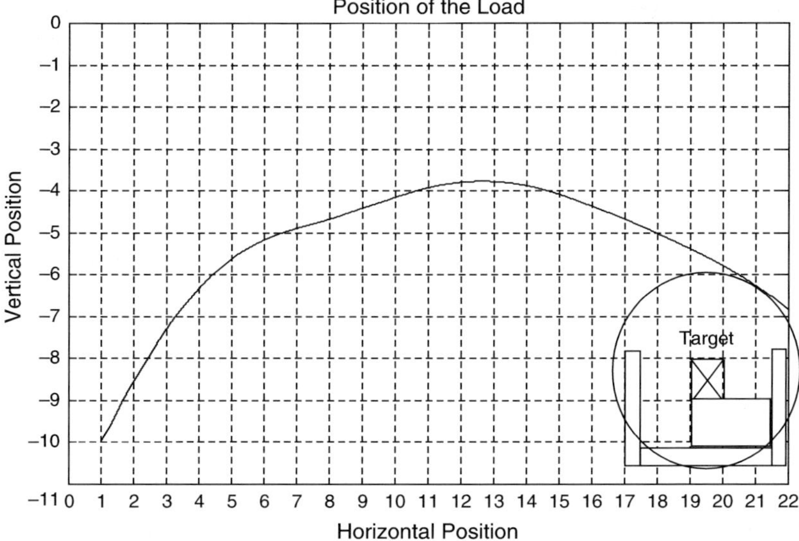

Fig. 18 Fuzzy controller that is trained with modified Table Look Up Scheme

5 Summary and Conclusion

Fuzzy modeling of nonlinear systems is usually performed by learning from examples. However when a given system involves process noise, the samples (examples) should not be treated the same. This inherent randomness increases the level of complexity of system and deteriorates modeling performance while significantly increasing the number of rules. It is shown here that a stochastic-nonlinear system can be better approximated if the statistics of inconsistent and/or redundant rules are considered in the modeling process. This methodology is illustrated by a modified Table Look-up Scheme proposed as a deterministic learning method with stochastic approach. The method is expected to be suitable for fuzzy modeling of randomized complex systems.

This modified method has been tested in two different problems which are expected to have randomized nature. The first problem is time series prediction of a *randomized chaotic* Mackey-Glass time series. The second problem is modeling human control strategy as a highly complex and stochastic system tested on over head crane control problem. The advantage of the modified Table Look up Scheme is shown for both problems; specifically, the proposed modified learning algorithm is considerably superior when noise levels are significant or the system is to be learned with a lower number of fuzzy rules.

Modeling human behavior, similar to the first problem, is also a function approximation problem, where the issue is approximating the complex mapping between sensory inputs and control outputs of the human acting as a controller. The learning problem implemented when a human controls a computer simulation of an overhead crane. Overhead crane is a system with relatively slow though highly nonlinear dynamics, which can be robustly controlled by a human operator. Therefore a computer simulator for manual control of the crane is developed here and used for learning human control strategy using the modified Table Look-up Scheme. Simulation results show higher similarity between the learned model performance and a sample of a successful manual control of the crane by a human operator.

References

1. J. Sjoberg, et al., "Nonlinear Black-Box Modeling in System Identification: a Unified Overview," *Automatica*, Vol.51, No. 12, pp. 1691–1724, 1995.
2. I. Rojas, et al. "Self-Organized Fuzzy System Generation from Training Examples," *IEEE Transactions on Fuzzy Systems*, Vol. 8, No. 1, pp. 23–36, Feb 2000.
3. Y. Jin, "Fuzzy Modeling of High-Dimensional Systems: Complexity Reduction and Interpretability Improvement," *IEEE Transactions on Fuzzy Systems*, Vol. 8, No. 2, pp. 212–221, Apr 2000.
4. S. Abe and M. Lan, "Fuzzy Rule Extraction Directly from Numerical Data for Function Approximation," *IEEE Transactions on Systems, Man and Cybernetics*, Vol 25, No 1, pp. 119–129 Jan 1995.
5. L.-X. Wang and J. M. Mendel, "Generating Fuzzy Rules by Learning from Examples," *IEEE Transactions on Systems, Man and Cybernetics*, Vol 22, No. 6, pp. 1414–1427, Nov 1992.
6. J. H. Nie and T. H. Lee, "Rule-based modeling: Fast construction and optimal manipulation," *IEEE Transactions on Systems, Man and Cybernetics-Part A*, Vol. 26, pp.728–738, Nov 1996.
7. A. H. Meghdadi and M-R. Akbarzadeh-T, "Fuzzy Modeling of Nonlinear Stochastic Systems by Learning from Examples," Proceedings of the IFSA World Congress and 20th NAFIPS International Conference, pp. 2746–2751, July 2001, Vancouver, Canada.
8. P. Carmona, J. Castro and J. Zurita, "Strategies to Identify Fuzzy Rules Directly from Certainty Degrees: A Comparison and a Proposal," *IEEE Transactions on Fuzzy Systems*, Vol. 12, No. 5, pp. 631–640, October 2004.
9. P. Carmona, J. Castro, and J. Zurita, "FRIwE: Fuzzy Rule Identification with Exceptions," *IEEE Transactions on Fuzzy Systems*, Vol. 12, No. 1, pp. 140–151, Feb 2004.
10. D. Kukolj, E. Levi, "Identification of complex systems based on neural and Takagi-Sugeno fuzzy model," *IEEE Transactions on Systems, Man and Cybernetics-Part B*, vol. 34, issue 1, pp. 272–282, Feb. 2004.
11. J. R. Jang, "ANFIS: Adaptive-Network Based Fuzzy Inference Systems," *IEEE Transactions on Systems, Man and Cybernetics*, Vol. 23, No.03, pp. 665–685, May 1993.
12. G. A. Carpenter, et all, "Fuzzy ARTMAP: A Neural Network Architecture for Incremental Supervised Learning of Analog Multidimensional Maps," *IEEE Transactions on Neural Networks*, Vol. 3, Issue:5, pp. 698–713, Sept 1992.
13. S.-K. Oh, W. Pedrycz and H.-S. Park, "Genetically Optimized Fuzzy Polynomial Neural Networks," *IEEE Transactions on Fuzzy Systems*, pp. 125–144, Vol. 14, No. 1, February 2006.
14. M. Kumar, R. Stoll, and N. Stoll, "Deterministic Approach to Robust Adaptive learning of Fuzzy Models," *IEEE Transactions on Systems, Man, and Cybernetics-Part B: Cybernetics*, pp. 767–780, Vol. 36, No. 4, August 2006.
15. L.-X. Wang, J. Mendel, "Fuzzy Basis Functions, Universal Approximation, and Orthogonal Least-Squares Learning," *IEEE Transactions on Neural Networks*, 814, Sept. 1992.
16. L.-X. Wang and C. Wei, "Approximation Accuracy of Some Neuro-Fuzzy Approaches," *IEEE Transactions on Fuzzy Systems*, Vol. 9, No. 4, Aug 2000.
17. L.-X. Wang, *A Course in Fuzzy Systems and Control*, Prentice Hall, 1997.
18. W. Pedrycz, M. Reformat, "Evolutionary Fuzzy Modeling," *IEEE Transactions on Fuzzy Systems*, Vol. 11, No 5, Oct 2003.

19. S. Kang, et al. "Evolutionary Design of Fuzzy Rule Base for Nonlinear System Modeling and Control," *IEEE Transactions on Fuzzy Systems*, Vol. 8, No. 1, Feb 2000.
20. W. E. Combs and J. Andrews, "Combinatorial Rule Explosion Eliminated by a Fuzzy Rule Configuration," *IEEE Transactions on Fuzzy Systems*, Vol 6 No. 1, Feb 1998.
21. M-Y. Chen, "Establishing Interpretable Fuzzy Models from Numeric Data," Proceedings of the 4th World Congress on Intelligent Control and Automation, pp. 1857–1861, June 2002.
22. M.-R. Akbarzadeh-T and A. H. Meghdadi, "Fuzzy Modeling of Human Control Strategy for Over Head Crane," Proceedings of the 10th IEEE International Conference on Fuzzy Systems, pp. 1076–1079, 2–5 Dec 2001, Melbourne, Australia.
23. M. C. Nechyba, *Learning and Validation of Human Control Strategies*, PhD Thesis, Carnegie Mellon University, 1998.
24. M. C. Nechyba and Y. Xu, "Stochastic Similarity for Validating Human Control Strategy Models," *IEEE Trans. on Robotics and Automation*, Vol. 14, no. 3, pp. 437–51, 1998.
25. B. Kimiaghalam, A. Homaifar, M. Bikdash, "Hybrid Fuzzy-PD Control for a Dock Mounted Pantry Crane," NASA URC Conference, 1999.
26. B. Kimiaghalam, A. Homaifar, M. Bikdash, "Using Genetic Algorithm for Optimal Crane Control," NASA URC Conference, Feb. 1998.
27. B. Kimiaghalam, PhD Thesis, Pre-published copy, North Carolina A&T University, 2001.

Conceptual Soft-Computing based Web search: FIS-CRM, FISS Metasearcher and GUMSe Architecture

José A. Olivas, Pablo J. Garcés, Javier de la Mata, Francisco P. Romero and Jesús Serrano-Guerrero

SMILe-ORETO Research Group (Soft Management of Internet e-Laboratory)

Abstract In this work it is presented a brief summary of FIS-CRM (**F**uzzy **I**nterrelations and **S**ynonymy **C**onceptual **R**epresentation **M**odel), its first application FISS (**F**uzzy **I**nterrelations and **S**ynonymy **S**earcher), and a new platform based on agents GUMSe (**GUM Se**arch), developed with the aim of improving the semantic capabilities of Web search engines.

Key words: Soft Computing · Internet Search · Agents

1 Introduction

Due to the increase of Internet and the consequent exponential growth in the information contained, the searching process has become complicated and ineffective, overwhelming the users with irrelevant information. The reason is that most of the search engines base their search on the co-occurrence of keywords within the text, dealing only with lexical and sometimes syntactic aspects (Ricarte & Gomide, 2001). That is, search systems carry out a "word" matching process instead of a "concept" matching process. It is important to differentiate this kind of systems from the one proposed by Zadeh (2003) which is focused on the development of question-answering systems, a very interesting point of view in information retrieval problems.

Nowadays, soft computing techniques have taken an important role (Pasi, 2002; Herrera-Viedma & Pasi, 2003) in order to provide a way for optimizing web search results. In this sense, lots of approaches have been proposed in the recent last years. Some of them are leaded to the construction of flexible adaptive sites (based on web patterns, user profiles, access patterns, user behaviour patterns···) using data mining techniques (Martin-Bautista, Vila, Kraft & Chen, 2001; Perkovitz & Etzioni, 2000; Tang & Zhang 2001; Cooley et al, 1997). Some others are focused on the organization of the retrieved documents into groups, being important to point out the ones based on dynamic clustering (Zamir & Etzioni, 1999) in contrast to those supported by predefined thematic groups. We can also find search systems based on flexible query languages (Herrera-Viedma, 2001; Choi, 2001), or systems based on fuzzy association rules that help the user to find new terms to be used in the query (Delgado et al, 2003).

Another different group of approaches centre their studies in the definition of models for representing documents, most of them based on extensions of the original vector space model (Salton et al, 1995). We can find systems based on term interrelations stored in ontologies (not fuzzy) such as WordNet, a semantic net of word groups (Miller, 1995). Gonzalo et al (1998) propose a system based on WordNet in which vector elements have three values (to identify the corresponding tree in the net and the sense it is used in the document). This kind of systems require a special matching mechanism, like the ontomatching algorithm proposed by Kiryakov & Simov (1999) when comparing the concepts associated to the words. Whaley (1999) propose other corpus-based search system that uses the probability that certain concepts co-occur together to disambiguate meanings. Leacock and Chodorow (1998) approach the sense disambiguation problem by studying the local context of the words and comparing them to the habitual context words of each one of the word senses. This system requires the usual context words to be stored in a repository. Loupy and El-Bèze (2002) also uses WordNet and propose a disambiguation system based on training the system with corpus of documents. Ramakrishnan et al (2004) study the disambiguation from a soft point of view. In this case, words are not disambiguated to an only sense, but rather to a set of senses with their corresponding relevance degrees (obtained by Bayesian Belief Networks). Lafourcade (2001, 2003) introduce the concept of relative synonymy to define a model of concept-based vectors. In this model, a term can be represented by a conceptual vector which is obtained by the linear combination of the definitions of the whole set of concepts. This system requires a concept repository.

In this work, it is presented FIS-CRM (**F**uzzy **I**nterrelations and **S**ynonymy **C**onceptual **R**epresentation **M**odel), and its first application FISS (**F**uzzy **I**nterrelations and **S**ynonymy **S**earcher), with conceptual manipulation possibilities (developed within the SMILe[1] framework), and with the acquired experience, a new test platform based on agents, GUMSe (**GUM Se**arch) is also proposed.

2 The FIS-CRM Model

FIS-CRM is a model for representing the concepts contained in any kind of document. It can be considered an extension of the vector space model (VSM) (Salton, Wang & Yang, 1975). Its main characteristic is that it is fed on the information stored in a fuzzy synonymy dictionary and various fuzzy thematic ontologies. The dictionary stores the synonymy degree between every pair of recognized synonyms.

---

[1] SMILe-ORETO Research Group (**S**oft **M**anagement of **I**nternet **e**-**L**aboratory)

Each ontology stores the generality degree between every word and its more general words. The way of calculating this value is the one proposed in (Widyantoro & Yen, 2001).

The key of this model is first to construct the base vectors of the documents considering the number of occurrences of the terms (what we call VSM vectors) and afterwards readjust the vector weights in order to represent concept occurrences, using for this purpose the information stored in the dictionary and the ontologies.

The readjusting process involves sharing the occurrences of a concept among the synonyms which converge to the concept and give a weight to the words that represent a more general concept than the contained ones.

2.1 FIS-CRM Basis. The Concept of 'Concept'

One of the aspects that the current version of the model differs from the original one is on the format of the dictionary of synonyms which stores the synonymy degrees. Now, the dictionary manages separately the different meanings of the words as Fernández proposes (Fernández, 2000).

The record format of the dictionary is: $< termA, N, termB, SD(termA, termB) >$, where N is the number that identifies a sense of *termA* and SD represents the synonymy degree between the pair (term A, sense s_N) and the pair (term B, sense s_x) and is obtained from the expression below. Note that if a term-sense has m synonyms it will have m entries in the dictionary.

$$SD(As_1, Bs_2) = \frac{\text{Number of synonyms held in common by As}_1 \text{ and Bs}_2}{\text{Number of total synonyms of As}_1 \text{ and Bs}_2} \quad (1)$$

The extended dictionary, which FIS-CRM is supported on, stores entries for each pair of recognized term-sense synonyms, but does not need to store the synonymy degree of all the combinations of meanings of the two involved terms. That is, assuming that term 'A' and term 'B' both have several meanings, the entries stored in the dictionary are the following ones:

$$SD(A_i, B) = MAX_{j=1}^{b}\left(SD(A_i, B_j)\right)$$
$$SD(B_j, A) = MAX_{i=1}^{a}\left(SD(B_j, A_i)\right) \quad (2)$$

Where the first expression is evaluated for the a senses of A, and the second one for the b senses of B. The aim of this calculation is to disambiguate the correct sense thanks for which both terms are synonymous. That is, in the first expression of Fig. 2, assuming that the term-sense 'Ai' is in fact synonym of only one of the meanings of 'B', the MAX function provides the synonymy degree of the right sense of 'B' that makes these words synonyms. The second expression of the figure makes the same operation in order to disambiguate the right sense of 'A'.

It is very interesting to realize that the second term of each term pair of the dictionary does not need to be explicitly disambiguated because its right sense will be the one that satisfies the expression below, in which b represents the number of meanings of the word B and A_i represents the i-meaning of the word A.

$$SD(A_i, B) = MAX_{j=1}^{b}(SD(A_i, B_j)) \qquad (3)$$

The way of calculating the synonymy degree between two term-meaning pairs fits the definition of FIS-CRM concept. A concept in FIS-CRM is not an absolute concept that has a meaning itself (there is not any kind of concept definition set or concept index). In FIS-CRM a concept is dynamically managed by means of the semantic areas of different words.

For example, when examining a single word (t_1) of a document, if it is related to other more general word (t_2) by means of a generality interrelation, the semantic area (SA_1) of the first one will be included in the semantic area (SA_2) of the second one. In this case we can assume that SA_1 is included in SA_2 with a membership degree equal to the generality degree between both terms ($GD(t_1,t_2)$). In this case we talk about one occurrence of the concept referred by t_1 and a number of occurrences of the concept referred by t_2 equal to $GD(t_1,t_2)$.

On the other hand, if two synonymous words co-occur in a query or a in a document, having a synonymy degree less than 1 because of their different semantic shades, they will have a common semantic area (the common semantic shades of these words). This common area defines the concept referred by the co-occurrence of these words, that is, the intersection of these areas represents an only concept.

Considering a concept, obtained from the occurrences of various synonyms, as a fuzzy set, we can define the membership degree of each one of the words that form the concept to the concept itself. So, assuming that m words (synonyms each other) co-occur in a document or in a query, the membership degree of each term t_i to the concept C, which they converge to, is calculated by expression 3:

$$\mu(t_i, C) = MIN_{j=1}^{m}(SD(t_i, t_j)) \qquad (4)$$

Once this value is defined, it is possible to define the number N of occurrences of a concept C (formed by the co-occurrence of m synonyms) in a query or document by the expression 4, in which w_i is the weight of the term t_i in the document, that in this case and in order to simplify, is the number of occurrences of the term t_i.

$$N = \sum_{i=1}^{m} w_i * \mu(t_i, C) \qquad (5)$$

2.2 *The Construction of FIS-CRM Vectors*

When obtaining the vector weights, what we pursue is to share the number of occurrences of each concept among the words of the set of synonyms whose semantic area is more representative of the semantic area of that concept.

To achieve this aim, the construction of the vector of documents and queries takes two steps:

1) Representing them by their base weight vectors (based on the occurrences of the contained words). This vector is what we called in this document the VSM vector.
2) Weights readjusting (obtaining FIS-CRM vectors based on concept occurrences). Thus, a word may have a weight in the new vector even if it is not contained in it, as long as the referenced concept underlies the document. The weights readjusting process undertakes two tasks:

   a) Giving a weight to the words that represent a more general concept that the contained one.
   b) Sharing the occurrences of a concept among the representative set of fuzzy synonyms of the concept.

To share the occurrences of a concept among its representative synonyms involves a problem: Which are the synonyms of a concept? The answer to this question is based on the fact that the words which converge to a concept will belong to several sets of synonyms (semantic areas). In this case it would be possible to get a fuzzy measurement of membership of the concept to every set of synonyms, but what we really want is to share the occurrences of the concept among as many synonyms as possible, in order to be able to increase the similarity of the document (or query) where the concept occurs respect to other documents containing similar concepts. So, the set of synonyms taken as the representative set of synonyms of the concept is the one with more cardinality of those containing the converging words to the concept.

The main handicap of the sharing process is managing weak words (words with several meanings). So, sense disambiguation of this kind of words is fundamental for carrying out this process successfully.

Sense disambiguation of weak words is implicitly carried out by the sharing process (as it will be later on described). Although the essence of the sharing process is unique, we distinguish three situations depending on the implication of weak or strong words (with only one meaning). So, there are three types of synonymy sharing:

- Type-1: When one or several strong synonyms co-occur in a document or query.
- Type-2: When one or several weak synonyms co-occur and the occurrences are shared among their strong synonyms. It requires word sense disambiguation.
- Type-3: When one or several weak synonyms co-occur and they have not strong synonyms to share the occurrences. It also requires word sense disambiguation.

2.2.1 Sharing Type-1: Readjustment among Strong Words

Let us consider a piece of the VSM vector of a query or document containing several occurrences of m strong synonyms. (The case of $m = 1$ is included in this situation as the one of $n = m$ is).

$$t_1 \quad t_2 \quad \cdots \quad t_m \quad t_{m+1} \cdots t_n$$

VSM vector

$$\boxed{\mathbf{w_1}\ \mathbf{w_2}\ \cdots\ \mathbf{w_m}\ \mathbf{0}\ \cdots\ \mathbf{0}}$$

Where w_i is reflects the number of occurrences of the term t_i.

We assume that these synonyms converge to a concept C whose most suitable set of synonyms is formed by n strong terms. It is very important to emphasize that in this situation only strong synonyms receive a part of weight. The FIS-CRM vector for this query or document would be this one:

$$t_1 \quad t_2 \quad \cdots \quad t_m \quad t_{m+1} \cdots t_n$$

FIS-CRM vector

$$\boxed{\mathbf{w'_1}\ \mathbf{w'_2}\ \cdots\ \mathbf{w'_m}\ \mathbf{w'_{m+1}}\ \cdots\ \mathbf{w'_n}}$$

Where w'_i is the weight of the term t_i.

To calculate each w'_i, the number of occurrences (N, obtained by the expression 5) of the concept C which the m terms converge to, is shared among the n strong synonyms.

The premise of the sharing process is to guarantee that if we compare (with a simple product of vectors) the piece of vector containing the n elements with other piece of vector containing the same number of occurrences of the concept, the matching value should be N^2, that is, the N occurrences in a document multiplied by the N occurrences in the other document. Thus, the new weights must satisfy this expression:

$$\sum_{i=1}^{n} w'^2_i = N^2 \tag{6}$$

In order to assign a weight for each term which is proportional to its membership degree to the concept C contained in the document, every new weight must satisfy this other expression:

$$w'_i = \alpha * \mu(t_i, C) \tag{7}$$

Solving these equations we find a value of α that satisfies these two expressions, which is:

$$\alpha = N * \sqrt{\frac{1}{\sum_{i=1}^{n} \mu(t_i, C)^2}} \tag{8}$$

And finally we can obtain each new weight by this expression:

$$w'_i = N * \sqrt{\frac{1}{\sum_{i=1}^{n} \mu(t_i, C)^2}} * \mu(t_i, C) \tag{9}$$

We can observe that, if we consider the particular case in which all the involved synonyms have each other a synonymy degree of 1, the membership degree of all the terms to the concept will be 1, and so, the number of occurrences of the concept will be the sum of occurrences of each term, and we could finally obtain expression 10, which is the one proposed in the first version of FIS-CRM.

$$w'_i = \sum_{i=1}^{n} w_i * \sqrt{\frac{1}{n}} \tag{10}$$

For example, let us consider a piece of the VSM vector of a document like the one below. We suppose that the terms A and B (synonyms) co-occur in a document with 2 and 3 occurrences respectively and that the most suitable set of synonyms to the concept they converge to contains three other words (C, D and E). We also assume that the synonymy degrees all among these terms are: SD(A,B)=0.9, SD(A,C)=0.8, SD(A,D)=0.7, SD(A,E)=0.6, SD(B,C)=0.7, SD(B,D)=0.8, SD(B,E)=0.9, SD(C,D) =0.5, SD(C,E)=0.6 and SD(D,E)=0.9.

	A	B	C	D	E
VSM vector	2	3	0	0	0

In this case, the number of the occurrences (N, obtained by the expression 5) of the concept formed by the co-occurrence of A and B is 4,5 (despite existing 5 occurrences of the terms A and B). The corresponding FIS-CRM vector will be this one:

	A	B	C	D	E
FIS-CRM vector	2.35	2.35	1.83	1.83	1.56

We can check that when comparing this piece of vector with other piece of vector equal to this, the matching value would be 20.25 (2.35*2.35 + 2.35*2.35 + 1.83*1.83 + 1.83*1.83 + 1.56*1.56) which is equal to 4.5^2 as we expected to get.

2.2.2 Disambiguating Weak Terms

The disambiguation of a weak term consists on identifying its right semantic area, therefore, in our model it consists on choosing its most suitable set of synonyms. The disambiguation process can take up to three phases:

Phase 1: If some other strong synonym of the weak term is contained in the document, its most suitable set of synonyms will be the bigger one that contains all these terms (the weak one and the strong one or ones). The sense disambiguation process would finish at this point.

Phase 2: If some other weak synonym of the weak term is contained in the document, its most suitable set of synonyms will be the bigger one that contains all these terms: the weak one to disambiguate and the other weak ones (also disambiguated for the same reason). The sense disambiguation process would finish at this point.

Phase 3: In any other situation, we disambiguate the weak term by its closer context. So, the first thing to do is to identify the most suitable ontology to the piece of document which the weak word is contained in. The piece of document is formed by the x significant terms before and after the word. The size of this window must be big enough to capture the context of the term, but not as big as to embrace occurrences of the word with other meaning. A window of 10-size terms has been tested to be enough for this purpose. The most suitable ontology for a set of terms would be the one that gets the maximum value for the expression 11.

$$MAX_{k=1}^{z} \left(\sum_{i=1}^{n_d} \sum_{j=1}^{m_k} GD_k(t_i, t_j) \right) \tag{11}$$

Where: z is the number of ontologies

n_d is the number of terms included in the piece of document

m_k is the number of terms of the ontology k

The cost of this calculation could be highly reduced by considering only a set of representative terms of each one of the ontologies.

Once obtained the most suitable ontology for the context of the weak term, the next step is to choose the set of synonyms of the weak term which has more similarity to this ontology. A reference measurement of the similarity of a set of synonyms S to an ontology k can be the one shown below, which is based on the same idea that expression above:

$$SIM(S, k) = \sum_{i=1}^{n_s} \sum_{j=1}^{m_k} GD_k(t_i, t_j) \tag{12}$$

Where: n_s is the number of terms included in the set of synonyms S

m_k is the number of terms of the ontology k

It is important to point out that these values could be obtained only once (as long as the ontologies and the dictionary do not change) and so could be stored in a repository.

2.2.3 Sharing Type-2: Sharing Occurrences from Weak Terms to Strong Words

This type of sharing is carried out when one or several weak synonyms co-occur in a document or a query and the occurrences are shared among their strong synonyms (leaving the weak terms without any weight). This type of readjustment requires disambiguating the sense of the weak words as explained in the previous section.

Let us consider a piece of the VSM vector of a query or document containing several occurrences of m weak synonymous words. And let us consider the g strong synonyms belonging to the set of synonyms of the disambiguated sense of these weak terms.

If m is equal to 1, the disambiguation process could possibly need the third phase of it (if no strong synonyms co-occur), but if $m > 1$ the disambiguation process could be resolved in the second phase (or even in the first one if strong words co-occur). To embrace the more general situation of this type of readjustment, let us consider that f strong synonyms of them co-occur in the document (f could be equal to 0). The vector could be represented this way:

$$
\begin{array}{c}
\text{VSM vector}
\end{array}
\quad
\begin{array}{ccccccccc}
t_1 & t_2 & \cdots & t_m & t_{m+1} & \cdots & t_n & t_{n+1} & \cdots & t_p \\
\hline
w_1\, w_2 & & \cdots & w_m\, w_{m+1} & \cdots & & w_n & 0 & \cdots & 0
\end{array}
$$

Where: w_i is the number of occurrences of the term t_i

   The first m terms are the weak ones contained in the document.

   The next f (n-m) terms are the strong ones contained in the document.

   The last g (p-n) terms are the strong terms of the set of synonyms not contained in the document.

In this case, the number of occurrences (N) of the concept, which the n synonyms converge to, is shared among the strong synonyms (from t_{m+1} to t_p).

It is important to point out that in order to calculate the number N, when managing the synonymy degree between two weak words, we must take into consideration the number that identifies the sense obtained by the disambiguation process. In the case of the synonymy degree (SD) between a strong word and a weak word it is implicitly disambiguated taking the value $SD(t_{strong}, t_{weak})$ as explained in Sect. 2.1.

The weights of the strong synonyms (from t_{m+1} to t_p) of the corresponding FIS-CRM vector are calculated with the formula of expression 9, assigning 0 weight to the first m terms (the weak ones).

2.2.4 Sharing Type-3: Sharing Occurrences Among Weak Terms

This type of sharing is carried out when one or several weak synonyms co-occur and they have not strong synonyms to share the occurrences of the concept they converge to. It also requires word sense disambiguation.

Let us consider a piece of the VSM vector of a query or document containing several occurrences of m weak synonymous words. And let us consider the set of n synonyms of the right disambiguated sense (all of them are weak terms).

$$t_1 \quad t_2 \quad \cdots \quad t_m \quad t_{m+1} \cdots t_n$$

VSM vector

$$\boxed{w_1 \; w_2 \quad \cdots \quad w_m \quad 0 \quad \cdots \quad 0}$$

In this case, the number of occurrences of the concept to which the m weak terms converge is shared among all the synonyms of its right set of synonyms. In this case we should take the same considerations as the ones explained in the previous section about the identification of the number of the senses involved.

2.2.5 Readjustment Due to the Generality Interrelation

The aim of this type of readjustment is to give a weight to the words that represent a more general concept than the contained one, as described in S9ect. 2.4 by means of the semantic areas involved.

This makes possible that a word A, not contained in a document, gets a weight if a word B (narrower than A) is present in the document. In this case, the readjustment made is linear and proportional to the generality degree between A and B.

The vector below stores the weights before this type of readjustment is made.

$$A \qquad B$$

$$\boxed{\mathbf{0} \quad \mathbf{w_B}}$$

In this case, the weight of the term A would be readjusted to this one:

$$w'_A = w_B \,^* GD(B, A) \tag{13}$$

Thus, this document could be retrieved using a query with the word A (never retrieved using VSM vectors).

This type of readjustment does not require any disambiguation process as it is implicitly made when obtaining the most suitable ontology to the document (assuming that all the entries of a word in an ontology have the same sense).

3 Implementing FIS-CRM in a Web Search System

One of the most interesting uses of FIS-CRM is to integrate it in the crawler/indexer subsystem of a search engine. This way, if we apply FIS-CRM to the indexed web pages and if we also apply it to the user query, when comparing the corresponding vectors we should get a matching value based on concept co-occurrences instead of term-occurrences as it happens when indexing with a standard VSM vector.

Integrating FIS-CRM in the web crawler/indexer does not mean increasing the efficiency of the search process, as indexing is carried out in an offline process.

Moreover, the matching mechanism required if using FIS-CRM vectors is exactly the same as it uses standard VSM vectors, that is, it could be one based on the product of the vectors or some other complex measurement as the vectors cosine.

So, the web search system is clearly divided in two subsystems: the offline subsystem (web crawler and indexer) and the online subsystem, whose functionality is showed in Fig. 1.

The online subsystem undertakes the following functions:

- Constructing the query vector/s from the user query.
- Matching process.
- Representing the retrieved snippets using the FIS-CRM model.
- Grouping the retrieved snippets according to the contained concepts.

The last three tasks of the online subsystem are the same than the ones implemented in the FISS metasearcher (later on described in Sect. 4).

The construction of the vector (or vectors) of the query is carried by applying FIS-CRM, and in this case the disambiguation of the weak words contained in the query plays an essential role. In order to disambiguate the sense of the words in the query some considerations must be taken into account. The only mandatory entry is the query expression, which contains the terms of the query and it could also contain logical operators (to simplify, the AND operator is assumed in the examples). The user can optionally select the desired thematic ontology (from the offered ones). If entered, the ontology will be used to disambiguate the sense of the weak terms in the query. If the query contains weak terms and the user does not select any ontology, the system will offer the user the set of synonyms of every sense of every weak term (the ones selected will determine their right senses). In the case that the query contains weak terms and the user does not either enter any ontology or select any sense of the weak terms, the system will build various query vectors (so many as the result of combining all the senses of the not disambiguated weak terms). For example, if the query contains two weak terms, A and B, with two meanings each one, identified by two different sets of synonyms each: $S_1{}^A$, $S_2{}^A$ for term A and $S_1{}^B$,

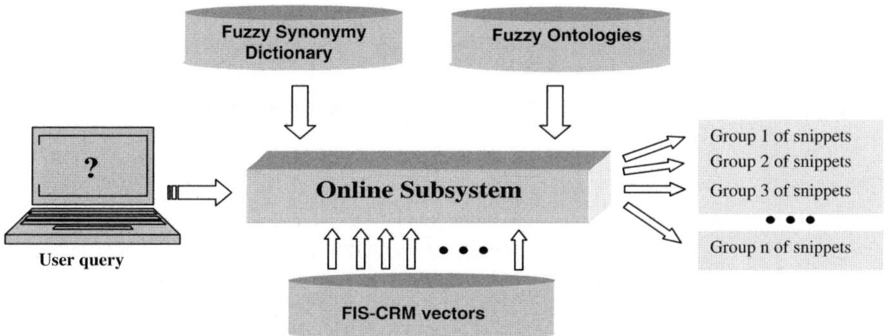

Fig. 1 Online subsystem

$S_2{}^B$ for term B, and if they have not been disambiguated, the system will generate four query vectors. In each vector, the occurrence of each term will be shared among the synonyms of the corresponding set. If we consider that each term in the query has a membership degree to each set of synonyms they belong, we can obtain a fuzzy measure of the compatibility of each one of the vectors with the query. That is, if a vector is built from the $S_1{}^A$ set and the $S_1{}^B$ set, the compatibility degree of this vector with the query could be obtained by the minimum of the values $\mu(A, S_1{}^A)$ and $\mu(B, S_1{}^B)$. Generalizing the expression, the compatibility degree of a vector v to a query Q of n terms can be obtained by this expression:

$$CD(v, Q) = MIN_{i=1}^{n}(\mu(t_i, S^i)) \tag{14}$$

This value can be used to rank the retrieved documents by each one of these query vectors.

3.1 Example of the Capabilities Retrieving

To easily test the effectiveness of this system, let us consider some example web pages and the results produced when evaluating some queries using this system and using a standard search engine. Let us take these pieces of web pages (Fig. 2), all of them containing the word *stack*:

Matching a query with the word *stack* and the VSM vectors for these web pages (only considering the most relevant contained words), the matching values could be the ones shown in Fig. 3, and so, the three pages would be retrieved.

This is the typical case of queries with words with several meanings. In this case, users are used to refining their query including another word. For example, if the user is interested in the *data structure* related sense he could probably add this compound word. In this case, making a query with both words *stack* and *"data structure"*, the matching values for the three VSM vectors would be 0 (as no one contains both words), not retrieving any, and worst of all, not retrieving the first web page which should match this query (according to the concepts contained in it). Something similar could happen if the user was interested in the *fire* related sense and tried to refine the query with the word *fireplace*. In this other case, no vector contains the words *stack* and *fireplace*, so no pages would be retrieved.

Now let us consider what would happen if using FIS-CRM vectors (when indexing) instead of VSM vectors. The set of synonyms of the different senses

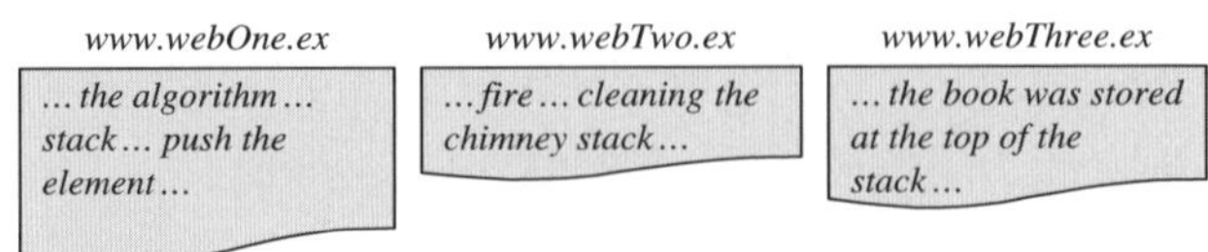

Fig. 2 Web pages examples

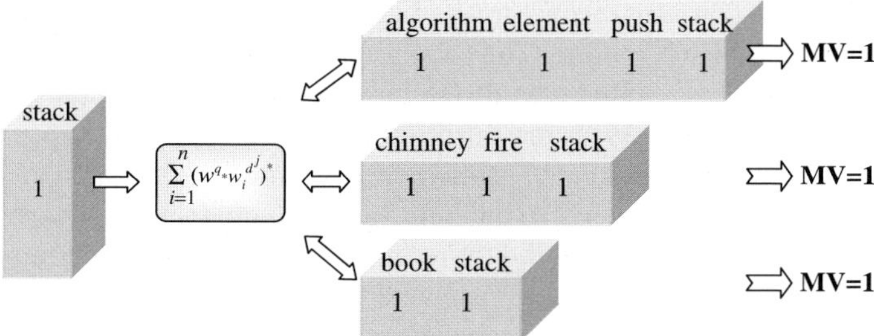

Fig. 3 Matching process with VSM vectors

of the word *stack* could be these three ones (referred in the three documents respectively):

stack$_1$ → (stack) : Assuming that there is no other synonym for the sense related to the *data structure* sense.

stack$_2$ → (stack, flue): For the sense related to *chimneys*.

Stack$_3$ → (stack, shelves): For the sense related to *books keeping*.

And let us suppose that the user disambiguates the sense of *stack* in the query by selecting a *programming* ontology (the stack$_1$ set of synonyms will define the concept). In this case, in the query vector the term *"data structure"* gets a weight because of its fuzzy generality interrelation with the term *stack,* but the term *stack* can not share its occurrences because its set of synonyms only include this term.

Something similar happens to the term *"data structure"* in the first vector, but in this case the disambiguation of *stack* was made at indexing time due to the fact that the *software* ontology was the most suitable one for this document.

In the second vector, the word *chimney* has shared its weight with its synonym *fireplace* and the term *stack* is disambiguated to its second sense (*stack$_2$*). In this case, the occurrences of *stack* are given to its strong synonym *flue*.

In the third vector, the term *stack* is disambiguated to the third sense (*stack$_3$*). Assuming that *shelves* is also a weak term, the occurrences are shared between the two terms of the set.

Thus, with these FIS-CRM vectors the matching values would be the ones shown in Fig. 4 (the new vector weights are calculated using SD reference values as nowadays only a Spanish dictionary is integrated in the system).

Now we can notice that now the matching value of the query and the first web

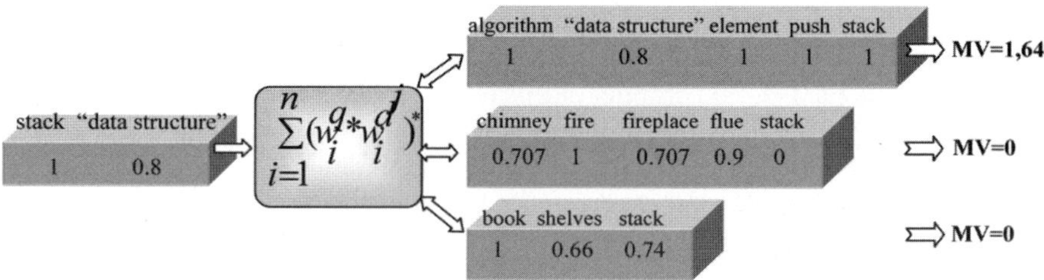

Fig. 4 Matching process with FIS-CRM vectors

page increases substantially respect to the one previously obtained, as we desired, getting a relevant matching value. The second page obtains a 0 matching value, as we desired, because the occurrences of the term *stack* in the page are totally given to its strong synonym *flue*. With the third page something similar happens. In this case, although the term *stack* keeps some weight (as it can not be given to any strong synonym), the matching value is 0, as there is no any matching between the term *"data structure"* in the query and the term *"data structure"* in the document.

We can also realize that with these new vectors a simple query with the word *fireplace* will retrieve the second web page (impossible to retrieve with VSM vectors), that is, despite the web page does not contain the word *fireplace*, it will be retrieved as the related concept does. Something similar happens if we make a query with the word *shelves*. In this case, as we desired, the only page retrieved will be the third one (never retrieved using the VSM vectors).

4 Implementing FIS-CRM in the FISS Metasearcher

The FISS metasearcher was the first system integrating FIS-CRM. The main difference of this system and the one described in Sect. 2 is that FISS uses the services of a commercial search engine and so it is fed up from the web pages indexed by this system.

The main characteristics of FISS are:

- It generates alternative queries (from the original query) using the fuzzy synonymy dictionary to do it.
- It uses the services of other searcher (Google in this case) to look for links to web pages.
- Taking the snnipets returned by Google, the fuzzy synonymy dictionary and the fuzzy ontologies and using FIS-CRM to represent the documents, it produces a conceptual vision of the retrieved documents.
- It dynamically groups the documents obtained using a soft clustering algorithm that considers the co-occurrence of concepts.

The detailed search process (Fig. 5) is divided into the following steps:

1. Generation of new queries. Alternative queries are generated from the synonyms of the words in the query. Each one of these queries has a compatibility degree with the original query that is calculated from the synonymy degree between the words included in it and the words specified in the original query. The value is obtained using a T-norm (product in this case).
2. Processing queries and obtaining snippets. The generated queries are sent to Google and afterwards the snippets of the documents are extracted from the returned HTML code.
3. Constructing an index of words that includes all the words included in the snippets. At this step, words are pre-processed (excluding the ones in a stop list, converting into lower case form, changing into to singular form, etc. . .).

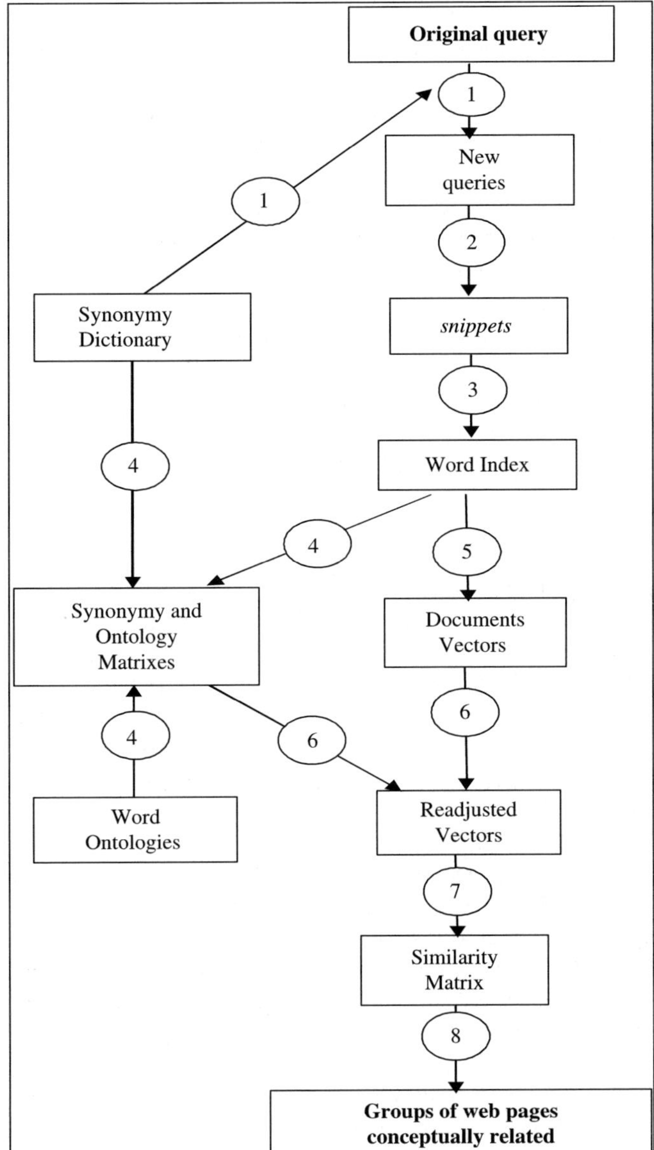

Fig. 5 Detailed search process

4. Constructing the synonymy and the ontology matrixes, which store the synonymy degree and the generality degree between every pair of words in the index.
5. Representing the *snippets* as vectors.
6. Readjustment of the vectors weights using FIS-CRM.
7. Generating the similarity matrix that stores the similarity degree between every pair of documents in the collection.

The similarity function contemplates the co-occurrence of words and phrases, documents size, rarity of words and the co-occurrence of concepts (in an implicit way).

8. Generation of groups of web pages conceptually related. In this step, groups are also labelled with the words that represent the concepts inside the groups documents.

We have implemented a clustering algorithm that, although it is based on the one proposed in (King & Ravikumar, 2001), has been adapted to optimize the resulting groups and to be able to get a hierarchical structure of groups.

Concerning the characteristics of the obtained results, the whole collection of documents retrieved can be considered as a fuzzy set that is formed by the documents that would have been obtainedusing the original query and the ones obtained from the generatedqueries. The first ones' membership degree to the fuzzy set is 1, and the second ones have a membership degree that is equal to the compatibility degree of the query that produced them with the original query.

Besides, the resulting clusters can be considered as fuzzy sets, so each one of the documents retrieved has a membership degree (obtained from an average similarity degree) to each one of these clusters. To make the solution easy to understand, the documents, that have a membership degree to a cluster lower than the similarity threshold (obtained automatically), are not included in this cluster.

5 The GUMSe Architecture

GUMSe (GUM Search) is a meta-search engine also based in FIS-CRM. Some typical tendencies in the use of search engines limit the effectiveness of the search process. Added to search based on key-words problems, there is a usual lack of experience in the users when using search engines. Meta-search engines appear as a promising alternative for trying to relieve this low precision of the current search engines. Nowadays, many of them only index a small portion of the information contained in Internet.

An increase in the semantic capability of user queries is necessary to improve the relevance of the search results. For this purpose, it is usually used the query expansion technique (Efthimiadis, 1996) with terms semantically related with the terms introduced by the user. There are different strategies that can be adopted to expand a query each one with its own characteristics. Basically, these mechanisms can be classified in manual, automatic and interactive query expansions. The *automatic algorithms* try to incorporate automatically new terms semantically related with those ones introduced by the user, to obtain a set of documents closer to the user search intention. But these approximations often are not easy to do, due to the different problems that vocabulary presents. Different approaches have been developed, for trying to get the correct meaning of the terms of a query and to focus the search for obtaining better results, in an automatic fashion. These algorithms are usually named Word Sense Disambiguation (WSD) algorithms. The *interactive query expansion* method consists of requesting the help of the user. By means of this mechanism the system suggest a series of terms among which the user has to choose. This type of systems frequently uses a tree structure from more general concepts towards more specific terms. It can make the process slow and uncomfortable because usually the user needs to answer several questions.

GUMSe meta-searcher tries to combine the most interesting aspects of those mechanisms of query expansion. The best results of one of another depend on the search circumstances and the quality of the existing knowledge structures. The intention is to develop a mechanism that complements the deficiencies of the rest of the searchers but that includes in certain degree some aspects of the different approaches. In GUMSe there is given priority to the *automatic algorithms* but taking into account the *interactive expansion*, since in case of not having enough knowledge to disambiguate the meaning of the terms of the query, the user is requested for help. Obviously it is possible to do the *manual query expansion*, doing a new query with new terms.

On the other hand, the query expansion is done by the help of a series of knowledge structures, which initially will be WordNet but it will also be based on the ontologies created by the system itself. This evolution relies on the *support to search agents* to store the user queries, and taking into account the more relevant documents obtained, tries to build or improve the ontologies that the query expansion stage uses.

There are many meta-search engine architectures (Glover, 1999; Kerschberg, 2001; Li, 2001) proposed in which there are several common components. This is the case for the components that make queries to be put to different search engines, as well as those components which calculate the relevance of the retrieved documents.

The GUMSe architecture is also based on the use of a series of agents with different functions which communicate with each other through a net, as shown in Fig. 6. Each agent develops a specific task, working cooperatively with other agents and reducing the system complexity. However, a communication language known by all the system agents is necessary to work cooperatively. This common language is known as *Agent Communication Language* (ACL).

WordNet's use is another remarkable aspect of GUMSe. It is the initial base of search process. Whereas dictionaries are organized by alphabetical order, WordNet (Miller, 1990, 1995) is organized on the basis of the semantic relations among different words. The basis of WordNet are the synonymy and hyponymy relations and other similar ones. In this way, sets of synonymous words can be found grouped in sets called synsets, and a polysemic word can belong to different synsets. The hyponymy relation (and the hypernymy one) is established among different terms, making a hierarchy based on the hyponymy. So, if dog "is a type of" *canine*, then *canine* is a hyponymy of *dog*. These relations provide important information that allows the expansion of the query terms to incorporate semantic information into the search process, as well as a mechanism to identify the correct meanings of the terms involved in a query.

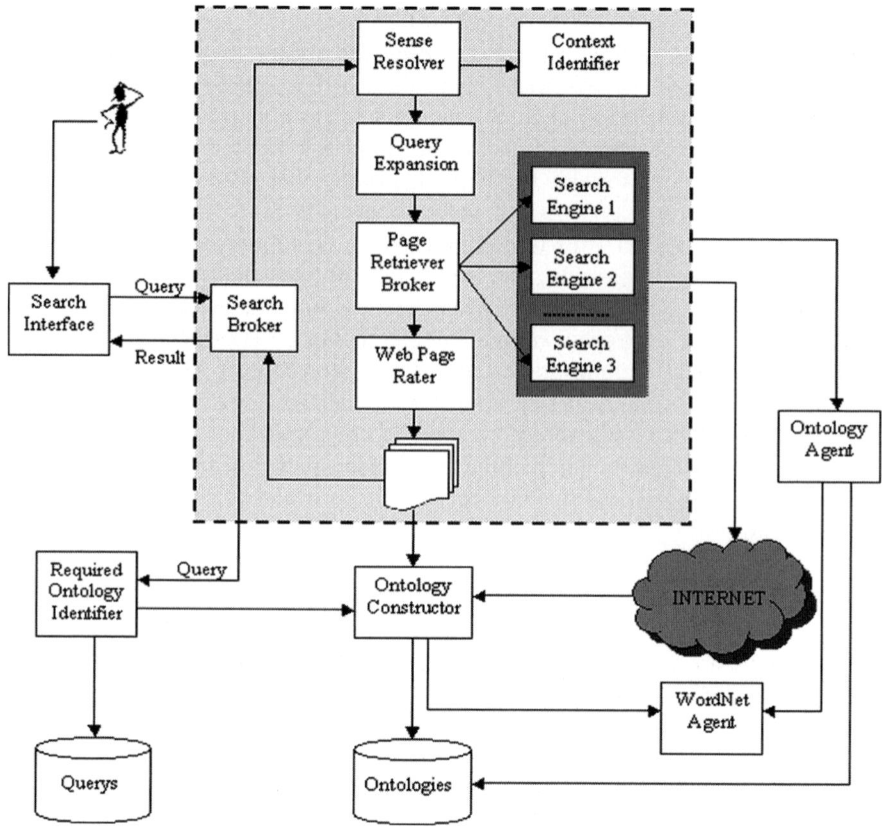

Fig. 6 GUMSe Architecture

The agents which configure the architecture are the following:

1. *Search Interface:* It is in charge of interacting with the user through a web interface.
2. *Search Broker:* It coordinates the searching process and receives the user's demands through the interface.
3. *Sense Resolver:* It determines the most similar meaning to the searching intentions of the user in case that the query has polysemic terms. It tries to determine the correct meaning without the user intervention by means of the Context Identifier Agent. If it can not do it, it shows to the user the meanings of the terms. In other case it obtains the correct meanings with the user intervention.
4. *Context Identifier:* It is in charge of deciding automatically in which sense the terms introduced by the user fit in. To do this, some kind of semantic relation between the terms of the query such as hyponymous or synonyms must exist. The approach used is to construct a linguistic context for each term, and to measure the number of common words shared by the term contexts. The linguistic context is constructed using WordNet.

5. *Query Expansion:* It builds the new queries which will increase the semantic content of the search and also calculates the relevance factor of each one to be subsequently considered in the number of admitted documents. Each term has a semantically related linguistic context obtained by WordNet and the ontologies. With this set of terms, new queries are built combining new terms appearing in a linguistic context and different search operators.

6. *Page Retriever Broker:* Its mission is to send the queries to the search engines we use and get the set of documents which will be subsequently evaluated. Nowadays Google and Altavista are used, although a rise in the number of searching engines is expected in order to improve the relevance of the searches.

7. *Web Page Rater:* It gives a measure of the degree of relevance of each document. For getting this measure, ontologies, synonymy and hyponymy relations are used. Factors like the appearance of terms strongly related to rejected meanings, which entail a reduction in the degree of relevance also count.

8. *Ontology Constructor:* The function of this agent is the improvement of the searching process through the construction of new ontologies which allow the expansion of the queries sent to the rest of searching engines. According to Widyantoro 2001, the idea of semantic relation is based on the co-occurrence of terms between different documents. In GUMSe a narrowing-down towards this idea based on physical proximity between terms is proposed. In other words, two terms t_i and t_j are physically closer if they are in the same sentence (paragraph) than if they are in different paragraphs. The explanation of this idea is that the terms in the same sentence are more likely to be strongly attached by predicates such as "is a" or "is part of", or even be part of axioms inside the ontology than two terms which are in different paragraphs that talk about different things. The search results are also used to improve and feed back the ontology. This feedback allows to progressively increase of the quality of the search.

9. *Required Ontology Identifier:* It checks the queries made by the users and tries to identify the most demanded concepts to build the ontologies which give them a support and improve the searching process. It groups terms in related sets, using WordNet, and shows the group with the highest number of related terms. Then the results of the search with those terms are used to construct the ontologies.

10. *Ontology Agent:* It obtains the semantical relations using both WordNet (with the intervention of the WordNet agent) and the constructed ontologies.

11. *WordNet Agent:* It is used as a broker for the different searching needs which allows the inclusion of semantic charge, meanings identification, synonyms and hyponymous relations. In case of not exist an ontology that supports a concrete query, the use of WordNet will be the only way to improve the search (Brezeale, 2000).

5.1 Searching Process

The searching process is coordinated by the *Search Broker Agent*. The semantic load of the query is successively refined to allow the expansion of its results with conceptually related documents. The four stages proposed are the following (Fig. 7):

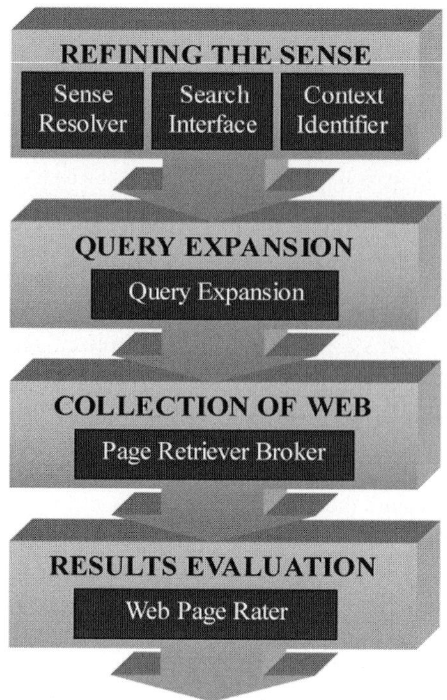

Fig. 7 Search stages and agents

Stage 1: Refining the sense of the search.

The agents involved for refining the sense of the search are the *Search Interface* which allows interaction with the user, and *Sense Resolver* and *Context Identifier* agents which are in charge of the refining of the sense of the search. The concept of meaning or sense will be used to focus the search and to improve the results. In order to refine the queries it is necessary to interact with the user. This mechanism allows an improvement in the relevance of the results (Bruza, 2000). On the other hand, the run time of the query management increases.

In addition, the discarded meanings will be used in the phase of evaluation as negative indicators. The relevance of a document is penalized if it has some negative terms. This procedure improves the reliability calculation of the retrieved documents relevance. Figure 8 shows the interactive interface of GUMSe to ask the user the suitable meaning.

Stage 2: Query expansion.

In this stage the data acquired in the previous phase is used to contribute a greater semantic load to the set of queries. The idea is to construct a set of additional queries that complement the original one. In these queries new terms semantically related to those of the original one are included. It is the case of synonymous, terms used to define the correct sense and also the information stored in ontologies. The main agent involved in this stage is the *Query Expansion Agent*.

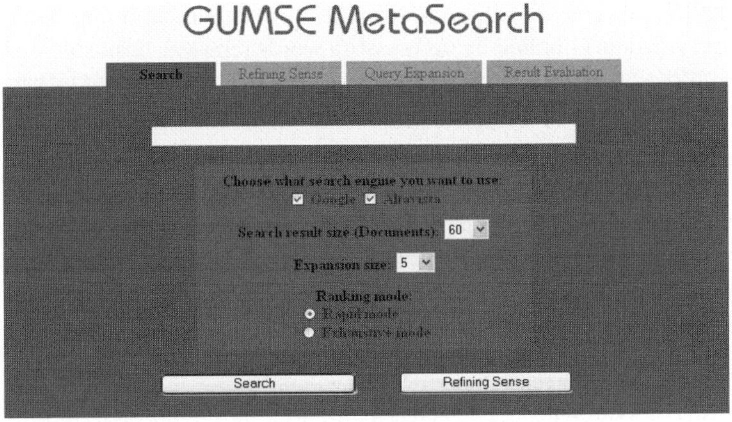

Fig. 8 GUMSe interactive search interface

For each new query a conceptual similarity factor is calculated (Abásolo, 2002). The result depends on the degree of conceptual similarity of its terms with original terms. It also depends on the used search operators. This factor determines the number of pages that will be later on evaluated.

An example: Search about "salt" (the condiment). If we only introduce the term "salt" in the search engine (i.e. Google) we get references about the Olympic Games held in Salt Lake City, the Society for Applied Learning Technology (SALT) and only one reference linked to the selected concept in the top ten references. To improve the relevance of the query one can turn to the synonyms sent back by WordNet (table salt, common salt) to expand the query to "salt table common", which increases the relevance of the sent back documents since the first 10 refer to the desired concept. It is necessary that the user selects the correct sense in the *refining sense stage* or takes other terms in the query that permit an automatic sense disambiguation.

Stage 3: Collection of the web.

The next step involves launching the different queries into a series of search engines to retrieve the documents. This task is carried out by the *Page Retrieve Broker* agent. The following step consists of sending new queries simultaneously to multiple Web search engines. The result is a collection of snippets of the most relevant documents. A proportional number of documents to the calculated in the previous step factor will be selected to be evaluated for each query. The selection is based on the property that search engines usually give back documents ordered by their relevance. the original query will be the one that *a priori* has great relevance.

One query could be expanded adding related terms. This is the traditional method. But GUMSe also includes new methods of query expansion that try to take advantage of the new characteristics of actual search engines. A meta-search engine depends on the underlying search engines to provide a reasonable set of results (Glover, 1999). Each existing search engine has its own characteristics and its concrete search operators. When the same query is used, different search engines will have different behaviors depending on their previous assumptions. For example, due to logic operators applied by defect, stop words removal, term suffix expansion and term order sensitivity. Due to this phenomenon, it is necessary to isolate the new expanded queries construction of a concrete search engine.

Stage 4: Results evaluation.

Finally, the obtained pages are retrieved and shown to the user sequenced under a unique ranked list. This process of evaluation is carried out by the *Rater Web Page Agent*. Due to the fact that the query results come from different sources, it is necessary to group them only in one collection of documents, eliminating repeated documents. Later, the relevancy of each document is calculated to determine the order in which they will appear. Finally, it is necessary to show the user the recovered pages classified by their relevance.

The terms or synonyms not selected by the user have a negative weight. A negative weight for a term penalizes any document that contains that term. However, documents containing the term still could appear in the results, because other terms may contribute to the document score. In this way, the documents with the selected meaning are chosen more frequently, increasing the degree of relevance of the document retrieved set.

$$Ranking(D) = \alpha P_D - (1 - \alpha) N_D \tag{15}$$

$$P_D = \frac{\sum_{t=1}^{NTP} f_{tD}}{\sum_{t=1}^{NTP} 1/SD_{tD}} \qquad N_D = \frac{\sum_{t=1}^{NTN} f_{tD}}{\sum_{t=1}^{NTN} 1/SD_{tD}} \tag{16}$$

In formula 16, P is the weight of positive terms in the document and N is for negative terms. One term is negative if exist a semantic relation (obtained from WordNet) with an incorrect sense of some query term, and positive if the relation exists with the correct sense. f_{tD} is the frecuency of the term t in the document D and SD_{tD} is the semantic distance respect to the related meaning. NTP is the number of positive terms and NTN is the number of negative terms in D. *Rater Web Page Agent* carries out this process of evaluation.

This method is very slow because it have to download many pages. For this reason, GUMSe implements other algorithm of evaluation call *rapid evaluation*. This method does not analyse the content of the document and it is faster. It is based in the next four criteria:

- *Conceptual Similarity Measure (CSM):* This value indicates the similarity of the query i with the original query.
- *The Position* of the document D in the query i.
- *The Search Trust Grade (STG):* It is the weight of the results from the different search engines.
- *The Chorus Effect:* It assumes that different search engines usually retrieve the same relevant documents in the first positions.

The ranking of each document D is calculated using the formula 17 where DQ is the number of queries that retrieve the document D and P_i is the position of the document D in the query i where 1 is the best position.

$$Ranking_{rapid}(D) = \frac{DQ}{\sum_{i=1}^{DQ} \sqrt[\alpha]{STG_i} * P_i * CSM_i} \tag{17}$$

This stage is necessary because the query results received can be manipulated. Most search engines rank pages using the frequency of appearance of the query word(s) within a document. Term-based methods are very susceptible to corruption (spamming). Some people try to elicit higher rankings for their Web sites by analyzing ranking algorithms. This practice, called SEO (Search Engine Optimization), is a major headache for search engines. With the proposed evaluation, we tried to avoid this problem, as well as to improve the relevance of the search.

5.2 Ontologies Construction

The *Ontology Constructor Agent* is specialized in the construction of those ontologies suggested by the *Required Ontology Identifier*. These ontologies are used to improve the semantic load in the query expansion stage about certain subjects in which the user emphasizes making multiple queries. The *Search Broker Agent* detects the concepts more frequently consulted. The stages followed in the construction of the ontologies process are the following (Fig. 9):

1. *Information Acquisition*: Retrieval of the most relevant files according to the *Web Page Rater* for the construction of the ontology.
2. *Syntactic analyzer (parser)*: A natural language parser which simplifies the sentences of the files selected in the previous stage through the use of *stoplist*, stemming algorithms (Porter, 80) and WordNet, this last one to study the different word classes which configure a sentence (nouns, verbs, prepositions) in order to take only the profitably processable knowledge. The simplifications of the sentences have the form: Concept Infinitive Concept (Serrano-Guerrero, 2004).
3. *Making of the data-vector*: Through the vector model technique we create a terms dictionary from the concepts taken in the previous stage, and every term is given a numerical value referred to each of the terms of the ontology based on the co-occurrence of terms in the same document and depending on the physical distance between terms whose basic unit is the sentence.
4. *Discovery of the implicit knowledge and elimination of the ambiguity*: Through WordNet, the updated knowledge can be refined. We can also simplify the terms dictionary eliminating redundant terms such as synonyms, since, for example, there is no point in storing terms such as "look like" and "resemble". If we take a

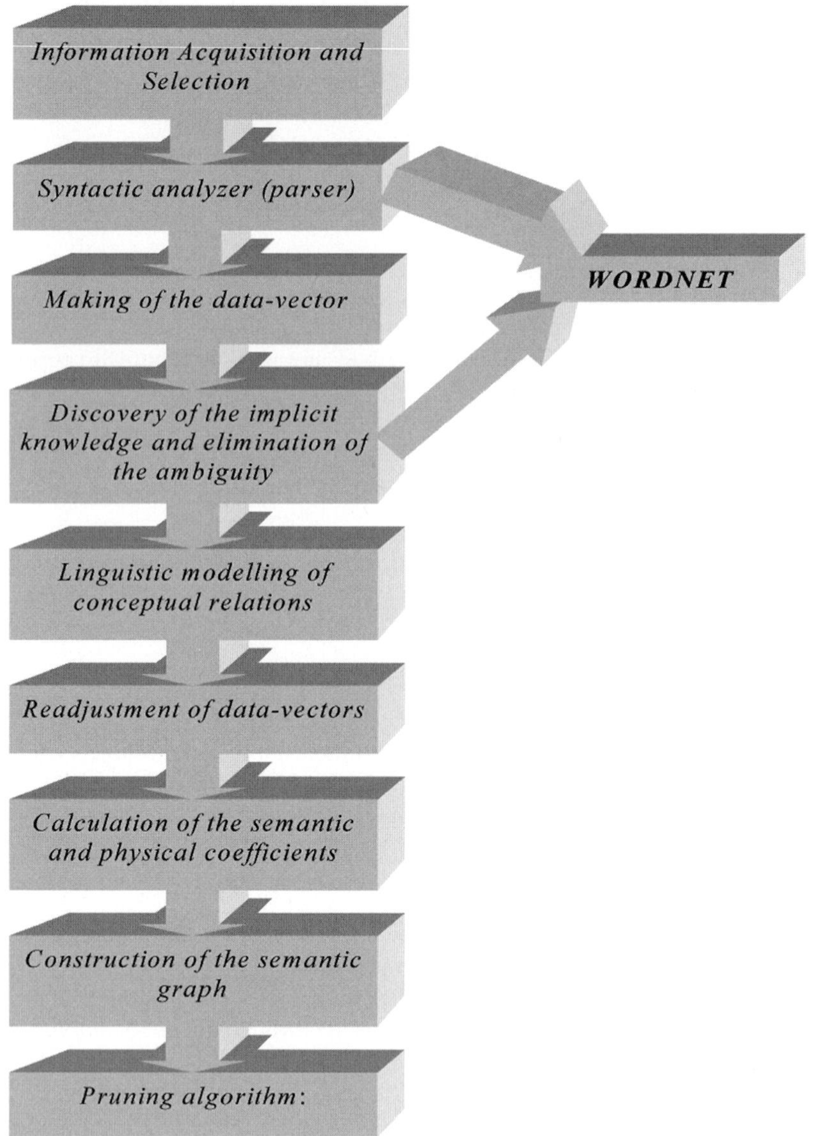

Fig. 9 Summary of the stages (Ontology Constructor Agent)

definition from WordNet in which an interpreter and a composer are both referred to as musicians we can refine the design as shown in Fig. 10:

5. *Linguistic modelling of the conceptual relations*: Discovery of the relations ("it is a", "it is part of", etc.) between different terms and their codification through first order logic. There are many different verbs that end up meaning the same, so they can be grouped in different kinds of relations. This linguistic modelling

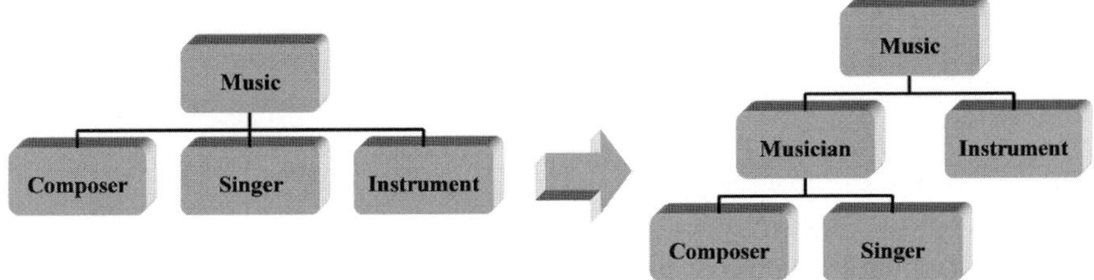

Fig. 10 Refining the Knowledge

is being developed for the extraction of a greater knowledge from the documents as well as for the making of possible deductions about them.

6. *Readjustment of data-vectors*: The readjustment occurs adding new terms to the dictionary ("Musician" in the preceding example), and summing "1" to the parameters X and Y of the data vector according to those terms ("Composer", "Interpreter") with which it co-occurs in WordNet definition, as they are physically in the same sentence.

7. *Calculation of the semantic and physical coefficients*: The construction of the semantic graph is made according to two relations: *physical dependence* (number of attributes two terms have in common) and *semantic dependence* (semantic proximity between two concepts). The physical coefficient is calculated according to the physical proximity between two terms as it is more probable that two very close terms are related through predicates like "is a" or "is part of" while the semantic coefficient is based on the concurrence of terms between documents and in the same document.

8. *Construction of the semantic graph*: Elaboration of a graph structure where all the knowledge is related.

9. *Pruning algorithm*: Creation of a syntactic tree in which only the most syntactically relevant terms appear. We can use three criteria:

   - *Physical power*: It is based on the physical coefficient. Once the root node of the tree is chosen, only the terms with the highest degree of pure physical relation will hang from it.
   - *Semantic power*: It is based on the semantic coefficient. Once the root node of the tree is chosen, only the terms with the highest degree of semantic relation will hang from it.

These two first algorithms work in the way proposed by Widyantoro and Yen (2001) but considering each coefficient in each case.

   - *Processing in layers*: Entails abstracting some semantic characteristics of the graph to create a higher abstraction level concept by pruning the information that deals with highly specific characteristics, keeping only the main and most general concepts.

6 Conclusions

At this moment, these systems are only prototypes. So, the evaluation with standard collections of documents such as TREC or CACM is not possible yet. Nevertheless, numerous testing exercises with particular collections of documents (retrieved *ad hoc*) have been carried out, obtaining excellent results.

To get an idea of the retrieving capability of this system, let us consider the following data: At present, Google search engine retrieves more than 16 million of links to web pages containing the word *stack*. From these, over one million are related to the *"data structure"* meaning (this is an estimated value), but from these ones only 300.000 links exactly contain the words *stack* and *"data structure"*, that is, there are over 700.000 links related to the *"data structure"* meaning but not containing this word, so none of these pages would be retrieved by a word-matching IR system, but the concept-matching IR systems presented would.

Anyway, the search systems and the FIS-CRM model are capable of being improved in some aspects such as the management of another type of interrelations and the study of their influences in the sense disambiguation method. Concerning the deduction capability when managing queries with "qualifiers" (as the example query of "high frequency CPU" mentioned in the introduction), future approaches will take into consideration the possibility of defining labelled fuzzy sets related to some user specified attributes, providing an approximation to this problem.

As a final conclusion and to sum up the proposed systems characteristics, considering the interesting results produced by FISS and GUMSe metasearchers and considering the quality of the results obtained by implementing FIS-CRM in the web crawler component, and taking into account that any search engine could manage FIS-CRM vectors without having to adapt its matching mechanism keeping its efficiency, we sincerely think that the proposed approach provides a very useful and easy implementing contribution to a very complex problem.

References

Abásolo, C. and Gómez, M. (2002): A framework for meta-search based on numerical aggregation operators. In Procceedings of the Congrés Català d'Intel.ligència Artificial, CCIA 2002.

Bordogna, G., Carrara, P. Pasi, G. (1995). Fuzzy approaches to extend boolean information retrieval, P. Bosc, J. Kacprzyk Eds. Fuzziness in database management-systems, Germany: Physica-Verlag, 231–274.

Brezeale, D. (1999). The Organization of Internet Web Pages Using Wordnet And Self-Organizing Maps. Masters thesis, University of Texas at Arlington, August 1999.

Bruza, P.; McArthur, R. and Dennis S. (2000): Interactive Internet search: Keyword, directory and query reformulation mechanisms compared. In Proceedings of the 23rd Annual International ACM SIGIR Conference on Research and Development in Information Retrieval. 280–287.

Choi, D. (2001). Integration of document index with perception index and its application to fuzzy query on the Internet, Proceedings of the BISC Int. Workshop on Fuzzy Logic and the Internet, 68–72.

Cooley, R., Mobasher, B., Srivastaba, J. (1997). Grouping web page references into transactions for mining world wide web browsing patterns, Technical report TR 97-021, University of Minnesota, Minneapolis.

Delgado, M., Martin-Bautista, M.J., Sánchez, D., Serrano, J.M., Vila, M.A. (2003). Association rules and fuzzy associations rules to find new query terms, Proc. of the Third Conference of the EUSFLAT, 49–53.

Deerwester, Scott, Dumais, S.T., Furnas, G.W., Landauer, T.K., Harshman, R. (1990). Indexing by latent semantic analysis, Journal of the American Society for Information Science, 41(6), 391–407.

Efthimiadis, E. N. (1996). Query Expansion. ARIST, v31, 121–187.

Fernandez, S. (2001). A contribution to the automatic processing of the synonymy using Prolog, PhD Thesis, University of Santiago de Compostela, Spain.

Fernández, S., Sobrino, A.(2000). Hacia un tratamiento computacional de la sinonimia. In Revista de la Sociedad Española para el Procesamiento del Lenguaje Natural (SEPLN), Nº 26, September 2000, 89–95.

Glover, E. J.; Lawrence, S.; Birming- ham, W. P. & Giles C. L. (1999): Architecture of a metasearch engine that supports user information needs. In Eighth International Conference on Information and Knowledge Management (CIKM'99) ACM, 210 216.

Gonzalo, J., Verdejo, F., Chugur, I., Cigarran, J. (1998). Indexing with WordNet synsets can improve retrieval, Proc. of the COLING/ACL Work. on usage of WordNet in natural language processing systems.

Herrera-Viedma, E. (2001). An information retrieval system with ordinal linguistic weighted queries based on two weighting elements, Int. J. Uncertainty, Fuzziness and Knowledge-Based Systems, 9, 77–88.

Herrera-Viedma, R., Pasi, G. (2003). Fuzzy approaches to access information on the Web: recent developments and research trends, Proc. of the Third Conference of the EUSFLAT, 25–31.

Kerschberg, L.; Kim, W. and Scime, A. (2001): A Semantic Taxonomy-Based Personalizable Meta-Search Agent. Second International Conference on Web Information Systems Engineering (WISE'01).

King-Ip, L., Ravikumar, K. (2001). A similarity-based soft clustering algorithm for documents, Proc. of the Seventh Int. Conf. on Database Sys. for Advanced Applications.

Kiryakov, A.K., Simov, K.I. (1999). Ontologically supported semantic matching, Proceedings of "NODALIDA'99: Nordic Conference on Computational Linguistics", Trondheim.

Lafourcade, M., Prince, V. (2001). Relative Synonymy and conceptual vectors, Proceedings the Sixth Natural Language Processing Pacific Rim Symposium, Japan, (202), 127–134.

Leacock, C., Chodorow, M. (1998), Combining local context and Wordnet similarity for word sense disambiguation, In WordNet, an Electronic Lexical Database, MIT Press, Cambridge Ma, 285–303.

Li, Z.; Wang, Y. and Oria, V. (2001): A New Architecture for Web Meta-Search Engines, in Proceedings of the 2001 Americas Conference on Information Systems, Boston, MA, 2001.

Loupy, C., El-Bèze, M. (2002). Managing synonymy and polysemy in a document retrieval system using WordNet, Proceedings of the LREC2002: Workshop on Linguistic Knowledge Acquisition and Representation.

Martin-Bautista, M.J., Vila, M., Kraft, D., Chen, J. (2001). User profiles and fuzzy logic in web retrieval, Proc. of the BISC Int. Workshop on Fuzzy Logic and the Internet, 19–24.

Miller, G. (ed.) (1990). WORDNET: An Online Lexical Database. International Journal of Lexicography, 3(4).

Miller, G.A. (1995). WordNet: A lexical database for English, Communications of the ACM 11, 39–41.

Miyamoto, S., Miyake, T. Nakayama, K. (1983). Generation of pseudothesaurus for information retrieval based on co-occurrences and fuzzy set operations, IEEE Transactions on Systems, Man and Cybernetics, Vol 13, 1, 62–70.

Miyamoto, S. (1990). Information retrieval based on fuzzy associations, Fuzzy Sets and Systems, 38 (2), 191–205.

Ohgaya, R., Takagi, T., Fukano, K., Taniguchi, K. (2002). Conceptual fuzzy sets- based navigation system for Yahoo!, Proc. of the 2002 NAFIPS annual meeting, 274–279.

Olivas, J.A., Garcés, P.J., Romero, F.P. (2003). An application of the FIS-CRM model to the FISS metasearcher: Using fuzzy synonymy and fuzzy generality for representing concepts in documents, Int. Journal of Approximate Reasoning 34, 201–219.

Pasi, G. (2002). Flexible information retrieval: some research trends, Mathware and Soft Computing 9, 107–121.

Perkovitz, M., Etzioni, O. (2000). Towards adaptive web sites: Conceptual framework and case study, Artificial Intelligence 118, 245–275.

Porter, M. F. (1980). An Algorithm for Suffix Stripping. Program, 14(3): 130–137.

Ramakrishnan, G., Prithviraj, B.P., Deepa E. et al, (2004). Soft word disambiguation, Second Global WordNet Conference, 2004.

Ricarte, I., Gomide, F. (2001). A reference model for intelligent information search, Proc. of the BISC International Workshop on Fuzzy Logic and the Internet, 80–85.

Romero, F. P.; Garcés, P.; Olivas, J. A. (2002). Improving Internet Intelligent Agents using Fuzzy Logic and Data Mining Techniques. Proc. of the International Conference on Artificial intelligence IC-AI'02, 225–230.

Salton, G., Wang, A., Yang, C.S.A. (1975). Vector space model for automatic indexing, Communications of the ACM 18, 613–620.

Serrano-Guerrero, J.; Olivas, J. A. (2004). Discovery of conceptual relations for ontology construction in GUMSe. 2004 Annual Meeting of the North American Fuzzy Information Processing Society Proceedings - NAFIPS, IEEE, Banff, Alberta, Canada, 647 – 651.

Sun, J., Shaban, K., Poddre, S., Karry, F., Basir, O., Kamel, M. (2003). Fuzzy Semantic measurement for synonymy and its application in an automatic question-answering system, IEEE Int. Conference on National Language Processing and Knowledge Engineering, Beijing.

Tang, Y., Zhang, Y. (2001). Personalized library search agents using data mining techniques, Proceedings of the 1st BISC Int. Workshop on Fuzzy Logic and the Internet, 119–124.

Takagi, T., Tajima, M. (2001). Proposal of a search engine based on conceptual matching of text notes, Proc. of the BISC Int. Workshop on Fuzzy Logic and the Internet, 53–58

Voorhees, E. (1998). Using WordNet for text retrieval, In WordNet: an electronic lexical database, MIT Press.

Whaley, J.M. (1999). An application of word sense disambiguation to information retrieval, Dartmouth College Computer Science Technical Report PCS-TR99-352.

Widyantoro, D., Yen, J. (2001). Incorporating fuzzy ontology of term relations in a search engine, Proceedings of the BISC Int. Workshop on Fuzzy Logic and the Internet, 155–160.

Zadeh, L. A. (1987). Fuzzy Sets and Applications (Selected Papers, edited by R. R. Yager, S. Ovchinnikov, R. M. Tong, H. T. Nguyen), John Wiley, Nueva York.

Zadeh, L. A. (2003). From search engines to Question-Answering System: The need for new tools, E. Menasalvas, J. Segovia, P.S. Szczepaniak (Eds.): Proceedings of the Atlantic Web Intelligence Conference - AWIC'2003. Lecture Notes in Computer Science (LNCS), Springer.

Zamir, O., Etzioni, O. (1999). Grouper: A dynamic clustering interface to web search results, Proceedings of the WWW8.

Mediative Fuzzy Logic: A New Approach for Contradictory Knowledge Management

Oscar Montiel, Oscar Castillo, Patricia Melin and Roberto Sepulveda

Abstract In this paper we are proposing a novel fuzzy method that can handle imperfect knowledge in a broader way than Intuitionistic (in the sense of Atanassov) fuzzy logic does (IFL). This fuzzy method can manage non-contradictory, doubtful, and contradictory information provided by experts, providing a mediated solution, so we called it Mediative Fuzzy Logic (MFL). We are comparing results of MFL, with IFL and traditional Fuzzy logic (FL).

1 Introduction

Uncertainty affects all decision making and appears in a number of different forms. The concept of information is fully connected with the concept of uncertainty; the most fundamental aspect of this connection is that uncertainty involved in any problem-solving situation is a result of some information deficiency, which may be incomplete, imprecise, fragmentary, not fully reliable, vague, *contradictory*, or deficient in some other way [1]. The general framework of fuzzy reasoning allows handling much of this uncertainty.

Nowadays, we can handle much of this uncertainty using Fuzzy logic type-1 or type-2 [2, 3], also we are able to deal with hesitation using Intuitionistic fuzzy logic, but what happens when the information collected from different sources is somewhat or fully contradictory. What do we have to do if the knowledge base changes with time, and non-contradictory information becomes into doubtful or contradictory information, or any combination of these three situations? What should we infer from this kind of knowledge? The answer to these questions is to use a fuzzy logic system with logic rules for handling non-contradictory, contradictory or information with a hesitation margin. Mediative fuzzy logic is a novel approach presented for the first time in [4] which is able to deal with this kind of inconsistent information providing a common sense solution when contradiction exists, this is a mediated solution.

There are a lot of applications where information is inconsistent. In economics for estimating the Gross Domestic Product (GDP), it is possible to use different variables; some of them are distribution of income, personal consummation expenditures, personal ownership of goods, private investment, unit labor cost, exchange

M. Nikravesh et al. (eds.), *Forging the New Frontiers: Fuzzy Pioneers II.*

rate, inflation rates, and interest rates. In the same area for estimating the exportation rates it is necessary to use a combination of different variables, for example, the annual rate of inflation, the law of supply and demand, the dynamic of international market, etc. [5]. In medicine information from experiments can be somewhat inconsistent because living being might respond different to some experimental medication. Currently, randomized clinical trials have become the accepted scientific standard for evaluating therapeutic efficacy, and contradictory results from multiple randomized clinical trials on the same topic have been attributed either to methodological deficiencies in the design of one of the trials or to small sample sizes that did not provide assurance that a meaningful therapeutic difference would be detected [6]. In forecasting prediction, uncertainty is always a factor, because to obtain a reliable prediction it is necessary to have a number of decisions, each one based on a different group, in [7] says: Experts should be chosen "whose combined knowledge and expertise reflects the full scope of the problem domain. Heterogeneous experts are preferable to experts focused in a single specialty".

2 Historical Background

Throughout history, distinguish good from bad arguments has been of fundamental importance to ancient philosophy and modern science. The Greek philosopher Aristotle (384 BC–322 BC) is considered a pioneer in the study of logic and, its creator in the traditional way. The Organon is his surviving collected works on logic [8]. Aristotelian logic is centered in the syllogism. In Traditional logic, a syllogism (deduction) is an inference that basically consists of three things: the major and minor premises, and the proposition (conclusion) which follows logically from the major and minor premises [9]. Aristotelian logic is "bivalent" or "two-valued", that is, the semantics rules will assign to every sentence either the value "True" or the value "False". Two basic laws in this logic are the law of contradiction (*p cannot be both p and not p*), and the law of the excluded middle (*p must be either p or not p*).

In the Hellenistic period, the stoics work on logic was very wide, but in general, one can say that their logic is based on propositions rather than in logic of terms, like the Aristotelian logic. The Stoic treatment of certain problems about modality and bivalence are more significant for the shape of Stoicism as a whole. Chrysippus (280 BC–206 BC) in particular was convinced that bivalence and the law of excluded middle apply even to contingent statements about particular future events or states of affairs. The law of excluded middle says that for a proposition, p, and its contradictory, $\neg p$, it is necessarily true, while bivalence insists that the truth table that defines a connective like 'or' contains only two values, true and false [10].

In the mid-19th century, with the advent of symbolic logic, we had the next major step in the development of propositional logic with the work of the logicians Augustus DeMorgan (1806–1871) [11] and, George Boole (1815–1864). Boole was primarily interested in developing special mathematical to replace Aristotelian syllogistic logic. His work rapidly reaps benefits, he proposed "Boolean algebra" that was used to form the basis of the truth-functional propositional logics utilized in computer design and programming [12, 13]. In the late 19th century, Gottlob Frege (1848–1925) claimed that all mathematics could be derived from purely logical principles and definitions and he considered verbal concepts to be expressible as symbolic functions with one or more variables [14].

L. E. J. Brouwer (1881–1966) published in 1907 in his doctoral dissertation the fundamentals of intuitionism [15], his student Arend Heyting (1898–1980) did much to put intuitionism in mathematical logic, he created the Heyting algebra for constructing models of intuitionistic logic [16]. Gerhard Gentzen (1909–1945), in (1934) introduces systems of natural deduction for intuitionist and classical pure predicate calculus [17], his cornerstone was cut-elimination theorem which implies that we can put every proof into a (not necessarily unique) normal form. He introduces two formal systems (sequent calculi) LK and LJ. The LJ system is obtained with small changes into the LK system and it is suffice for turning it into a proof system for intuitionistic logic.

Nowadays, Intuitionistic logic is a branch of logic which emphasizes that any mathematical object is considered to be a product of a mind, and therefore, the existence of an object is equivalent to the possibility of its construction. This contrasts with the classical approach, which states that the existence of an entity can be proved by refuting its non-existence. For the intuitionist, this is invalid; the refutation of the non-existence does not mean that it is possible to find a *constructive* proof of existence. Intuitionists reject the *Law of the Excluded Middle* which allows proof by contradiction. Intuitionistic logic has come to be of great interest to computer scientists, as it is a constructive logic, and is hence a logic of what computers can do.

Bivalent logic was the prevailing view in the development of logic up to XX century. In 1917, Jan Łkasiewicz (1878–1956) developed the three-value propositional calculus, inventing ternary logic [18]. His major mathematical work centered on mathematical logic. He thought innovatively about traditional propositional logic, the principle of non-contradiction and the law of excluded middle. Łkasiewicz worked on multi-valued logics, including his own three-valued propositional calculus, the first non-classical logical calculus. He is responsible for one of the most elegant axiomatizations of classical propositional logic; it has just three axioms and is one of the most used axiomatizations today [19].

Paraconsistent logic is a logic rejecting the principle of non-contradiction, a logic is said to be *paraconsistent* if its relation of logical consequence is not explosive. The first paraconsistent calculi was independently proposed by Newton C. da Costa (1929-) [20] and Jas'kowski, and are also related to D. Nelson's ideas [21]. Paraconsistent logic was proposed in 1976 by the Peruvian philosopher Miró Quesada, it is a non-trivial logic which allows inconsistencies. The modern history of paraconsistent logic is relatively short. The expression "paraconsistent logic" is at present time well-established and it will make no sense to change it. It can be interpreted in many different ways which correspond to the many different views on a logic which permits to reason in presence of contradictions. There are many different paraconsistent logics, for example, *non-adjunctive, non-truth-functional, many-valued,* and *relevant.*

Fuzzy sets, and the notions of inclusion, union, intersection, relation, etc, were introduced in 1965 by Dr. Lofti Zadeh [2], as an extension of Boolean logic. Fuzzy logic deals with the concept of partial truth, in other words, the truth values used in Boolean logic are replaced with degrees of truth. Zadeh is the creator of the concept Fuzzy logic type-1 and type-2. Type-2 fuzzy sets are fuzzy sets whose membership functions are themselves type-1 fuzzy sets; they are very useful in circumstances where it is difficult to determine an exact membership function for a fuzzy set [22].

K. Atanassov in 1983 proposed the concept of Intuitionistic fuzzy sets (IFS) [23], as an extension of the well-known Fuzzy sets defined by Zadeh. IFS introduces a new component, degree of nonmembership with the requirement that the sum of membership and nonmbership functions must be less than or equal to 1. The complement of the two degrees to 1 is called the hesitation margin. George Gargov proposed the name of intuitionistic fuzzy sets with the motivation that their fuzzification denies the law of excluded middle, wish is one of the main ideas of intuitionism [24].

3 Mediative Fuzzy Logic

Since knowledge provided by experts can have big variations and sometimes can be contradictory, we are proposing to use a Contradiction fuzzy set to calculate a mediation value for solving the conflict. Mediative Fuzzy Logic is proposed as an extension of Intuitionistic fuzzy Logic (in the sense of Atanassov [9]). Mediative fuzzy logic (MFL) is based in traditional fuzzy logic with the ability of handling contradictory and doubtful information, so we can say that also it is an intutitionistic and paraconsistent fuzzy system.

A traditional fuzzy set in X [8], given by

$$A = \{(x, \mu_A(x)) | x \in X\} \tag{1}$$

where $\mu_A : X \to [0, 1]$ is the membership function of the fuzzy set A.

An intuitionistic fuzzy set B [9] is given by

$$B = \{(x, \mu_B(x), \nu_B(x)) | x \in X\} \tag{2}$$

where $\mu_B : X \to [0, 1]$ and $\nu_B : X \to [0, 1]$ are such that

$$0 \leq \mu_B(x) + \nu_B(x) \leq 1 \tag{3}$$

and $\mu_B(x); \nu_B(x) \in [0, 1]$ denote a degree of membership and a degree of nonmembership of $x \in A$, respectively.

For each intuitionistic fuzzy set in X we have a "hesitation margin" $\pi_B(x)$, this is an intuitionistic fuzzy index of $x \in B$, it expresses a hesitation degree of whether x belongs to A or not. It is obvious that $0 \leq \pi_B(x) \leq 1$ for each $x \in X$.

$$\pi_B(x) = 1 - \mu_B(x) - v_B(x) \tag{4}$$

Therefore if we want to fully describe an intuitionistic fuzzy set, we must use any two functions from the triplet [10].

1. Membership function
2. Non-membership function
3. Hesitation margin

The application of intuitionistic fuzzy sets instead of fuzzy sets, means the introduction of another degree of freedom into a set description, in other words, in addition to μ_B we also have v_B or π_B. Fuzzy inference in intuitionistic has to consider the fact that we have the membership functions μ as well as the non-membership functions v. Hence, the output of an intuitionistic fuzzy system can be calculated as follows [11]:

$$IFS = (1 - \pi)FS_\mu + \pi FS_v \tag{5}$$

where FS_μ is the traditional output of a fuzzy system using the membership function μ, and FSv is the output of a fuzzy system using the non-membership function v. Note in (6), when $\pi = 0$ the IFS is reduced to the output of a traditional fuzzy system, but if we take into account the hesitation margin of π the resulting IFS will be different.

In a similar way, a contradiction fuzzy set C in X is given by:

$$\zeta_C(x) = \min(\mu_C(x), v_C(x)) \tag{6}$$

where $\mu_C(x)$ represents the agreement membership function, and for the variable $v_C(x)$ we have the non-agreement membership function.

We are using the agreement and non-agreement, instead membership and non-membership, because we think these names are more adequate when we have contradictory fuzzy sets.

We are proposing three expressions for calculating the inference at the system's output, these are

$$MFS = \left(1 - \pi - \frac{\zeta}{2}\right) FS_\mu + \left(\pi + \frac{\zeta}{2}\right) FS_v \tag{7}$$

$$MFS = \min\left(((1-\pi)^* FS_\mu + \pi^* FS_v), \left(1 - \frac{\zeta}{2}\right)\right) \tag{8}$$

$$MFS = ((1-\pi)^* FS_\mu + \pi^* FS_v)^* \left(1 - \frac{\zeta}{2}\right) \tag{9}$$

In this case, when the contradictory index ζ is equal to zero, the system's output can be reduced to an intituionistic fuzzy output or, in case that $\pi = 0$, it can be reduced to a traditional fuzzy output.

4 Experimental Results

For testing the system we dealt with the problem of population control. This is an interesting problem that can be adapted to different areas. We focused in controlling the population size of an evolutionary algorithm by preserving, killing or creating individuals in the population. Dynamic population size algorithms attempt to optimize the balance between efficiency and the quality of solutions discovered by varying the number of individuals being investigated over the course of the evolutionary algorithm's run. We used Sugeno Inference system to calculate FS_μ and FS_ν, so the system is divided in two main parts: the inference system of the agreement function side, and the inference system of the non-agreement function side.

At the FS_μ side, we defined the variable percentage of cycling (pcCy-cling) with three terms, Small, Medium and Large. The universe of discourse is in the range [0,100]. We used a Sugeno Inference System, which in turn have three variables for the outputs: MFSCreate, MFSKill and MFSPreserve. They correspond to the amount of individuals that we have to create, kill and preserve in the population. Each output variable has three constant terms, so we have:

1. For the MFSCreate variable, the terms are: Nothing = 0, Little = 0.5, and Many = 1.
2. For FSKill we have: Nothing = 0, Little = 0.5, All = 1.
3. For MFSPreserve we have: Nothing = 0, More or Less=0.5, All = 1.

The rules for the $FS\mu$ side are:

if (pcCycled is small) then (create is nothing)(kill is nothing)(preserve is all)
if (pcCycled is medium) then (create is little)(kill is little)(preserve is moreOrLess)
if (pcCycled is large) then (create is many)(kill is all)(preserve is nothing)

At the side FS_ν, we defined the input variable NMFpcCycled with three terms: NoSmall, NoMedium, and NoLarge, they are shown in Fig. 2. They are applied to a Sugeno Inference System with three output variables: nCreate, nKill, and nPreserve. In similar way, they are contributing to the calculation of the amount of individuals to create, kill and preserve, respectively. Each output variable has three constant terms, they are:

1. For nCreate we have the output terms: Nothing = 0, Little = 0.5, Many=1.
2. For nKill we have: Nothing = 0, Little = 0.5, and All = 1.
3. For nPreserve we have: Nothing = 0, More or Less=0.5, and All = 1.

The corresponding rules are:

if (NMFpcCycled is Nsmall) then (create is nothing)(kill is nothing)(preserve is all)

if (NMFpcCycled is Nmedium) then (create is little)(kill is nothing)(preserve is moreOrLess)
if (NMFpcCycled is Nlarge) then (create is many)(kill is all)(preserve is nothing)

Using the agreement function ($FS\mu$) and the non-agreement functions (FS_ν) we obtained the hesitation fuzzy set and the contradictory fuzzy set.

We performed experiments for the above-mentioned problem obtaining results for traditional and intuitionistic fuzzy systems. Figs. 5, 6, and 7 show results of a traditional fuzzy system (FS), and in Figs. 8, 9, and 10 we have the intuitionistic fuzzy outputs (IFS). Moreover, we did experiments for calculating the meditative fuzzy output using (7), (8), and (9).

Experiment #1. Using (7)

Equation (7) is transofrmed into (10), (11), and (12) for the three different outputs *MFSCreate*, *MFSKill*, and *MFSPreserve*. Figures 11, 12 and 13 correspond to these outputs.

$$MFSCreate = \left(1 - \pi - \frac{\zeta}{2}\right) FS_{Create} + \left(\pi + \frac{\zeta}{2}\right) FS_{nCreate} \qquad (10)$$

$$MFSKill = \left(1 - \pi - \frac{\zeta}{2}\right) FS_{Kill} + \left(\pi + \frac{\zeta}{2}\right) FS_{nKill} \qquad (11)$$

$$MFSPreserve = \left(1 - \pi - \frac{\zeta}{2}\right) FS_{Preserve} + \left(\pi + \frac{\zeta}{2}\right) FS_{nPreserve} \qquad (12)$$

In Fig. 3 we are showing the hesitation fuzzy set for the Membership functions of Figs. 1 and 2, they are the agreement and non-agreement membership function respectively. Fig. 3 shows the hesitation fuzzy set obtained using (4). Fig. 4 shows the contradiction fuzzy set obtained using (6). Figs. 5, 6, and 7 correspond to the outputs *MFSCreate*, *MFSKill*, and *MFSPreserve*, we can see in these figures that the hesitation and contradiction fuzzy set did not impact the output. Figs.8, 9 and 10 show that the corresponding outputs were impacted by the hesitation fuzzy set and they were calculated using the IFS given in (5). The outputs in Figs. 11 to 19 were impacted by the hesitation and contradiction fuzzy set, they were calculated using MFL. We used (7) to calculate the output in Figs. 11, 12, and 13. Equation (8) was used to obtain Figs. 14, 15 and 16.

Finally, Figs. 17, 18, and 18 were obtained using (9). In general, we observed that the best results were obtained using (7), with this equation we obtained a softer inference output in all the test that we made, this can be observed comparing Fig.13 against Figs. 16 and 19.

Fig. 1 Membership functions in a traditional fuzzy system (FS)

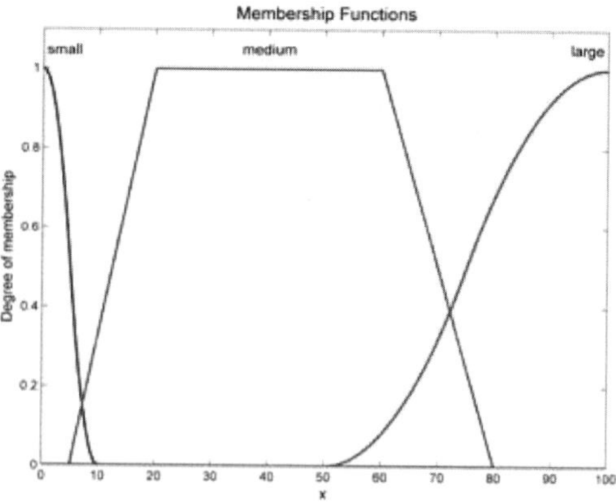

Fig. 2 Non-agreement membership functions for Mediative Fuzzy System (MFS)

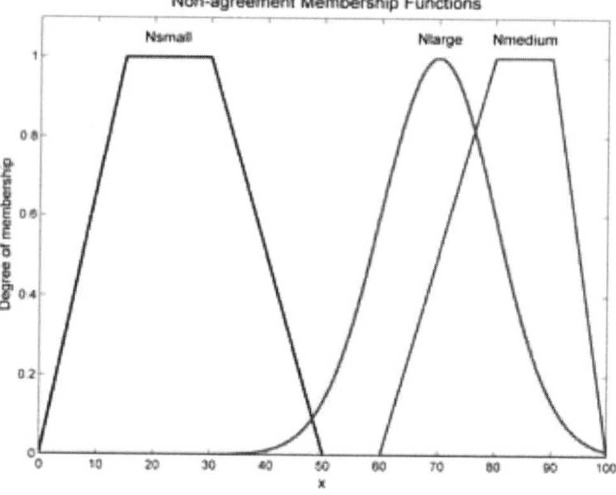

Fig. 3 Hesitation fuzzy set. We applied equation (4) to eachs complementary subset of membership and non-membership function, in this case agreement and non-agreement membership functions

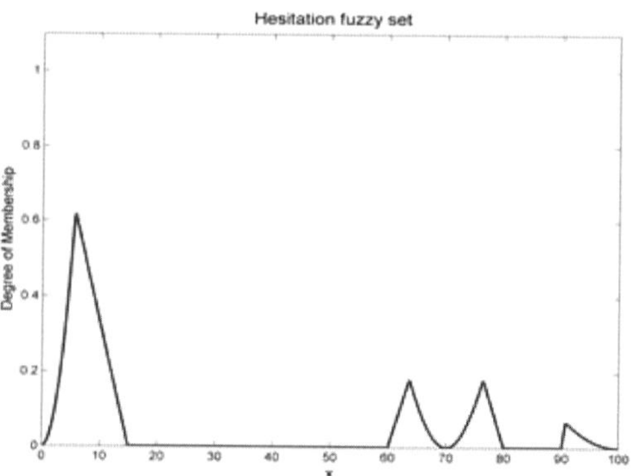

Fig. 4 Contradiction fuzzy
set. We obtained this set
applying equation (6) to each
subset of agreement and
non-agreement membership
functions

Fig. 5 Traditional FS for the
output Create. We can
observe that the system is
inferring that we have to
create 50% more individual
in the actual population size
when we have a percentage of
cycling between 12 and 55

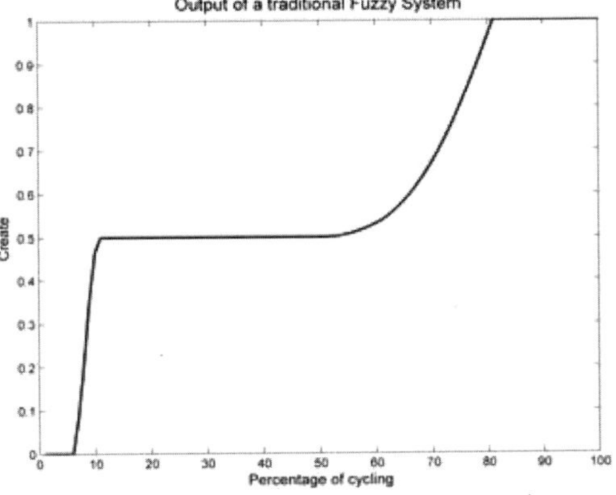

Fig. 6 The output Kill of FS,
says that we have to remove
50% of the less fit individuals
when we have more or less a
percentage of cycling
between 12 and 55

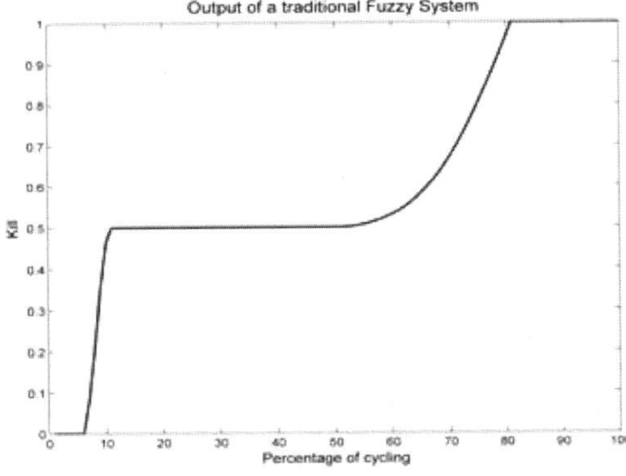

Fig. 7 The output Preserve of traditional FS says how many individuals we have to preserve, this is depending on the degree of cycling. Note that this result is in accordance with Figs.6 and 7

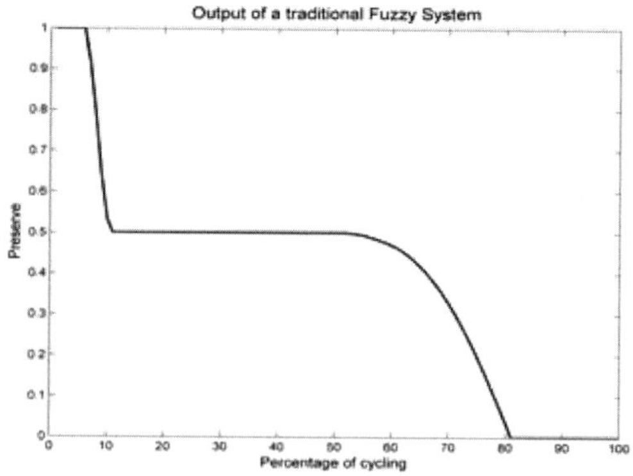

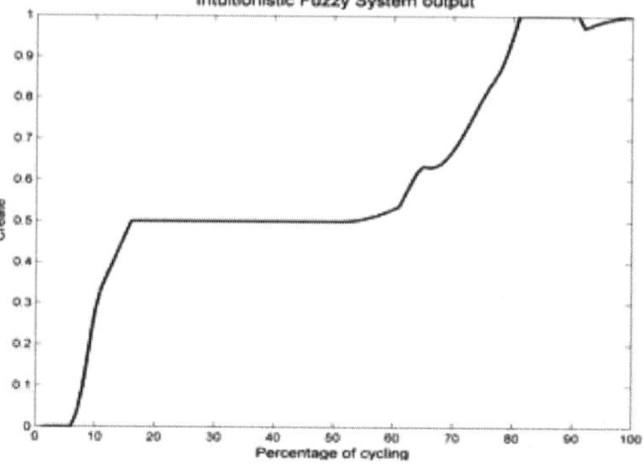

Fig. 8 WE can observe that although there is contradictory knowledge, we have a softener trasition in the lower part of Percentage of cycling, but when contradiction increases we cannot say the same. WE used (5), with FS_{Create} and FSn_{Create}

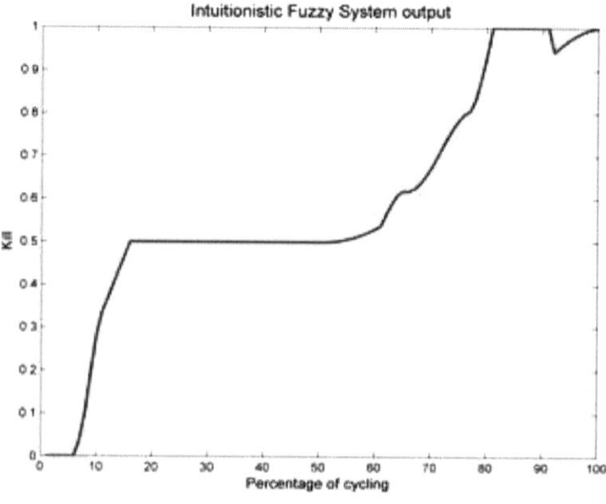

Fig. 9 In fact, a comparison using contradictory knowledge in IFS is not feared since the idea of this logic is not to use this kind of knowledge, but it is interesting to plot the inference output to compare results with MFS

Fig. 10 IFS do not reflect contradictory knowledge at the systems' output

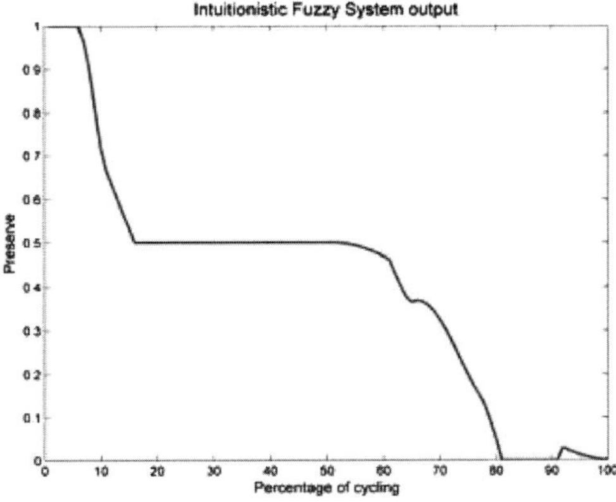

Fig. 11 MFS reflects contradictory knowledge at the systems' output. Note that here we have a softener transition values in the range between 50 and 100. We used (7) for plotting Figs. 11, 12 and 13. Experiment #1

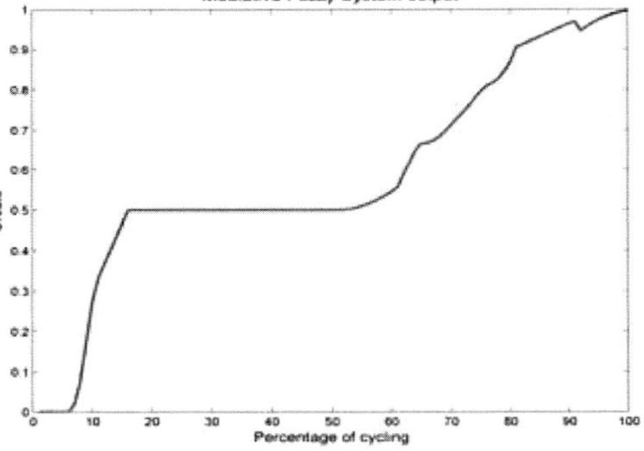

Fig. 12 Although, we have the highest degree of contradiction around the value 80, inference gives for this region reasonably good output values. Experiment #1

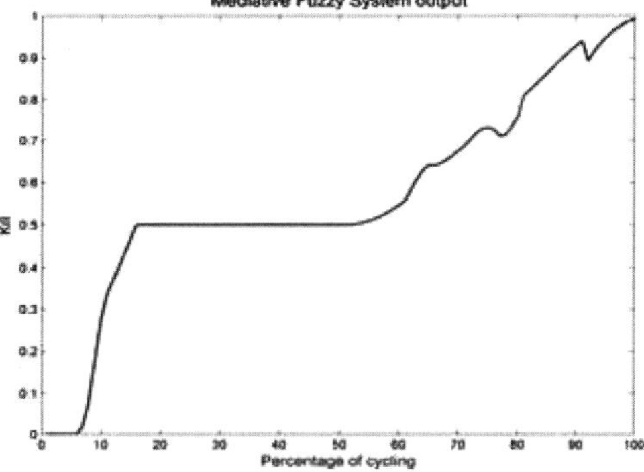

Fig. 13 Comparing results of MFS against FS and IFS, we can see that MFS can gives a softener transition when we have hesitation and contradiction fuzzy sets.. We used (7) in experiment#1

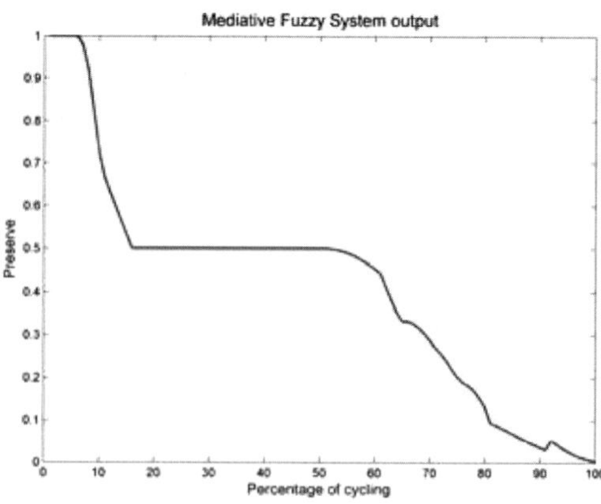

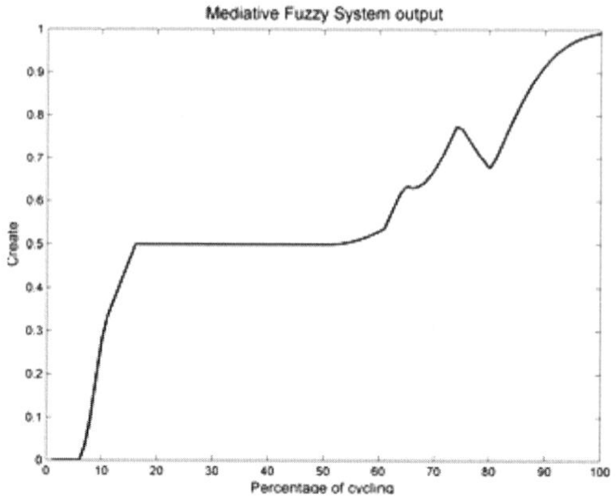

Fig. 14 MFS for the output *MFSCCreate*. Figs. 14. 15 and 16 were plotted using (8) as base. Experiment#2

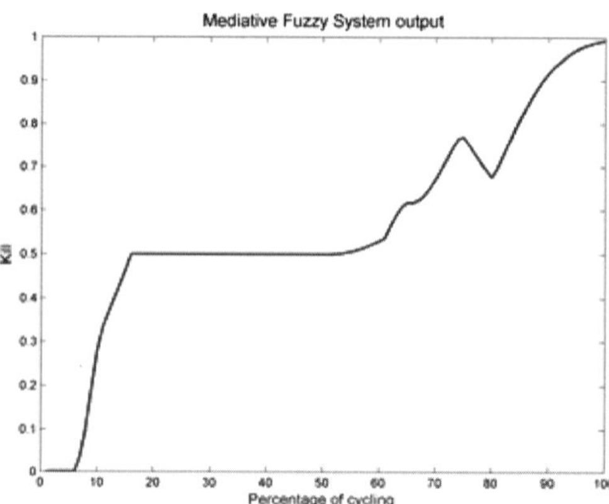

Fig. 15 MFS for the output *MFSkill*. Experiment #2

Fig. 16 Output for the variable *MFSPreserve*. Experiment #2

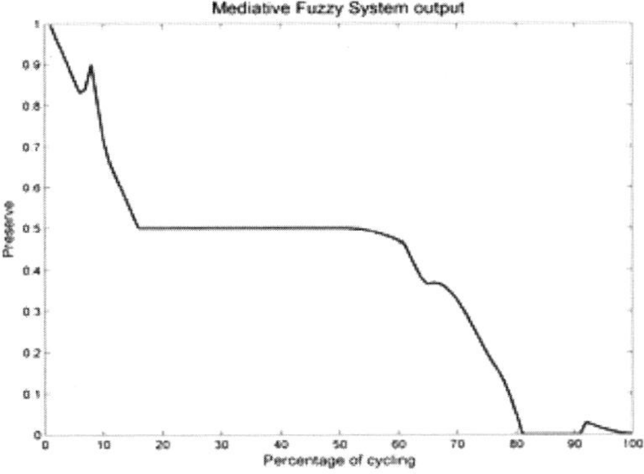

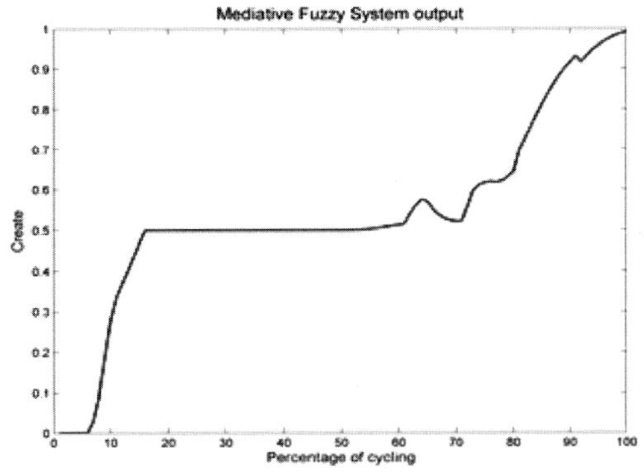

Fig. 17 Output for the variable *MFSCreate*. Figs. 17, 18 and 19 were plotted using (9) as base. Experiment #3

Fig. 18 Output for the variable *MFSKill*. Experiment # 3

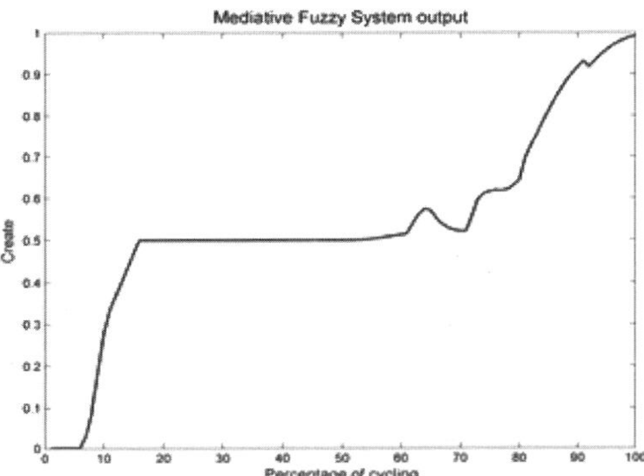

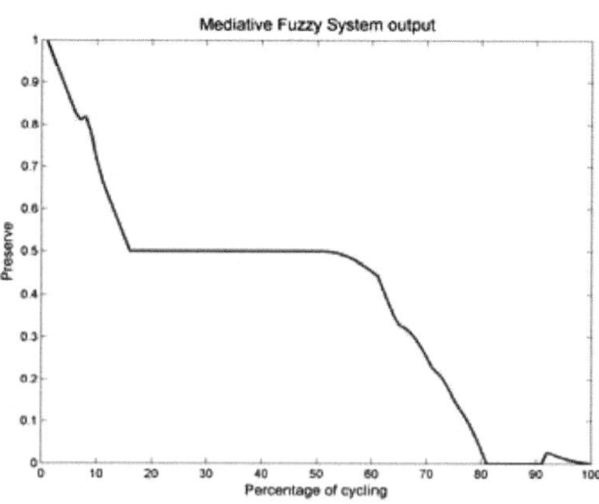

Fig. 19 Output for the variable Preserve in Experiment #3

5 Conclusions

Through time fuzzy logic type-1 and type-2 have demonstrated their usefulness for handling uncertainty in uncountable applications. Intuitionistic fuzzy logic is relatively a new concept which introduces the degree of nonmbership as a new component, this technology also have found several application niches. Mediative fuzzy logic is a novel approach that enables us to handle imperfect knowledge in a broader way than traditional and intuitionistic fuzzy logic can do. MFL is a sort of paraconsistent fuzzy logic because it can handle contradictory knowledge using fuzzy operators. MFL provides a mediated solution in case of a contradiction, moreover it can be reduced automatically to intuitionistic and traditional fuzzy logic in an automatized way, this is depending on how the membership functions (agreement and non-agreement functions) are established. We introduced three equations to perform the meditative inference. In this experiment we found the best results using (7) that is an extension of (5), i.e. it is an extension of the formula to calculate the intuitionistic fuzzy output. MFL is a good option when we have knowledge from different human experts, because it is common that experts do not fully agree all the time, so we can obtain contradiction fuzzy sets to represent the amount of disagree with the purpose of impacting the inference result. Traditional FL, and IFL will not impact the output when we have contradictory knowledge.

References

1. George J. Klir, Bo Yuan, *Fuzzy Sets and Fuzzy Logic: Theory and Applications*, Ed. Prentice Hall USA,1995.
2. L. A. Zadeh, "Fuzzy Sets", Information and Control, Vol. 8, pp. 338–353, 1965.
3. Jerry M. Mendel, Uncertain Rule-Based Fuzzy Logic Systems, Introduction and new directions, Ed. Prentice Hall, USA, 2000.

4. O. Montiel, O. Castillo, P. Melin, A. Rodríguez Días, R. Sepúlveda, ICAI-2005
5. Donald A. Bal, Wendell H. McCulloch, JR., "International Business. Introduction and Essentials", Fifth Edition, pp. 138-140, 225. USA. 1993.
6. Horwitz R. I.: Complexity and contradiction in clinical trial research. Am. J. Med., 82: 498–510,1987.
7. J. Scott Armstrong, "Principles of Forecasting. A Handbook for researchers and Practitioners", Edited by J. Scott Armstrong, University of University of Pennsylvania, Wharton School, Philadelphia, PA., USA, 2001.
8. Aristotle, The Basic Works of Aristotle, Modern Library Classics, Richard McKeon Ed., 2001.
9. Robin Smith, Aristotle's logic, Stanford Encyclopedia of Philosophy, 2004, http://plato.stanford.edu/entries/aristotle-logic/.
10. Dirk Baltzly, Stanford Encyclopedia of Philosophy, 2004, http://plato.stanford.edu/entries/stoicism/
11. J J O'Connor and E F Robertson, Augustus DeMorgan, MacTutor History of Mathematics: Indexes of Biographies (University of St. Andrews), 2004, http://www-groups.dcs.st-andrews.ac.uk/~history/Mathematicians/De_Morgan.html
12. George Boole, The Calculus of Logic, Cambridge and Dublin Mathematical Journal, *Vol. III* (1848), 1848, pp. 183–98
13. J J O'Connor and E F Robertson, George Boole, MacTutor History of Mathematics: Indexes of Biographies (University of St. Andrews), 2004, http://www-groups.dcs.st-andrews.ac.uk/~history/Mathematicians/Boole.html
14. J J O'Connor and E F Robertson ,Friedrich Ludwig Gottlob Frege, MacTutor History of Mathematics: Indexes of Biographies (University of St. Andrews), http://www-groups.dcs.st-andrews.ac.uk/~history/Mathematicians/Frege.html
15. J J O'Connor and E F Robertson , Luitzen Egbertus Jan Brouwer, MacTutor History of Mathematics: Indexes of Biographies (University of St. Andrews), http://www-groups.dcs.st-andrews.ac.uk/~history/Mathematicians/Brouwer.html
16. J J O'Connor and E F Robertson, Arend Heyting, *MacTutor History of Mathematics: Indexes of Biographies* (University of St. Andrews), 2004, http://www-history.mcs.st-andrews.ac.uk/Mathematicians/Heyting.html
17. J J O'Connor and E F Robertson , Gerhard Gentzen, MacTutor History of Mathematics: Indexes of Biographies *(University of St. Andrews)*, 2004, http://www-history.mcs.st-andrews.ac.uk/Mathematicians/Gentzen.html
18. J J O'Connor and E F Robertson, Jan Lukasiewicz, MacTutor History of Mathematics: Indexes of Biographies (University of St. Andrews), 2004, http://www-history.mcs.st-andrews.ac.uk/Mathematicians/Lukasiewicz.html
19. Wikipedia the free encyclopedia (Biography of Jan Lukasiewicz), http://en.wikipedia.org/wiki/Jan_Lukasiewicz
20. Wikipedia the free encyclopedia (Biography of Newton da Costa), http://en.wikipedia.org/wiki/Newton_da_Costa
21. How to build your own paraconsistent logic: an introduction to the Logics of Formal (In) Consistency. W. A. Carnielli. In: J. Marcos, D. Batens, and W. A. Carnielli, organizers, *Proceedings of the Workshop on Paraconsistent Logic* (WoPaLo), held in Trento, Italy, 5–9 August 2002, as part of the 14th Euro-pean Summer School on Logic, Language and Information (ESSLLI 2002), pp.58–72.
22. Jerry M. Mendel and Robert I. Bob John, Type-2 Fuzzy Sets Made Simple, IEEE Transactions on Fuzzy Systems, Vol. 10, No. 2, April 2002.
23. Atanassov, K., "Intuitionistic Fuzzy Sets: Theory and Applications", Springer-Verlag, Heidelberg, Germany. 1999.
24. Mariana Nikilova, Nikolai Nikolov, Chris Cornelis, Grad Deschrijver, "Survey of the Research on Intuitionistic Fuzzy Sets", In: Advanced Studies in Contemporary Mathematics **4**(2), 2002, p. 127–157.
25. O. Castillo, P. Melin, *A New Method for Fuzzy Inference in Intuitionistic Fuzzy Systems*, Proceedings of the International Conference NAFIPS 2003, IEEE Press, Chicago, Illinois, USA, Julio 2003, pp. 20-25.

Fuzzy Cognitive Maps Structure for Medical Decision Support Systems

Chrysostomos D. Stylios and Voula C. Georgopoulos

Abstract Fuzzy Cognitive Maps (FCMs) are a soft computing technique that follows an approach similar to human reasoning and human decision-making process, considering them a valuable modeling and simulation methodology. FCMs can successfully represent knowledge and experience, introducing concepts for the essential elements and through the use of cause and effect relationships among the concepts Medical Decision Systems are complex systems consisting of irrelevant and relevant subsystems and elements, taking into consideration many factors that may be complementary, contradictory, and competitive; these factors influence each other and determine the overall diagnosis with a different degree. Thus, FCMs are suitable to model Medical Decision Support Systems and the appropriate FCM structures are developed as well as corresponding examples from two medical disciplines, i.e. speech and language pathology and obstetrics, are described.

1 Introduction

Fuzzy Cognitive Maps (FCMs) are a modeling and simulation methodology based on an abstract conceptual representation of any system. In fact, they are a computational intelligence modeling and inference methodology suitable for modeling complex processes and systems that are systems consisted of a great number of highly related and interconnected elements and subsystems. FCMs can successfully represent knowledge and experience, introducing concepts for the essential elements and through the use of cause and effect relationships among the concepts. They are used to develop models of aggregated behavior and inferring models that govern the components and interaction from large amount, possibly incomplete and uncertain data (Kosko 1992; Jang et al. 1997; Stylios and Groumpos 2000).

Medical Decision Systems have to consider a high amount of data and information from interdisciplinary sources (patient's records and information, doctors' physical examination and evaluation, laboratory tests, imaging tests etc) and, in addition to this, medical information may be vague, missing or not available. Furthermore the Medical Diagnosis procedure is a complex one, taking into consideration a variety of inputs in order to infer the final diagnosis. Medical Decision Systems are complex systems consisting of irrelevant and relevant subsystems and elements, taking into consideration many factors that may be complementary, contradictory, and competitive; these factors influence each other and determine the overall diagnosis with a different degree. It is apparent that Medical Decision Support Systems require a modeling tool that can handle all these challenges and at the same time to be able to infer a decision. An Advanced Medical Decision Support System must be capable of extracting causal knowledge from the appropriate medical domain, building a causal knowledge base, and making inference with it. FCM is as a major vehicle of causal knowledge representation and inference (Lee and Kim 1998).

Kosko (1986) first introduced expanded cognitive maps in the engineering area to describe the cause and effect between concepts. This primitive FCM used crisp values {-1, 0, 1} to describe causality and used concepts and dis-concepts in order to describe positive or negative concepts. The same period a first attempt to develop a generic system FCM for decision analysis proposed the POOL2 where both negative and positive assertions are weighted and kept separately based on the negative-positive-neutral (NPN) interval [-1,1] (Zhang et al. 1989), (Zhang et al. 1992).

FCMs attracted the interest of many researchers from different areas. FCMs were used to represent knowledge (Taber 1991), to model complex dynamical systems, such as social and psychological processes and organizational behavior (Craiger et al. 1996), and as an advanced artificial intelligence approach for engineering applications, (Jain 1997). FCMs were used for fault detection (Pelaez and Bowles 1996), and modelling process control and supervision of distributed systems (Stylios et al. 1999; Stylios and Groumpos 2004;). Other research efforts introduced FCMs to analyze urban areas (Xirogiannis et al. 2004), to represent the management of relationships among organizational members in airline service (Kang and Lee 2004) and to modeling software development project (Stach and Kurgan 2004; Stach et al. 2004). FCMs have been used for web-mining inference amplification (Lee et al. 2002) (Kakolyris et al. 2005). Finally, FCMs have been used successfully in the medical diagnosis and decision area; specifically, they have been used to model the complex process of radiotherapy (Papageorgiou et al. 2003), for differential diagnosis of specific language impairment (Georgopoulos et al. 2003) and for diagnosis and characterization for tumor grade (Papageorgiou et al. 2006).

An FCM is an interconnected network of concepts. Concepts represent variables, states, events, trends, inputs and outputs, which are essential to model a system. The connection edges between concepts are directed and they indicate the direction of causal relationships while each weighted edge includes information on the type and the degree of the relationship between the interconnected concepts. Each connection is represented by a weight which has been inferred through a method based on fuzzy rules that describes the influence of one concept to another. This influence can be positive (a promoting effect) or negative (an inhibitory effect). The FCM development method is based on Fuzzy rules that can be either proposed by human experts and/or derived by knowledge extraction methods (Stylios et al. 1999).

This chapter describes FCM structures suitable for Medical Decision Support Systems as well as corresponding examples from medical disciplines. First, the Competitive FCM and its applicability in differential diagnosis of two language disorders is presented. Next a distributed m-FCM is described and an example for the differential diagnosis of a speech disorder is discussed. Finally, a hierarchical structure for FCMs is presented and the usage of this hierarchical approach in an obstetrics decision support problem supporting the obstetricians on how to proceed during labor is analyzed.

2 Fuzzy Cognitive Maps

FCMs are a soft computing technique that follows an approach similar to human reasoning and the human decision-making process. Soft computing methodologies have been investigated and proposed for the description and modeling of complex systems. An FCM looks like a cognitive map, it consists of nodes (concepts) that illustrate the different aspects of the system's behavior. These nodes (concepts) interact with each other showing the dynamics of the model. In the case of system modeling, the FCM is developed by human experts who operate/supervise/know the system and its behavior under different circumstances in such a way that the accumulated experience and knowledge are integrated in a causal relationship between factors/characteristics/components of the process or system modeled (Stylios and Groumpos 2004).

Fuzzy Cognitive Maps can be constructed either by actual experts or based on transforming expert knowledge from the literature. It is a knowledge-based methodology that utilizes the knowledge and experience of experts. It is accepted that the perceptions of experts create a subjective rather than objective model of the system. The main concern is to determine and describe which elements constitute the model and what elements influence other elements as well as the degree of this influence. There is an inference mechanism that describes the relations among elements as fuzzy causal relationships. Different values of influence are recommended and accepted; this is the main strength of this method. FCMs are ideal for knowledge and conceptual representation of complex systems in a soft computing way where the concepts of the system and their relationships are mainly fuzzy and not precisely estimated.

Experts design and develop the fuzzy graph structure of the system, thus the FCM describes the perception of experts about the system. Experts determine the structure and the interconnections of the network using fuzzy conditional statements. Experts' concern is to describe whether one concept influences another. Cause and effect relations among concepts are the basis of expectations and this is important in every system trying to model and replicate brain-like intelligence. Experts use linguistic variables in order to describe the relationship among concepts, and then all the linguistic variables are combined and so the weights of the causal interconnections among concepts are concluded. The simplest FCMs act as asymmetrical networks of threshold or continuous concepts and converge to an equilibrium point or limit cycles. At this level, they differ from Neural Networks in the way they are developed as they are based on extracting knowledge from experts. FCMs have non-linear structure of their concepts and differ in their global feedback dynamics (Papageorgiou et al. 2004).

Given two events that are represented by two concepts A and B, in FCM terms, the main questions that have to be answered are:

i) Does event A cause B or vice versa?
ii) What is the strength of the causal relationship?

Causality plays a key role in any knowledge-based system. The issue of how to understand and interpret the existing cause and effect relations is central to any effort to design systems that have some human like intelligence. The main problems concerning causality in that context are the extraction and elicitation of causal knowledge, its representation and its use. In the general approach, causal information emerges from statistical data and information, by looking at data that occur simultaneously, but it is clear that the co occurrence of data, although most likely correlated, does not always mean that the data are causally linked.

2.1 Mathematical Representation of Fuzzy Cognitive Maps

The graphical illustration of an FCM is a signed directed graph with feedback, consisting of nodes and weighted arcs. Nodes of the graph stand for the concepts that are used to describe the behavior of the system and they are connected by signed and weighted arcs representing the causal relationships that exist between the concepts (Fig. 1).

Each concept is characterized by a number A_i that represents its value and it results from the transformation of the fuzzy real value of the system's variable, for which this concept stands, in the interval [0,1]. Between concepts, there are three possible types of causal relationships that express the type of influence from one concept to the others. The weights of the arcs between concept C_i and concept C_j could be positive ($W_{ij} > 0$) which means that an increase in the value of concept C_i leads to the increase of the value of concept C_j, and a decrease in the value of concept C_i leads to the decrease of the value of concept C_j. Or there is negative causality ($W_{ij} < 0$) which means that an increase in the value of concept C_i leads to the decrease of the value of concept C_j and vice versa.

The value A_i of concept C_i expresses a degree, which is related to its corresponding physical value. Ateach simulation step, the value A_i of a concept C_i is

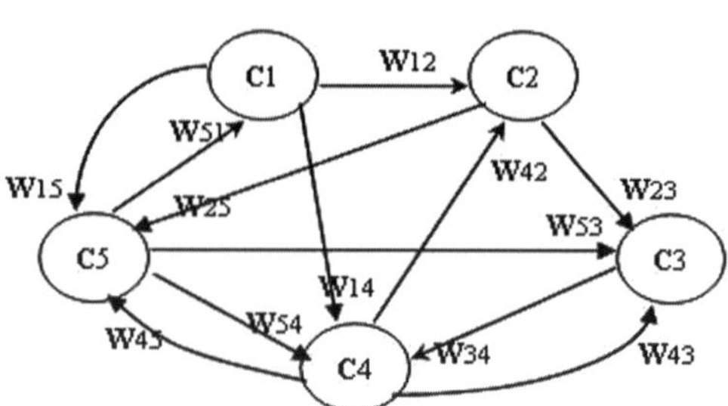

Fig. 1 The Fuzzy Cognitive Map model

calculated by computing the influence of other concepts C_j's on the specific concept C_i following the calculation rule:

$$A_i^{(k+1)} = f\left(A_i^{(k)} + \sum_{\substack{j\neq i \\ j=1}}^{N} A_j^{(k)} \cdot w_{ji}\right) \tag{1}$$

where $A_i^{(k+1)}$ is the value of concept C_i at simulation step $k+1$, $A_j^{(k)}$ is the value of concept C_j at simulation step k, w_{ji} is the weight of the interconnection from concept C_j to concept C_i and f is the sigmoid threshold function:

$$f = \frac{1}{1 + e^{-\lambda x}} \tag{2}$$

where $\lambda > 0$ is a parameter that determines its steepness. In this particular approach, the value $\lambda = 1$ has been used. This function is selected since the values A_i of the concepts, lie within [0, 1].

2.2 Method to Construct Fuzzy Cognitive Maps

The development and construction method of FCMs has great importance to sufficiently model any system. The proposed method is dependent on a group of experts who operate, monitor, supervise the system. This methodology extracts the knowledge from the experts and exploits their experience of the system's model and behavior (Stylios and Groumpos 2000).

The group of experts determines the number and kind of concepts that comprise the FCM. An expert from his/her experience knows the main factors that describe the behavior of the system; each of these factors is represented by one concept of the FCM. Experts know which elements of the systems influence other elements; for the corresponding concepts they determine the negative or positive effect of a concept on the others, with a fuzzy degree of causation. In this way, an expert transforms his/her knowledge in a dynamic weighted graph, the FCM. Experts describe the existing relationship between the concepts and thus, justify their suggestions. Each expert determines the influence of one concept on another as "negative" or "positive" and then evaluates the degree of influence using a linguistic variable, such as "strong influence", "medium influence", "weak influence", etc.

More specifically, the causal interrelationships among concepts are declared using the variable *Influence* which is interpreted as a linguistic variable taking values in the universe U=[−1, 1]. Its term set *T(influence)* is suggested to comprise nine variables. Using nine linguistic variables, an expert can describe in detail the influence of one concept on another and can discern between different degrees of influence. The nine variables used here are: *T(influence)*={negatively very strong, negatively strong, negatively medium, negatively weak, zero, positively weak,

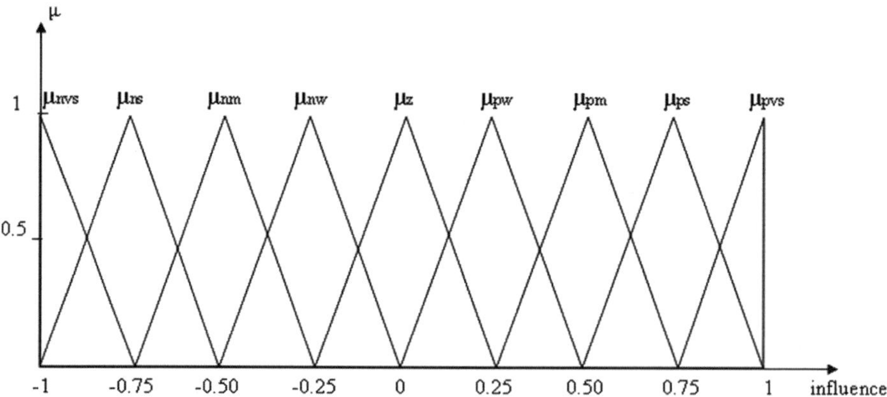

Fig. 2 Membership functions of the linguistic variable *Influence*

positively medium, positively strong and positively very strong}. The corresponding membership functions for these terms are shown in Fig. 2 and they are μ_{nvs}, μ_{ns}, μ_{nm}, μ_{nw}, μ_z, μ_{pw}, μ_{pm}, μ_{ps} and μ_{pvs}.

Thus, every expert describes each interconnection with a fuzzy linguistic variable from the set, which correspond to the relationship between the two concepts and determines the grade of causality between the two concepts. Then, all the proposed linguistic variables suggested by experts, are aggregated using the SUM method and an overall linguistic weight is produced, which with the defuzzification method of Center Of Gravity (COG) (Lin & Lee 1996), is transformed to a numerical weight W_{ji}, belonging to the interval $[-1, 1]$. A detailed description of the development of FCM model is given in (Stylios and Groumpos 2004).

This FCM development approach utilizes the knowledge and experience of experts asking them to describe the existing causal relationship using Fuzzy rules. Since a knowledge base reflects its sources, it is critical to identify suitable knowledge sources, i.e. domain experts taking into account perceived expertise level. Each expert draws an FCM and these are easily combined, leading to the combined FCM being potentially stronger than an individual FCM because the information is derived from a multiplicity of sources, making point errors less likely (Stylios et al. 1999).

3 Medical Decision Support Systems Based on Fuzzy Cognitive Maps

When medical experts are called upon to make a decision they take into consideration a variety of factors (concepts) giving each one a different degree of importance (weight). The description, characteristics and information of the factors they use may be complementary, similar, conflicting, vague, or even incomplete (Zeleznikow and Nolan 2001). Medical experts have a conceptual model in mind by which they process these factors and their degrees of importance, making comparisons, integrating the available information, and differentiating their importance, thus, finally reaching a decision out of a number of alternative potential decisions. A well known approach for designing a medical decision support system involves the process of mapping experts' knowledge concerning the decision into a computer program's knowledge base. One can create a representation of the experts' knowledge using causal concept maps, which are developed by considering experts as the creators of the "map" that explicitly represents their expert knowledge drawn out as a diagram. In essence, this is an integrated interactive, graphic diagram of each expert's mental model of the procedure to reach a decision. Concepts of the map are factors that are usually considered to reach a decision, as well as the potential decisions. In the graphical form of a cognitive map the concepts are the nodes. The "causal" component of these maps refers to the cause-effect relationships that hold between factors involved in the decision and the possible decisions and between different the factors themselves. The cause-effect relationships are connections between the nodes and are depicted in the graphical form as signed directed edges from one node (the causing concept) to another node (the affected concept). Causal knowledge generally involves many interacting concepts that make them difficult to deal with, and for which analytical techniques are inadequate (Park and Kim 1995). A cognitive map is a technique adequate for dealing with interacting concepts (Chaib-draa and Desharnais 1998).

The type of cognitive maps proposed here for developing the medical decision support systems are Fuzzy Cognitive Maps. Where the values of the nodes themselves and the weightings of the connections are expressed using a fuzzy (linguistic value), such as those described in the previous section. This is an appropriate modeling technique for the medical decision support system since the weighting in a human reasoning decision process almost never carries an exact numerical value.

The area of Medical Diagnosis and Medical Decision Support is characterized by complexity requiring the investigation of new advanced methods for modeling and development of sophisticated systems. Medical Decision Support Systems (MDSSs) have attracted the interest of many researchers and still considerable efforts are under way. Especially for MDSS, FCMs have been successfully applied (Georgopoulos et al. 2003; Papageorgiou et al. 2003; Georgopoulos and Stylios 2005).

4 Competitive FCM for Medical Diagnosis

A specific type of MDSS for differential Diagnosis has been proposed where a new structure, the Competitive Fuzzy Cognitive Map (CFCM) is presented (Georgopoulos et al. 2003). The CFCM introduced the distinction of two main kinds of concepts: decision-concepts and factor-concepts. Figure 3 illustrates an example CFCM model which is used to perform medical decision/diagnosis, and includes both types of concepts of the FCM and the causal relations among them. All the

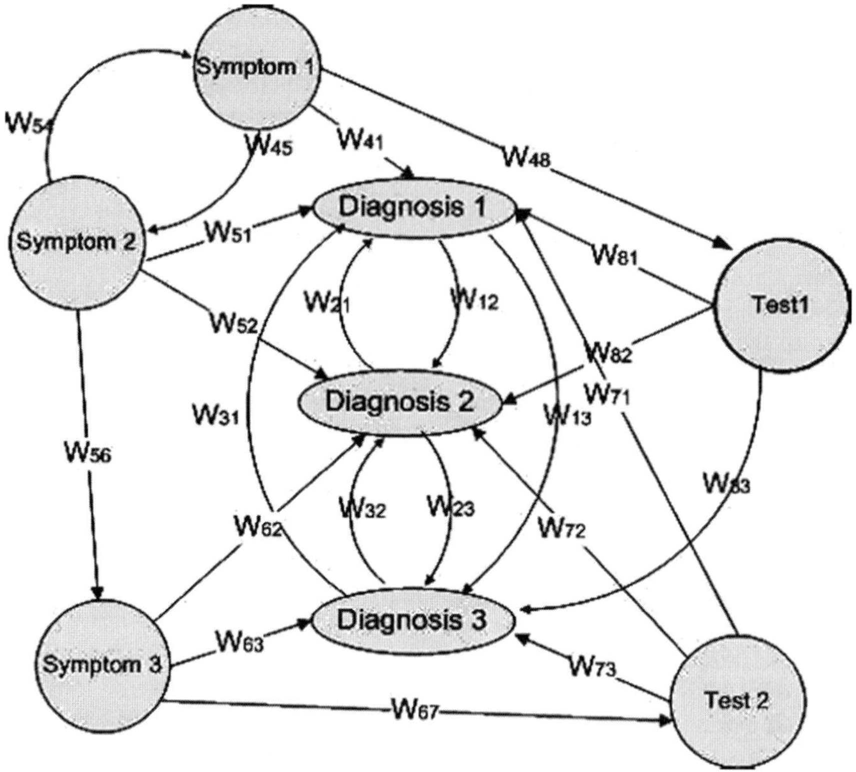

Fig. 3 A conceptual model for Medical Diagnosis

concepts can interact with each other and determine the value of diagnosis concepts that interest us thus indicating the final diagnosis.

In the CFCM model, each decision concept represents a single decision/diagnosis, which means that the decision concepts must be mutually exclusive because the MDSS intention is to infer always only one diagnosis. This is the case of most medical applications, where, according to symptoms, medical professionals have to conclude to only one diagnosis and then must determine, accordingly, the treatment. It is well known that the medical diagnosis procedure is a complex process that has to take under consideration a variety of interrelated factors, measurements and functions. This is the case of any real world diagnosis problem, where many different factors are taken into consideration. In carrying out any diagnosis procedure, some of these factors are complementary, others are similar and others conflicting, and most importantly, factors influence other factors.

The factor-concepts can be considered as inputs to the MDSS such as patient data, observed symptoms, patient records, experimental and laboratory tests etc, which can be dynamically updated based on the system interaction, whereas the decision-concepts are considered as outputs where their estimated values outline the possible diagnosis for the patient. The factor-concepts can be interrelated and they partially influence the diagnosis. For such a situation, FCM are suitable as their strength is their ability to describe systems and handle situations where there are feedback relationships and relationships between the factor concepts. Such interconnections are shown in Fig. 3 where the "competitive" interconnections between the diagnosis concepts are also illustrated.

4.1 CFCM for Dyslexia and Specific Language Impairment

Dyslexia and Specific Language Impairment (SLI) are frequent developmental disorders that may have a serious impact on an individual's educational and psychosocial life. Both are considered as important public health problems since they affect the lives of many individuals. Prevalence studies report percentages between 3% and 15% for dyslexia and 3% to 10% for SLI.

In general terms, developmental dyslexia is identified if a child has poor literacy skills despite adequate intelligence and opportunities to learn. SLI is diagnosed when oral language lags behind other areas of development for no apparent reason. Although, these two developmental disorders have separate and distinct definitions, they share many similar symptoms and characteristics that can make it difficult for clinicians to differentiate between them.

Specifically, in several studies that have investigated the reading skills of children with SLI, literacy problems (high incidence of reading difficulties) have been documented at an early age of these children. Similarly, it has also been found that many dyslexic children show a history of language impairment.

In the current differential diagnosis model there are two diagnosis concepts, i.e. the two disorders that are studied: Concept 1 Dyslexia and Concept 2 Specific Language Impairment (SLI). Two types of factors are the factor-concepts that are considered as measurements that determine the result of the diagnosis in this model and they are:

- Concept 3 Reduced Lexical Abilities
- Concept 4 Decreased MLU
- Concept 5 Problems in Syntax
- Concept 6 Problems in Grammatical Morphology
- Concept 7 Impaired or Limited Phonological development
- Concept 8 Impaired Use of Pragmatics
- Concept 9 Reading Difficulties
- Concept 10 Problems in Writing and Spelling
- Concept 11 Reduced Ability of Verbal Language Comprehension
- Concept 12 Difference between Verbal and Nonverbal IQ
- Concept 13 Heredity
- Concept 14 Impaired Sociability
- Concept 15 Impaired Mobility
- Concept 16 Attention Distraction
- Concept 17 Reduced Arithmetic Ability

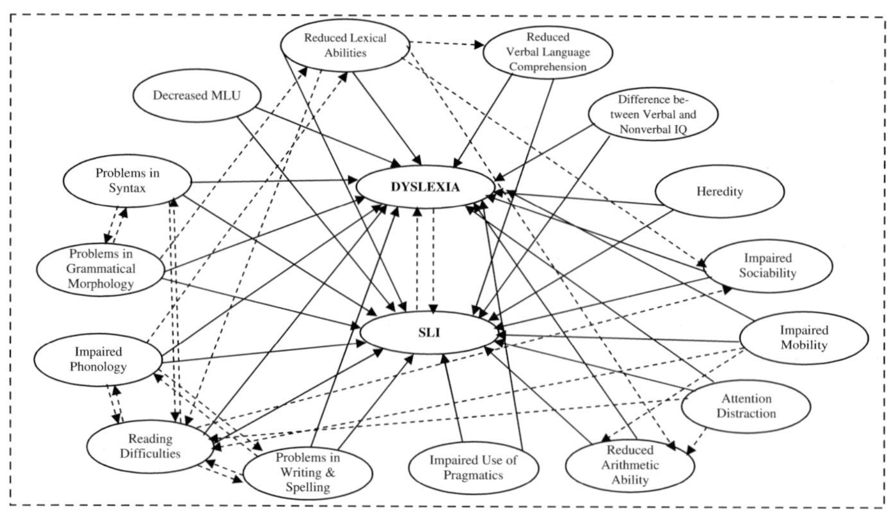

Fig. 4 Fuzzy Cognitive Map Differential Diagnosis of Dyslexia and SLI

The connections between the concepts are shown in Fig. 4 by arcs. However, due to limited space the sign and weights of the connections are not shown in Fig. 4, but can be determined from Table 1 (Malandraki and Georgopoulos 2006).

Four case studies from the literature are examined here, two on Dyslexia (Psychcorp 2005; Pierson 1999) and two on SLI (Van der Lely 1997; McGregor and Appel 2002), as experimental clinical cases that were used to run the differential diagnosis model. Three of the cases were school-age children and one was a preschool child.

Table 1 Weights between Factor and Disorder concepts

	C1	C2	C3	C4	C5	C6	C7	C8	C9	C10	C11	C12	C13	C14	C15	C16	C17
C1		-1															
C2	−1																
C3	+ M-H	+ VVH							+L		+L			+L			+L
C4	+ L-M	+VH															
C5	+ M-H	+ VVH				+L			+L								
C6	+ M-H	+ VVH			+L				+L								
C7	+VVH	+ H	+L						+L	+L							
C8	NONE *	+M-H															
C9	+VVH	+M-H			+L		+L			+L				+L			
C10	+VVH	+M					+L		+L								
C11	+ L	+M															
C12	+VVH	+VVH															
C13	+VH	+M-H															
C14	+M	+M-H															
C15	CD	+M							+L								+L
C16	+M-H	+M															
C17	CD	+M-H															

*No consistent and clear relationship was reported in the literature regarding the pragmatic aspects of language of children with dyslexia

Details on the history and the assessment results of these cases can be found in Psychcorp 2005, Pierson, 1999, Van der Lely 1997 and McGregor and Appel 2002.

In Table 2 the factors used by the model in the diagnosis of each case are presented. In addition, the degree of occurrence of each factor in each case study is denoted with similar qualitative degrees of very-very high, very-high, high, medium, low, very low, and 0. The designation of weight "NR" in Table 2 indicates that the factor is not reported in the particular case and a value of zero is used in the computational model.

Table 2 Initial Factor Concept Fuzzy Values for Four Cases

Factor-Concepts	Case 1	Case 2	Case 3	Case 4
Reduced Lexical Abilities	Very very high	Very very high	Medium	Very very high
Decreased MLU	NR	Very very high	NR	NR
Problems in Syntax	Very very high	High	Medium	Very high
Problems in Grammatical Morphology	Very high	Very high	Medium	NR
Impaired or Limited Phonological development	0	Low	Very very high	Very very high
Impaired Use of Pragmatics	Low	Very very high	0	0
Reading Difficulties	0	NR	Very very high	Very very high
Difficulties in writing and Spelling	0	NR	Very very high	Very very high
Reduced Ability of Verbal Language Comprehension	0	Very high	High	High
Difference between Verbal and Nonverbal IQ	High	Very very high	0	Very high
Heredity	High	0	NR	NR
Impaired Sociability	- Medium	0	Medium	0
Impaired Mobility	0	0	Medium to high	NR
Attention Distraction	0	0	Very very high	NR
Reduced Arithmetic Ability	0	NR	Medium	NR

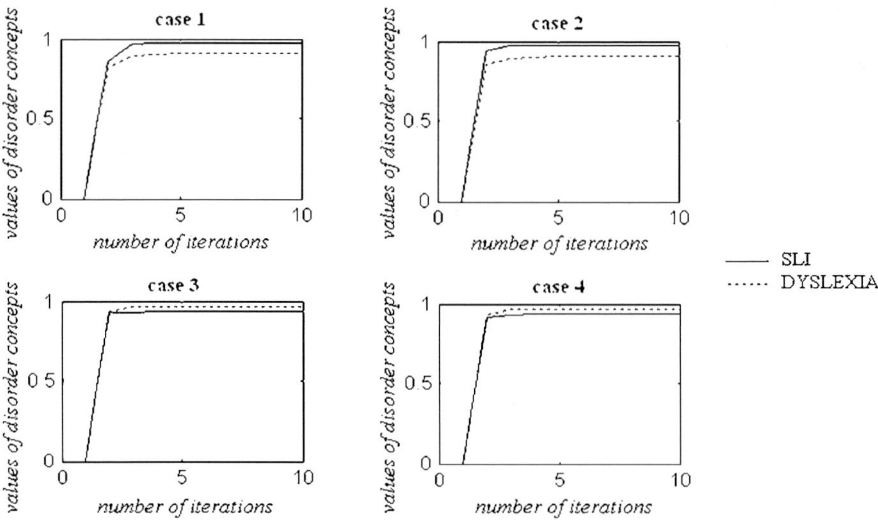

Fig. 5 Output nodes (disorder concepts) of differential diagnosis FCM for Dyslexia and SLI for four known cases

Results showed that for all four cases, even though some of the information was incomplete, the outcome given by the model agreed with the published diagnosis. That is in all four cases, the correct diagnosis was concluded: SLI, SLI, Dyslexia, and Dyslexia, respectively (Fig. 5). In the two cases of Dyslexia the largest-final diagnosis, even though correct, differed by a relatively small amount from the other diagnosis (SLI) which points out the difficulty in differential diagnoses of the two disorders.

5 Distributed m-FCM for Medical Diagnosis

For the case of a large complex system a common known approach is the decomposition into subsystems, which is a well known technique that has been used extensively on conventional approaches (Mesarovic et al. 1970), (Siljak 1979). But this decomposition is not easily applicable when subsystems have common elements that prohibit the simplified approach of summing up the individual components behavior. We follow the same direction in using FCMs to model complex medical decision support systems. With the proposed perspective for the modeling and analysis of complex systems, each component of the infrastructure constitutes a part of the intricate web that forms the overall infrastructure (Stylios 2002).

The case where multiple infrastructures are connected as "systems of systems" is considered. A Fuzzy Cognitive Map models each subsystem and the complex system is modeled with the interacting Fuzzy Cognitive Maps. FCMs communicate with each other as they operate in a common environment, receiving inputs from other FCMs and transmitting outputs to them. The linkage between two FCMs has the meaning that one state-concept of one FCM influences or is correlated to the state-concept of the other. This distributed multiple m-FCM is shown in Fig. 6. FCMs are connected at multiple points through a wide variety of mechanisms, representing by bi-directional relationship existing between states of any pair of FCMs, that is, FCM_k depends on FCM_l through some links, and probably FCM_l depends on FCM_k through other links. There are multiple connections among FCMs such as feedback and feed forward paths, and intricate and branching topologies. The connections create an intricate web, depending on the weights that characterize the linkages. Interdependencies among FCMs increase the overall complexity of the "system to systems".

Figure 6 illustrates an combined Distributed Fuzzy Cognitive Map, which aggregates five FCM models for the five subsystems of the complex system. Among the subsystems and thus, among the FCM models, there are interdependencies that are illustrated as interconnections between concepts belonging to different FCMs, where each FCM can be easily modeled (Stylios and Groumpos 2004).

5.1 Distributed m-FCM for Differential Diagnosis of Dysarthria and Apraxia of Speech

Dysarthria is the term to describe a group of disorders of oral communication resulting from disturbances in muscle control over the speech production mechanism due to damage to the central or peripheral nervous system (Darley et al. 1969a, 1969b). Dysarthrias are associated with certain neurologic and very debilitating disorders, such as Parkinson's disease, Huntington's disease, Multiple Sclerosis,ALS

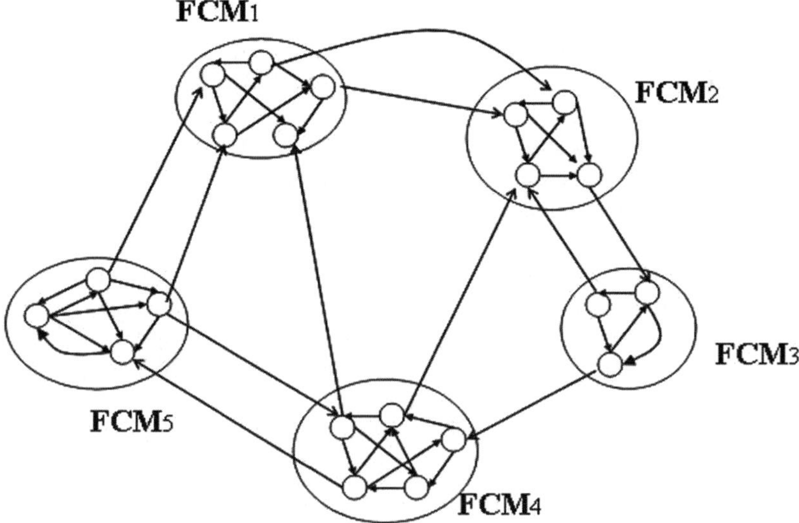

Fig. 6 The Distributed m-FCM model

(Lou Gehrig's Disease) Disease, Cerebral Palsy, Brain Tumors, Stroke, Certain Types of Brain Surgery, etc. Neurological impairment in the form of paralysis, weakness, or lack of coordination of the muscles that support speech production, can result in different forms of dysarthria. Darley et al. (Darley et al. 1969a, 1969b) identified seven forms of dysarthria: spastic, flaccid, ataxic, hypokinetic, hyperkinetic chorea, hyperkinetic dystonia, and mixed dysarthrias. The most common types of mixed dysarthrias can be: flaccid-spastic, ataxic-spastic, hypokinetic-spastic, ataxic-flaccid-spastic, and hyperkinetic-hypokinetic.

Apraxia of speech is defined as "a neurogenic speech disorder resulting from impairment of the capacity to program sensorimotor commands for the positioning and movement of muscles for the volitional production of speech. It can occur without significant weakness or neuromuscular slowness, and in the absence of disturbances of conscious thought or language" (Duffy 1995). Although apraxia of speech is controversial, most definitions of the disorder refer to impairment in programming, planning, or sequencing the movements of speech.

The differentiation between the dysarthria types can be a challenging task for a speech and language pathologist (SLP), since many speech and oral motor characteristics of the dysarthrias are overlapping. Additionally, despite the fact that the distinction between AOS (Apraxia of Speech) and dysarthrias is usually an easier process, differentiation between AOS and ataxic dysarthria or the establishment of a co-occurrence of both AOS and a dysarthria type can be challenging as well (Duffy 1995) . One of the most widely used and accepted systems for the differential diagnosis of the dysarthria types is the DAB system or the Darley, Aronson and Brown (Darley et al. 1969a, 1969b) system which has some difficulties associated with its use since there are too many parameters to remember, overlapping symptoms etc.

Characteristics of the specific type of dysarthria is important in treatment design, since clinicians working with patients with dysarthria must make a variety of decisions including: what aspects of the disorder will be responsive to treatment, what type of intervention and how much intervention is needed and when to undertake intervention, etc.

For the differential diagnosis of dysarthia an m-FCM system is developed, it is consisted of 89 factors. These factors are divided: 31 concepts represent oral-motor characteristics and 58 the speech characteristics (See (Duffy 1995) for a complete set of the factors used). But some of these factors can be grouped together since they represent separate assessment procedures, thus certain FCM subsystems can be developed and so the most suitable approach is the distributed m-FCM diagnosis model. For example, "voice quality assessment" can include nasality of speech, hoarseness, breathiness, voice tremor, strained voice, voice breaks, diplophonia. A fuzzy cognitive map subsystem with these factors can provide a value for the concept voice quality in the FCM of Fig. 7. Similarly, the concept "voice pitch" is represented by another FCM subsystem with concepts such as low pitch, high pitch, pitch breaks, and monopitch. Thus, the distributed m-FCM model for the Differential Diagnosis System of Dysarthria and Apraxia of Speech, shown in Fig. 7 is a hierarchical one where the results of subsystem FCMs used for various assessments are fed into the supervisor "upper FCM". The connection between the output concepts of the lower FCMs to the output concepts of the upper ones can have values of

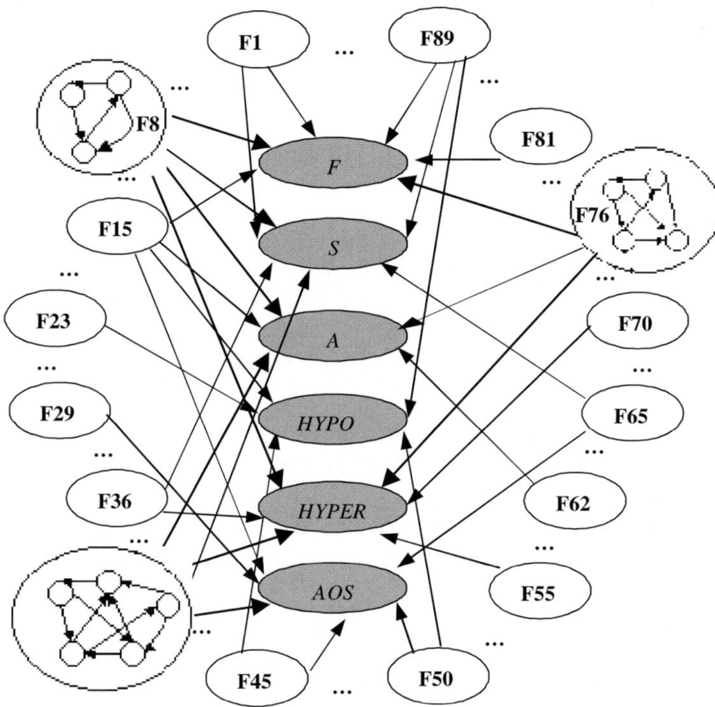

Fig. 7 Diagram of Differential Diagnosis Hierarchical System of Dysarthria and Apraxia of Speech

low (L), medium (M), high (H), near zero (0) etc. Table 3 illustrates an example of some of the weights between factors and diagnoses since it is not possible to show the whole table with the 89 factors and their connection to each of the 7 possible diagnoses would be quite complex. It is mentioned that the diagnosis m-FCM here is not a CFCM because there is not co-occurrence of more than one dysarthria, as well as dysarthria and apraxia. This can be observed in Table 4 where there is a comparison of diagnosis provided by a speech and language pathologist (SLP) and the Dysarthria-Apraxia distributed m-FCM DSS for four patient cases.

Table 3 Examples Of Fuzzy Values Of Weights Between Factor Concepts And Diagnosis Concept

Factor	Flaccid Dys.	Spastic Dys.	Ataxic Dys.	Hypokinetic Dys.	Hyperkinetic Dys.	Apraxia of Speech
Dysphagia	M	M	0	M	M	0
Hyperactive gag	0	H	0	0	0	0
Voice quality	M to H	M to H	L to M	M to H	M to H	0
Distorted vowels	0	0	H	0	H	M
Irregular AMRs	0	0	H	0	H	0
. . ..	. . ..	. . ..	. . ..	. . ..	. . ..	. . ..

Table 4 Comparison of Diagnosis Provided by SLP and Dysarthria-Apraxia FCM DSS

Initial Diagnosis of Case by SLP	Output Values of Differential Diagnostic System Resulting Diagnosis					
	Flaccid Dys.	Spastic Dys.	Ataxic Dys.	Hypok. Dys.	Hyperk. Dys.	Apraxia of Speech
Case 1 Ataxic Dys.	0.5622	0.8081	*0.9170*	0.5000	0.8355	0.6225
Case 2 Flaccid Dys.	*0.9284*	0.6900	0.5156	0.6514	0.5312	0.5312
Case 3 AOS	0.5467	0.7432	0.8936	0.7186	0.8727	*0.9975*
Case 4 Mixed Dys.	0.5101	*0.9272*	*0.9487*	0.6934	0.8222	0.5248

7 Hierarchical Structure for Medical Decision Support System

A knowledge based methodology is more suited to accomplish complex tasks when the nature of the tasks, systems, problems and solutions is not well defined or not known beforehand. In medical applications there are situations involving a significant number of variable factors such as changing characteristics, unexpected disturbances, different combinations of fault and alarm situations, where the approach of knowledge based system has certain advantages and flexibility which make such method particularly attractive and suitable for complex systems.

Medical Decision Support systems may have a large number of operating rules and constraints requiring complex logic methods. Knowledge based systems have considerable potential for successful applications requiring synthetic and abstract logic, such as decision making, diagnosis, alarm management.

A hierarchical structure is proposed where the m-FCM can be used to model the supervisor, which is the Medical Decision Support Systems (Fig. 8). The m-FCM consists of concepts representing each one of the FCM modeling various discipline sources (patient's records and information, doctors' physical examination and evaluation, laboratory tests, imaging tests etc). In addition there are other concepts representing issues for emergency behavior, estimation and overall decision and etc. The m-FCM is an integrated, aggregated and abstract model of the complex system and it represents the relationships among the subsystems and the procedure for inferring the final decision by evaluating all the information from them.

Consequently, the m-FCM system has a generic purpose, it receives information from all the subsystems in order to accomplish a task, it makes decisions and it can plan strategically. This m-FCM uses a more abstract representation, general knowledge and adaptation heuristics.

7.1 Two-level Structure for Decision Support During Labor

During the crucial period of labor, obstetricians evaluate the whole situation, they take into consideration a variety of factors, they interpret and evaluate the FHR signal and they continuously reconsider regarding the procedure of the delivery. Obstetricians have to determine whether they will proceed with a Caesarian section

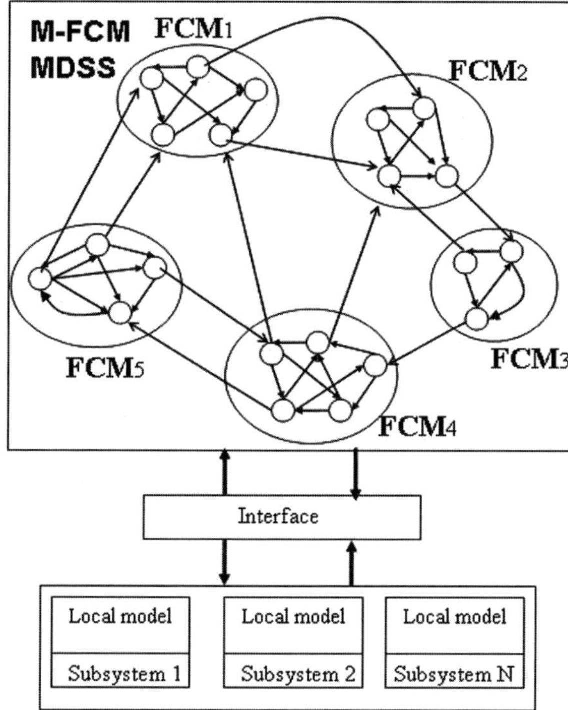

Fig. 8 The hierarchical structure with the m-FCM for Medical Decision Support Systems

or a natural delivery based on the physical measurements, and the intepretation of Fetal Heart Rate (FHR) and other essential indications and measurements.

Cardiotocography was introduced into obstetrics practice and it has been widely used for antepartum and intrapartum fetal surveillance. Cardiotocogram (CTG) consists of two distinct signals, i.e. the recording of instantaneous Fetal Heart Rate (FHR) and Uterine Activity (UA), which are two biosignals. FHR variability is believed to reflect the interactions between the sympathetic nervous system (SNS) and the parasympathetic nervous system (PSNS) of the fetus. Considerable research efforts have been made to process, evaluate and categorise FHR either as suspecious, or pathological or normal. There have been proposed integrated methods based on Support Vector Machines, Wavelets and other computational intelligence techniques to interpet the FHR (Georgoulas et al., 2006a; 2006b).

Here, the development of a Fuzzy Cognitive Map to model the way by which the obstetrician makes a decision for a normal delivery or a caesarean section is investigated. This is an online procedure where the obstetrician evaluates whether either the woman or the fetus are at serious risk and thus, he/she has to intervene, stopping the physiological delivery and perform a caesarean section or to continue with natural delivery.

The main parameters that the obstetrician evaluates are:

a) Interpretation of CTG

Cardiotocogram (CTG) has great importance and it is an essential main factor in the decision system and it is represented in the FCM as a concept for FHR and a concept for uterine contractions UA. The interpretation and classification of the FHR is essential; various advanced techniques have been proposed to classify FHR based on Computational Intelligence Techniques such as Wavelets and Support Vector Machines (Georgoulas et al. 2006b). The FHR is classified as normal, suspicious or pathological.

b) Bishop score

The Bishop score describes the potential and condition of the woman to deliver naturally. Having a cervical dilatation less than 3 and a pathological FHR means that the physician does not expect this woman to have a normal labor.

c) Uterine contractions

Uterine contractions (UCs) are the second parameter that obstetrician takes into consideration. The Obstetrician evaluates UCs' intensity and frequency and whether UCs are automatically or induced by the medicine oxytocine. The decelerations of the FHR are evaluated in conjuction with the uterine contractions. Obstetricians adjust the pharmaceutical dose of oxytocine or even to stop it when there are prolonged, severe or repetitive decelerations as a first action before thinking to perform a caesarean section.

d) Presence of meconium

The presence of meconium is a physical measurement. It is an indication of stress to the fetus and it leads many times to abnormal FHR

e) Duration of labor

Duration of labor is also a critical factor, as it is known that labor is a continuous stressful situation for the fetus. We can reduce the duration of the labor adjusting the pharmaceutical dose of oxytocine when contractions of the uterine are not satisfactory or there is no improvement in the Bishop score taking into account that the FHR monitoring is normal.

f) Oxytocine

This is the quantity of the medicine Oxytocine, that the pregnant woman has received.

In case of a suspicious FHR, obstetricians have to stop the oxytocine, wait for 20 minutes and if the FHR is strongly pathological, they continue with other examinations, testing the pH of fetal head. Because fetal scalp blood sampling for pH and blood gas assessment is an invasive technique, its use is not widespread and obstetricians usually prefer in the case of a pathological FHR to perform a Caesarean section. This is the critical point where the clinician needs an expert system to help him to distinguish between physiological stress and pathological distress and to decide whether he can wait for a normal labor or must immediately perform a Caesarean section.

FHR evaluation is a main concern during labor but another main concern for obstetricians is the Bishop score. The Bishop score describes the physiological findings from the fetus and the mother and it describes the labor in five stages. The Bishop score represents another main concept on the Fuzzy Cognitive Map.

The wide use of oxytocine during labor interferes with the Bishop score and the uterine contractions, although in women with hypertension during pregnancy or labor, oxytocine is eliminated. It is also clear, though, that we adjust the pharmaceutical dose (oxytocine) to the uterine contractions.

The duration of labor is also a critical point to the experts, as it is known that labor is a continuous state of stress to the fetus. As the time go by, obstetricians expect that the Bishop score is improved by the contractions of the uterine or the use of the oxytocine, otherwise they cannot wait especially when the FHR is a suspicious one and obstetricians have to perform a Caesarian section in order not to put fetus at risk.

Another factor that is taking under consideration is the presence of Meconium. When the outcome of the FHR is pathological, the presence of Meconium is highly considered in the decision of proceed to a Caesarian section. Thus another factor-concept of the Fuzzy Cognitive Map is the presence and quantity of Meconium.

Experienced obstetricians take into consideration all these factors during labor. Obstetricians exploiting their clinical experience developed the following Fuzzy Cognitive Map that is depicted on Fig. 9. Their knowledge on managing the labor was utilized and representing in the concepts of the FCM and the weighted interconnections among them.

The FCM model consists of 9 concepts:

Concept 1	Decision for Normal Delivery
Concept 2	Decision for Caesarian section
Concept 3	Fetus Heard Rate (FHR)
Concept 4	Presence of Meconium
Concept 5	Time duration of labor
Concept 6	Bishop score
Concept 7	Oxytocine
Concept 8	Contractions of the uterine
Concept 9	Hypertension

The relationships among concepts are represented by the corresponding weights. So the influence from concept C_i towards concept C_j is presented by the weight W_{ij}.

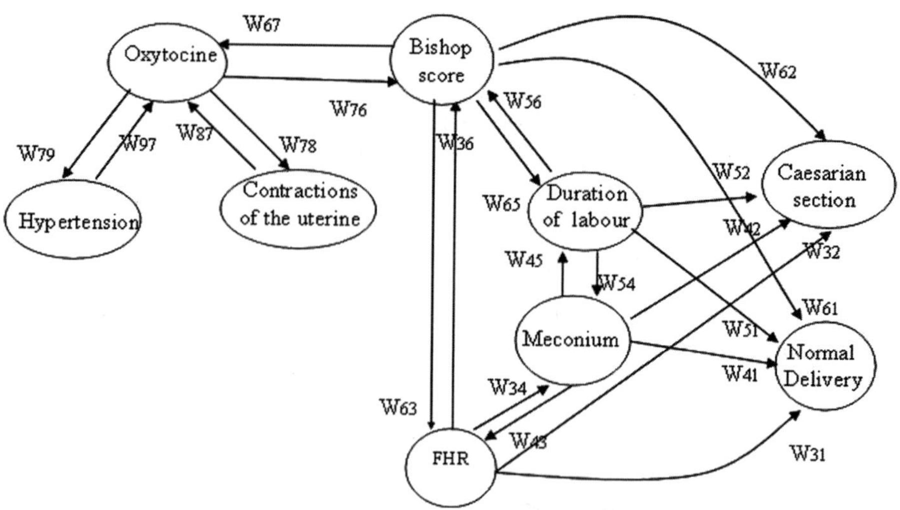

Fig. 9 Fuzzy Cognitive Map model for decision during labor

Experienced obstetricians have estimated the degree of influence from one concept to another and it is presented in Table 5. Linguistic values of interconnections are suggested by experts and are transformed in numerical weights.

At each step, values of concepts are calculated according to the influence from interconnected concepts. Some concepts can have only external input such as the concept C_3 (FHR), which stand for the evaluation and classification of FHR, which is performed at the lower level by the Support Vector Machine (Georgoulas et al. 2006a). The interactions among concepts change values of concepts. New values of some concepts may require some action from the obstetrician; as an example, a new value for oxytocine concept means descrease or increase of the pharmaceutical action to the woman. When the system reaches the steady state, the value of the concept for Natural delivery and value of the concept for Caesarian section have to

Table 5 Relationships among concepts representing by fuzzy values in obstetrics example

	C_1	C_2	C_3	C_4	C_5	C_6	C_7	C_8	C_9
C_1	-	-	-	-	-	-	-	-	-
C_2	-	-	-	-	-	-	-	-	-
C_3	very high (normal)	very high (pathological)	-	-	-	high	-	-	-
C_4	low	high	-	-	High	-	-	-	-
C_5	high (<8h)	high (>8h)	-	medium	-	medium	-	-	-
C_6	medium	high	-	-	medium	-	medium	-	-
C_7	-	-	-	-	-	medium	-	medium	medium
C_8	-	-	-	-	-	-	medium	-	-
C_9	-	-	-	-	-	-	low	-	-

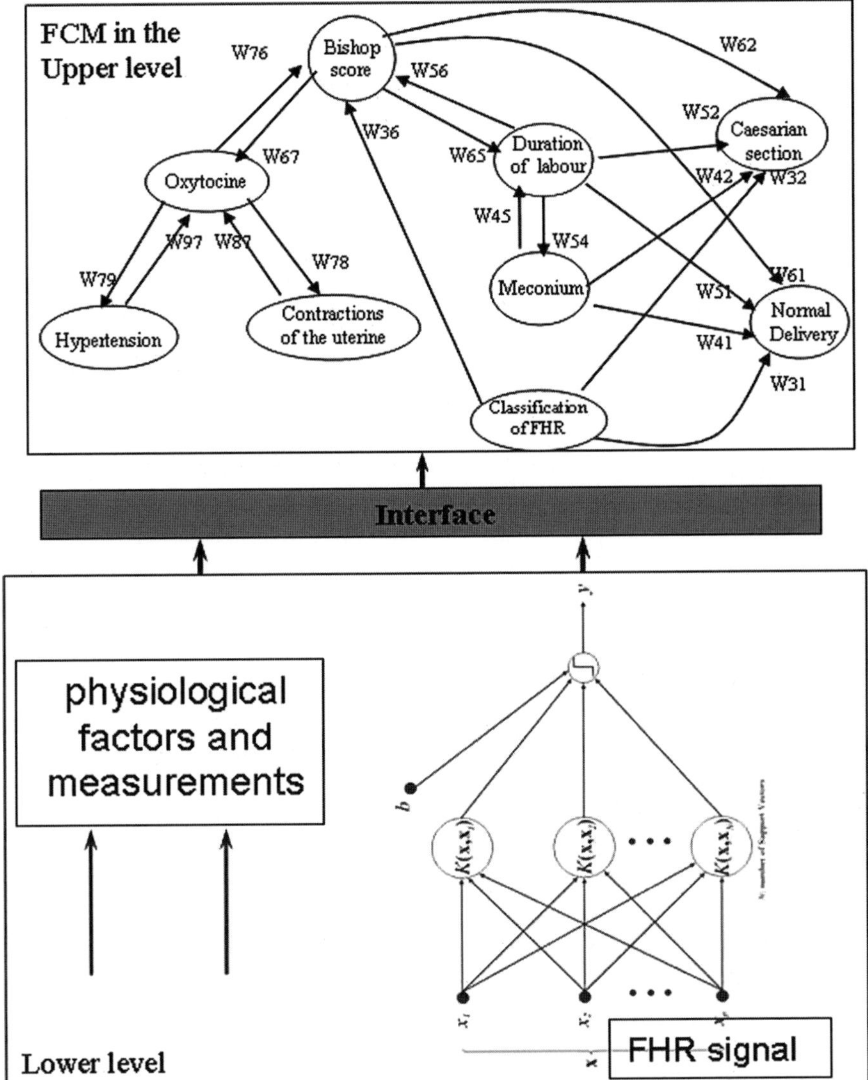

Fig. 10 The two-level structure for Decision Support during labour

be mutually exclusive and only one suggestion will be the outcome of the system. Thus, the FCM is the upper level is a CFCM, as shown in Fig. 10.

It is apparent that labor is a complex situation where the obstetrician has to consider a variety of factors and to make an on line decision on how to proceed with the delivery. It is a crucial decision, for the health of both the fetus and the mother. On the other hand, it is extremely hard to make decisions. Thus a two-level structure is proposed, where at the lower level there are either simple sensors or more advanced systems such as the FHR classification system based on Support

Vector Machines. Information from the lower level is transformed in suitable form through the interface and this information is transmitted to the FCM on the upper level. This supervisor FCM will infer a final suggestion to the obstetrician on how to proceed with the labor.

7 Conclusion

This chapter describes three novel types of Fuzzy Cognitive Map (FCM) structures suitable for Medical Decision Support Systems. The three structures are: a) the Competitive FCM, suitable when a single out of different possible diagnoses must be reached, b) a distributed m-FCM as complex medical decision support system where a large number of interacting factors are involved, and c) a hierarchical structure where the m-FCM receives information from all the subsystems in order to accomplish the task of making decisions. For each structure a corresponding example of the FCM is described performing medical decision support. The real examples presented here, are successful applications of the proposed methodologies and structures in the fields of speech pathology, language pathology, and obstetrics.

Acknowledgments The Project is co-funded by the European Social Fund and National Resources – (EPEAEK-II) **ARCHIMIDIS I**.

References

Chaib-draa B, Desharnais J (1998) A Relational Model of Cognitive Maps. International Journal of Human-Computer Studies 49:181–200

Craiger JP, Goodman DF, Weiss RJ, Butler A. (1996) Modeling organizational behavior with fuzzy cognitive maps. Journal of Computational Intelligence and Organizations 1:120–123

Darley FL, Aronson AE, Brown JR (1969) Differential diagnostic patterns of dysarthria. Journal of Speech and Hearing Research 12:246–269

Darley FL, Aronson AE, Brown JR (1969) Clusters of deviant speech dimensions in the dysarthrias. Journal of Speech and Hearing Research 12:462–496

Dickerson J, Kosko B (1994) Fuzzy Virtual Worlds. AI Expert 25–31

Duffy JR (1995) Motor speech disorders: substrates, differential diagnosis, and management Mosby-Year Book, St. Louis

Georgopoulos VC, Malandraki GA, Stylios CD (2003) A Fuzzy Cognitive Map Approach To Differential Diagnosis of Specific Language Impairment. Journal of Artificial Intelligence in Medicine 29:261–278

Georgopoulos VC, Stylios CD (2005) Augmented Fuzzy Cognitive Maps Supplemented with Case Base Reasoning for Advanced Medical Decision Support. In: Nikravesh M, Zadeh LA, Kacprzyk J (eds.) Soft Computing for Information Processing and Analysis Enhancing the Power of the Information Technology. Studies in Fuzziness and Soft Computing, Springer-Verlag, pp. 389–398.

Georgopoulos VC, Malandraki GA (2005) A Fuzzy Cognitive Map Hierarchical Model for Differential Diagnosis of Dysarthrias and Apraxia of Speech. Proceedings of the 27th Annual International Conference of the IEEE Engineering in Medicine and Biology Society, 1–4 September 2005, Shanghai, China.

Georgoulas G, Stylios C, Groumpos PP (2006a) Feature Extraction and Classification of Fetal Heart Rate Signals using Wavelet Analysis and Support Vector Machines. International Journal of Artificial Intelligent Tools, 15: 411–432

Georgoulas G, Stylios C, Groumpos PP (2006b) Novel Methodology for Fetal Heart Rate Signal Classification During the Intrapartum Period. IEEE Transactions on Biomedical Engineering 53: 875–884.

Jain L (1997) Soft Computing Techniques in Knowledge-Based Intelligent Engineering Systems: Approaches and Applications. Studies in Fuzziness and Soft Computing, 10, Springer Verlag

Jang J, Sun C, Mizutani E. (1997) Neuro-Fuzzy and Soft Computing. Prentice-Hall, NJ

Kakolyris A, Stylios G., Georgopoulos V (2005) Fuzzy Cognitive Maps application for Webmining. Proceedings of the 4th Conference of the European Society for Fuzzy Logic and Technology EUSFLAT-LFA 2005, 7–9 September 2005 Barcelona, Spain

Kang I, Lee S (2004) Using fuzzy cognitive map for the relationship management in airline service. Expert Syst. Appl. 26:545–555

Khan MS, Khor S,Chong A (2004). Fuzzy cognitive maps with genetic algorithm for goal-oriented decision support. Int. J. Uncertainty, Fuzziness and Knowledge-based Systems. 12:31–42.

Kosko B (1992) Neural Networks and Fuzzy Systems. Prentice-Hall, Englewood Cliffs, NJ

Kosko B (1997). Fuzzy Engineering. Prentice-Hall, Englewood Cliffs, NJ

Lee KC, Kim HS (1998) A Causal Knowledge-Driven Inference Engine for Expert System. In Proceedings of the annual Hawaii international conference on system science 284–293.

Lee KC, Kin JS, Chung NH, Kwon SJ (2002) Fuzzy cognitive map approach to web-mining inference amplification. Journal of Experts Systems with Applications 22:197–211.

Lin CT, Lee CSG (1996) Neural Fuzzy Systems: A Neuro-Fuzzy Synergism to Intelligent Systems. Prentice Hall, Upper Saddle River N.J.

Malandraki GA, Georgopoulos VC (2006) An Artificial Intelligence System for Differential Diagnosis of Dyslexia and Specific Language Impairment. Chapter to appear In: Dyslexia in Children: New Research, Nova Science Publishers, Inc., Hauppauge, NY

McGregor KK, Appel A (2002) On the relation between mental representation and naming in a child with specific language impairment. Clinical Linguistics and Phonetics, 16:1–20

Mesarovic MD, Macko D, Takahara Y (1970) The theory of Hierarchical Multilevel Systems. Academic Press, New York.

Mitchell CM, Sundstrom GA, (1997) Human interaction with complex systems: design issues and research approaches. IEEE Transactions on Systems, Man, and Cybernetics. 27:265–273

Muata K, Bryson O (2004) Generating consistent subjective estimates of the magnitudes of causal relationships in fuzzy cognitive maps. Computers and Operations Research 31: 1165–1175

Papageorgiou EI, Stylios CD, Groumpos PP (2003) An Integrated Two-Level Hierarchical Decision Making System based on Fuzzy Cognitive Maps (FCMs). IEEE Transactions on Biomedical Engineering 50: 1326–1339

Papageorgiou EI, Stylios CD, Groumpos PP (2004) Active Hebbian Learning Algorithm to train Fuzzy Cognitive Maps, in International Journal of Approximate Reasoning, Vol. 37, Issue 3, pp. 219–247.

Papageorgiou EI, Spyridonos P, Stylios CD, Ravazoula R, Groumpos PP, Nikiforidis G (2006) A Soft Computing Method for Tumour Grading Cognitive Maps. Artificial Intelligence in Medicine 36:58–70.

Park KS, Kim SH (1995) Fuzzy cognitive maps considering time relationships. International Journal of Human-Computer Studies 42:157–168.

Pelaez CE, Bowles JB (1996) Using Fuzzy Cognitive Maps as a System Model for Failure Modes and Effects Analysis. Information Sciences 88:177–199.

Pierson JM (1999) Transforming engagement in literacy instruction: The role of student genuine interest and ability. Annals of Dyslexia 49:307–29.

Psychcorp, Harcourt Assessment, Inc. (2005) Case Study, No.4.

Siljak DD (1979) Overlapping Decentralized Control. In Singh M, Titli A (eds) Large Scale Systems Engineering Applications, North-Holland, New York

Stach W, Kurgan L (2004) Modeling Software Development Projects Using Fuzzy Cognitive Maps. Proceedings of the 4th ASERC Workshop on Quantitative and Soft Software Engineering (QSSE'04), Banff, AB, 55–60

Stach, W., Kurgan, L., Pedrycz, W., and Reformat M., (2004) "Parallel Fuzzy Cognitive Maps as a Tool for Modeling Software Development Projects" NAFIPS 2004

Stylios CD, Groumpos PP, Georgopoulos VC (1999) An Fuzzy Cognitive Maps Approach to Process Control Systems. Journal of Advanced Computational Intelligence 3:409–417.

Stylios C, Groumpos P (2000) Fuzzy Cognitive Maps in Modelling Supervisory Control Systems. J Intell Fuzz Syst 8:83–98.

Stylios CD (2002) The Knowledge Based Technique of Fuzzy Cognitive Maps for Modeling Complex Systems. Proc. of 16th European Meetings on Cybernetics and Systems Research (EMCSR) April 2 - 5, 2002, University of Vienna, Austria, 524–529.

Stylios CD, Groumpos PP (2004) Modeling Complex Systems Using Fuzzy Cognitive Maps. IEEE Transactions on Systems, Man and Cybernetics: Part A Systems and Humans 34: 155–162..

Taber R (1991) Knowledge processing with fuzzy cognitive maps. Expert Systems with applications 2:83–87.

Van der Lely HKJ (1997) Language and cognitive development in a grammatical SLI boy: Modularity innateness. Journal of Neurolinguistics10:75–107.

Xirogiannis G, Stefanou J, Glykas M (2004) A fuzzy cognitive map approach to support urban design. Expert Systems with Applications 26:257–268

Zeleznikow J, Nolan J (2001) Using soft computing to build real world intelligent decision support systems in uncertain domains. Decision Support Systems 31:263–285

Zhang WR, Chen SS, Bezdek JC (1989) Pool 2: a generic system for cognitive map development and decision analysis. IEEE Transactions on Systems, Man and Cybernetics 19:31–39

Zhang WR, Chen SS, Wang W, King RS (1992) A cognitive-map-based approach to the coordination of distributed cooperative agents. IEEE Transactions on Systems, Man and Cybernetics 22:103–114.

Fuzzy-Neuro Systems for Local and Personalized Modelling

Nikola Kasabov, Qun Song and Tian Min Ma

Abstract This chapter presents a review of recently developed fuzzy-neuro models for local and personalised modelling and illustrates them on a real world case study from medical decision support. The local models are based on the principles of evolving connectionist systems, where the data is clustered and for each cluster a separate local model is developed and represented as a fuzzy rule, either of Takagi-Sugeno, or Zadeh-Mamdani types. The personalised modelling techniques are based on transductive reasoning and include a model called TWNFI. It is also illustrated on medical decision support problem where a model for each patient is developed to predict an outcome for this patient and to rank the importance of the clinical variables for them. The local and personalised models are compared with statistical, neural network and fuzzy-neuro global models and show a significant advantage in accuracy and explanation.

Key words: Fuzzy-neuro systems · Local modelling · Transductive reasoning · Personalised modelling

1 Introduction

The complexity and the dynamics of real-world problems, such as adaptive speech recognition and language acquisition [11, 13, 20], adaptive intelligent prediction and control systems [1], intelligent agent-based systems and adaptive agents on the Web [43], mobile robots [10], visual monitoring systems and multi-modal information processing [16, 25], large Bio-informatics data processing, and many more [2, 3], require sophisticated methods and tools for building on-line, adaptive, knowledge-based intelligent systems (IS). Such systems should be able to: (1) learn fast from a large amount of data (using fast training); (2) adapt incrementally in an on-line mode; (3) dynamically create new modules – have open structure; (4) memorise information that can be used at a later stage; (5) interact continuously with the environment in a "life-long" learning mode; (6) deal with knowledge (e.g. rules), as well as with data; (7) adequately represent space and time [2, 6, 14, 16, 18, 30, 33].

M. Nikravesh et al. (eds.), *Forging the New Frontiers: Fuzzy Pioneers II.*
© Springer-Verlag Berlin Heidelberg 2008

2 Local Learning in ECOS

Evolving connectionist systems (ECOS) are modular connectionist-based systems that evolve their structure and functionality in a continuous, self-organised, on-line, adaptive, interactive way from incoming information; they can process both data and knowledge in a supervised and/or unsupervised way [17].

ECOS learn local models from data through clustering of the data and associating a local output function for each cluster. Clusters of data are created based on similarity between data samples either in the input space (this is the case in some of the ECOS models, e.g. the dynamic neuro-fuzzy inference system DENFIS [21]), or in both the input space and the output space (this is the case in the EFuNN models [17]). Samples that have a distance to an existing cluster center (rule node) N of less than a threshold Rmax (for the EfuNN models it is also needed that the output vectors of these samples are different from the output value of this cluster center in not more than an error tolerance E) are allocated to the same cluster Nc. Samples that do not fit into existing clusters, form new clusters as they arrive in time. Cluster centers are continuously adjusted according to new data samples, and new clusters are created incrementally.

The similarity between a sample $S = (x,y)$ and an existing rule node $N = (W_1, W_2)$ can be measured in different ways, the most popular of them being the normalized Euclidean distance:

$$d(S, N) = \left[\Sigma_{(i=1,...,n)} (x_i - W_1(i))^2 \right] / n, \tag{1}$$

where n is the number of the input variables.

ECOS learn from data and automatically create a local output function for each cluster, the function being represented in the W_2 connection weights, thus creating local models. Each model is represented as a local rule with an antecedent – the cluster area, and a consequent – the output function applied to data in this cluster, e.g.:

$$IF\ (data\ is\ in\ cluster\ Nc)$$

$$THEN\ (the\ output\ is\ calculated\ with\ a\ function\ Fc) \tag{2}$$

Implementations of the ECOS framework require connectionist models that support these principles. Such model is the evolving fuzzy neural network (EFuNN).

3 The Evolving Fuzzy Neural Network (EFuNN) Model

3.1 The Principles and Architecture of Evolving Fuzzy Neural Networks (EFuNNs)

EFuNNs adopt some known techniques from [19], but here all nodes in an EFuNN are created /connected during learning. The nodes representing *membership functions* (*MF*) can be modified during learning. In EFuNNs, each input variable is represented here by a group of spatially arranged neurons to represent a fuzzy quantization of this variable. For example, three neurons can be used to represent *"small"*, *"medium"* and *"large"* fuzzy values of the variable. Different *MFs* can be attached to these neurons. New neurons can evolve in this layer if, for a given input vector, the corresponding variable value does not belong to any of the existing *MFs* to a degree greater than a set threshold. A new fuzzy input neuron, or an input neuron, can be created during the adaptation phase of an EFuNN. An optional short-term memory layer can be used through feedback connections from the rule node layer (see Fig. 1). The layer of feedback connections could be used if temporal relationships between input data are to be memorised structurally. The third layer contains rule nodes that evolve through supervised / unsupervised learning. The rule nodes represent prototypes of input-output data associations, graphically represented as an association of hyper-spheres from the fuzzy input and fuzzy output spaces. Each rule node r is defined by two vectors of connection weights – $W_1(r)$ and $W_2(r)$, the latter being adjusted through supervised learning based on the output error, and the former being adjusted through unsupervised learning based on a similarity measure within a local area of the problem space. The fourth layer of neurons represents fuzzy quantization for the output variables, similar to the input fuzzy neurons representation. The fifth layer represents the real values for the output variables.

The evolving process can be based on two assumptions, that either no rule nodes exist prior to learning and all of them are created during the evolving process, or

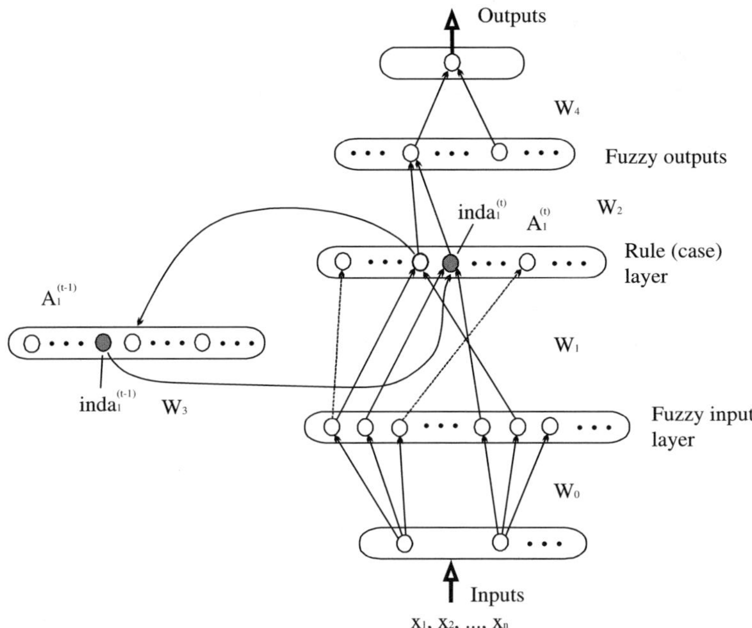

Fig. 1 An EFuNN architecture with a short term memory and feedback connections (adapted from [17, 21])

there is an initial set of rule nodes that are not connected to the input and output nodes and become connected through the learning process.

Each rule node (e.g., r_1) represents an association between a hyper-sphere from the fuzzy input space and a hyper-sphere from the fuzzy output space (see Fig. 2), the $W_1(r_j)$ connection weights representing the co-ordinates of the centre of the sphere in the fuzzy input space, and the $W_2(r_j)$ – the co-ordinates in the fuzzy output space. The radius of an input hyper-sphere of a rule node is defined as $(1 - Sthr)$, where *Sthr* is the *sensitivity threshold* parameter defining the minimum activation of a rule node (e.g., r_1) to an input vector (e.g., (Xd_2, Yd_2)) in order for the new input vector to be associated to this rule node.

Two pairs of fuzzy input-output data vectors $d_1 = (Xd_1, Yd_1)$ and $d_2 = (Xd_2, Yd_2)$ will be allocated to the first rule node r_1 if they fall into the r_1 input sphere and in the r_1 output sphere, i.e. the local normalised fuzzy difference between Xd_1 and Xd_2 are correspondingly smaller than the radius r and the local normalised fuzzy difference between Yd_1 and Yd_2 is smaller than an *error threshold (Errthr)*.

The local normalised fuzzy difference between two fuzzy membership vectors d_{1f} and d_{2f} that represent the membership degrees to which two real values d_1 and d_2 data belong to the pre-defined *MF* are calculated as $D(d_{1f}, d_{2f}) = \sum |d_{1f} - d_{2f}| / \sum (d_{1f} + d_{2f})$. For example, if $d_{1f} = [0, 0, 1, 0, 0, 0]$ and $d_{2f} = [0, 1, 0, 0, 0, 0]$, then $D(d_{1f}, d_{2f}) = (1 + 1)/2 = 1$ which is the maximum value for the local normalised fuzzy difference. If data example $d_1 = (Xd_1, Yd_1)$ where

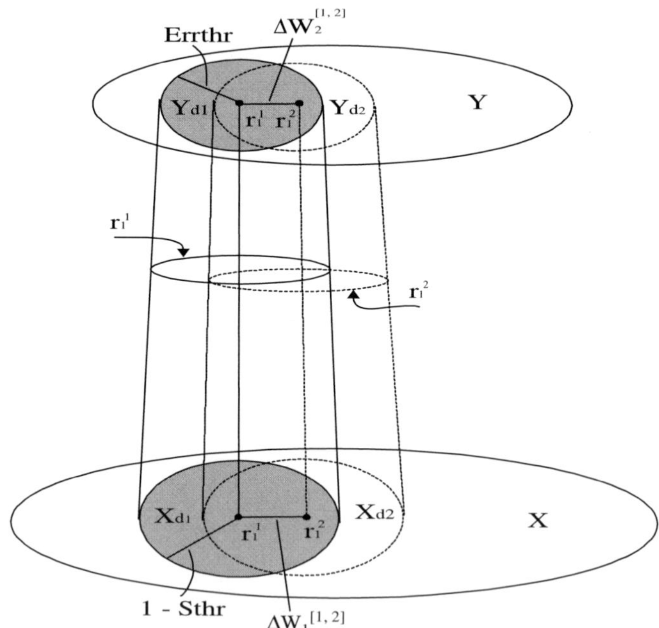

Fig. 2 Each rule created during the evolving process associates a hyper-sphere from the fuzzy input space to a hyper-sphere from the fuzzy output space. Through accommodating new nodes the centre of the rule node moves slightly (adapted from [17, 21])

Xd_1 and Yd_1 are correspondingly the input and the output fuzzy membership degree vectors, and the data example is associated with a rule node r_1 with a centre $r_1{}^1$, then a new data point $d_2 = (Xd_2, Yd_2)$, that is within the shaded area as shown in Fig. 2, will be associated with this rule node too.

Through the process of associating (learning) new data points to a rule node, the centre of this node hyper-sphere is adjusted in the fuzzy input space depending on a *learning rate lr_1* and in the fuzzy output space depending on a *learning rate lr_2*, as it is shown in Fig. 2 on two data points. The adjustment of the centre $r_1{}^1$ to its new position $r_1{}^2$ can be represented mathematically by the change in the connection weights of the rule node r_1 from $W_1(r_1{}^1)$ and $W_2(r_1{}^1)$ to $W_1(r_1{}^2)$ and $W_2(r_1{}^2)$ as it is presented in the following vector operations:

$$W_1(r_1{}^2) = W_1(r_1{}^1) + lr_1{}^* Ds(Xd_1, Xd_2) \tag{3}$$

$$W_2(r_1{}^2) = W_2(r_1{}^1) + lr_2{}^* Err(Yd_1, Yd_2)^* A_1(r_1{}^1) \tag{4}$$

where: $Err(Yd_1, Yd_2) = Ds(Yd_1, Yd_2) = Yd_1 - Yd_2$ is the signed value rather than the absolute value of difference vector; $A_1(r_1{}^1)$ is the activation of the rule node $r_1{}^1$ for the input vector Xd_2.

The idea of dynamic creation of new rule nodes over time for a time series data is graphically illustrated in Fig. 3. While the connection weights from W_1 and W_2 capture spatial characteristics of the learned data (centres of hyper-spheres), the temporal layer of connection weights W_3 from Fig. 1 captures temporal dependences between consecutive data examples. If the winning rule node at the moment $(t - 1)$ (to which the input data vector at the moment $(t - 1)$ was associated) was $r_1 = inda_1(t - 1)$, and the winning node at the moment t is $r_2 = inda_1(t)$, then a line between the two nodes is established as follows:

$$W_3(r_1, r_2)^{(t)} = W_3(r_1, r_2)^{(t-1)} + lr_3{}^* A_1(r_1)^{(t-1)*} A_1(r_2)^{(t)} \tag{5}$$

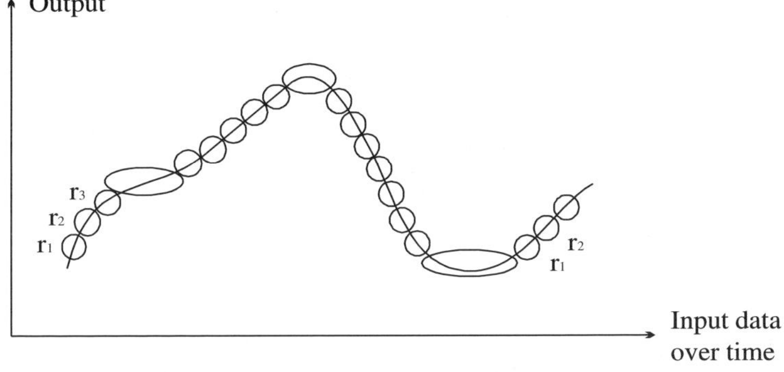

Fig. 3 The rule nodes in an EFuNN evolve in time depending on the similarity in the input data

where: $A_1(r)^{(t)}$ denotes the activation of a rule node r at a time moment (t); lr_3 defines the degree to which the EFuNN associates links between rules (clusters, prototypes) that include consecutive data examples (if $lr_3 = 0$, no temporal associations are learned in an EFuNN).

The learned temporal associations can be used to support the activation of rule nodes based on temporal, pattern similarity. Here, temporal dependences are learned through establishing structural links. These dependences can be further investigated and enhanced through synaptic analysis (at the synaptic memory level) rather than through neuronal activation analysis (at the behavioural level). The ratio (spatial – similarity) / (temporal – correlation) can be balanced for different applications through two parameters Ss and Tc such that the activation of a rule node r for a new data example d_{new} is defined as the following vector operations:

$$A_1(r) = f(Ss^* D(r, d_{new}) + Tc^* W_3(r^{(t-1)}, r)) \tag{6}$$

where: f is the activation function of the rule node r, $D(r, d_{new})$ is the normalised fuzzy difference value and $r^{(t-1)}$ is the winning neuron at time moment $(t - 1)$.

Several parameters were introduced so far for the purpose of controlling the functioning of an EFuNN. Some more parameters will be introduced later, that will bring the EFuNN parameters to a comparatively large number. In order to achieve a better control of the functioning of an EFuNN structure, the three-level functional hierarchy is used, namely: genetic level, long-term synaptic level, and short-term activation level.

At the genetic level, all the EFuNN parameters are defined as genes in a chromosome, these are:

(a) Structural parameters, e.g., number of inputs, number of MF for each of the inputs, initial type of rule nodes, maximum number of rule nodes, number of MF for the output variables, number of outputs.

(b) Functional parameters, e.g., activation functions of the rule nodes and the fuzzy output nodes; mode of rule node activation ('one-of-n', or 'many-of-n') depending on how many activation values of rule nodes are propagated to the next level); learning rates lr_1, lr_2 and lr_3; sensitivity threshold ($Sthr$) for the rule layer; error threshold ($Errthr$) for the output layer; forgetting rate; various pruning strategies and parameters, as explained in the EFuNN algorithm below.

3.2 The Basic EFuNN Algorithm

In an EFuNN, a new rule node r_n is connected and its input and output connection weights are set. The EFuNN algorithm, to evolve EFuNNs from incoming examples, is given below as a procedure of consecutive steps [16]. Vector and matrix operation expressions are used to simplicity of presentation.

1. Initialise an EFuNN structure with maximum number of neurons and no (or zero-value) connections. Initial connections may be set through inserting fuzzy rules in the structure. If initially there are no rule nodes connected to the fuzzy input and fuzzy output neurons, then create the first node $r_n = 1$ to represent the first example d_1 and set its input $W_1(r_n)$ and output $W_2(r_n)$ connection weight vectors as follows:
   <create a new rule node>: $W_1(r_n) = EX$; $W_2(r_n) = TE$, where TE is the fuzzy output vector for the current fuzzy input vector EX.

2. WHILE <there are examples in the input stream> DO
   Enter the current example (Xd_i, Yd_i), EX denoting its fuzzy input vector. If new variables appear in this example, which are absent in the previous examples, create new input and / or output nodes with their corresponding membership functions.

3. Find the normalised fuzzy local distance between the fuzzy input vector EX and the already stored patterns (prototypes, exemplars) in the rule (case) nodes, r_j, $r_j = r_1, r_2, \ldots, r_n$,

$$D(EX, r_j) = \Sigma |EX - W_1(j)/2| / \Sigma(W_1(j))$$

4. Find the activation $A_1(r_j)$ of the rule (case) r_j, $r_j = r_1, r_2, \ldots, r_n$. Here, radial basis, *radbas*, activation, or a saturated linear one, *satlin*, can be use, i.e.

$$A_1(r_j) = radbas(D(EX, r_j)), \ or A_1(r_j) = satlin(1 - D(EX, r_j))$$

The former may be appropriate for function approximation tasks, while the latter may be preferred for classification tasks. In case of the feedback variant of an EFuNN, the activation is calculated as explained above:

$$A_1(r_j) = radbas(Ss^*D(EX, r_j) - Tc^*W_3), \ or$$
$$A_1(r_j) = satlin(1 - Ss^*D(EX, r_j) + Tc^*W_3)$$

5. Update the pruning parameter value for the rule nodes, e.g. *age, average activation* as pre-defined in the EFuNN chromosome.

6. Find all case nodes r_j with an activation value $A_1(r_j)$ above a sensitivity threshold *Sthr*.

7. If there is no such case node, then <create a new rule node> using the procedure from step 1.
   ELSE

8. Find the rule node $inda_1$ that has the maximum activation value (e.g., $maxa_1$).

9. There are two modes: 'one-of-n' and 'many-of-n'.

   (a) In case of 'one-of-n' EFuNNs, propagate the activation $maxa_1$ of the rule node $inda_1$ to the fuzzy output neurons.

$$A_2 = satlin(A_1(inda_1))^* W_2(inda_1))$$

    (b) In case of '*many-of-n*' mode, the activation values of all rule nodes that are above an activation threshold of *Athr* are propagated to the next neuronal layer.

10. Find the winning fuzzy output neuron $inda_2$ and its activation $maxa_2$.
11. Find the desired winning fuzzy output neuron $indt_2$ and its value $maxt_2$.
12. Calculate the fuzzy output error vector: $Err = TE - A_2$.
13. IF ($inda_2$ is different from $indt_2$) or ($D(A_2, TE) > Errthr$), <create a new rule node>
    ELSE
14. Update: (a) the input, (b) the output, and (c) the temporal connection vectors (if such exist) of the rule node $k = inda_1$ as follow:

    (a) $Ds(EX, W_1(k)) = EX - W_1(k); W_1(k) = W_1(k) + lr_1{}^*Ds(EX, W_1(k))$, where lr_1 is the learning rate for the first layer;

    (b) $W_2(k) = W_2(k) + lr_2{}^*Err^*maxa_1$, where lr_2 is the learning rate for the second layer;

    (c) $W_3(l, k) = W_3(l, k) + lr_3{}^*A_1(k)^*A_1(l)^{(t-1)}$, here l is the winning rule neuron at the previous time moment $(t-1)$, and $A_1(l)^{(t-1)}$ is its activation value kept in the short term memory.

15. Prune rule nodes j and their connections that satisfy the following fuzzy pruning rule to a pre-defined level:
    IF (a rule node r_j is *OLD*) AND (average activation $A_{1av}(r_j)$ is *LOW*) AND (the density of the neighbouring area of neurons is *HIGH* or *MODERATE* (i.e. there are other prototypical nodes that overlap with j in the input-output space; this condition apply only for some strategies of inserting rule nodes as explained in a sub-section below) THEN the probability of pruning node (r_j) is *HIGH*.
    The above pruning rule is fuzzy and it requires that the fuzzy concepts of *OLD, HIGH*, etc., are defined in advance (as part of the EFuNN's chromosome). As a partial case, a fixed value can be used, e.g. a node is *OLD* if it has existed during the evolving of an EFuNN from more than 1000 examples. The use of a pruning strategy and the way the values for the pruning parameters are defined depends on the application tasks.
16. Aggregate rule nodes, if necessary, into a smaller number of nodes (see the explanation in the following subsection).
17. END of the while loop and the algorithm.
18. Repeat steps 2 to step 17 for a second presentation of the same input data or for Eco training if needed.

With good dynamic characteristics, the EFuNN model is a novel efficient model especially for on-'line tasks. The EFuNN model has the following major strong points: (1) incremental, fast learning (possibly 'one pass'); (2) on-line adaptation; (3) 'open' structure; (4) allowing for time and space representation based on biological plausibility; (5) rule extraction and rule insertion.

4 Dynamic Evolving Neural Fuzzy Inference System (DENFIS)

4.1 General Principle of DENFIS

The Dynamic Evolving Neural-Fuzzy Inference Systems (DENFIS) is also based on the ECOS principle and motivated by EFuNNs. DENFIS has an approach similar to EFuNNs especially similar to EFuNNs' *m-of-n* mode. DENFIS is a kind of dynamic Takagi-Sugeno type fuzzy inference systems. An evolving clustering method (ECM) is used in DENFIS models to partition the input space for creating the fuzzy rules. DENFIS evolve through incremental, hybrid (supervised/unsupervised), learning and accommodate new input data, including new features, new classes, etc. through local element tuning. New fuzzy rules are created and updated during the operation of the system. At each time moment the output of DENFIS is calculated through a fuzzy inference system based on m-most activated fuzzy rules which are dynamically selected from the existing fuzzy rule set. As the knowledge, fuzzy rules can be inserted into DENFIS before, or during its learning process and, they can also be extracted during the learning process or after it. The fuzzy rules used in DENFIS are indicated as follows:

$$R_l : \quad \text{if } x_1 \text{ is } F_{11} \text{ and } x_2 \text{ is } F_{12} \text{ and } \ldots \text{ and } x_P \text{ is } F_{1P},$$
$$\text{then } y_l = b_{l0} + b_{l1}x_1 + b_{l2}x_2 + \ldots + b_{lP}x_P \tag{7}$$

where "x_j is F_{lj}", $l = 1, 2, \ldots m$; $j = 1, 2, \ldots P$, are $M \times P$ fuzzy propositions that form m antecedents for m fuzzy rules respectively; x_j, $j = 1, 2, \ldots, P$, are antecedent variables defined over universes of discourse X_j, $j = 1, 2, \ldots, P$, and F_{lj}, $l = 1, 2, \ldots M$; $j = 1, 2, \ldots, P$ are fuzzy sets defined by their fuzzy membership functions $\mu_{Flj} : X_j \to [0, 1]$, $l = 1, 2, \ldots M$; $j = 1, 2, \ldots, P$. In the consequent parts of fuzzy rules, y_l, $l = 1, 2, \ldots m$, are the consequent variables defined by linear functions.

In DENFIS, F_{lj} are defined by the following *Gaussian* type membership function

$$Gaussian MF = \alpha \exp\left[-\frac{(x - m)^2}{2\sigma^2} \right] \tag{8}$$

When the model is given an input-output pair (x_i, d_i), it calculates the following output value:

$$f(x_i) = \frac{\sum_{l=1}^{M} y_l \prod_{j=1}^{P} \alpha_{lj} \exp\left[-\frac{(x_{ij} - m_{lj})^2}{2\sigma_{lj}^2} \right]}{\sum_{l=1}^{M} \prod_{j=1}^{P} \alpha_{lj} \exp\left[-\frac{(x_{ij} - m_{lj})^2}{2\sigma_{lj}^2} \right]} \tag{9}$$

The goal is to design the system from (8) so that the following objective function is minimized:

$$E = \frac{1}{2}\left(f(\boldsymbol{x_i}) - d_i\right)^2 \tag{10}$$

For optimizing the parameters b_{lj}, m_{lj}, α_{lj} and σ_{lj} in DENFIS, the steepest descent algorithm can be used:

$$\varphi(k+1) = \varphi(k) - \eta_\varphi \frac{\partial E}{\partial \varphi} \tag{11}$$

here, η is the learning rate and φ can represent b, m, α or σ respectively. DENFIS has following characteristics:

- Building a Takagi-Sugeno fuzzy inference engine dynamically.
  The Takagi-Sugeno fuzzy inference engine is used in both on-line and off-line modes of DENFIS. The difference between them is that for forming a dynamic inference engine, only first-order Takagi-Sugeno fuzzy rules are employed in DENFIS on-line mode and both first-order Takagi-Sugeno fuzzy rules and expanded high-order Takagi-Sugeno fuzzy rules are used in DENFIS off-line modes. To build such a fuzzy inference engine, several fuzzy rules are dynamically chosen from the existing fuzzy rule set depending on the position of current input vector in the input space.
- Dynamic creation and updating of fuzzy rules.
  All fuzzy rules in the DENFIS on-line mode are created and updated during a 'one-pass' training process by applying the *Evolving Clustering Method* (ECM) and the *Weighted Recursive Least Square Estimator with Forgetting Factors* (WRLSE).
- Local generalisation.
  Similar to EFuNNs, DENFIS model has local generalisation to speed up the training procedure and to decrease the number of fuzzy rules in the system.
- Fast training speed.
  In the DENFIS on-line mode, the training is a 'one-pass' procedure and in the off-line modes, WLSE and small-scale MLPs are applied, which lead DENFIS to have the training speed for complex tasks faster than some common neural networks or hybrid systems such as multi-layer perceptron with back-propagation algorithm (MLP-BP) and *Adaptive Neural-Fuzzy Inference System* (ANFIS), both of which adopt global generalisation.
- Satisfactory accuracy.
  Using DENFIS off-line modes, we can achieve a high accuracy especially in non-linear system identification and prediction.

4.2 Dynamic Takagi-Sugeno Fuzzy Inference Engine

The Takagi-Sugeno fuzzy inference engine [40] utilised in DENFIS is a dynamic inference model. In addition to dynamically creating and updating fuzzy rules in the DENFIS on-line mode, the major differences between such inference engine and the general Takagi-Sugeno fuzzy inference engine are described as follows:

First, depending on the position of the current input vector in the input space, different fuzzy rules are chosen from the fuzzy rule set, which has been estimated during the training procedure, for constructing an inference engine. If there are two input vectors very close to each other, especially in DENFIS off-line modes, two identical fuzzy inference engines are established and they may be exactly the same. In the on-line mode, however, although sometimes two inputs are exactly same, their corresponding inference engines are probably different. This is because these two inputs come into the system from the data stream at different moments and the fuzzy rules probably have been updated during this interval.

Secondly, also depending on the position of current input vector in the input space, the antecedents of fuzzy rules, which have been chosen from the fuzzy rule set for forming an inference engine, may be different. An example is illustrated in Fig. 4, where two fuzzy rule groups, FG_1 and FG_2, are estimated depending on two input vectors x_1 and x_2 respectively in a 2-D input space. We can know from this example that, for instance, the region C represents a linguistic meaning 'large' in FG_1 on the X_1 axis but it represents a linguistic meaning 'small' on that in FG_2. Also, the region C is presents as different membership functions respectively in FG_1 and FG_2.

4.3 Fuzzy Rule Set, Rule Insertion and Rule Extraction

Fuzzy rules in a DENFIS are created during a training procedure, or come from rule insertion. In the on-line mode, the fuzzy rules in the rule set can also be updated as new training data appear in the system [22].

As the DENFIS uses a Takagi-Sugeno fuzzy inference engine the fuzzy rules inserted to or extracted from the system are Takagi-Sugeno type fuzzy rules. These rules can be inserted into the rule set before or during the training procedure and they can also be exacted from the rule set during or after the training procedure.

The inserted fuzzy rules can be the rules that are extracted from a fuzzy rule set created in previous training of DENFIS, or they can also be general Takagi-Sugeno type fuzzy rules. In the latter, the corresponding nodes of the general Takagi-Sugeno fuzzy rules have to be found and located in the input space. For an on-line learning mode, their corresponding radiuses should also be defined. The region can be obtained from the antecedent of a fuzzy rule and the centre of this region is taken as the node corresponding with the fuzzy rule. A value of $(0.5 \sim 1)Dthr$ can be taken as the corresponding radius.

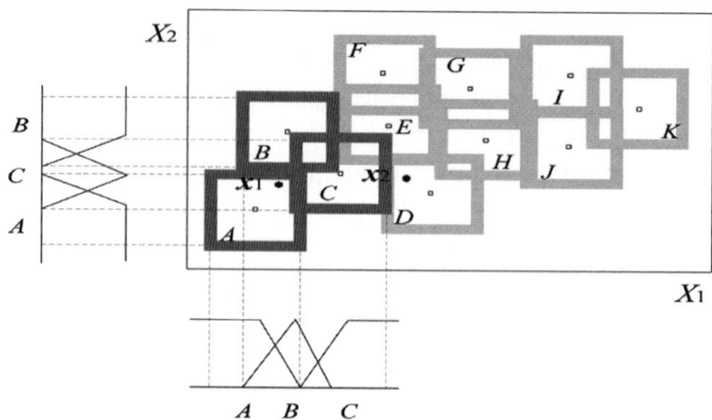

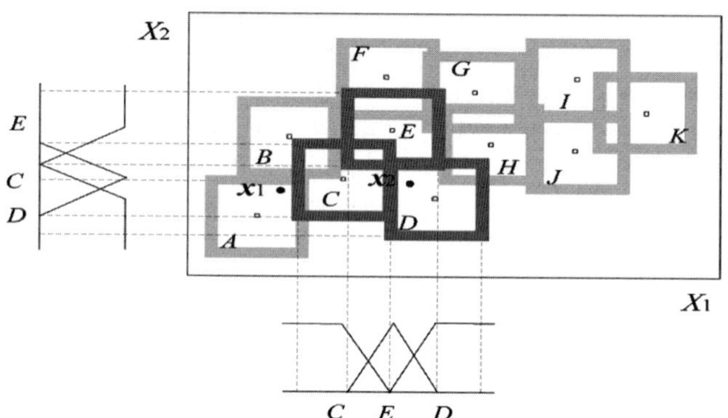

Fig. 4 Two fuzzy rule groups corresponding to input vectors x_1 and x_2 in a 2-D space (from [21])

4.4 Comparison of DENFIS On-line Mode and EFuNN

Similar to EFuNN, the DENFIS on-line mode applies a one-pass, on-line training algorithm and adopts local generalisation that make both DENFIS performs its training procedure very fast.

The EFuNN is a fuzzified neural network and it updates the system by using a method similar to an on-line gradient descent method [4, 9, 27, 28, 32]. The EFuNN algorithm is a simple one so that the rule insertion, rule extraction and rule aggregation are easy to realize, however, the EFuNN normally requires more rules than a DENFIS on-line mode for achieving the similar result.

The DENFIS model is a neural fuzzy inference system with capability of learning. The DENFIS on-line mode updates the system by using a weighted recursive least square estimation algorithm which makes the DENFIS more effective and more accurate than EFuNN. Instead of the data space partitioning in both input space and output space, the DENFIS on-line mode applies the ECM that performs a partitioning in the input space. DENFIS has been used to develop an adaptive decision support system for renal function evaluation [26], as explained the last section of the chapter and shown in Fig. 8.

5 Transductive Reasoning and Weighted Data Normalization for Personalised/Individual Modelling

5.1 Transductive versus Inductive Modelling

Most of learning models and systems in artificial intelligence developed and implemented so far are based on inductive methods, where a model (a function) is derived from data representing the problem space and this model is further applied on new data [12, 24, 5]. The model is usually created without taking into account any information about a particular new data vector (test data). An error is measured to estimate how well the new data fits into the model. The inductive learning and inference approach is useful when a global model ("the big picture") of the problem is needed even in its very approximate form. In contrast to the inductive learning and inference methods, transductive inference methods estimate the value of a potential model (function) only in a single point of the space (the new data vector) utilizing additional information related to this point [41]. This approach seems to be more appropriate for clinical and medical applications of learning systems, where the focus is not on the model, but on the individual patient. Each individual data vector (e.g.: a patient in the medical area; a future time moment for predicting a time series; or a target day for predicting a stock index) may need an individual, local model that best fits the new data, rather then a global model, in which the new data is matched without taking into account any specific information about this data.

An individual model M_i is trained for every new input vector x_i with data use of samples D_i selected from a data set D, and data samples $D_{0,i}$ generated from an existing model (formula) M (if such a model is existing). Data samples in both D_i and $D_{0,i}$ are similar to the new vector x_i according to defined similarity criteria.

Transductive inference is concerned with the estimation of a function in single point of the space only [41]. For every new input vector x_i that needs to be processed for a prognostic task, the N_i nearest neighbours, which form a sub-data set D_i, are derived from an existing data set D. If necessary, some similar vectors to vector x_i and their outputs can also be generated from an existing model M. A new model M_i is dynamically created from these samples to approximate the function in the point x_i - Figs. 5 and Figs. 6. The system is then used to calculate the output value y_i for this input vector x_i.

5.2 Weighted Data Normalization

In many neural network and fuzzy models and applications, raw (not normalized) data is used. This is appropriate when all the input variables are measured in the

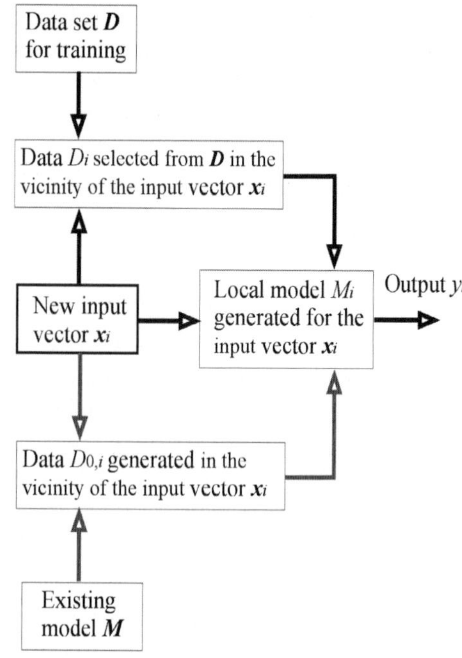

Fig. 5 A block diagram of a *transductive* reasoning system

same units. Normalization, or standardization, is reasonable when the variables are in different units, or when the variance between them is substantial. However, a general normalization means that every variable is normalized in the same range, e.g. [0, 1] with the assumption that they all have the same importance for the output of the system [5].

For many practical problems, variables have different importance and make different contribution to the output(s). Therefore, it is necessary to find an optimal normalization and assign proper importance factors to the variables. Such a method can also be used for feature selection or for reducing the size of input vectors through

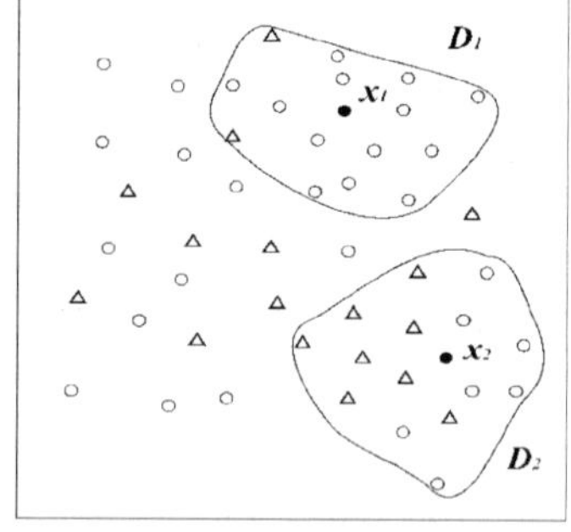

Fig. 6 In the centre of a transductive reasoning system is the new data vector (here illustrated with two of them – x_1 and x_2), surrounded by a fixed number of nearest data samples selected from the training data D and generated from an existing model M

keeping the most important ones [34]. This is especially applicable to a special class of neural networks or fuzzy models – the clustering based models (or also: distance-based; prototype-based) such as: RBF [31], ART [7], ECOS [15, 17, 21]. In such systems, distance between neurons or fuzzy rule nodes and input vectors are usually measured in Euclidean distance, so that variables with a wider normalization range will have more influence on the learning process and vice versa.

5.3 Transductive Neuro- Fuzzy System with Weighted Data Normalization

5.3.1 Principles and Structure of the TWNFI

TWNFI is a dynamic neural-fuzzy inference system with a local generalization, in which, either the *Zadeh-Mamdani* type fuzzy inference engine [44, 45] is used, or the *Takagi-Sugeno* fuzzy inference [40] is applied. Here, the former case is introduced. The local generalization means that in a sub-space of the whole problem space (local area) a model is crated that performs generalization in this area. In the TWNFI model, Gaussian fuzzy membership functions are applied in each fuzzy rule for both the antecedent and the consequent parts. A steepest descent (BP) learning algorithm is used for optimizing the parameters of the fuzzy membership functions [24, 42]. The distance between vectors x and y is measured in TWNFI in *weighted normalized Euclidean distance* defined as follows (the values are between 0 and 1):

$$\|x - y\| = \left[\frac{1}{P} \sum_{j=1}^{P} w_j \left| x_j - y_j \right|^2 \right]^{\frac{1}{2}} \tag{12}$$

where: $x, y \in R^P$ and w_j are weights.

To partition the input space for creating fuzzy rules and obtaining initial values of fuzzy rules, the ECM (Evolving Clustering Method) is applied [21, 35] and the cluster centers and cluster radiuses are respectively taken as initial values of the centres and widths of the Gaussian membership functions. Other clustering techniques can be applied as well. A block diagram of the TWNFI is shown in Fig. 7.

5.3.2 The TWNFI Learning Algorithm

For each new data vector x_q an individual model is created with the application of the following steps:

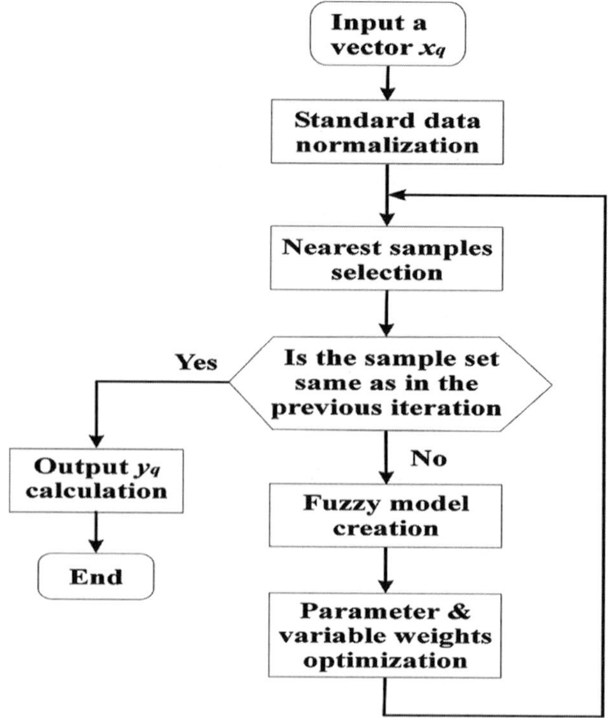

Fig. 7 A block diagram of the TWNFI

1. Normalize the training data set and the new data vector x_q (the values are between 0 and 1) with value 1 as the initial input variable weights.

2. Search in the training data set in the input space to find N_q training examples that are closest to x_q using *weighted normalized Euclidean distance* defined as (1). The value of N_q can be pre-defined based on experience, or - optimized through the application of an optimization procedure. Here we assume the former approach.

3. Calculate the distances d_i, $i = 1, 2, \ldots, N_q$, between each of these data samples and x_q. Calculate the vector weights $v_i = 1 - (d_i - \min(d))$, $i = 1, 2, \ldots, N_q$, $\min(d)$ is the minimum value in the distance vector $d = [d_1, d_2, \ldots, d_{N_q}]$.

4. Use the ECM clustering algorithm to cluster and partition the input sub-space that consists of N_q selected training samples.

5. Create fuzzy rules and set their initial parameter values according to the ECM clustering procedure results; for each cluster, the cluster centre is taken as the centre of a fuzzy membership function (Gaussian function) and the cluster radius is taken as the width.

6. Apply the steepest descent method (back-propagation) to optimize the weights and parameters of the fuzzy rules in the local model M_q following (16–23).

7. Search in the training data set to find N_q samples (the same to Step 2); if the same samples are found as the last search, the algorithm goes to Step 8, otherwise, to Step 3.

8. Calculate the output value y_q for the input vector x_q applying fuzzy inference over the set of fuzzy rules that constitute the local model M_q.
9. End of the procedure.

The weight and parameter optimization procedure is described below:

Consider the system having P inputs, one output and M fuzzy rules defined initially through the ECM clustering procedure, the l-th rule has the form of:

R_l: If x_1 is F_{l1} and x_2 is F_{l2} and $\ldots x_P$ is F_{lP}, then y is G_l.

Here, F_{lj} are fuzzy sets defined by the following Gaussian type membership function:

$$GaussianMF = \alpha \, \exp\left[-\frac{(x-m)^2}{2\sigma^2} \right] \tag{13}$$

and G_l are logarithmic regression functions:

$$G_l = bl0\,x_1^{bl1}\,x_2^{bl2}\ldots x_P^{blp} \tag{14}$$

Using the Modified Centre Average defuzzification procedure the output value of the system can be calculated for an input vector $x_i = [x_1, x_2, \ldots, x_P]$ as follows:

$$f(x_i) = \frac{\sum\limits_{l=1}^{M} G_l \prod\limits_{j=1}^{P} \alpha_{lj} \exp\left[-\dfrac{w_j^2 \left(x_{ij} - m_{lj} \right)^2}{2\sigma_{lj}^2} \right]}{\sum\limits_{l=1}^{M} \prod\limits_{j=1}^{P} \alpha_{lj} \exp\left[-\dfrac{w_j^2 \left(x_{ij} - m_{lj} \right)^2}{2\sigma_{lj}^2} \right]} \tag{15}$$

Here, w_j are weights of the input variables.

Suppose the TWNFI is given a training input-output data pair $[x_i, t_i]$, the system minimizes the following objective function (a weighted error function):

$$E = \frac{1}{2} v_i \left[f(x_i) - t_i \right]^2 \tag{16}$$

(v_i are defined in Step 3)

The steepest descent algorithm (BP) is used then to obtain the formulas for the optimization of the parameters b_{lj}, α_{lj}, m_{lj}, σ_{lj} and w_j such that the value of E from (6) is minimized:

$$b_{l0}(k+1) = b_{l0}(k) - \frac{\eta_b}{b_{l0}(k)} G_l(k) v_i \left[f^{(k)}(x_i) - t_i \right] \Phi(x_i) \tag{17}$$

$$b_{lj}(k+1) = b_{lj}(k) - \eta_b G_l(k) \ln(x_{ij}) v_i \left[f^{(k)}(x_i) - t_i \right] \Phi(x_i) \tag{18}$$

$$\alpha_{lj}(k+1) = \alpha_{lj}(k) -$$
$$\frac{\eta_\alpha v_i \Phi(\boldsymbol{x}_i)}{\alpha_{lj}(k)} \left[f^{(k)}(\boldsymbol{x}_i) - t_i \right] \left[G_l(k) - f^{(k)}(\boldsymbol{x}_i) \right] \tag{19}$$

$$m_{lj}(k+1) = m_{lj}(k) -$$
$$\frac{\eta_m w_j^2(k) v_i \Phi(\boldsymbol{x}_i)}{\sigma_{lj}^2(k)} \left[f^{(k)}(\boldsymbol{x}_i) - t_i \right] \left[G_l(k) - f^{(k)}(\boldsymbol{x}_i) \right] \left[x_{ij} - m_{lj}(k) \right] \tag{20}$$

$$\sigma_{lj}(k+1) = \sigma_{lj}(k) -$$
$$\frac{\eta_\sigma w_j^2(k) v_i \Phi(\boldsymbol{x}_i)}{\sigma_{lj}^3(k)} \left[f^{(k)}(\boldsymbol{x}_i) - t_i \right] \left[G_l(k) - f^{(k)}(\boldsymbol{x}_i) \right] \left[x_{ij} - m_{lj}(k) \right]^2 \tag{21}$$

$$w_j(k+1) = w_j(k) -$$
$$\frac{\eta_w w_j(k) v_i \Phi(\boldsymbol{x}_i)}{\sigma_{lj}^2(k)} \left[f^{(k)}(\boldsymbol{x}_i) - t_i \right] \left[f^{(k)}(\boldsymbol{x}_i) - G_l(k) \right] \left[x_{ij} - m_{lj}(k) \right]^2 \tag{22}$$

here,

$$\Phi(\boldsymbol{x}_i) = \frac{\prod_{j=1}^{P} \alpha_{lj} \exp\left\{ -\frac{w_j^2(k) \left[x_{ij} - m_{lj}(k) \right]^2}{2\sigma_{lj}^2(k)} \right\}}{\sum_{l=1}^{M} \prod_{j=1}^{P} \alpha_{lj} \exp\left\{ -\frac{w_j^2(k) \left[x_{ij} - m_{lj}(k) \right]^2}{2\sigma_{lj}^2(k)} \right\}} \tag{23}$$

where: η_b, η_α, η_m, η_σ η_w and are learning rates for updating the parameters b_{lj}, α_{lj}, m_{lj}, σ_{lj} and w_j respectively.

In the TWNFI training–simulating algorithm, the following indexes are used:

- Training data samples: $i = 1, 2, \ldots, N$;
- Input variables: $j = 1, 2, \ldots, P$;
- Fuzzy rules: $l = 1, 2, \ldots, M$;
- Training epochs: $k = 1, 2, \ldots$.

6 Fuzzy- Neuro Systems for Local and Personalized Modeling: A Real World Case Study on Renal Function Evaluation

A real data set from a medical institution is used here for experimental analysis [36, 37, 38, 39]. The data set has 447 samples, collected at hospitals in New Zealand and Australia. Each of the records includes six variables (inputs): age, gender, serum creatinine, serum albumin, race and blood urea nitrogen concentrations, and one output - the glomerular filtration rate value (GFR).

All experimental results reported here are based on 10-cross validation experiments with the same model and parameters and the results are averaged. In each experiment 70% of the whole data set is randomly selected as training data and another 30% as testing data.

For comparison, several well-known methods are applied on the same problem, such as the MDRD logistic regression function widely used in the renal clinical practice [23], MLP neural network [29], adaptive neural fuzzy inference system (AN-FIS) [12], and a dynamic evolving neural fuzzy inference system (DENFIS) [21], along with the proposed TWNFI. Results are presented in Table 1. The results include the number of fuzzy rules (fuzzy models), or neurons in the hidden layer (MLP), the testing RMSE (root mean square error), the testing MAE (mean absolute error), and the weights of the input variables (the upper bound for the variable normalization range).

Two experiments with TWNFI are conducted. The first one applies the transductive NFI without WDN: all weights' values are set as '1' and are not changed during the learning. Another experiment employs the TWNFI learning algorithm.

Table 1 Experimental results on GFR data

Model	Neurons or rules	Testing RMSE	Testing MAE	Weights of input variables					
				Age w1	Sex w2	Scr w3	Surea w4	Race w5	Salb w6
MDRD (global model)	—	7.74	5.88	1	1	1	1	1	1
MLP (global model)	12	8.44	5.75	1	1	1	1	1	1
ANFIS (global model)	36	7.49	5.48	1	1	1	1	1	1
DENFIS (local model)	27	7.29	5.29	1	1	1	1	1	1
TNFI (personalised)	**6.8** (average)	**7.12**	**5.14**	1	1	1	1	1	1
TWNFI (personalised and weighted variables)	**6.8** (average)	**7.04**	**5.11**	0.82	0.80	1	0.63	0.21	0.36

Table 2 A TWNFI model of a single patient (one sample from the GFR data)

Input variables	Age 58.9	Sex Female	Scr 0.28	Surea 28.4	Face White	Salb 38
Weights of input variables (TWNFI)	0.94	0.83	1	0.52	0.31	0.29
Results	GFR (desired) 18.0		MDRD 14.9		**TWNFI** 16.4	

The experimental results illustrate that the TWNFI method results in a better accuracy and also depicts the average importance of the input variables represented as the calculated weights.

For every patient sample, a personalised model will be created and used to evaluate the output value for the patient and to also estimate the importance of the variables for this patient as shown in Table 2. The TWNFI not only results in a better accuracy for this patient, but shows the importance of the variables for her/him that may result in a more efficient personalised treatment.

The transductive neuro-fuzzy inference with weighted data normalization method (TWNFI) performs a better local generalisation over new data as it develops an individual model for each data vector that takes into account the new input vector location in the space, and it is an adaptive model, in the sense that input-output pairs of data can beadded to the data set continuously and immediately made available for transductive inference of local models. This type of modelling can be called "personalised", and it is promising for medical decision support systems. As the TWNFI

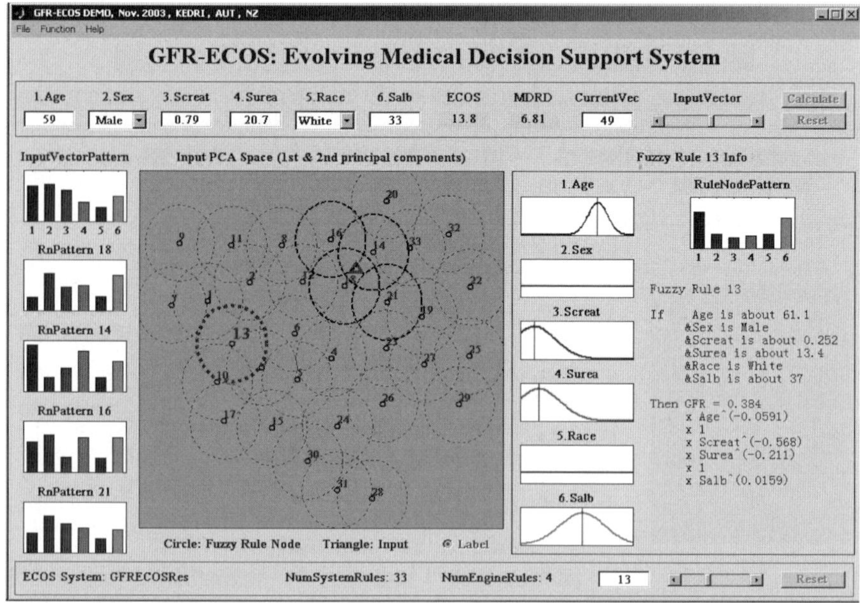

Fig. 8 Local, adaptive GFR Renal Function Evaluation System based on DENFIS: GFR-DENFIS (adapted from [26])

creates a unique sub-model for each data sample, it usually needs more performing time than an inductive model, especially in the case of training and simulating on large data sets.

7 Future Research

Further directions for research include: (1) TWNFI system parameter optimization such as optimal number of nearest neighbours; (2) Transductive feature selection; (3) Applications of the TWNFI method for other decision support systems, such as: cardio-vascular risk prognosis; biological processes modelling and prediction based on gene expression micro-array data [15].

Acknowledgments We are grateful to Dr. Masoud Nikravesh for his invitation to prepare this chapter. The research is funded by the New Zealand Foundation for Research, Science and Technology under grant NERF/AUTX02-01, and the Top Achiever Doctoral Scholarship from the Tertiary Education Commission (TEC) of New Zealand. The authors would like to thank Dr. Mark R. Marshall for his great contributions to the case study of renal function evaluation. The New Zealand National Kidney Foundation under grant GSN-12, and the Auckland Medical Research Foundation also funded the research.

References

1. Albus JS (1975) A new approach to manipulator control: The cerebellar model articulation controller (CMAC), Trans. of the ASME: Journal of Dynamic Systems, Measurement, and Control 220–227
2. Amari S, Kasabov N (1998) Brain-like Computing and Intelligent Information Systems. Springer Verlag, NewYork
3. Arbib M. (1995) The .andbook of Brain Theory and Neural Networks. The MIT Press, Cambridge, MA, U.S.A.
4. Biehl M, Freking A, Holzer M, Reents G, Schlosser E (1998) On-line learning of Prototypes and Principal Components. Sadd D (eds) On-line learning in Neural Networks, Cambridge University Press, Cambridge, UK, pp 231–249
5. Bishop C (1995) Neural networks for pattern recognition. Oxford University Press, Oxford, UK
6. Blanzieri E, Katenkamp P (1996) Learning radial basis function networks on-line In: Morgan Kaufmann (eds) Proc. of Intern. Conf. On Machine Learning, Bari , Italy, pp. 37–45
7. Carpenter G, Grossberg S (1991) Pattern recognition by self-organizing neural networks. Massachusetts. The MIT Press, Cambridge, MA, U.S.A.
8. Chakraborty S, Pal K, Pal NR (2002) A neuro-fuzzy framework for inferencing. Neural Networks 15: 247–261
9. Freeman J, Saad D (1997) On-line learning in radial basis function networks. Neural Computation 9(7)
10. Fukuda T, Komata Y, Arakawa T (1997) Recurrent Neural Networks with Self-Adaptive GAs for Biped Locomotion Robot In: Proceedings of the International Conference on Neural Networks ICNN'97, IEEE Press
11. Furlanello C, Giuliani D, Trentin E (1995) Connectionist speaker normalisation with generalised resource allocation network. In: Toretzky D, Tesauro G, Lean T (eds) Advances in NIPS7 MIT Press, Cambridge, MA, U.S.A., 1704–1707

12. Jang R (1993) ANFIS: adaptive network based fuzzy inference system. IEEE Trans. on Syst.Man, and Cybernetics 23(3) 665–685
13. Kasabov N (1998) A framework for intelligent conscious machines utilising fuzzy neural networks and spatial temporal maps and a case study of multilingual speech recognition. In: Amari S, Kasabov N (eds) Brain-like computing and intelligent information systems. Springer Verlag, Singapore, pp 106–126
14. Kasabov N (1998) ECOS: A framework for evolving connectionist systems and the ECO learning paradigm. In: S.Usui and T.Omori (eds) Proc. of ICONIP'98, IOS Press, Kitakyushu, Japan, pp 1222–1235
15. Kasabov N (2002) Evolving connectionist systems: Methods and Applications in Bioinformatics, Brain study and intelligent machines. Springer Verlag, London, New York, Heidelberg
16. Kasabov N (1998) Evolving Fuzzy Neural Networks - Algorithms, Applications and Biological Motivation. In: Yamakawa, T. and G.Matsumoto (eds) Methodologies for the conception, design and application of soft computing, World Scientific, pp 271–274
17. Kasabov N (2001) Evolving fuzzy neural networks for on-line supervised/unsupervised, knowledge–based learning. IEEE Trans. SMC – part B, Cybernetics 31(6): 902–918
18. Kasabov N (1996) Foundations of Neural Networks, Fuzzy Systems and Knowledge Engineering. The MIT Press, CA, MA
19. Kasabo N, Kim JS, Watts M, Gray A (1997) FuNN/2- A Fuzzy Neural Network Architecture for Adaptive Learning and Knowledge Acquisition. Information Sciences - Applications 101(3-4): 155–175
20. Kasabov N, Kozma R, Kilgour R, Laws M, Taylor J, Watts M, Gray A (1999) A Methodology for Speech Data Analysis and a Framework for Adaptive Speech Recognition Using Fuzzy Neural Networks and Self Organising Maps. In: Kasabov N, Kozma (eds) Neuro-fuzzy techniques for intelligent information systems, Physica, Springer Verlag
21. Kasabov N, Song Q (2002) DENFIS: Dynamic, evolving neural-fuzzy inference systems and its application for time-series prediction. IEEE Trans. on Fuzzy Systems 10:144–154
22. Kasabov N, Woodford B (1999) Rule Insertion and Rule Extraction from Evolving Fuzzy Neural Networks: Algorithms and Applications for Building Adaptive, Intelligent Expert Systems. In: Proc. of IEEE Intern. Fuzzy Systems Conference, Seoul, Korea, pp. 1406–1411
23. Levey AS, Bosch, JP, Lewis JB, Greene T, Rogers N, Roth D for the Modification of Diet in Renal Disease Study Group (1999) A More Accurate Method To Estimate Glomerular Filtration Rate from Serum Creatinine: A New Prediction Equation. Annals of Internal Medicine 130: 461–470
24. Lin CT, Lee CSG (1996) Neuro Fuzzy Systems. Prentice Hall.
25. Massaro D, Cohen M (1983) Integration of visual and auditory information in speech perception Journal of Experimental Psychology: Human Perception and Performance 9: 753–771
26. Marshall MR, Song Q, Ma TM, MacDonell S, Kasabov N (2005) Evolving Connectionist System versus Algebraic Formulae for Prediction of Renal Function from Serum Creatinine. Kidney International 67:1944–1954
27. Moody J Darken C (1988) Learning with localized receptive fields. In: Touretzky D, Hinton G, Sejnowski T (eds) Proceedings of the 1988 Connectionist Models Summer School, San Mateo, Carnegie Mellon University, Morgan Kaufmann
28. Moody J, Darken C (1989) Fast learning in networks of locally-tuned processing units Neural Computation 1: 281–294
29. Neural Network Toolbox User's Guide (2001). The Math Works Inc., ver. 4.
30. Platt, J (1991) A resource allocating network for function interpolation. Neural Computation 3: 213–225
31. Poggio T (1994) Regularization theory, radial basis functions and networks. In: From Statistics to Neural Networks: Theory and Pattern Recognition Applications. NATO ASI Series, No.136, pp. 83–104
32. Rummery GA, Niranjan M (1994) On-line Q-learning using connectionist system. Cambridge University Engineering Department, CUED/F-INENG/TR , pp166
33. Schaal S, Atkeson C (1998) Constructive incremental learning from only local information. Neural Computation 10: 2047–2084

34. Song Q, Kasabov N (2003) Weighted Data Normalizations and feature Selection for Evolving Connectionist Systems Proceedings. In: Lovell BC, Campbell DA, Fookes CB, Maeder AJ (eds) Proc. of The Eighth Australian and New Zealand Intelligence Information Systems Conference (ANZIIS2003), Sydney, Australia, pp. 285–290

35. Song, Q, Kasabov N (2001) ECM - A Novel On-line, Evolving Clustering Method and Its Applications. In: Kasabov N, Woodford BJ (eds) Proceedings of the Fifth Biannual Conference on Artificial Neural Networks and Expert Systems (ANNES2001), Dunedin, New Zealand, pp. 87–92

36. Song Q, Kasabov N (2004) TWNFI – Transductive Neural-Fuzzy Inference System with Weighted Data Normalization and Its Application in Medicine. IEEE Neural Networks. (accepted)

37. Song Q, Ma TM, Kasabov N (2003) A Novel Generic Higher-order TSK Fuzzy Model for Prediction and Applications for Medical Decision Support. In: Lovell BC, Campbell DA, Fookes CB, Maeder AJ (eds) Proc. of The Eighth Australian and New Zealand Intelligence Information Systems Conference (ANZIIS2003), Sydney, Australia, pp. 241–245

38. Song Q, Ma TM, Kasabov N (2004) LR-KFNN: Logistic Regression-Kernel Function Neural networks and the GFR-NN Model for Renal Function Evaluatio. In: Mohammadian M (eds) Proc. Of International Conference on Computational Intelligence for modelling, Control and Automation 2004 (CIMCA2004), Gold Coast, Australia, pp. 946–951

39. Song Q, Ma TM, Kasabov N (2005) Transductive Knowledge Based Fuzzy Inference System for Personalized Modeling. Lecture Notes of Artificial Intelligence 3614: 528–535

40. Takagi T, Sugeno M (1985) Fuzzy Identification of systems and its applications to modeling and control. IEEE Trans. on Systems, Man, and Cybernetics 15: 116–132

41. Vapnik V(1998) Statistical Learning Theory. John Wiley & Sons Inc.

42. Wang LX (1994) Adaptive Fuzzy System And Control: Design and Stability Analysis. Englewood Cliffs, NJ: Prentice Hall

43. Woldrige M, Jennings N (1995) Intelligent agents: Theory and practice. The Knowledge Engineering review (10).

44. Zadeh LA (1988) Fuzzy Logic. IEEE Computer 21: 83–93

45. Zadeh LA (1965) Fuzzy Sets. Information and Control 8: 338–353

Dynamic Simulation of a Supply Chain with and without Flexibility in the Lead Time Agreements

Kaj-Mikael Björk, * Anton Sialiauka and Christer Carlsson

Abstract While supply chain efficiency has remained a vital field of research during several decades, the development of improvement schemes has accelerated greatly. However, schemes like Vendor Managed Inventory are not easily implemented in a supply chain because of the great responsibility the scheme will give to the producer in the supply chain. Other, simpler schemes to improve supply chain efficiency may include Flexible Lead Times (FLT), a concept presented in Björk and Carlsson (2006). The total effect of the improvement scheme needs to be shown in a comprehensive way to establish the concept's advantage. This will be done through a simulation study of tissue supply chain, including a producer and several distributors. The simulation model takes dynamically the orders, the production schedule and the delivery decisions into account simultaneously. Numerical optimization is therefore needed in each simulation step to model the decisions for the producer and the distributors. While the distributor decision model is a simple one and solvable analytically, the producer is faced with a combined production distribution problem that is solved successfully with a genetic algorithm and the solution from the tissue supply chain example will give us insight in the possible savings a FLT concept may provide.

Key words: Supply Chain Management · Simulation · Optimization · Genetic Algorithms

1 Introduction

Supply chain improvement schemes are an essential part of supply chain reengineering projects. For instance, Glosuch (McCullen and Towill 2001), Dell (Kapuscinski et al. 2004) and Philips (Kok et al. 2005) are great examples of how the concepts can be actualized and implemented in a real supply chain. The development of the schemes can be carried out in many ways, but most of them are born intuitively since

---

* Rheinisch Westfälische Technische Hochschule, Aachen / DE-252074 Aachen / Germany / Email: Anton.Sialiauka@rwth-aachen.de

M. Nikravesh et al. (eds.), *Forging the New Frontiers: Fuzzy Pioneers II.*

they often are logical concepts in the given context. The verification of them is much more difficult, however, especially if they are to be analyzed quantitatively before an actual implementation. But the improvement of supply chain performance is of utmost importance to last in the stiff competition today (Houlihan 1985). Any improvement herein will also have great impact on the bottom line result, which evokes the reengineering process conducted in the supply chains. The Bullwhip effect is one major cause for unsatisfactory performance within a supply chain. Forrester (1961) is often referred to as the first source in the literature, where the Bullwhip effect is presented. He investigated the dynamics of the supply chain and specially the expanding variance in the demand patterns upstream in a supply chain. Interestingly enough he found very large demand variations which resulted in both rush-orders and a total drop in demand for the manufacturer. Lee et al. (1997a, 1997b) were the first to show that the Bullwhip effect exists because of the players' rationality. Strangely enough it seems to be so that the individual optimization problem each player carries out is the driver to the problem and not the cure to it (Carlsson and Fuller 1999, 2002).

In order to make the supply chain smoother, simpler, the idea of FLT (Flexible Lead Times) was proposed in a research project called SmartBulls (http://iamsr. abo.fi/smartbulls) to overcome one of the more significant costs from the Bullwhip effect: the variability in the production. FLT means that there is flexibility towards the producer and not towards the customers or distributors. This is contrary to many research papers concerning flexibility and supply chain efficiency: most often contributions in this field of research try to make the production more flexible to adapt to the fluctuating demand. For the specific business context of this paper and in the SmartBulls research project it is not possible, however, because of the capital intensive continuous production facilities that are based on a large-scale chemical process. The concept of FLT is just one way of sharing the flexibility, but proposed in the SmartBulls project because it is very easy to implement. Another way would be to have flexibility in quantities, like in Carlsson and Fuller (1999). However, it seems that flexibility in lead times (of perhaps only a few days) is easier to be accepted by the distributors (the customers of the producers) than uncertainty in quantities. In order to investigate a supply chain improvement suggestion (such as FLT) before an actual implementation, simulation models have been turned to quite frequently (Dejonckheere et al. 2002, 2003, 2004; Higuchi and Troutt 2004). Due to the many and influential sources of stochastic variation and interdependencies within the supply chain, simulation has shown to be a highly effective tool to help practitioners and researchers find operationally and economically optimal solutions (Simchi-Levi et al. 2003). Among the techniques supporting the solving of the multi-decisional problems, as a supply chain is, simulation plays an important role, above all for its main property to provide what-if analysis and to evaluate quantitatively benefits from operating in certain cooperative environments (Terzi and Cavalieri 2004). Therefore, an extensive simulation effort to investigate the effect of FLT was conducted in the last stage of the SmartBulls project.

Many decisions within supply chain management are based on rational assumptions and can be captured with optimization models. These models can be stochastic (Lee et al. 1997a, 1997b), deterministic (basic Economic Order Quantity models,

EOQ) or based on fuzzy numbers (Björk and Carlsson 2005), for instance. The solution approaches for optimization models in supply chain decision making situations have often been analytical, but also numerical approaches are being used extensively. The use of numerical optimization models is necessary, since the complexity of SCM (Supply Chain Management) is high if the models are desired to capture the reality without major simplifications. Therefore genetic algorithms can be used to solve the complex decision making problem, see Björk and Carlsson (2006) for instance. Also, the purpose of the models can be different: prescriptive models to ensure excellence in performance if the suggestions of the models are followed (such as Carlsson and Rönnqvist 2005) or descriptive optimization models such as the ones in Lee et al. (1997a, 1997b). The same optimization models can be used descriptively as well as prescriptively. Especially in a simulation effort, the logical and rational decisions made in a supply chain can be modeled with optimization models. Simulation models are also often using optimization routines in the calculations, especially if the environment is very complex and dynamic. This is the case in many applications, for instance the one presented in Perea-Lopez et al. (2003). They used a Model Predictive Control (MPC) approach to model a supply chain with many distribution channels and combined simulation with optimization in an interesting way. The work in this chapter uses a similar approach, but the optimization models are different and the simulated supply chain is different than the one used in Perea-Lopez et al. (2003). The idea of a rolling-planning simulation-optimization model is not new, but found useful and robust, and therefore this approach was taken also in the simulation of a tissue supply chain presented in the next sections of this chapter.

The SmartBulls project dealt with typical process industry supply chain applications. This context is very different from assembly line industry applications, for which most research contributions regarding flexibility and supply chain efficiency seem to target. Production can be made flexible quite easily in an assembly line compared to the inflexible production plant of the tissue paper. The production efficiency aspect needs to be addressed specifically, dynamically and in a detailed way in the effort of making a paper supply chain efficient, whereas the production and distribution decisions may be decoupled activities in an assembly line based supply chain. Also the market is different for computers than tissue paper; the customer can wait to receive the ordered goods (the computer) whereas the customer of tissue paper seldom wants to wait up to several weeks to receive a bag of toilet paper! The concept of FLT is, in this context, interesting since it will force the distributors in the supply chain to offer some flexibility to the activities, where it is most needed; the production planning. However, the concept needed to be investigated thoroughly in a setting where the production planning, inventory decisions and the distribution is executed dynamically and in a realistic way. Aggregates for the production are not sufficient, but actual production plans need to be considered. Earlier work in this specific context is given by Björk and Carlsson (2006) where the joint production and distribution planning optimization problem was addressed with several numerical models (with rigorous linear programming models and genetic algorithms), i.e. the optimization problem from the producer's perspective. The genetic algorithm was improved in the conference proceedings by Björk and Sialiauka (2006). The distributor's perspective (the customers of the producer) can be modeled according to analytical economic order quantity problems in the conference proceedings by Björk and Carlsson (2005), where the fuzzy optimization problem was solved analytically for the proposed problem. The simulation framework, where both the producer and a number of distributors are modeled simultaneously, is briefly found in the conference proceedings by Sialiauka and Björk (2005) and also found in the Master thesis Sialiauka (2005). But there is no coherent presentation of both the simulation effort with the results and the optimization models used in the simulation model to capture the rational decisions made by the actors of the supply chain. This chapter will give the full picture and present the dynamic link between the optimization models, the simulation effort and the simulation results exemplified with data given for a specific tissue supply chain in one of the Nordic countries. Finally the simulation results will answer the question whether the concept of FLT may be beneficial for an entire supply chain and also quantify the savings.

The outline of this chapter is the following: First the optimization models from the producer's perspective (Model 1 & 2) solved with a genetic algorithm will be presented along with the optimization problem from the distributors' perspective (analytical economic order quantity model, Model 3). Then the simulation framework will be presented with the supply chain structure and the use of the optimization models in the simulation framework will be explained. Finally the simulator will be used to simulate an actual tissue supply chain with one producer and a number of distributors, all distributing several products. This example will show that the concept of FLT can in fact save money for the entire supply chain, if the actors are making rational decisions according to the optimization models presented.

2 The Optimization Models

2.1 The Producer's Perspective

In order to present the rational decision making problem for the producer, we will present two models, one without FLT (meaning the producer is not given any slack or flexibility in the lead time agreement) and one where the order fulfilment do not need to be on time, but is allowed to be either somewhat later than planned or somewhat earlier than planned. In this second model, the producer is given the flexibility to choose the delivery date within the limits of the flexibility agreement. The producer can therefore use the flexibility gained to make more efficient production-delivery plans. In Björk and Carlsson (2006) a set of optimization models were presented for these kinds of problems. The models were solved with a MILP-solver (Mixed Integer Linear Programming), but the solver could not solve the instances of the problems needed in the tissue supply chain example later on, since the size of the problem grew fast. Therefore, the genetic algorithm, presented in Björk and Carlsson (2006), was used to get a good solution rapidly. The problems need to be solved quickly, since they will occur in each simulation step, leaving the MILP-solver unsuitable for the task. First we will present how the combined production planning and inventory management optimization problem without FLT (normal crisp lead times) can be solved with a genetic algorithm. This is referred to Model A2 in Björk and Carlsson (2006), but we will call it Model 1.

2.1.1 Model 1

Given the total time horizon (number of periods), a set of products, a set of distributors, shortage penalty costs as well as unit holding costs for the producer, production setup costs, the orders (demand) from the distributors, maximal production per time period (one product is only produced at one time period) and an extra inventory requirement in the last time period (it is needed because it is not wise to have the optimization procedure drive the inventories to zero in the final time period), we can solve the optimization problem with a genetic algorithm. First the indexes for the problem need to be defined:

$i = [0, 1, 2, \ldots, I]$, time period index
$j = [1, 2, \ldots, J]$, product index
$k = [1, 2, \ldots, K]$, distributor index

The parameters in the problem are the following:

π_j, shortage penalty costs for negative inventory (annual)
Φ_j, unit holding costs for positive inventory (annual)
$\Psi_{j',j}$, production setup cost (per change)
Γ_j, extra inventory in the last time period
$\Delta_{i,j,k}$, demand from the distributors
M_j, maximal production per time period

The variables to be evaluated for each chromosome are the following:

$R_{i,j} \geq 0$, the amount of product j that is produced in period i *(0 or max)*
$x_{i,j}$, the inventory status (both + and -) of product j in period i
$V_{i,j',j}$, whether product j needs to be set up in the production in period i or not

The parameters are given a priori, but the variables need to be obtained from the optimization problem. In order to solve this planning problem we need to define a chromosome structure. Therefore consider the following chromosome.

The chromosome structure in Fig. 1 represents a production schedule for one production line (we focus on a single machine optimization problem) with 5 products in 13 periods. Each gene in the chromosome represents a production for a certain period, i.e. product 1 will be produced the first two periods, and then product 3 will be produced for 3 periods, and so on. Evaluating the chromosome will be relatively easy, since we assume deterministic demand for the producer. A step by step procedure is given in the following. Given the parameters in the beginning of this chapter

Fig. 1 Example of a chromosome

1	1	3	3	3	4	5	5	5	2	2	2	2

and the chromosome in Fig. 1, we can calculate the variable $R_{i,j}$ · $R_{i,j}$ is set to be M_j if the time period (i) and the corresponding gene in the chromosome has a match (i.e. the value in the gene is equal to j) otherwise it is set to zero. Then the inventory status for all products and time periods can be calculated according to

$$x_{i,j} = x_{i-1,j} + R_{i,j} + \sum_{k=1}^{K} \Delta_{i,j,k} \tag{1}$$
$$, \forall j = 1, \ldots, J, i = 1, \ldots, I, x_{0,j} = start\ inv$$

When the inventory position is set (negative inventory means backorders) we can calculate the total cost for the chromosome (i.e. the fitness value) according to

$$lfitness = \sum_{j=1}^{J} \sum_{i=1}^{I} \pi_j \cdot \min(x_{i,j}, 0)$$
$$+ \sum_{j=1}^{J} \sum_{i=1}^{I} \phi_j \cdot \max(x_{i,j}, 0) + \sum_{i=1}^{I} \sum_{j'}^{J} \sum_{j}^{J} \Psi_{j',j} V_{i,j',j} \tag{2}$$

where $V_{i,j',j}$ is a variable set to 1 if there is a change from product j' to j in time i, otherwise it is 0. $V_{i,j',j}$ is extractable directly from the chromosome.

After the fitness of the chromosome is calculated, all chromosomes will be ranked, and both mutation and crossover are allowed. In the examples in this chapter, only a single point crossover operator is used as well as a mutation operator that could be characterized as a "bit inversion" operator. Still, since the chromosome is not composed of binary variable genes, the altered value will be random. For the parent selection rule for the crossover operator, we have found that a good strategy could be to give the best chromosome the altered fitness value of $1/2$, and the second best the value of 1/3, and so on. The chance of a parent being selected is then in proportion to its altered fitness value (linear ranking, Baker 1985). The simplicity of the chromosome makes the genetic algorithm able to find close-to-optimal solutions very fast; in Björk and Carlsson (2006) it was concluded that for a small example problem the genetic algorithm found the global optimum in 5 out of 10 test cases.

2.1.2 Model 2

The Model 2 has the feature added (from Model 1) that there is flexibility in the lead time agreement. Flexibility in lead times means in the example case in the next section that the producer is able to make the delivery one period ahead of schedule, on time or one period after the schedule without extra costs. The decision is the producer's alone. Otherwise the models are similar to each other. Comparison of results between the Models 1 and 2 will give an insight of how the extra flexibility in the lead times will help the producer to find better production schedules and make better inventory decisions. However, in order to model the flexibility with the genetic algorithm above, we need an extended chromosome, where the first part consists of the integers $-1, 0, 1$ (when flexibility of only one time period is considered). The first gene shows how many periods the first incoming order is shifted (assuming that each order is given a number from the sequence $1, 2, \ldots$, the total number of orders for the total time horizon), the second gene shows the second order, and so on. Recall that we have a deterministic demand for the period, and that there are a finite number of orders already placed when the planning task takes place. The second part represents the production schedules, and will be the same as in the previously presented chromosome.

In Fig. 2 an example chromosome is given with 8 incoming orders and a planning horizon of 13 periods. The example chromosome is given only to illustrate the chromosome structure. To evaluate this chromosome we need an extra variable to calculate the altered shipping dates. The continuous variables $D_{i,j,k}$ presents the altered demand of product j in period i for distributor k (after a shift has been done according to the flexibility chosen by the first part of the chromosome in Fig. 2.) Evaluating the second chromosome will differ from the one previously presented. First the actual demand will be altered with shifts presented in the first part of the chromosome; after this the evaluation of the chromosome will be similar to the evaluation of the first chromosome for the Model 1 (the parameter $\Delta_{i,j,k}$ is replaced with $D_{i,j,k}$ in (1). The solution space has increased with the flexibility and the genetic algorithm will need more time to improve the objective value compared to the cases with Model 1. Therefore, the genetic algorithm of Model 2 was improved with a heuristic approach in Björk and Sialiauka (2006). The heuristic was shown to obtain better solutions faster than with a plain genetic algorithm of the Model 2, c.f. Björk and Carlsson (2006).

2.2 The Distributors' Perspective, Model 3

The decision making process for the customers of the producer, the distributors, is less complex, since the production aspects do not need to be considered. It is worth noticing that the decisions of the distributors and the producer are independent of each other. No advanced coordination scheme such, as Vendor Managed Inventory, is suitable according to the tissue producer. The optimization problem for the distributors just concern the time and quantity of the replenishment order

-1	-1	1	1	0	0	1	0	1	1	3	3	3	4	5	5	5	2	2	2	2
				Part 1										Part 2						

Fig. 2 Example of another chromosome

(the distributors are replenishing their inventories when a certain threshold in the inventory level is reached). Therefore the basic EOQ-model is considered here to capture the distributors' optimization problem well enough. Each distributor is facing a stochastic demand with uncertainty in both lead times and the demand, for each product (every distributor has all the products under consideration). The distributors will replenish their inventories according to an EOQ and the time for replenishment will be given by a threshold in the inventory level, given by a safety stock and a lead time demand.

2.2.1 The Order Quantity

The answer to the question of how much to order can be answered with the basic EOQ-formula or some of its extensions (such as the one with back-order costs). The main idea is to capture the right quantity given the trade-off between the inventory costs and the order setup costs. Then the inventory will have a seesaw effect as in Fig. 3.

The optimal order quantity for each distributor is calculated using the classical EOQ-model (Harris 1913), which is given by

$$\Delta_{i,j,k} = \sqrt{\frac{2Delta_{annual,j,k}\Gamma_{j,k}}{\phi_{j,k}}} \tag{3}$$

if the inventory level has reached the threshold otherwise the order amount is 0 for the specific time period. The parameters and variables in (3) are the following: $\Delta_{i,j,k}$ – economical order quantity (which is also an instance of the demand experienced by the producer), $\Delta_{annual,j,k}$ – demand in the simulation period, $\Gamma_{j,k}$ – ordering cost, $\phi_{j,k}$ – unit holding cost. There are several extensions of the basic EOQ-model that capture stochastic demand, fuzzy demand, back-orders, fuzzy lead times (c.f. Björk and Carlsson, 2005, for instance).

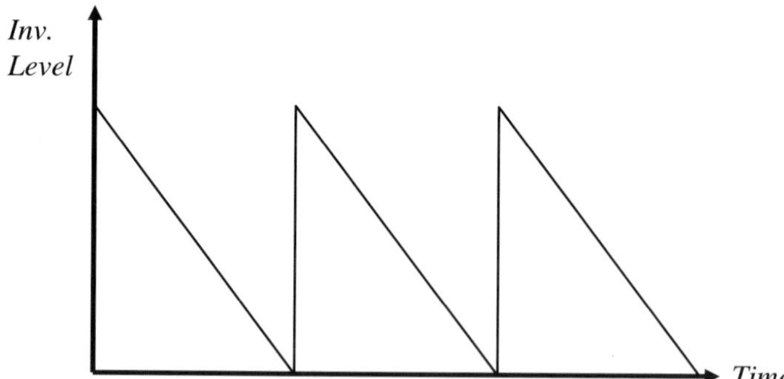

Fig. 3 The representation of the basic EOQ-model

2.2.2 Threshold Level of the Inventory

A periodic review model with stochastic demand and lead times was applied in order to find the reorder point (i.e. to find out when the replenishment decision neds to be done). For each product and distributor, the reorder point is calculated separately and corresponds to the sum of the average demand during the lead time plus some safety stock. At the end of each day this amount is compared with the current system inventory and a decision about the replenishment is made. The system inventory level means an inventory level on hand plus ordered quantity in transit. For the considered stochastic model (no possibilistic model is known for the safety stock calculations to the authors' knowledge today), where both the demand and the lead times are allowed to be uncertain, the reorder point is calculated according to the following formula (Kapuscinski et al. 2004; Stevenson 2002):

$$R = \mu_\Delta \mu_{lt} + SS = \mu_\Delta \mu_{lt} + Z\sqrt{\mu_\Delta{}^2 \sigma_{lt}{}^2 + \mu_{lt} \sigma_\Delta{}^2} \tag{4}$$

, where: μ_{lt} – mean lead time, μ_Δ – mean demand, SS – safety stock, Z – safety factor, $\sigma_{lt}{}^2$ – lead time variance, $\sigma_\Delta{}^2$ – demand variance.

The models for the replenishment decisions for the independent distributors are kept simple. There are several extensions that could be used, for instance the EOQ-model with fuzzy backorders by Björk and Carlsson (2005), but since there is no theory of fuzzy safety stock levels in the literature, the models are kept simple and probabilistic. However, a future research track is to develop a fuzzy safety stock measure, which would enable the use of the models in Björk and Carlsson (2005) in the simulation framework presented in the next section.

3 The Simulation Framework to Investigate FLT

It is important to carry out the optimization problems for the producer as well as the distributors. However, the target with this research effort was not to focus on a single actor in the supply chain, but to investigate the effect of FLT on the supply chain as an entity. Therefore, we need to include the distributors as well as the production of the tissue paper into the model. In addition, several products needed to be modeled, c.f. Fig. 4. In this figure, the circles represent an instance of the demand the distributors experience for a number of products. The distributors are then making replenishment decisions according to some optimal policy (Model 3). The producer knows about this policy and gets information about the demand experienced by the distributors. He uses this information to plan the production and the delivery-schedules (Model 1 - the case without flexibility and the Model 2 - the case with FLT).

The nature of the FLT agreement is such that the producer will benefit from the flexibility gained through them; if the producer is given more degrees of freedom, he will be able to coordinate his actions in a better way. Still the goal with

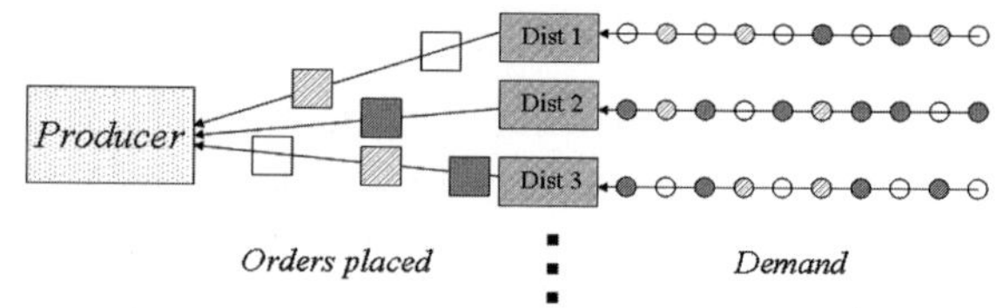

Fig. 4 The simulated supply chain with many products and several distributors

this research effort is to provide the supply chain with solutions that can reduce total supply chain costs and not just move the costs from one echelon to the next. Therefore, it was necessary to show how FLT will affect the distributors; to quantify the increase of the inventory costs, due to the increased safety stock to meet theflexibility used by the producer and if the savings for the producer will compensate the extra costs for the distributors. To study these aspects in a more realistic setting, certain criteria should be fulfilled. The study should be done in a changing environment with uncertainties in demand, which will result in uncertain lead times (other than the possible flexibility agreement under study), due to the limited production capacity at the producer. Several products as well as several distributors need to be included in the study in order to model correctly the outcome of FLT. The simulation study was done with data from a Nordic tissue producer and the study included 8 products (all produced in one machine) and 5 distributors, each distributing all 8 products. With the help of simulation the supply chain is compared when there is a flexibility agreement between the distributors and the producer and when there is not. The customers of the distributors, which are outside of the simulated supply chain, should not experience any difference in service level or price and the total costs for all distributors and the producer is compared for the different cases.

3.1 The Setup of the Simulation Framework

The general context of tissue production was chosen in this particular study for several reasons. First, after discussions with the case company in SmartBulls, it came to light, that the customers of the producer (the distributors) valued especially the service level commitment and the short lead times they have been promised. In fact, two days lead times were offered to most customers regardless of the size of the order. To get some insights in this particular problem, the simulation study was conducted for several cases. However, four basic cases were chosen to be investigated more thoroughly: two days lead time with no flexibility, two days lead time with one day flexibility (in each direction), four days lead time with no flexibility and four days lead time with two days flexibility (in each direction), c.f. Fig. 5.

The other aspects of the supply chain were the same for each of the four different cases. The distributors faced the same (stochastic) demand and applied the same inventory policies for each case: they used order-up-to policies for the safety stock to

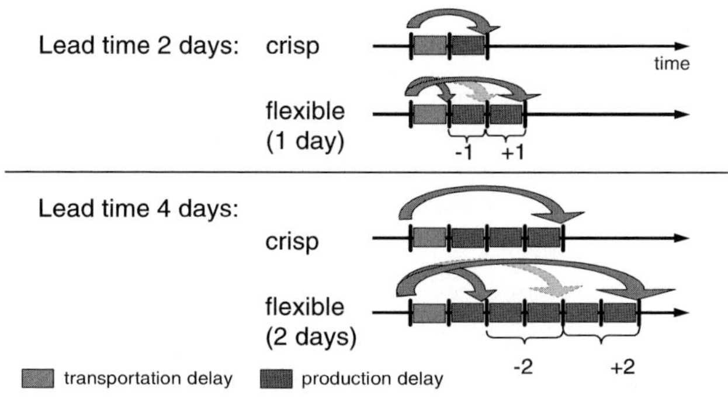

Fig. 5 The four main cases in the simulation study

meet the uncertainties in both demand and lead times described in the previous section. The safety stock was somewhat increased for the flexible lead time cases, since the lead time variances, σ_{lt}^2, were higher. In the simulation experiment they shared both inventory data and the demand data (as well as information about the used order-up-to policy, which in fact can be extracted from the inventory, demand and order data) with the producer. Any other kind of collaboration was not assumed. The producer uses the information to plan both production and delivery schedules some periods ahead. Depending on the case (flexibility or not) this combined production and inventory management problem (with the orders from the distributors and estimated orders based on estimated demand in the future) are solved with Model 1 and 2, respectively. After the optimal solution is found for the production-delivery problem, it is implemented for the next periods. However, the parameters used in the optimization problem become more and more uncertain the further away in the future the planning goes. This fact enforces the need of a new optimization of the schedules whenever updated demand- and order-information is available. Therefore the numerical optimization (production and delivery planning) is performed each time step in the simulation framework according to a rolling horizon approach, c.f. Fig. 6.

3.2 The Simulator

In this simulation framework the use of optimization is inevitable to model smart decision making. The inventory problem of the distributor can be described with safety stock and economic order quantity calculations with analytical solutions (Model 3). The optimal production and delivery problem is harder to model in a similar manner. Numerical optimization is the only way out (Model 1 & 2), even if analytical models would ease up the calculations in the simulation process significantly. The simulation process can be described as follows (i.e. pseudo code for the simulation of one period, day i):

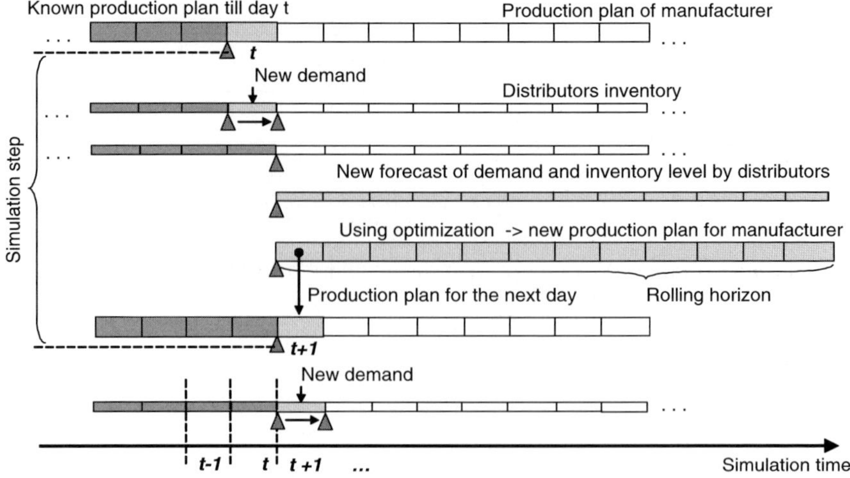

Fig. 6 Simulation framework with rolling horizon optimization

Start the Simulation, step(i)
Get the production plan (i) from the numerical optimization routine
(Models 1 and 2 respectively)
Fill the producer inventory of the product produced (i-1)
Fill the distributor inventory (i) with the products arrived in (i-1)
Meet the demand for the Distributors (i)
Review the inventories of the Distributors (i)
If system inventory level < Reorder point (Model 3)
make new Order with:
order quatity := EOQ (Model 3)
date := i
realization date := i + lead time
put this Order in the in-Transit-Queue
Fulfill the orders according to the solution from the combined produc-
tion and distribution problem as well as the FIFO rule in the in-Transit-
Queue(i)
Calculate the virtual inventory at the producer (negative or positive).
This inventory level is used in the production-distribution optimization
problem for the next step
End Simulation Step (i)

In the pseudo-code for one simulation step, the dynamic link between the optimization problems and how their solutions are used in the simulation framework is
exemplified in the pseudo code above. The setup and the path through the simulation
problem are given by the following pseudo-code:

Start Simulation ( )
Initialize model by setting the time period to 0

Parameterize the simulation model (such as length of simulation etc)
Parameterize the Genetic algorithm (number of iterations, number of chromosomes etc)
Simulate demand for each distributor and product for the entire simulation period by the demand-models extracted from the data given by the tissue producer (5)
Predict the demand and orders for the producer and for the time interval (1, rolling_horizon) (for i=0)
Optimize production plan for the producer and for the first day (for i=1)
Simulate according to the pseudo-code for a simulation step from time-period 1 to the end of simulation
Evaluate total costs and service levels
End simulation

Based on the logic and the object of simulation a set of input parameters has to be declared. The total supply chain cost is a major evaluation parameter in the model; the following cost parameters are included into the cost model:

Distributor costs:

- Unit holding cost
- Unit penalty cost
- Order cost (per order)

Producer costs:

- Unit holding cost
- Unit penalty cost
- Order cost (per order)
- Setup cost matrix (asymmetric)

All these costs are necessary in the optimization problems or in the evaluation of the simulation results. The penalty costs for the producer are not used in the evaluation of the simulation results since the producer and the distributors are assumed as one complete entity in the simulation model. The penalty costs are necessary in the individual optimization problem carried out in each time step, however.

4 The Tissue Supply Chain – Simulation Results

There are quite many local actors producing tissue paper (soft paper) compared to other paper markets (such as the fine paper market). The mills are not necessarily small, but generally taken, well spread out over a large geographical region. The reason is found in the high transportation costs per ton as the density is low. The general conditions for supply chain collaboration in the tissue market are good compared to the fine paper market: significant measures to coordinate actions have been taken already and the distributor and retailer have quite advanced ICT-capabilities. Therefore it may be assumed that information sharing is possible. For instance, some of the supermarkets are equipped with POS data handling, which makes it possible to collect real sales data in the distributors' databases. The coordination of the actions taken among the actors in the supply chain is a tougher issue, however. We believe that FLT is a change in the right direction to, not only share information, but also create smart action plans accordingly.

The simulator presented in the previous section is useful, not only for the investigation of FLT, but also for other kinds of "what-if" analysis. In this section the focus will be set on the effect of FLT, but doing so, other aspects, such as capacity utilization, demand variation and production setup cost structure need to be taken into account. Therefore, a somewhat detailed presentation of the results from the simulation experiments is given. Also, once the simulator had been built, it is of course preferable that it will be used to give a rigorous and complete picture of the supply chain under study (the Nordic tissue supply chain). Therefore many simulation runs have been made for slightly different settings and in the end the final conclusions will be drawn to answer the question whether FLT is beneficial for the entire supply chain. All examples reflect the results of a simulation period of 130 days for the four lead time cases described earlier. The capacity utilization for the producer is assumed to be 92% for all cases. The first analysis is given in Fig. 6, where the results for all lead time cases are given for different demand variations. This sensitivity analysis is interesting, since there are significant differences in the demand variations for different products. But first let us consider how the demand (towards the distributors, note that the demand to the producer is given by individual orders from the distributors) is modeled. The distributors are assumed to face a stochastic demand given by the equation:

$$D_i = d + \rho D_{i-1} + u_i, \tag{5}$$

where D_i is the demand at time i, d is the solid part of the demand, ρ – a connection factor between the different periods (smoothing parameter), u – stochastic demand variation (this is the basic demand model, which is an AR(1)-model, used in many research papers, c.f. Lee et al. 1997a, 1997b; Lee et al. 2000). The parameters in (5) are obtained from real distributor demand data. The notation of 100% corresponds to a standard deviation of the mean demand of 58%, because 100% denotes the variation for the uniform distribution of the stochastic variable u in (5). This demand variation (100%) is quite common for the considered context, but since the demand variation is not the same for all products, simulation runs with different demand variations were done. As one can expect the minimum costs are accomplished by 0% demand variation (steady demand) and maximal flexibility in lead time as well as the longer (4 days) lead time case. In Fig. 7, this minimum cost value is set as the reference level 100%. The results in Fig. 7 are interesting in the sense that for all simulated demand variations, except 0% (which is an unnatural case for the real supply chain), the flexibility of one day within the lead time of two days will give savings in the total supply chain costs of at least 5 %. Flexibility in

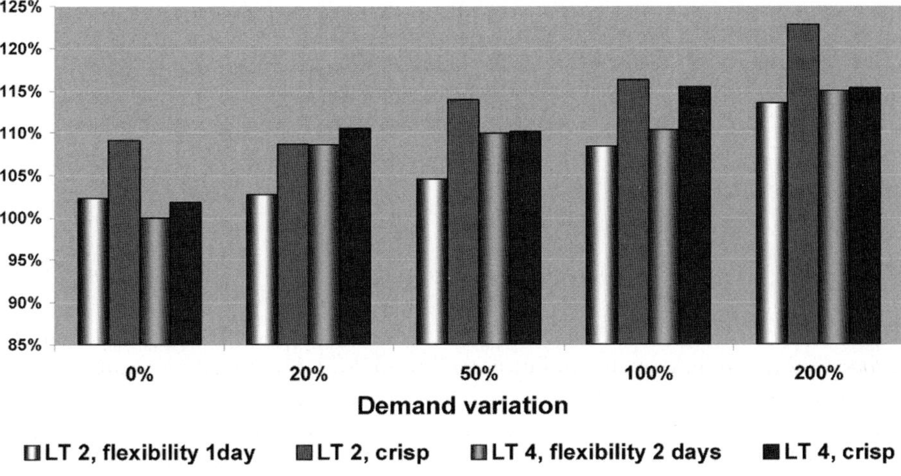

Fig. 7 The total supply chain costs by different demand variations

the four days lead time case does not reduce the total costs significantly, however. It is also worth noticing that an increase of the lead times from 2 to 4 days can be profitable within the fixed lead time framework: by the greatest demand variation – case 200 % - the total supply chain costs are less for the longer lead time.

From Fig. 7 it can be concluded that the lead times should be held short, but flexibility should be introduced in the lead time agreements (except for the steady demand case). The general expression from the simulation runs were that if there is flexibility in the lead time agreements, there will be fewer production setups. In a specific simulation run (compare Figs. 8 to 9, the red dots), the number of times

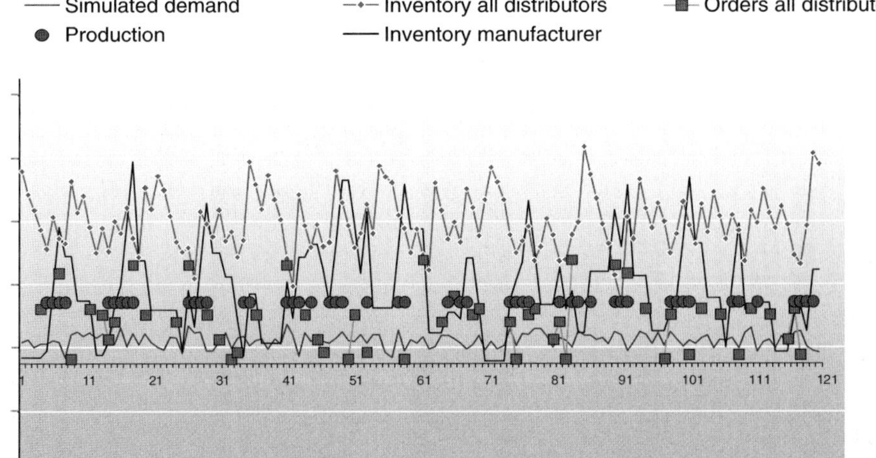

Fig. 8 An example of the simulation supply chain dynamics for one product, case crisp lead times (2 days), demand variation 100 %

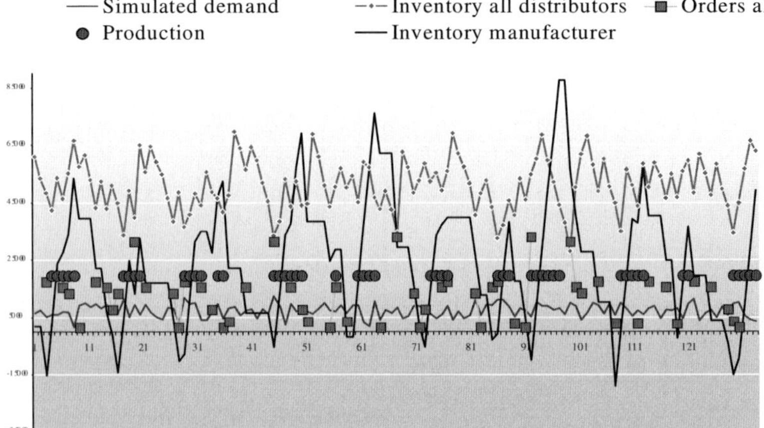

Fig. 9 An example of the simulation supply chain dynamics for one product, case FLT (one day in each direction) lead times - 2 days, demand variation - 100 %

a specific product was produced was reduced from 20 to 13. Still the same amount of the product was produced, but in longer batches. From the examples in Figs. 8 and 9 it can also be noticed that the FLT will increase the inventory level for the distributors (in Figs. 8, 9, the notation "all distributors" is used for the sum of all distributor inventories in the distributor echelon). Still, the increase was moderate.

Since the number of production setups are, in general, decreasing when FLT is introduced, the overall production setup costs will be reduced. This was also a trend in most simulation runs made. At the same time the inventory costs for the distributors were somewhat increased. In Fig. 10, a comparison of the different cost structures for some simulation runs are made. For the 2 days lead time case we found that the savings were only in the production setup costs and that the inventory costs are increased both at the distributors and the producer. For the 4 days lead time case the savings from the production setup costs does barely cover the increase in

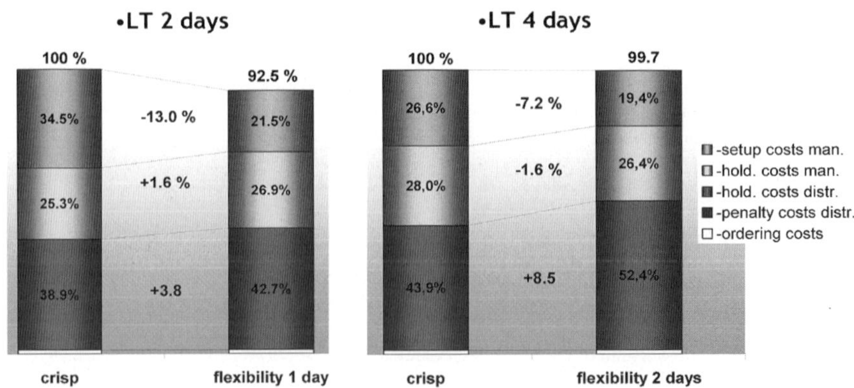

Fig. 10 Cost structure in percentage of the total supply chain costs, Case: demand variation 200%

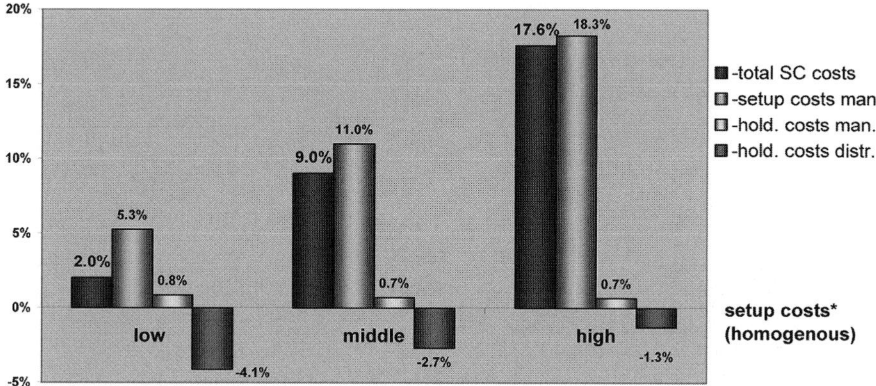

***- low - 33%, middle - 100%, high - 300% of maximum of current setup costs**

Fig. 11 Savings in the total supply chain costs by the different setup costs scenarios, demand variation - 100 %

theinventory costs. It is also worth noticing that for this specific case, the flexibility in the lead time agreements were not significantly improving the total costs for the LT 4 days' case.

The results given in this section have been based on a setup costs matrix given by the Tissue producer. This setup matrix is heterogeneous and asymmetric, meaning that the cost to change the production from product 1 to 2 need not be the same as the cost of changing the production from product 2 to 1. In addition, the variations in setup costs are high. In order to investigate the sensitivity of the total cost savings given different setup costs scenarios, some extra simulation runs were made (c.f. Fig. 11). Through the experiments with a genetic algorithm it was recognized that with a symmetric and homogenous setup matrix the reliable results in the near of optimum can be reached in a few simulation runs. Such a setup matrix was used in order to check the former results and also to investigate the impact of increasing setup costs on savings through the flexibility. The middle case in the following figure represents the smoothened current situation and gives approximately the same results as were given before. Further, the increase in the setup costs of the value of 200% can give the savings for the whole supply chain up to 18%. The setup cost analysis is interesting since the trend in the paper industry is to build larger, more expensive and more efficient paper machines. This trend will increase the importance of the reduction of the costly setup time, making the flexibility concept even more important for future applications.

Figure 12 shows, in a more visible way, the estimated savings due to flexibility for both lead time cases (2 and 4 days), respectively. The savings, if FLT are implemented for the 2-days lead time case, are more significant than the savings given by FLT in the 4-days lead time case. In the chart 100% corresponds to the total supply chain costs with the fixed lead times respectively. It is worth noticing that in all cases, the impact of introducing flexibility in the lead time agreement will reduce the total costs, even if the reduction is insignificant for some specific cases. Also

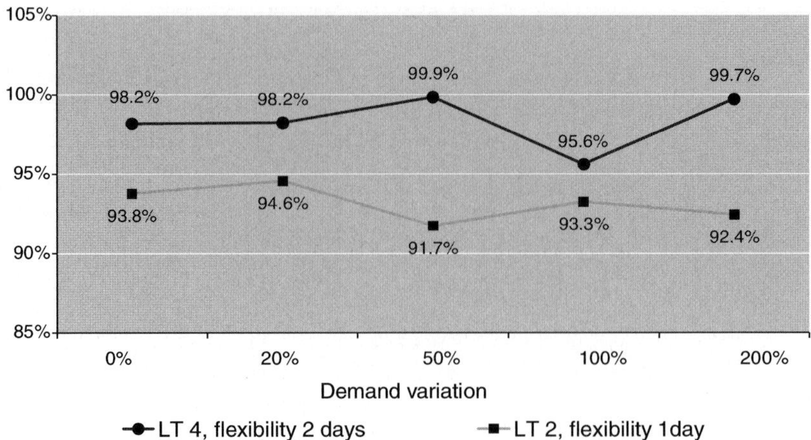

Fig. 12 The ratio of the total supply chain costs in the flexible cases compared with the crisp cases

the additional conclusions that could be drawn from these experiments were that, in general, the lead times should be kept short (which is contrary to what we predicted in the beginning of this project), but the lead times should indeed be flexible towards the producer. The savings in this study was then on average of 7.5% when the capacity utilization was fairly high. In addition, due to the sensitivity of the production setup costs, other forest industry applications may experience an even higher degree of cost reductions with an appropriate FLT-implementation. Fine paper machines, for instance, are more sophisticated (and expensive) than tissue paper machines. At the point when this study was conducted, the time- and data availability restrictions made it impossible to investigate the fine paper supply chain further. This track is left for future research.

5 Conclusions and Future Research

While SCM and supply chain efficiency remain to be appealing topics of research, the means of investigation have become more sophisticated during the last decade. The necessity to improve the performance of the supply chains is undoubtedly an issue that many managers, directors and researcher dwell over today. The way of investigating proposed improvement schemes may vary, but in order to find better credibility, researchers today tend to lean on mathematical modeling instead of a pure conceptual modeling approach. Mathematical models have the advantage of giving some insight about the efficiency of an improvement scheme before it is implemented in a real supply chain. Therefore, mathematical models were also used in this research work to investigate a new improvement scheme, not applied anywhere before to the authors' knowledge; the idea of flexibility in the lead time agreements. The idea was tested in a comprehensive simulation-optimization framework and an example of the efficiency of FLT was shown in a tissue producing context. The simulation framework are using both numerical optimization with a genetic algorithm and a analytical optimization model to describe the rational decision making processes in the simulated two-level supply chain. Production aspects as well as inventory aspects must be considered simultaneously and the simulation results for the tissue supply chain showed that the concept of FLT introduce a better balance, reducing the total costs with about 7.5 % for a specific case. The advantage with the new concept is that it would be relatively easy to implement in comparison to other initiatives such as VMI (Vendor Managed Inventory).

Even if the results in this research work give a promising and interesting answer to the question regarding FLT, much remains to do. In fact, many questions have been raised during the process, of which the perhaps most important one is how to share the wealth gained by better production planning through the flexibility agreements. In the suggested framework, the simulation results indicate that the distributors jointly will pay for the gained flexibility through increased inventory costs but the gain for the producer will be greater than the increase in cost for the distributors. The question is how the producer would motivate the distributor to share some of their flexibility. One idea would be to have the producer to purchase the flexibility for a fixed price for a given range of slack. This scenario is very close to the assumptions made in this simulation study. An interesting alternation of the FLT concept would be that the distributor would supply the producer with the necessary flexibility, but the producer would need to pay a premium for each occurrence of the realized flexibility. In other words, the producer could purchase both the possibility to make early shipments and late shipments for a predetermined price per day flexibility. This would imply an altered version of the optimization problem for the producer, where the used flexibility comes with a penalty cost. It would be an interesting version, since the sharing of the wealth in the supply chain would come very naturally with these models. A further improvement, where the price would be determined dynamically, is also very appealing, because of the fact that the distributor occasionally will have extra flexibility to share cheaply (for instance in a situation, where the distributor have enough inventory to accept a late shipment) or when the flexibility need to be shared expensively (no inventory at hand). The list of possible improvements is endless; especially if adaptive, collaborative and automated supply chain improvements need to be considered. Evolving systems where the flexibility could be determined and controlled by transparent multi-agent systems balancing the objectives and the optimization problems for every member in the supply web, with many actors and products are surely a challenge to create, maintain or develop, but probably the right way to go in the future research efforts regarding supply chain efficiency.

References

Baker JE (1985) Adaptive Selection Methods for Genetic Algorithms. Proceedings of an International Conference on Genetic Algorithms and their Application, Hillsdale, New Jersey, USA: Lawrence Erlbaum Associates. pp. 101–111

Björk K-M. Carlsson C (2005) The Outcome of Imprecise Lead Times on the Distributors, Proceedings of the 38th Annual Hawaii International Conference on System Sciences (HICSS'05) - Track 3. pp 81–90

Björk K-M. Carlsson C (2006) The Effect of Flexible Lead Times on a Paper Producer. Accepted with conditions to International Journal of Production Economics

Björk K-M. Sialiauka A (2006) A Heuristic to Improve the Solution of a Combined Production and Inventory Planning Problem. Accepted to IPMU 2006

Carlsson C. Fuller R (1999) Soft computing and the Bullwhip effect. Economics and Complexity 2: pp. 1–26

Carlsson C. Fuller R (2002) A Position Paper for Soft Decision Analysis. Fuzzy Sets and Systems 131: 3–11

Carlsson D. Rönnqvist M (2005) Supply chain management in forestry – case studies at Södra Cell AB. European Journal of Operational Research 163: 589–616

Dejonckheere J. Disney SM. Lambrecht MR. Towill DR (2002) Transfer Function Analysis of Forecasting Induced Bullwhip in Supply Chains. International Journal of Economics 78: 133–144

Dejonckheere J. Disney SM. Lambrecht MR. Towill DR (2003) Measuring and Avoiding the Bullwhip Effect: A Control Theoretic Approach. European Journal of Operational Research 147: 567–590

Dejonckheere J. Disney SM. Lambrecht MR. Towill DR (2004) The Impact of Information Enrichment on the Bullwhip Effect in the Supply Chains: A Control Engineering Perspective. European Journal of Operational Research 153: 727–750

Forrester J (1961) Industrial dynamics. Cambrigde, MA: MIT Press

Harris FW (1913) How many parts to make at once. The Magazine of Management 10: 135–136

Higuchi T. Troutt MT (2004) Dynamic Simulation of the Supply Chain for a Short Life-Cycle Product – Lessons from the Tamagotchi Case. Computers and Operations Research 31: 1097–1114

Houlihan J (1985) International Supply Chain Management. International Journal of Physical Distribution and Materials Management 15: 22–38

Kapuscinski R. Zhang RQ. Carbonneau P. Moore R. Reeves B (2004) Inventory Decisions in Dell's Supply Chain. Interfaces 34: 191–205

Kok T. Janssen F. van Doremalen J. van Wachem E. Clerkx M. Peeters W (2005) Philips Electronics Synchronizes Its Supply Chain to End the Bullwhip Effect. Interfaces 35: 37–48

Lee H. Padmanabhan V. Whang S (1997a) Information Distortion in a Supply Chain: The Bullwhip Effect. Management Science 43: 546–558

Lee H. Padmanabhan V. Whang S (1997b) The Bullwhip Effect in Supply Chains. Sloan Management Review 38: 93–102

McCullen P. and Towill DR (2001) Practical ways of reducing bullwhip: The case of the Glosuch Global Supply Chain. Control December/January, pp. 24–39

Perea-Lopez E. Ydstie E. Grossmann I (2003) A model predictive control strategy for supply chain optimization. Computers and Chemical Engineering 27: 1201–1218

Sialiauka A (2005) The Impact of Flexible Lead Times in a Forest Industry Supply Chain: A Simulation Study. Master thesis for Rheinisch-Westfälische Technischen Hochschule, Aachen

Sialiauka A. Björk K-M (2005) The Impact of Flexible Lead Times in a Forest Industry Supply Chain: A Simulation Study. Conference proceedings for Modeling Simulation and Optimization, pp. 73–78

Simchi-Levi D. Kaminsky P. Simchi-Levi E (2003) Designing and Managing the Supply Chain: Concepts, Strategies, and Case Studies, 2nd edition, McGraw-Hill, New York

Terzi, S. Cavalieri S (2004) Simulation in the supply chain context: a survey. Computers in Industry 53: 3–16

Robust Statistics and Fuzzy Industrial Clustering

Barbara Diaz and Antonio Morillas

Abstract The search for groups of important sectors in an economy has been and still is one of the more recurrent themes in input-output analysis. But a sector can probably be important for some questions at the same time, to a different degree. In this direction, a multidimensional fuzzy clustering analysis gives as a result a classification of sectors illustrating the different roles that each one plays in the economy. But multivariate outliers, witch have not been studied in the literature in previous applications of clustering techniques to input-output studies, have even worst effects in clustering than the univariate ones, due to their influence on the correlation matrices, and because their presence can mask the real clusters. We will show how fuzzy clustering and robust statistics should work together in this kind of studies, so the clustering can benefit from the use of robust statistics in data preparation, identification and computation of dissimilarities or deciding the best number of clusters and specially avoiding the dangerous effects coming from the presence of multivariate outliers.

1 Introduction

The search for the important sectors in an economy started, within the input-output analysis, in a very elementary way. The buy and sell direct relationships among sectors or the use of the multiplier based on the inverse of Leontief's model would not be considered nowadays sufficient criteria to identify key sectors. Nevertheless, Hirschman (1958) considered the seminal work by Chenery and Watanabe (1958) and Rasmussen (1956) as a way to measure what he called linkages in his development theory. Probably the discussion over causality in economics (Simon, 1952; Wold, 1954) contributed to the idea that linkages effects, using causality or hierarchy as a criterion, were a determinant of the economic relevance of a sector.

However, in most of recent works one can already glimpse the importance of the industries networks development, where a growth impulse does not so much follow from the hierarchy of autonomous driving sectors but rather from the recursive interdependence of clusters or groups of industries. The studies on industrial clusters are almost as numerous as diverse in their considerations on what should be taken into

M. Nikravesh et al. (eds.), *Forging the New Frontiers: Fuzzy Pioneers II.*
© Springer-Verlag Berlin Heidelberg 2008

account. The seminal works by Czamanski (1974) and, subsequently, Porter (1990) caused a growing interest in industrial clustering from different areas of knowledge. The applications in the input-output context have made use of very different analysis methods including some multivariate statistical techniques as well.

Regarding these last ones, one of the first applications of multivariate analysis to study industrial groups through an input-output table was Czamanski (1974), in which a principal component analysis is carried out on the US input-output table for 1963, on the basis of their buy and sell profiles, with the aim of grouping sectors. Roepke et al. (1974) made a similar application to the Ontario (1965) tables. Latham (1976) and Harrigan (1982) criticised these methods considering that they only grouped vertically linked sectors. Ó hUallacháin (1984) proposes a posterior analysis of the components found in the inter-industrial flux matrix, to verify the complementarities of the grouped sectors, supporting the suitability of these statistical methods. Feser and Bergman (2000) differentiate in a group what they consider key and secondary sectors, attending to their factorial load,[1] in an application that aims to identify the industrial clusters in the manufacturing sector of the US, using the 1987 table and Czamanski's coefficients. In later works Feser et al. (2001a, b) carry out a study of the spatial concentration of employment for each group.[2]

Rey and Mattheis (2000) elaborate a consensus methodology to combine the results of some principal components and clustering techniques and incorporate, for the first time, a non-hierarchical clustering to determine industrial clusters. Lainesse and Poussart (2005) introduce a hybrid proposal, based on the former one, which adds a supplementary analysis of indicators with the aim of including or excluding sectors according to their complementarities.

Lately, fuzzy clustering[3] has been applied, looking for similarity between sectors (Dridi and Hewings, 2003) or identifying key sectors from a strategic, dynamic and multivariate approach (Díaz, Moniche and Morillas, 2006). Fuzzy clustering relaxes the grouping made and gives to each element a membership value to each cluster, allowing an enrichment of posterior analysis.

In this last work, a relational fuzzy clustering method allowing the use of ordinal and nominal variables was applied, and it was pointed out that most of the sectors with difficulties to be classified could be considered as outliers. Furthermore, it was realized that the mega clusters usually obtained as a result in this kind of studies

---

[1] It could be considered the first antecedent of our proposal to use fuzzy clustering to get non-mutually exclusive groups.

[2] A good summary of these works is given by Lainesse and Poussart (2005).

[3] Fuzzy cluster analysis has its origins in Ruspini (1969) In Zadeh (1977) there is an outline of a conceptual framework for cluster analysis and pattern classification based on the theory of fuzzy sets, which he initiated in his seminal work, Zadeh (1965), since at that time there wasn't still a unified theory for them. Bezdek (1981) subsequently generalized Dunn (1973) work, introducing fuzzy c means. In Kaufman and Rousseeuw (1990) an objective function in which the dissimilarity or distance measure is a L1 norm (not squared) is used, which makes it more robust than fuzzy c means (Rousseeuw, 1995). Hathaway et al. (2000) present a generalization of fuzzy c means to allow the use of different distances (norms).

had never been considered related with the existence of these outliers. The masking effect of the outliers could be causing the low number of clusters observed in the literature from different authors.

In this work a study of multivariate outliers is carried out, in order to evaluate the results from a previous fuzzy industrial clustering. Given the distortions that all these outliers cause in the clustering procedure, multivariate detection of outliers is considered a necessary previous step for any multivariate study carried out in an input-output environment.

2 Antecedents and Previous Results

Hathaway *et al.* (1989) show that there is a dual relational problem of object fuzzy *c* means for the special case in which the relational data coincide with the Euclidean distances among their elements. Their method is called RFCM (relational fuzzy *c* means). Later, Hathaway and Bezdek (1994) introduce a non Euclidean relational fuzzy clustering (NERFCM). Their objective is to modify RFCM to be effective with arbitrary relational data that show dissimilarities among objects. In this direction, the fuzzy analysis clustering in Kaufman and Rousseeuw (1990) can be used with both kinds of input data, since the distances in the objective function are distances among objects and not to the centres of the clusters. Dave and Sen (2002) propose another robust version of fuzzy analysis clustering called robust non Euclidean fuzzy relational clustering (robust-NE-FRC).

In NERFCM, the objective function to minimize is:

$$\min \sum_{i=1}^{k} \frac{\sum_{j=1}^{n} \sum_{l=1}^{n} u_{ij}^m u_{il}^m r_{jl}}{2 \sum_{t=1}^{n} u_{it}^m}$$

where N is the number of cases (sectors), $m > 1$ is a parameter controlling the amount of fuzziness in the partition, k is the number of clusters, u_{ij} is the membership of case j to cluster i, with $u_{ij} \in [0, 1]$ and $\sum_{i=1}^{k} u_{ij} = 1$. Memberships cannot be negative and each sector has a constant total membership, distributed over the different clusters. By convention, this total membership is normalized to one, and r_{jl} is the relational data measuring the relationship between cases j and l.

The algorithm for the NERFCM proceeds in several steps. In a first stage the membership matrix, for a given number of clusters k, is initialized. Then, with those memberships, the *mean vectors*[4] for each one of the k clusters are determined. In a third step, distances for each element to the *mean vector* of each cluster are calculated. It is in this step where a modification from RFCM takes part, so that these

---

[4] It is basically calculated as the proportion of the membership of each element j (elevated to m) to cluster i in relation to the sum of memberships (elevated to m) of all elements to cluster i.

distances are not negative[5]. In a fourth step, the memberships are updated. This process is carried out until a convergence rate in the membership matrix is reached.

The variables used, their statistical sources and respective definitions, are shown in Annex 1. Three concepts or blocks are distinguished. The economic integration block contains 4 variables. It captures both the intensity and the quality of the sectoral relationships. The variables are: the multiplier derived from the relative influence graph,[6] its dispersion as measured by the variation coefficient, the cohesion grade[7] and a topologic integration index,[8] obtained for each sector. The last two variables expressed qualitative aspects related with the relative position of a sector in the exchange structure. The economic weights block (with 4 variables) summarized the relative participation in the national output, in exports and in wage rents and employment generation. The economic potential block (with 5 variables) included a dynamic vision of the former variables through the observed trends plus the technological innovation capacity along with the information given by Eurostat(1998a, b) and the high technology indicators of the Spanish Statistics National Institute (INE) given in Annex 2.

The base for the study was the last available Spanish 1995 input-output table. The fact that its initial 70 branches (sectors) were reduced to 66 in that work were the inexistence of technology data for the first three branches, which belong to the primary sector and the inexistence of relationships with the other sectors of branch 70 (*Homes with domestic personnel*).

Several previous steps were necessary to take before applying the NERFCM algorithm. The first one was a pre-processing of the variables, while step two provided the relational matrix on which the NERFCM operated. Step four is fulfilling the need for an initialization of the membership values for the NERFCM objective function, which is carried out using a robust crisp clustering method. These four steps are exposed in more detail in the next paragraphs.

As a first step, the subjacent principal components were extracted, both to avoid the possible effects of multicolineality on the similarity matrix, and to provide a

---

[5] This modification is active whenever some negative value for the distances is encountered,, allowing the use of non-Euclidean matrices. This is the difference between RFCM and NERFCM. For details on this modification called β-spread transformation, see Hathaway and Bezdek, (1994), p.431. For a detailed description of NERFCM see pp. 433–434.

[6] This multiplier comes from the relative influence graph associated to the distribution coefficients matrix and can be interpreted in terms of elasticities (Lantner, (1974), (2001); Mougeot et al., (1977); Morillas, 1983a; and de Mesnard, 2001). Its expression is: $\hat{x}^{-1}\Delta x = (I - D)^{-1}\hat{e}\hat{y}^{-1}\Delta y$. Since it is weighted by the export rate from outside the graph structure ($\hat{e}$ = final demand coefficient), it could be said that it is very close to the net multiplier mentioned by Oosterhaven (2002, 2004). The interindustry matrix has been deprived of self-consumption, to reinforce the interdependencies analysis.

[7] The cohesion concept in an economic model was first introduced by Rossier 1980 and applied for the first time to input-output analysis by Morillas 1983a.

[8] See Morillas 1983b. Extending the global indicator proposed in that work, the integration index for a given sector is defined as $R_i = \frac{1}{2n-1} \sum_{\substack{j=1 \\ i \neq j}}^{n} \left(\frac{1}{e_{ij}} + \frac{1}{e_{ji}} \right)$ where e_{ij} is an element of the distances matrix.

better conceptual identification of the resulting clusters, reducing the number of variables for the classification task. In the rotated solution three components with an eigenvalue slightly greater than one explained about 88% of the variance (41%, 24% and 23% respectively). Considering only the more important factor loadings, the first of these components is mainly integrated by the variables multiplier, its dispersion index and the sector output. This component was termed *economic relevancy*. The second component (to be termed *integration*) mainly considered the two qualitative measures based on graph theory, i.e. cohesion and productive integration indexes. The third component, with exports and their trend as the more important variables, was identified as *export base*. The other variables seemed not to be of sufficient importance (or they were redundant) to explain the data cloud variability.

With these three numerical variables and the ordinal variable (termed *technology*) the fuzzy analysis clustering was carried out. Since NERFC operates on relational data, for the calculus of the distance matrix among sectors involving the principal components in step 1 (quantitative variables) and the technology variable (ordinal variable) the *daisy* method proposed by Kaufman and Rousseeuw (1990) was used. This method computes all pair wise dissimilarities among objects, allowing the use of different kind of variables.

Finally, the NERFC algorithm needed an initialization for the membership matrix. The results provided by the crisp *partitioning around medoids (pam)* clustering method (Kaufman and Rousseeuw, 1990), which was used in Rey and Mattheis (2000) were used as seeds[9]. This last one is a robust version of the well-known k means method (MacQueen, 1967) and it is less sensible to the presence of outliers. The *pam* method selects k representative or typical objects called *medoids* among the observations. A medoid is an object of a cluster whose average dissimilarity to all the objects in the cluster is minimal. Each observation is assigned to the closest medoid. The k medoids selected should minimize the objective function, the sum of the dissimilarities of all objects j to their nearest medoids m_i[10]. The *pam* method sequentially selects k centrally located objects, used as initial medoids. Then, it computes the swapping cost of interchanging a selected object with an unselected one. If the objective function is reduced with the new medoid, and this is done until there is no benefit in changing the latest medoids.

To select the best number of clusters k for the partitioning around medoids method the silhouette width method, proposed in Rousseeuw (1987)[11] was used.

---

[9] The results from the partitioning around medoids method are used to initialize the membership matrix assigning for each element j a membership equal to one for the cluster where it belongs and a membership of zero to the rest of clusters. It is thus a binary matrix with zeros and ones.

[10] Since the matrix we are using is already a dissimilarity matrix, these are the distances used within the algorithm. In other cases the metric to compute these distances needs to be specified.

[11] The silhouette width for a given object (sector) j is calculated as the difference between the minimum average distance of j to objects in other clusters (between dissimilarity, b_j) and the average distance between j and all the other observations in its cluster i (within dissimilarity, aj), divided by the maximum between aj and bj. The silhouette width has values between - 1 and 1, $-1 \leq s_j \leq 1$. If $s_j = 1$, the within dissimilarity is much smaller than the between dissimilarity, so the object j has been assigned to an appropriate cluster, if $s_j = 0$, it means $a_j \approx b_j$ and it is not clear whether j should be assigned to A or B. It can be considered as an intermediate case. If $s_j = -1$, object j is

The best option given by the silhouette plot in the PAM method had three clusters with 32, 23 and 11 sectors respectively. This option was subsequently used as an input into NERFCM algorithm. The average silhouette width indicated an ill-delimited classification. There was a lot of fuzziness, which prompted fuzzy clustering.

A membership value equal to or greater than 0.5 in a cluster was considered as indicative of the priority (although not exclusively) of classifying such a sector to a certain cluster. In this situation were 24 sectors of cluster 1, 15 of cluster 2, and 9 of cluster 3 (see Table 1). Even accepting this threshold, it means that the other 18 sectors (i.e. more than 27% of the sectors) could be considered as sectors that could be classified simultaneously from different perspectives. This reasonable result would not have been obtained if any other crisp (as opposed to fuzzy) clustering technique[12] was used.

Table 1 Clusters and Membership values

CLUSTER	Membership ≥ 0.5			Membership < 0.5			
	Sectors			Sectors	Membership	Alternative cluster	Membership
Cluster 1	5	18	43	58	0.4993	Cluster 2	0.3769
Local base:	6	19	46	48	0.4744	Cluster 2	0.4175
– Economic Relevancy	7	21	51	23	0.4619	Cluster 2	0.3897
– Integrated	8	27	52	42	0.4475	Cluster 2	0.4366
– Low Export Base	14	35	55	4	0.4462	Cluster 2	0.4242
	15	38	56	44	0.4445	Cluster 2	0.3021
	16	39	57	41	0.4357	Cluster 2	0.3361
	17	40	59	37	0.3786	Cluster 2	0.3739
Cluster 2	2	50		61	0.4965	Cluster 1	0.3106
Services, Final Consumption:	3	60		1	0.4781	Cluster 1	0.4037
– Low Integration	9	62		22	0.4743	Cluster 1	0.4090
– Low Economic Relevancy	10	63		11	0.4727	Cluster 1	0.3276
– Low Export Base	12	64		24	0.4614	Cluster 1	0.3723
	13	65		25	0.4522	Cluster 1	0.4322
	36	66		26	0.4093	Cluster 1	0.3623
	49			45	0.3993	Cluster 1	0.3708
Cluster 3	28	34		20	0.4600	Cluster 1	0.2942
Export Base:	29	47				Cluster 2	0.2458
– Technological, Exporting	30	53		33	0.3900	Cluster 2	0.3233
– Low Economic Relevancy	31	54				Cluster 1	0.2867
– Medium-High Integration	32						

---

misclassified. The average silhouette width for each cluster and an overall average silhouette width are calculated. The overall average silhouette width is the average of s_j for all objects in the whole dataset. The largest overall average silhouette indicates the best clustering. Therefore, the number of clusters with maximum overall average silhouette width is considered the optimal number of clusters.

[12] Only sectoral grading on the basis of the intensity of correlations as suggested by Czamanski's method or on the basis of the factorial loads as applied in Feser and Bergman (2000) can approach this result to some extent.

Taking into account the sectors with higher membership value in cluster 1, it would have the highest mean for "economic relevancy" and "integration", but the lowest for "export base". All the sectors with higher membership value in this cluster had a low technological level. These sectors could be considered to form the local base of the Spanish economy. Among them were those with more economic relevancy and the ones showing a greater interdependence.

Cluster 2 was basically distinguished for being less integrated, with "economic relevancy" and "export base" below the mean (below zero in the respective components). Its technological level was also low. The keys to interpret this cluster were its lack of integration and slight relevancy as driving force for the economy, both for the poor economic boost that they are able to generate and for their low exporting capacity. The best-classified sectors in cluster 2 were those oriented to final demand, especially to public and private consumption.

The principal features of cluster 3 were that it had the largest exports and the highest technological levels. Its score on "integration" was above average, but it was the least relevant in terms of "economic relevancy". The keys to interpret them was that they are exporting sectors with medium and high technology. It could be concluded that they were the most important group of sectors from the perspective of a growth based in foreign demand and innovation, a very desirable scenario of development for the Spanish economy.

3 The Outliers Masking Effect

Many observations in the clusters which were difficult to classify in a specific group, as can be seen looking at their membership values, have been identified as univariate outliers. For example, in cluster 1, sector 37 (Construction), being clearly a member, because its "economic relevancy" is the largest one, presented serious difficulties to be classified. Curiously, among the sectors with highest level of membership in cluster 1, the two sectors with lowest possibility of belonging to this group were Construction (37) and Hotel and restaurant services (41), the first and second in the "economic relevancy" component, respectively. The reason for this could be that both are outliers, attending either to this component or the variables forming it, and they are playing a distorting role in the clustering process.

In clusters 2 and 3 similar problems can also be found. Just to mention some of them, among those with a low membership value were sector 11 (Prepared animal feeds and other food products) and 26 (Basic metals), which were erroneously classified in the silhouette plot. They ranked 4th and 6th respectively on "economic relevancy" and were among those with better exporting behaviour. They were outliers on "export base" and "economic relevancy".

Although univariate outliers being difficult to classify can be explaining the membership values of the above mentioned sectors, a deeper study of them shows this is not the only underlying problem.

Multidimensional outliers are observations considered to be rare not for their particular value in a given variable, but for the combination of values in the different variables they take. They are more difficult to identify than univariate outliers, since they cannot be considered "extreme values" as it is the case when there is only one variable under study (Gnanadesikan and Kettenring, (1972); Campbell, 1980). In a two dimensional space it is not difficult to think of data being multidimensional outliers even if they are not extreme values in any of the variables separately. For example, a couple of data points that not being univariate outliers are strongly distorting an otherwise linearly shaped cloud of points.

The presence of multidimensional outliers has even worst effects than the presence of one-dimensional outliers, because of the distortion[13] they cause on the orientation measures (decreasing the correlations)[14].

The idea of multivariate distance plays a key role in outlier identification. For determining the number of outliers, Rousseeuw and Van Zomeren (1990) propose the use of a "robust distance" to perform a discordance test. The test uses the Mahalanobis distances of all the points respect to the robust Minimum Covariance Determinant (MCD) location estimator[15]. That way $\bar{x}$ and $Cov(x)$ are substituted by the MCD robust estimators of location and covariance, so that they are not affected by the presence of outliers nor the masking effect, since in the computation of the mean and covariance sample matrices only the really core points are included, and they are not distorted by outliers.

After estimating for the original variables the robust location and covariance parameters by the MCD method, the robust Mahalanobis distances have been obtained for each one of the sectors. They can be seen in Fig. 1, where the 26 potential outliers[16], are identified by the number of the corresponding sector.

As it can be seen, sector number 37 (Construction) is a very extreme value in the cloud of data points. Even sectors 61 (Public administration and defence; compulsory social security) and 26 (Basic metals) are very far from the rest of sectors. Their relative large distances can be masking smaller ones that could be adequately featuring clusters. So, not too much confidence should be placed in the clustering result shown in Table 1, even if it comes from a robust estimation.

---

[13] An excellent exposition of this problem and alternative solutions can be seen in Barnett and Lewis (1994), Chap.), now in its third edition, that has become a classic in the subject.

[14] See Kempthorne and Mendel (1990) for related comments on the declaration of a bivariate outlier.

[15] Mahalanobis distance allows the computation of the distance of each point to the center of the points, taking into account the variability and the correlation among variables, $MD(x_i) = \sqrt{(x_i - \bar{x})' (Cov(x))^{-1} (x_i - \bar{x})}$, where x_i is p-dimensional, $\bar{x}$ is the arithmetic mean and $Cov(x)$ is the classical sample covariance matrix. As Rousseeuw and Van Driessen (1999, p. 212) point out, "Although it is quite easy to detect a single outlier by means of the Mahalanobis distances, this approach no longer suffices for multiple outliers because of the masking effect, by which multiple outliers do not necessarily have large Mahalanobis distances. It is better to use distances based on robust estimators of multivariate location and scatter". The Minimun Covariance Determinant consists in the search for the covariance matrix with minimum determinant. The reason is that the determinant of this matrix is inversely related to the intensity of the correlations.

[16] The maximum number of outliers allowed is: [n- [(n+p+1)/2]]=26

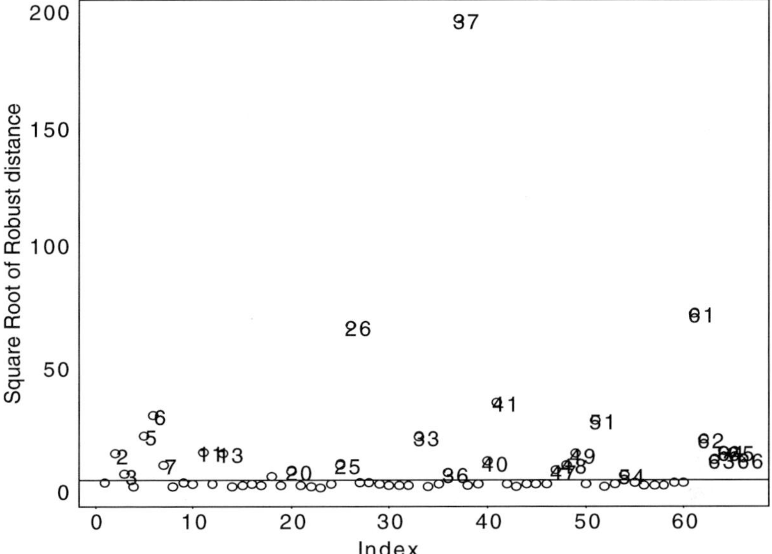

Fig. 1 Distance plot

Under multivariate normality hypothesis, these robust distances are asymptotically distributed as a χ^2 with p (number of variables) degrees of freedom[17]. Rousseeuw and Van Zomeren (1990) propose the consideration of the values over the 97.5 % quantile in the χ^2 distribution as outliers. Other authors propose, for outlier identification, an adaptive procedure searching for outliers specifically in the tails of the distribution, beginning at a certain chisq-quantile (see Filzmoser et al., 2005).

In this study, the hypothesis of normality does not hold. For this reason, the proposed (either χ^2 or F) distributions of the robust Mahalanobis distances cannot be considered as suitable ones. As Barnett and Lewis (1994), p. 461) point out *"the very presence of outliers may be masked by particularly strong real effects or by breakdowns in the conventional assumptions, such as normality, homoscedasticity, additivity and so on"*. Several interesting questions become apparent in Fig. 2, in relation with this subject.

In the first graph the Quantile-Quantile plot for the Mahalanobis distance shows how the Mahalanobis distances are far from following a Chi-square distribution, probably because normality is not present in the variables used in this study. Anyway, these types of distances should not be used because they present the masking effect, detecting very few outliers.

In the second graph, the robust distances, obtained by using the MCD mean and covariance matrices, are plotted against the square root of the quantiles of the

---

[17] The discordancy test uses this assumption, but some authors claim that an F distribution seems to be more appropriate. See Hardin and Rocke (2005) for a complete description of the distributions of robust distances coming from normally distributed data.

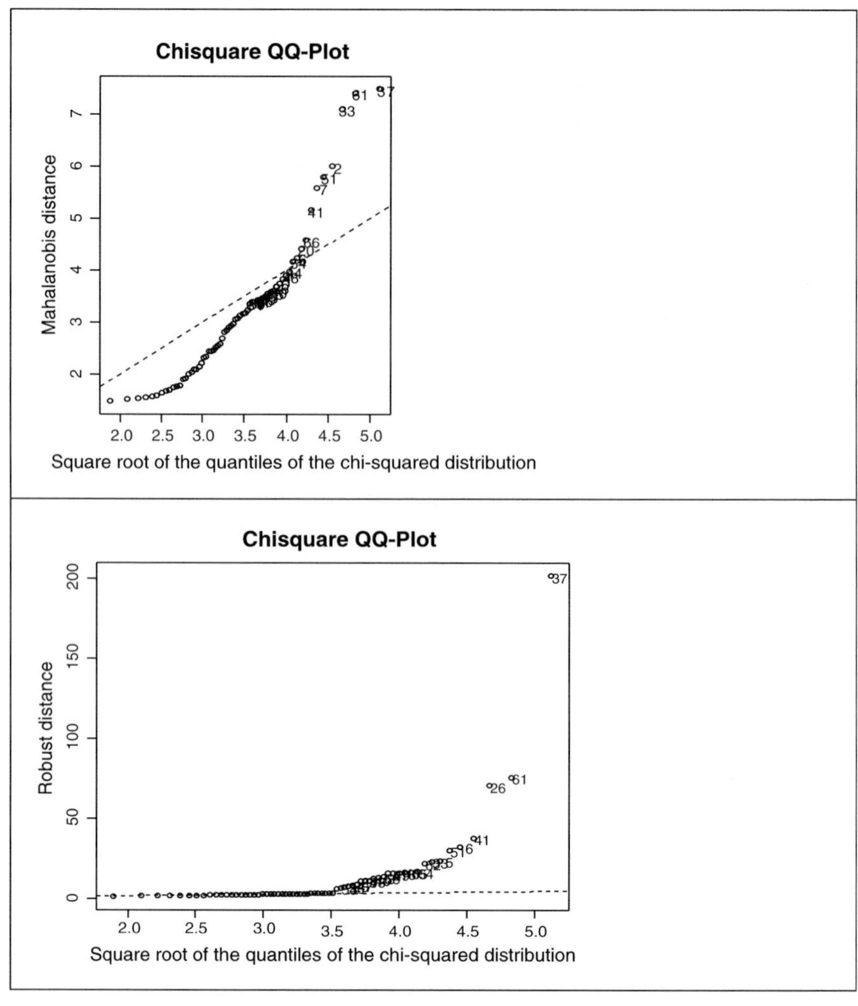

Fig. 2 Chisquare Q-Q plots for Mahalanobis and robust distances

chi-square distribution. The points that are included in the MCD sample (on the left, very close to the line) appear to have a chi-square distribution, but the points not included, are not distributed as a chi-square. There is an "elbow effect", as it was pointed out by Rousseeuw and Van Zomeren (1991), with the larger distances diverging in a systematic pattern. Hardin and Rocke (2005) give an explanation of the suitability of an F distribution for these points, when a normal multivariate distribution is present. In this case, the F distribution does not fit either, since there is not normality in the data.

It can be seen that, as expected, the Mahalanobis distances leads to the identification of far less points (only seven) as outliers, due to the masking effect. The use of of the 97.5% quantile in the chi-square distribution for the robust distances sometimes leads to the identification of too many points as outliers Hardin and Rocke

(2005), but they are often used, as it is also common to use this discordancy test that relies so heavily on the normality of the data, without even questioning the results[18].

In Fig. 3, two Distance Plots are shown. The one on the left displays the square root of the robust distances for each sector, while the one on the right displays the classical Mahalanobis distances. The solid horizontal lines represent values equal to the cutoff of the square root of the 97.5% quantile of the chi-squared distribution with p degrees of freedom. Points over these lines are considered outliers according to the discordancy test criterion.

In Fig. 3 some facts are highlighted. There are some sectors that should be considered as multidimensional outliers, regardless of the distance used. Those could be only the seven sectors considered as outliers when using the Mahalanobis distance, that is, sectors 37, 61, 33, 2, 51, 7 and 41, or many more, if the robust distance is taken instead.

But it is interesting to note that sector number 26, which is a clear outlier when considering the robust distance (it is in fact the sector with third higher robust distance) is not considered as a multidimensional outlier when using the Mahalanobis distance. We also note that sector 33 is the third one according to the Mahalanobis distance, although it is relegated to position number eight in the robust distances.

Sectors 2 and 7 are the 13th and 19th respectively according to the descending robust distance hierarchy. This is due to the fact that Mahalanobis distances and the robust distances are quite different because of the difference in the location of the

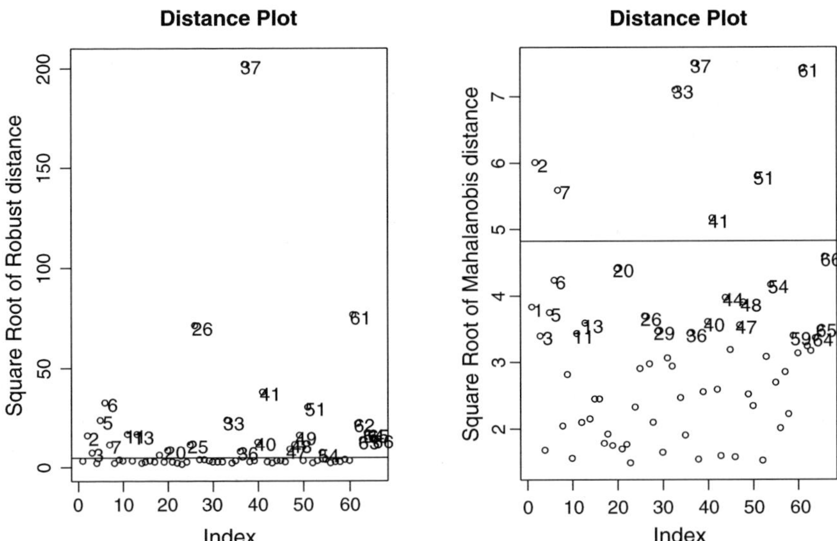

Fig. 3 Outliers from Mahalanobis and robust distances

---

[18] The inherent problems to the violation of the normality hypothesis represent an important issue that has not been definitely solved yet. In a recent Ph.D thesis (Dang, 2005) it can be seen that there is cause for concern.

center of the cloud of points in each of them and the covariation matrix. In fact, it can be observed when using the MCD estimator that estimations for the location shows a decrease respect to their original values of a 40% for the variable multiplier, a 28% for the cohesion index, 32% for production, 9% for exportations, 40% for employment, 8% for salaries, 9% for trend in the production and 0.5% in trend of the employment. On other hand, an increase of 12% for variable cohesion, 13% for integration and 25% for trend in exportations is also observed. This robust location estimation is more likely to be the real center of the cloud of points. Unfortunately it is not possible to have a display in this high dimension to visualize these shifts observed in the center when using one or other distance.

The use of one or the other distance is thus not trivial since not only the Mahalanobis distance will identify less outliers, they will also be different from the ones considered as such by the robust distance.

But, leaving aside theoretical problems and the search for an exact number of outliers, it is clear in Figs. 1, 2b) and 3 that using robust distances at least the three furthest sectors would clearly be considered as outliers.

Actually, they had already been considered difficult to classify due to their univariate outliers condition, being sector 37 (Construction) the one more difficult to be classified in cluster 1, having the same membership value both for clusters 1 and 2. Sector 61 and 26 were both considered belonging to cluster 2, being 26 (Basic metals) also erroneously classified in the silhouette plot. Sector 26 was 6th on "economic relevancy" and was among those with better exporting behaviour. It was an outlier on "export base" and "economic relevancy". Sector 61 (Public administration and defence; compulsory social security) is outlier both for its high integration, low exporting base, and low exporting base. It is a univariate outlier for all three components.

Anyway, it is not clear they are the only ones, because their extremely high values could be masking others important relative separations from the core of the cloud of points. Fig. 4 shows how when the three mentioned outliers are taken away, other possible outliers can still be seen.

All in all, there is coincidence in some of the multivariate outliers with those sectors considered difficult to classify (See Table 1). There were eighteen sectors with a membership value under 0.5 and several of those could be multivariate outliers. It is interesting to observe that some of the fuzziest ones, sectors 42, 4, 1, 22 and 45 are not included in the possible multivariate outliers list. As it was already pointed out, sectors 42 and 4 did not have a clear profile for any of the components and they are not multidimensional outliers either. Regarding sectors 1, 22 and 45 it seems that there is no reason for those sectors to be misclassified. They are univariate outliers in component 1 with a low economic relevancy, and component 3, with a low export base. But that does not make them rare points, they are not outliers in the multidimensional space. The fact that they are grouped in cluster 2 makes us think about the problems in this group.

It is remarkable that cluster number 2 is almost entirely integrated by multidimensional outliers. Cluster 2 had 23 sectors, eight of them with memberships below 0.5. Thirteen of the 26 points that could be considered outliers were in cluster 2, ten of them considered as correctly classified at that point. Sectors 62, 63, 64, 65 and

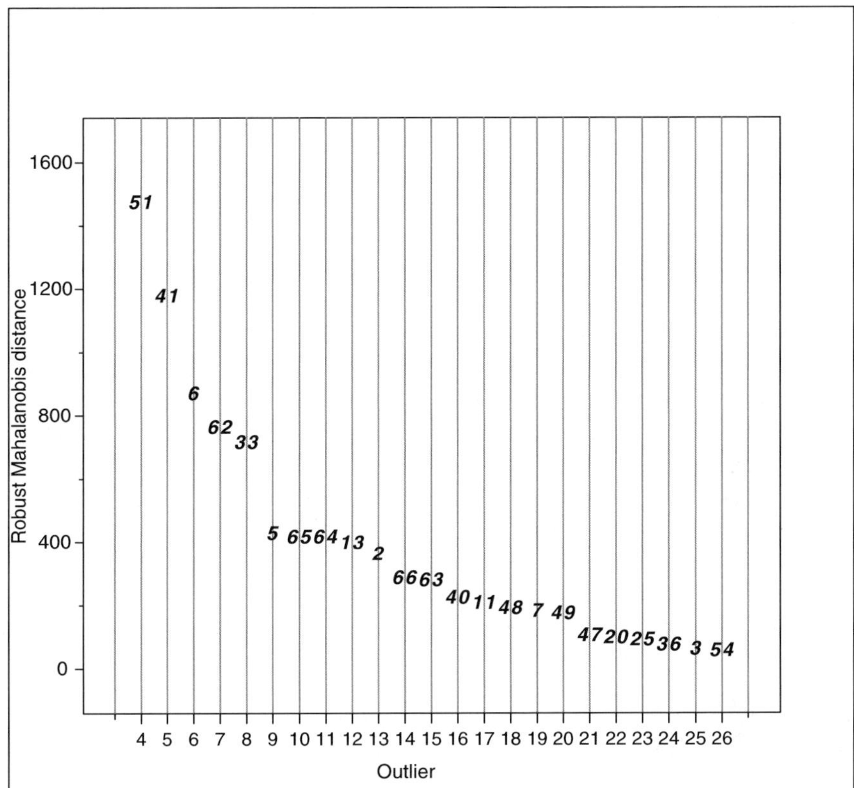

Fig. 4 Distance plot without the three furthest data

66 could all of them be multidimensional outliers according to the robust distance, as can be deducted from Fig. 4, and they all have low economic relevancy, low integration and low export base. Most probably, then, these sectors were weighting too much in this cluster. Unfortunately, the sample covariance matrixes are too sensitive to the presence of outliers and too often, a principal components or factorial analysis is "explaining" a structure that, observed in detail, has indeed been created by one or two anomalous data. The components extraction itself has to be questioned.

Finally, cluster three, which was the one considered as more important for the economic development, has as possible outliers sectors 33 and 20, which had problems for classification and could be affecting already the clustering results.

4 Conclusions

There hasn't been in the literature much emphasis on a multivariate approach to identify the industrial clusters or the key sectors for an economy. Undoubtedly, this multidimensional approach causes some complications not only because of the different measure scales of the variables that can be used, but also because the problem of multivariate outliers arises, being these more difficult to identify than the univariate ones.

The presence of outliers can be a really important problem, both for the extraction of the components, because of its incidence on the correlations matrix, and for the clusters definition. Being with no doubt a serious problem it had not been explored yet in this input-output analysis context. Surprisingly, the literature has never related the mega clusters results with them. Moreover, they not only cause effects on the cluster analysis but, with no doubt, they also have effects over the results of any multivariate application made before.

On the other hand, we have shown that the use of several robust techniques in combination with the relational fuzzy clustering (NERFCM) is an adequate treatment of the problem in some initial steps. These include a robust dissimilarity measure that allows the incorporation of different kind of variables, partitioning around medoids as an alternative to k-means for membership values initialization and silhouettes plots for selection of best number of clusters.

Finally, it is obligated to remark that the presence of multivariate outliers is completely distorting the clustering, and putting them aside would for sure improve the results. On the other hand, when theoretical hypothesis are not fitted, it seems obvious the outliers definition remain imprecise yet and for identifying them it should be always reasonable to take into account simultaneously other criteria regarding the context and objective of the analysis. A deeper research in this direction needs to be done in the future.

Acknowledgments This work has been partially supported by the Spanish State Secretary of Education and Universities and The Social European Fund under grant number EX2003-0563

References

V. Barnett and T. Lewis, Outliers*in Statistical Data*. Chichester, Wiley, 1994.

J.C. Bezdek, *Pattern Recognition with Fuzzy Objective Function Algorithms*. Plenum, New York, 1981.

N.A. Campbell, "Robust procedures in multivariate analysis I: Robust covariance estimation". *Applied Statistics*, Vol. 29, No. 3, pp. 231–237, 1980.

H. B. Chenery and T. Watanabe, "International comparison of the structure of production", *Econometrica*, Vol. XXVI, No. 26, pp. 487–521, 1958.

S. Czamanski, *Study of Clustering of Industries*, Dalhousie University, Halifax, 1974.

X. Dang, *Nonparametric multivariate outlier detection methods with applications*, University of Dallas at Texas, 2005.

R.J. Dave and S. Sen, "Robust Fuzzy Clustering of Relational Data". *IEEE Transactions on Fuzzy Systems,* Vol. 10, No. 6, pp. 713–727, 2002.

L. De Mesnard, "On Boolean Topological Methods of Structural Analysis", en Lahr, M.L. y Dietzenbacher, E. (eds.): *Input-Output Analysis: Frontiers and Extensions*, Palgrave, New York, Chap. 3, 2001.

B. Diaz, L. Moniche and A. Morillas, A., "A fuzzy clustering approach to the key sectors of the Spanish economy". *Economic systems Research*, Vol. 18, No. 3, to appear September 2006.

C. Dridi and G.J.D. Hewings, "Sectors associations and similarities in input-output systems: an application of dual scaling and fuzzy logic to Canada and the United States", *Annals of Regional Science*, Vol. 37, pp. 629–656, 2003.

J. C. Dunn, "A Fuzzy Relative of the ISODATA Process and Its Use in Detecting Compact Well-Separated Clusters". *Journal of Cybernetics*, Vol. 3, No. 3, pp. 32–57, 1973.

EUROSTAT, *Ressources humaines en Haute Technologie. Serie 'Statistiques en Bref'*, 1998a.

EUROSTAT, *Employment in high technology manufacturing sectors at the regional level*, 1998b.

E. J. Feser and E.M. Bergman, "National Industry Cluster Templates: a Framework for Applied Regional Cluster Analysis". *Regional Studies*, Vol. 34, No. 1, pp. 1–19, 2000.

E. J. Feser, J. Koo, H. Renski, and S.H. Sweeney, *Incorporating spatial analysis in applied industry cluster studies*. Working paper, 2001a. Available:http://www. planning.unc.edu/pdf/FeserPUBS/EDQ%20Revised.pdf

E.J. Feser, S.H. Sweeney and H.C. Renski, *A Descriptive Analysis of Discrete U.S. Industrial Complexes*. Prepared for the Journal of Regional Science, 2001b. Available: http://www.planning.unc.edu/pdf/FeserPUBS/Complexes.pdf

P. Filzmoser, R.G. Garrett and C. Reimann, "Multivariate outlier detection in exploration geochemistry". *Computers and Geosciences*, Vol. 31, No. 5, pp. 579–587, 2005.

R. Gnanadesikan and J.R. Kettenring, "Robust estimates, residuals and outlier detection with multiresponse data". *Biometrics*, Vol. 28, No. 1, Special Multivariate Issue (Mar., 1972) pp. 81–124, 1972.

J. Hardin and D.M. Rocke, "The distribution of robust distances". *Journal of Computational and Graphical Statistics*, Vol. 14, No. 4, pp. 1–19, 2005.

F. Harrigan, "The Relationship Between Industrial and Geographical Linkages". *Journal of Regional Science* Vol. 22, pp. 19–31, 1982.

R.J .Hathaway, J.C. Bezdek and Y. Hu, "Generalized Fuzzy c-Means Clustering Strategies Using Lp Norm Distances", *IEEE Transactions on Fuzzy Systems*, Vol. 8, No. 5, pp. 576–582, October 2000.

R.J. Hathaway and J.C. Bezdek, "NERF c-means: Non- Euclidean Relational Fuzzy Clustering", *Pattern Recognition*, Vol. 27, No. 3, pp. 429–437, 1994.

R.J. Hathaway, J. W. Davenport and J.C. Bezdek, "Relational duals of the c-means algorithms", *Pattern Recognition*, Vol. 22, No. 2, pp. 205–212, 1989.

A. O. Hirschman, *The Strategy of Economic Development*, 1958.

INE, Tablas input-output de la Economía Española de 1995, 2000. www.ine.es

INE a, Contabilidad Nacional. Base 1995. www.ine.es

INE b, INEbase en la metodología expuesta en Investigación y desarrollo tecnológico. Indicadores de alta tecnología. Available: http://www.ine.es/inebase/cgi/um?M= %2Ft14% 2Fp197&O=inebase&N=&L=

L. Kaufman and P.J. Rousseeuw, *Finding Groups in Data*. Wiley, New York, 1990.

P.J. Kempthorne and M.B. Mendel, M. B., "Comments on Rousseeuw and Van Zomeren (1990)". *Journal of the American Statistical Association*, Vol. 85, No. 411, pp. 647–648, 1990.

L. Lainesse and B. Poussart, " Méthode de repérage des filières industrielles sur le territoire québécois dasee sur les tableaux d'entrées-sorties", *L'economie du savoir. Institut de la Statistique du Québec, 2005. Available:* http://www.stat.gouv.qc.ca/ publications/savoir/methode_reperage_pdf.htm

R. Lantner, *Théorie de la Dominance Economique*. Paris. Dunod, 1974.

R. Lantner, "Influence Graph Theory Applied to Structural Analysis", in Lahr, M.L. y Dietzenbacher, E. (eds.): *Input-Output Analysis: Frontiers and Extensions*, Palgrave, New York, Chap. 15, 2001.

W.R. Latham, "Needless Complexity in the Identification of Industrial Complexes". *Journal of Regional Science* Vol. 16, pp. 45–55, 1976.

J. B. MacQueen, "Some Methods for Classification and Analysis of Multivariate Observations", *Proceedings of 5-th Berkeley Symposium on Mathematical Statistics and Probability"*, Berkeley, University of California Press, pp.281–297, 1967.

A. Morillas, *La teoría de grafos en el análisis input-output*. Secretariado de Publicaciones. Universidad de Málaga, 1983a.

A. Morillas, "Indicadores topológicos de las características estructurales de una tabla input-output. Aplicación a la economía andaluza". *Investigaciones Económicas*, No. 20, pp. 103–118, 1983b.

M. Mougeot, G. Duru and J.P. Auray, *La Structure Productive Francaise*, Económica, Paris, 1977.

B. Ó hUallachain, "The Identification of Industrial Complexes". *Annals of the Association of American Geographers* Vo. 74, No.3, 1984, pp. 420–436.

J. Oosterhaven and D. Stelder, "Net multipliers avoid exaggerating impacts: with a bi-regional illustration for the Dutch transportation sector", *Journal of Regional Science*, Vol. 42, pp. 533–543, 2002.

J. Oosterhaven, "On the definition of key sectors and the stability of net versus gross multipliers", *SOM Research Report 04C01*, Research Institute SOM, University of Groningen, The Netherlands, 2004. Available at: http://som.rug.nl.

M.E. Porter, "The Competitive Advantage of Nations". *Harvard Business Review*, Vol. 68, No.2, pp. 77–93, 1990.

N. P. Rasmussen, *Studies in Intersectoral Relations*, North Holland Publishing Company, Amsterdam, 1956.

S. Rey and D. Mattheis, *Identifying Regional Industrial Clusters in California*. Volumes I,II, III. Prepared for the California Employment Development Department., 2000. San Diego State University. Available: http ://irsr.sdsu.edu/ %7Eserge/node7.html

H. Roepke, D.Adams and R. Wiseman, "A New Approach to the Identification of Industrial Complexes Using Input-Output Data". *Journal of Regional Science* Vol. 14, pp. 15–29, 1974.

E. Rossier, *Economie Structural*, Económica, Paris, 1980.

P.J. Rousseeuw, "Discussion: Fuzzy Clustering at the Intersection". *Technometrics.* Vol. 37, No. 3, 1995, pp. 283–285.

P.J. Rousseeuw, "Silhouettes: a Graphical Aid to the Interpretation and Validation of Cluster Analysis". *Journal of Computational and applied Mathematics*, Vol. 20, pp. 53–65, 1987.

P.J. Rousseeuw and K. Van Driessen, "A fast algorithm for the minimum covariance determinant estimator". *Technometrics,* Vol. 41, No.3, pp. 212–223, 1999.

P.J. Rousseeuw and B.C. Van Zomeren, "Unmasking multivariate outliers and leverage points". *Journal of the American Statistical Association*, Vol. 85, No. 411, pp. 633–639, 1990.

P.J. Rousseeuw and B.C. Van Zomeren, "Robust Distances: Simulations and Cutoff Values," in *Directions in Robust Statistics and Diagnostics*, Part II, eds. W. Stahel and S. Weisberg. *The IMA Volumes in Mathematics and Its Applications*, Vol. 34, pp. 195–203, 1991.

E. Ruspini, "A new approach to clustering", *Information and control*, Vol. 15, No. 1, pp. 22–32, 1969.

H.A. Simon, "On the Definition of the Causal Relation", *Journal of Philosophy*, Vol. 49, 1952, pp. 517–528.

H. Wold, "Causality and Econometrics", *Econometrica*, Vol. 22, No. 2, pp. 162–177, 1954.

L.A. Zadeh, "Fuzzy sets and their application to pattern classification and clustering analysis", in *Classification and Clustering*, ed. J.V. Ryzin, 1977, pp. 251–282 & 292–299. Reprinted in *Fuzzy Sets, Fuzzy Logic and Fuzzy Systems, Selected papers by Lotfi A: Zadeh*, 1996, Ed. J. Klir, and B. Yuan, pp. 355–393.

L.A. Zadeh, "Fuzzy sets". *Information and Control*, Vol. 8, No. 3, pp. 338–353, 1965.

Annex 1 List of variables

VARIABLE	DEFINITION	COMMENTS
Economic integration block		
MULTIPLIC	Multiplier	Addition by columns of Ghosh's inverse matrix, Augustinovics' (1970) multiplier.
CVINVERS	Inverse of the multipliers' variation coefficient	Each sector is characterized by the inverse of the dispersion, represented by its column variation coefficient.
COHESIO	Cohesion Grade	Number of times that a sector is a path (from graphs theory.)
Ri	Total productive integration Index	A topological offer-demand integration index for each sector (see note 8).
Economic weights block		
PRODUC	Output	Output at basic price by product from the Spanish symmetric table.
EXPORT	Exports	The source is the Spanish symmetric table for 1995, INE(2000.)
EMPLEO	Full-time equivalent employment	Since this data is only available by industry (use table), the correction coefficient that presents the compensation of employees by products has been applied.
SALARIOS	Mean salaries	It is estimated with corrected paid employment (also with use table) and data of compensation of employees (from the symmetric table.)
Economic potential block		
TENPROD	Output trend	It is the real output variation rate for each sector for the period 1995-2000 (source: INE a).
TENEMPL	Employment trend.	It is the full-time equivalent employment variation rate for each sector respect to the initial period, from 1995 to 2000. The implicit sectoral deflator of the Gross Valued Added at basic prices (VABpb in the National Accounting of the Spanish National Institute of Statistics, INE a) is used.
TENSALAR	Mean salary trend	It is the real mean salary variation rate for each sector for the period 1995-2000. The consumption prices index (IPC) variation during that period (14,2) to deflate salaries for 2000 is used.
TENEXPO	Exportation trend	It is the real exportation variation rate for each sector for the period 1995-2000. The procedure used to deflate is the one used for output.
TECNO	Technological innovation capacity	Sectors classification attending to their technological level (high, medium or low technology.)

Annex 2 High and medium technology sectors

High and Medium-High Technology Sectors
Listed by the Spanish Institute of Statistics (INE)

CNAE-93 (NACE Rev. 1)	Sectors
	High Technology Manufacturing Sectors
244	Manufacture of pharmaceuticals, medicinal chemicals and botanical products
30	Manufacture of office machinery and computers
321	Electronic components
32–32.1	Manufacture of radio, television and communication equipment and apparatus excluding electronic components
33	Manufacture of medical, precision and optical instruments, watches and clocks
35.3	Manufacture of aircraft and spacecraft
	Medium-High Technology Manufacturing Sectors
24–24.4	Manufacture of chemicals and chemical products excluding manufacture of pharmaceuticals, medicinal chemicals and botanical products
29	Manufacture of machinery and equipment n.e.c.
31	Manufacture of electrical machinery and apparatus n.e.c.
34	Manufacture of motor vehicles, trailers and semi-trailers
35–35.3	Manufacture of other transport equipment excluding manufacture of aircraft and spacecraft
	High Technology Service Sectors
64	Post and telecommunications
72	Computer and related activities
73	Research and development

Source: INE. INE b), p.5.

Fuzzy Label Semantics for Data Mining

Zengchang Qin and Jonathan Lawry

Abstract This chapter gives a tutorial introduction on the label semantics framework for reasoning with uncertainty and several data mining models which are developed based on this framework. Modelling real world problems typically involves processing uncertainty of two distinct types. These are uncertainty arising from a lack of knowledge relating to concepts which, in the sense of classical logic, may be well defined and uncertainty due to inherent vagueness in concepts themselves. Traditionally, these two types of uncertainties are modeled in terms of probability theory and fuzzy set theory, respectively. Zadeh [31] recently argued that all the approaches for uncertainty modelling can be unified into a general theory of uncertainty (GTU). In this chapter, we will introduce an alternate approach for modelling uncertainties by using random set and fuzzy logic. This framework is referred to as label semantics where the labels could be discrete or fuzzy labels. Based on this framework, we proposed several new data mining models. These models not only give comparable accuracy to other well-known data mining models, but also high transparency by which we understand how classifications or predictions have been made instead of a black box.

1 Introduction

In some sense the world is not fuzzy. We can look out and see precisely a leaf falling from an old tree whose shadow lies on a green grassland, there are three people playing and laughing on the grassland, and not far away, there is a car parked on the side of road. Which of them are fuzzy? But this detail which we can see with our eyes is often unwanted precision when it comes to categorizing, classifying and clustering the real world into groups which we can label. We give labels to such objects as people, cars, grassland, trees and leaves, so that we can talk about these objects in terms of their common properties within their group. *Fuzzy Logic* was first proposed by Zadeh [28] as an extension of traditional binary logic. In contrast to a classical set, which has a crisp boundary, the boundary of a fuzzy set is blurred. This smooth transition is characterized by *membership functions* which give fuzzy sets flexibility in modeling linguistic expressions. In early research fuzzy logic was successfully applied in expert systems where the linguistic

M. Nikravesh et al. (eds.), *Forging the New Frontiers: Fuzzy Pioneers II.*
© Springer-Verlag Berlin Heidelberg 2008

interpretation fuzzy sets allowed for an interface between the human user and a computer system. Because our language is fuzzy, they share the uncertainty and impreciseness: One word has many different meanings and to describe one meaning, we could use many different words. Therefore, we may use fuzzy sets to model our language. This idea provides a good way of bridging the gap between human users and computing systems, and this motivates related research into Computing with Words [29].

Almost all the labels we give to groups of objects are fuzzy. For example, friends, pretty faces, tall trees etc. An object may belong to the set of objects with a certain label, with a certain *membership value*. In traditional set theory, this membership value only has two possible values, 1 and 0, representing the case where the object belongs to or does not belong to the set, respectively. We use a fuzzy term such as 'big' to label a particular group, because they share the property of objects within this group (i.e., they are big). The objects within this group will have different membership values varying from 0 to 1 qualifying the degree to which they satisfy the concept 'big'. An object with membership of 0.8 is more likely to be described as 'big' than an object with membership of 0.4. If we consider this problem in another way. Given an object, label 'big' can be used to describe this object with some appropriateness degrees. Follow this idea, we discuss a new approach based on random set theory to interpret imprecise concepts. This framework, first proposed by Lawry [10] and is referred to as *Label Semantics*, can be regarded as an approach to Modelling with Words[1] [11].

2 Label Semantics

Vague or imprecise concepts are fundamental to natural language. Human beings are constantly using imprecise language to communicate each other. We usually say 'John is tall and strong' but not 'John is exactly 1.85 meters in height and he can lift 100kg weights'. We will focus on developing an understanding of how an intelligent agent can use vague concepts to convey information and meaning as part of a general strategy for practical reasoning and decision making. Such an agent can could be an artificial intelligence program or a human, but the implicit assumption is that their use of vague concepts is governed by some underlying internally consistent strategy or algorithm. We may notice that *labels* are used in natural language to describe what we see, hear and feel. Such labels may have different degrees of vagueness (i.e., when we say Mary is *young* and she is *female*, the label *young* is more vague than

---

[1] According to Zadeh [30], Modeling with Words is a new research area which emphasis "modelling" rather than "computing", however, the relation between it and Computing with Words is close is likely to become even closer. Both of the research areas are aimed at enlarging the role of natural languages in scientific theories, especially, in knowledge management, decision and control. In this chapter, the framework is mainly used for modelling and building intelligent machine learning and data mining systems. In such systems, we use words or fuzzy labels for modelling uncertainty.

the label *female* because people may have more widely different opinions on being *young* than being *female*. For a particular concept, there could be more than one label that is appropriate for describing this concept, and some labels could be more appropriate than others. Here, we will use a random set framework to interpret these facts. *Label Semantics*, proposed by Lawry [10], is a framework for modelling with linguistic expressions, or labels such as *small, medium* and *large*. Such labels are defined by overlapping fuzzy sets which are used to cover the universe of continuous variables. Related to fuzzy sets is the theory of possibility, which can be seen as its numerical counterpart. It is possible to build bridges between probability and fuzzy sets where the latter are viewed as possibility distributions. In particular, we shall interpret possibility measures in the framework of random sets and belief function theory and we shall consider the problem of transforming a possibility distribution into a probability distribution and vice versa.

2.1 Mass Assignment on Fuzzy Labels

The underlying question posed by label semantics is how to use linguistic expressions to label numerical values. For a variable x into a domain of discourse Ω we identify a finite set of linguistic labels $\mathcal{L} = \{L_1, \cdots, L_n\}$ with which to label the values of x. Then for a specific value $x \in \Omega$ an individual I identifies a subset of $\mathcal{L}$, denoted D_x^I to stand for the description of x given by I, as the set of labels with which it is appropriate to label x. If we allow I to vary across a population V with prior distribution P_V, then D_x^I will also vary and generate a random set denoted D_x into the power set of $\mathcal{L}$ denoted by $\mathcal{S}$. We can view the random set D_x as a description of the variable x in terms of the labels in $\mathcal{L}$. The frequency of occurrence of a particular label, say S, for D_x across the population then gives a distribution on D_x referred to as a mass assignment on labels[2].

More formally,

Definition 1 (*Label Description*) *For $x \in \Omega$ the label description of x is a random set from V into the power set of $\mathcal{L}$, denoted D_x, with associated distribution m_x, which is referred to as mass assignment:*

$$\forall S \subseteq \mathcal{L}, \quad m_x(S) = P_V(\{I \in V | D_x^I = S\}) \tag{1}$$

where P_V is the prior distribution of population V. $m_x(S)$ is called the mass associated with a set of labels S and

---

[2] Since $\mathcal{S}$ is the power set of L, the logical representation $S \in \mathcal{S}$ can be written as $S \subseteq \mathcal{L}$. The latter representation will be used through out this chapter. For example, given $\mathcal{L} = \{L_1, L_2\}$, we can obtain $\mathcal{S} = \{\emptyset, \{L_1\}, \{L_2\}, \{L_1, L_2\}\}$. For every element in $\mathcal{S}$: $S \in \mathcal{S}$, the relation $S \subseteq \mathcal{L}$ will hold.

$$\sum_{S \subseteq \mathcal{L}} m_x(S) = 1 \tag{2}$$

Intuitively mass assignment is a distribution on appropriate label sets and $m_x(S)$ quantifies the evidence that S is the set of appropriate labels for x. Based on the data distribution $p(x)$, we can calculate the prior distribution of labels by summing up the mass assignment across the database as follows:

$$pm(S) = p(S) = \int_{\Omega} m_x(S) p(x) dx \Big/ \left(\sum_{S \subseteq \mathcal{L}} \int_{\Omega} m_x(S) p(x) \right) \tag{3}$$

However, the dominator equals to 1 according to the definition of mass assignment and (2), so that:

$$pm(S) = \int_{\Omega} m_x(S) p(x) dx \tag{4}$$

And in the discrete case:

$$pm(S) = \sum_{x \in \mathcal{D}} m_x(S) P(x) \tag{5}$$

For example, given a set of labels defined on the temperature outside: $\mathcal{L}_{Temp} = \{low, medium, high\}$. Suppose 3 of 10 people agree that '*medium* is the only appropriate label for the temperature of $15°$ and 7 agree 'both *low* and *medium* are appropriate labels'. According to def. 1,

$$m_{15}(medium) = 0.3 \text{ and } m_{15}(low, medium) = 0.7$$

so that the mass assignment for $15°$ is $m_{15} = \{medium\} : 0.3, \{low, medium\}: 0.7$. More details about the theory of mass assignment can be found in [1].

2.2 Appropriateness Degrees

Consider the previous example, can we know how appropriate for a single label, say *low*, to describe $15°$? In this framework, *appropriateness degrees* are used to evaluate how appropriate a label is for describing a particular value of variable x. Simply, given a particular value α of variable x, the appropriateness degree for labeling this value with the label L, which is defined by fuzzy set F, is the membership value of α in F. The reason we use the new term 'appropriateness degrees' is partly because it more accurately reflects the underlying semantics and partly to highlight the quite distinct calculus based on this framework [10]. This definition provides a relationship between mass assignments and appropriateness degrees.

Definition 2 (*Appropriateness Degrees*)

$$\forall x \in \Omega, \ \forall L \in \mathcal{L} \quad \mu_L(x) = \sum_{S \subseteq \mathcal{L}:L \in S} m_x(S)$$

Consider the previous example, we then can obtain $\mu_{medium}(15) = 0.7 + 0.3 = 1$, $\mu_{low}(15) = 0.7$. Based on the underlying semantics, we can translate a set of numerical data into a set of mass assignments on appropriate labels based on the reverse of definition 2 under the following assumptions: consonance mapping, full fuzzy covering and 50% overlapping [19]. These assumptions are fully described in [19] and justified in [12]. These assumptions guarantee that there is unique mapping from appropriate degrees to mass assignments on labels.

2.3 Linguistic Translation

It is also important to note that, given definitions for the appropriateness degrees on labels, we can isolate a set of subsets of $\mathcal{L}$ with non-zero masses. These are referred to as *focal sets* and the appropriate labels with non-zero masses as *focal elements*, more formally,

Definition 3 (*Focal Set*) *The focal set of $\mathcal{L}$ is a set of focal elements defined as:*

$$\mathcal{F} = \{S \subseteq \mathcal{L} | \exists x \in \Omega, m_x(S) > 0\}$$

Given a particular universe, we can then always find the unique and consistent translation from a given data element to a mass assignment on focal elements, specified by the function $\mu_L : L \in \mathcal{L}$. For example, Fig. 1 shows the universes of two variables x_1 and x_2 which are fully covered by 3 fuzzy sets with 50% overlap, respectively. For x_1, the following focal elements occur:

$$\mathcal{F}_1 = \{\{small_1\}, \{small_1, medium_1\}, \{medium_1\}, \{medium_1, large_1\}, \{large_1\}\}$$

Since $small_1$ and $large_1$ do not overlap, the set $\{small_1, large_1\}$ cannot occur as a focal element according to def. 3. We can always find a unique translation from a given data point to a mass assignment on focal elements, as specified by the function μ_L. This is referred to as *linguistic translation* and is defined as follows:

Definition 4 (*Linguistic Translation*) *Suppose we are given a numerical data set $\mathcal{D} = \{\langle x_1(i), \ldots, x_n(i)\rangle | i = 1, \ldots, N\}$ and focal set on attribute j: $\mathcal{F}_j = \{F_j^1, \ldots, F_j^{h_j} | j = 1, \ldots, n\}$, we can obtain the following new data base by applying linguistic translation to $\mathcal{D}$:*

$$\mathcal{D} = \{A_1(i), \ldots, A_n(i) | i = 1, \ldots N\}$$

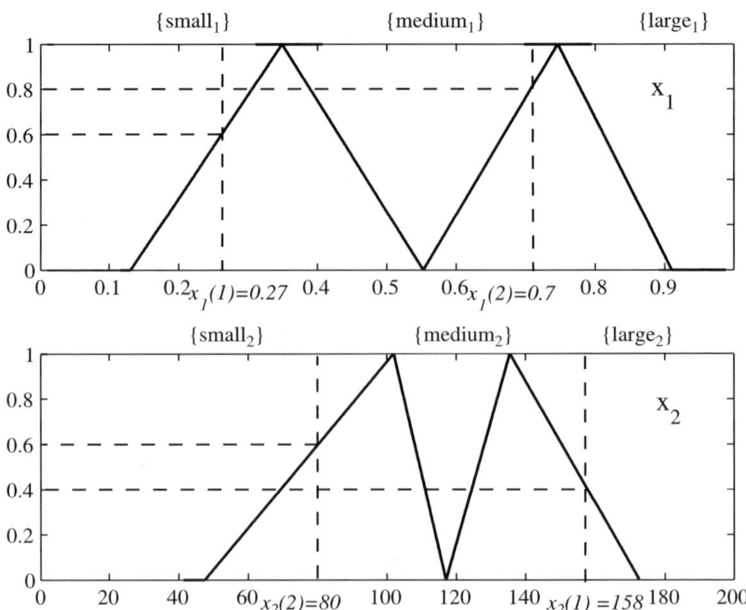

Fig. 1 A full fuzzy covering (discretization) using three fuzzy sets with 50% overlap on two attributes x_1 and x_2, respectively

$$A_j(i) = \{\langle m_{x_j(i)}(F_j^1), \ldots, m_{x_j(i)}(F_j^{h_j})\rangle\}$$

where $m_{x_j(i)}(F_j^r)$ is the associated mass of focal element F_j^r as appropriate labels for data element $x_j(i)$ where $r = 1, \ldots, h_j$ and $j = 1, \ldots, n$.

For a particular attribute with an associated focal set, linguistic translation is a process of replacing its data elements with the focal element masses of these data elements. For a variable x, it defines a unique mapping from data element $x(i)$ to a vector of associated masses $\langle m_{x(i)}(F^1), \ldots, m_{x(i)}(F^h)\rangle$.

See fig. 1. $\mu_{small_1}(x_1(1) = 0.27) = 1$, $\mu_{medium_1}(0.27) = 0.6$ and $\mu_{large_1}(0.27) = 0$. They are simply the memberships read from the fuzzy sets. We then can obtain the mass assignment of this data element according to def. 2 under the consonance assumption [19]: $m_{0.27}(small_1) = 0.4$, $m_{0.27}(small_1, medium_1) = 0.6$. Similarly, the linguistic translations for two data:

$$x_1 = \langle x_1(1) = 0.27\rangle, \langle x_2(1) = 158\rangle$$

$$x_2 = \langle x_1(2) = 0.7\rangle, \langle x_2(2) = 80\rangle$$

are illustrated on each attribute independently as follows:

$$\begin{bmatrix} x_1 \\ \hline x_1(1) = 0.27 \\ x_1(2) = 0.7 \end{bmatrix} \xrightarrow{LT} \begin{bmatrix} m_x(\{s_1\}) & m_x(\{s_1, m_1\}) & m_x(\{m_1\}) & m_x(\{m_1, l_1\}) & m_x(\{l_1\}) \\ \hline 0.4 & 0.6 & 0 & 0 & 0 \\ 0 & 0 & 0.2 & 0.8 & 0 \end{bmatrix}$$

$$\begin{bmatrix} x_2 \\ \hline x_2(1) = 158 \\ x_2(2) = 80 \end{bmatrix} \xrightarrow{LT} \begin{bmatrix} m_x(\{s_2\}) & m_x(\{s_2, m_2\}) & m_x(\{m_2\}) & m_x(\{m_2, l_2\}) & m_x(\{l_2\}) \\ \hline 0 & 0 & 0 & 0.4 & 0.6 \\ 0.4 & 0.6 & 0 & 0 & 0 \end{bmatrix}$$

Therefore, we can obtain:

$$x_1 \rightarrow \langle \{s_1\} : 0.4, \{s_1, m_1\} : 0.6 \rangle, \langle \{m_2, l_2\} : 0.4, \{l_2\} : 0.6 \rangle$$

$$x_2 \rightarrow \langle \{m_1\} : 0.2, \{m_1, l_1\} : 0.8 \rangle, \langle \{s_2\} : 0.4, \{s_2, m_2\} : 0.6 \rangle$$

3 Linguistic Reasoning

As a high-level knowledge representation language for modelling vague concepts, label semantics allows linguistic reasoning. This section introduces the linguistic reasoning mechanism for label semantics framework. Given a universe of discourse Ω containing a set of objects or instances to be described, it is assumed that all relevant expressions can be generated recursively from a finite set of basic labels $\mathcal{L} = \{L_1, \ldots, L_n\}$. Operators for combining expressions are restricted to the standard logical connectives of negation "$\neg$", conjunction "$\wedge$", disjunction "$\vee$" and implication "$\rightarrow$". Hence, the set of logical expressions of labels can be formally defined as follows:

Definition 5 (*Logical Expressions of Labels*) *The set of logical expressions, LE, is defined recursively as follows:*

(i) $L_i \in LE$ for $i = 1, \ldots, n$.
(ii) If $\theta, \varphi \in LE$ then $\neg\theta, \theta \wedge \varphi, \theta \vee \varphi, \theta \rightarrow \varphi \in LE$

Basically, we interpret the main logical connectives as follows: $\neg L$ means that L is not an appropriate label, $L_1 \wedge L_2$ means that both L_1 and L_2 are appropriate labels, $L_1 \vee L_2$ means that either L_1 or L_2 are appropriate labels, and $L_1 \rightarrow L_2$ means that L_2 is an appropriate label whenever L_1 is. As well as labels for a single variable, we may want to evaluate the appropriateness degrees of a complex logical expression $\theta \in LE$. Consider the set of logical expressions LE obtained by recursive application of the standard logical connectives in $\mathcal{L}$. In order to evaluate the appropriateness degrees of such expressions we must identify what information they provide regarding the the appropriateness of labels. In general, for any label expression θ we should be able to identify a maximal set of label sets, $\lambda(\theta)$ that are consistent with θ so that the meaning of θ can be interpreted as the constraint $D_x \in \lambda(\theta)$.

Definition 6 (λ-*function*) *Let θ and φ be expressions generated by recursive application of the connectives $\neg$, $\vee$, $\wedge$ and $\rightarrow$ to the elements of $\mathcal{L}$ (i.e. $\theta, \varphi \in LE$). Then the set of possible label sets defined by a linguistic expression can be determined recursively as follows:*

(i) $\lambda(L_i) = \{S \subseteq \mathcal{F} | \{L_i\} \subseteq S\}$

(ii) $\lambda(\neg\theta) = \overline{\lambda(\theta)}$

(iii) $\lambda(\theta \wedge \varphi) = \lambda(\theta) \cap \lambda(\varphi)$

(iv) $\lambda(\theta \vee \varphi) = \lambda(\theta) \cup \lambda(\varphi)$

(v) $\lambda(\theta \rightarrow \varphi) = \overline{\lambda(\theta)} \cup \lambda(\varphi)$

It should also be noted that the λ-function provides us with notion of logical equivalence '$\equiv_L$' for label expressions

$$\theta \equiv_L \varphi \iff \lambda(\theta) = \lambda(\varphi)$$

Basically, the λ-function provides a way of transferring logical expressions of labels (or linguistic rules) to random set descriptions of labels (i.e. focal elements). $\lambda(\theta)$ corresponds to those subsets of $\mathcal{F}$ identified as being possible values of D_x by expression θ. In this sense the imprecise linguistic restriction 'x is θ' on x corresponds to the strict constraint $D_x \in \lambda(\theta)$ on D_x. Hence, we can view label descriptions as an alternative to linguistic variables as a means of encoding linguistic constraints.

3.1 Appropriateness Measures

Based on definition 6, we can evaluate the appropriateness degree of $\theta \in LE$ is to aggregate the values of m_x across $\lambda(\theta)$. This motivates the following general definition of appropriateness measures.

Definition 7 (*Appropriateness Measures*) *$\forall \theta \in LE, \forall x \in \Omega$ the measure of appropriateness degrees of θ as a description of x is given by:*

$$\mu_\theta(x) = \sum_{S \in \lambda(\theta)} m_x(S)$$

Appropriateness degrees (def. 2) introduced at the beginning of this chapter are only a special case of the appropriateness measures where $\theta = L$ for $L \in \mathcal{L}$.

Example 1. Given a continuous variable x: $\mathcal{L} = \{small, medium, large\}$, $\mathcal{F} = \{\{small\}, \{small, medium\}, \{medium\}, \{medium, large\}, \{large\}\}$. Suppose we are told that "x is **not large** but it is **small or medium**". This constraint can be interpreted as the logical expression

$$\theta = \neg large \wedge (small \vee medium)$$

According to definition 6, the possible label sets of the given logical expression θ are calculated as follows:

$$\lambda(\neg large) = \{\{small\}, \{small, medium\}, \{medium\}\}$$

$$\lambda(small) = \{\{small\}, \{small, medium\}\}$$

$$\lambda(medium) = \{\{small, medium\}, \{medium\}, \{medium, large\}\}$$

So that we can obtain:

$$\lambda(\theta) = \lambda(\neg large \wedge (small \vee medium)) = \{\{small\}, \{small, medium\}, \{medium\}\}$$
$$\wedge (\{\{small\}, \{small, medium\}\} \vee \{\{small, medium\}, \{medium\}, \{medium, large\}\})$$
$$= \{\{small\}, \{small, medium\}, \{medium\}\}$$

If a prior distribution on focal elements of variable x are given as follows:

$$\{small\} : 0.1, \{small, med.\} : 0.3, \{med.\} : 0.1, \{med., large\} : 0.5, \{large\} : 0.0$$

The appropriateness measure for $\theta = \neg large \wedge (small \vee medium)$ is:

$$\mu_\theta(x) = \sum_{S \in \lambda(\theta)} m_x(S)$$
$$= m_x(\{small\}) + m_x(\{small, medium\}) + m_x(\{medium\})$$
$$= 0.1 + 0.3 + 0.1 = 0.5$$

3.2 Linguistic Interpretation of the Sets of Appropriate Labels

Based on the inverse of the λ-function (def. 6), a set of linguistic rules (or logical label expressions) can be obtained from a given set of possible label sets. For example, suppose we are given the possible label sets $\{\{small\}, \{small, medium\}, \{medium\}\}$, which does not have an immediately obvious interpretation. However by using the α-function, we can convert this set into a corresponding linguistic expression $(small \vee medium) \wedge \neg large$ or its logical equivalence.

Definition 8 (α-function)

$$\forall F \in \mathcal{F} \quad let \quad \mathcal{N}(F) = \left(\bigcup_{F' \in \mathcal{F}: F' \supseteq F} F' \right) - F \tag{6}$$

$$then \quad \alpha_F = \left(\bigwedge_{L \in F} L \right) \wedge \left(\bigwedge_{L \in \mathcal{N}(F)} \neg L \right) \tag{7}$$

We can then map a set of focal elements to label expressions based on the α-function as follows:

$$\forall R \in \mathcal{F} \quad \theta_R = \bigvee_{F \in R} \alpha_F \quad where \quad \lambda(\theta_R) = R \tag{8}$$

The motivation of this mapping is as follows. Given a focal element $\{s, m\}$ (i.e. $\{small, medium\}$) this states that the labels appropriate to describe the attribute are exactly *small* and *medium*. Hence, they include s and m and exclude all other labels that occur in focal sets that are supersets of $\{s, m\}$. Given a set of focal sets $\{\{s, m\}, \{m\}\}$ this provides the information that the set of labels is either $\{s, m\}$ or $\{m\}$ and hence the sentence providing the same information should be the disjunction of the α sentences for both focal sets. The following example gives the calculation of the α-function.

Example 2. Let $\mathcal{L} = \{very\ small\ (vs),\ small\ (s),\ medium\ (m),\ large\ (l),\ very\ large\ (vl)\}$ and $\mathcal{F} = \{\{vs, s\}, \{s\}, \{s, m\}, \{m\}, \{m, l\}, \{l\}, \{l, vl\}\}$. For calculating $\alpha_{\{l\}}$, we obtain

$$F' \in \mathcal{F} : F' \supseteq \{l\} = \{\{m, l\}, \{l\}, \{l, vl\}\} = \{m, l, vl\}$$

$$\mathcal{N}(\{l\}) = \left(\bigcup_{F' \in \mathcal{F}: F' \supseteq \{l\}} F' \right) - \{l\} = \{l, vl, m\} - \{l\} = \{vl, m\}$$

$$\alpha_{\{l\}} = \left(\bigwedge_{L \in F} L \right) \wedge \left(\bigwedge_{L \in \mathcal{N}(F)} \neg L \right) = (l) \wedge (\neg m \wedge \neg vl) = \neg m \wedge l \wedge \neg vl$$

Also we can also obtain

$$\alpha_{\{m,l\}} = m \wedge l \quad \alpha_{\{l,vl\}} = l \wedge vl$$

Hence, a set of label sets $\{\{m, l\}, \{l\}, \{l, vl\}\}$ can be represented by a linguistic expression as follows,

$$\theta_{\{\{m,l\},\{l\},\{l,vl\}\}} = \alpha_{\{m,l\}} \vee \alpha_{\{l\}} \vee \alpha_{\{l,vl\}} =$$

$$(m \wedge l) \vee (\neg m \wedge l \neg vl) \vee (l \wedge vl) \equiv_L large$$

where '$\equiv_L$' represents logical equivalence (see def. 6).

Basically, α-function provides a way of obtaining logical expressions from a random set description of labels. It is an inverse process of to the λ-function.

As a framework of reasoning with uncertainty, label semantics aims to model vague or imprecise concepts which can be used as a knowledge representation tool in high-level modelling tasks. We hope to develop models to be defined interms of linguistic expressions we can enhance robustness, accuracy and transparency. Transparent models should allow for a qualitative understanding of the underlying system in addition to giving quantitative predictions of behaviour. Based on label semantics, several new transparent data mining algorithms have been proposed. We found these algorithms have better transparency and comparable accuracy compared to other algorithms. These algorithms will be introduced in details in the following sections.

4 Linguistic Decision Tree

Tree induction learning models have received a great deal of attention over recent years in the fields of machine learning and data mining because of their simplicity and effectiveness. Among them, the ID3 [22] algorithm for decision trees induction has proved to be an effective and popular algorithm for building decision trees from discrete valued data sets. The C4.5 [24] algorithm was proposed as a successor to ID3 in which an entropy based approach to crisp partitioning of continuous universes was adopted. One inherent disadvantage of crisp partitioning is that it tends to make the induced decision trees sensitive to noise. This noise is not only due to the lack of precision or errors in measured features but is often present in the model itself since the available features may not be sufficient to provide a complete model of the system. For each attribute, disjoint classes are separated with clearly defined boundaries. These boundaries are 'critical' since a small change close to these points will probably cause a complete change in classification. Due to the existence of uncertainty and imprecise information in real-world problems, the class boundaries may not be defined clearly. In this case, decision trees may produce high misclassification rates in testing even if they perform well in training. To overcome this problems, many fuzzy decision tree models have been proposed [2, 9, 14, 15].

Linguistic decision tree (LDT) [19] is a tree-structured classification model based on label semantics. The information heuristics used for building the tree are modified from Quinlan's ID3 [22] in accordance with label semantics. Given a database of which each instance is labeled by one of the classes: $\{C_1, \cdots, C_M\}$. A linguistic decision tree with S consisting branches built from this database can be defined as follows:

$$T = \{\langle B_1, P(C_1|B_1), \cdots, P(C_M|B_1)\rangle, \cdots \langle B_S, P(C_1|B_S), \cdots, P(C_M|B_S)\rangle\}$$

where $P(C_k|B)$ is the probability of class C_k given a branch B. A branch B with d nodes (i.e., the length of B is d) is defined as: $B = \langle F_1, \cdots, F_d \rangle$, where $d \leq n$

and $F_j \in \mathcal{F}_j$ is one of the focal elements of attribute j. For example, consider the branch: $\langle\langle\{small_1\}, \{medium_2, large_2\}\rangle, 0.3, 0.7\rangle$. This means the probability of class C_1 is 0.3 and C_2 is 0.7 given attribute 1 can only be described as *small* and attribute 2 can be described as both *medium* and *large*.

These class probabilities are estimated from a training set $\mathcal{D} = \{x_1, \cdots, x_N\}$ where each instance x has n attributes: $\langle x_1, \cdots, x_n \rangle$. We now describe how the relevant branch probabilities for a LDT can be evaluated from a database. The probability of class C_k ($k = 1, \cdots, M$) given B can then be evaluated as follows. First, we consider the probability of a branch B given x:

$$P(B|x) = \prod_{j=1}^{d} m_{x_j}(F_j) \tag{9}$$

where $m_{x_j}(F_j)$ for $j = 1, \cdots, d$ are mass assignments of single data element x_j. For example, suppose we are given a branch $B = \langle\{small_1\}, \{medium_2, large_2\}\rangle$ and data x $= \langle 0.27, 158 \rangle$ (the linguistic translation of x_1 was given in Sect. 2.3). According to (9):

$$P(B|x) = m_{x_1}(\{small_1\}) \times m_{x_2}(\{medium_2, large_2\}) = 0.4 \times 0.4 = 0.16$$

The probability of class C_k given B can then be evaluated[3] by:

$$P(C_k|B) = \frac{\sum_{i \in \mathcal{D}_k} P(B|x_i)}{\sum_{i \in \mathcal{D}} P(B|x_i)} \tag{10}$$

where $\mathcal{D}_k$ is the subset consisting of instances which belong to class k. According to the Jeffrey's rule [13] the probabilities of class C_k given a LDT with S branches are evaluated as follows:

$$P(C_k|x) = \sum_{s=1}^{S} P(C_k|B_s)P(B_s|x) \tag{11}$$

where $P(C_k|B_s)$ and $P(B_s|x)$ are evaluated based on (9) and (10).

---

[3] In the case where the denominator is equals to 0, which may occur when the training database for the LDT is small, then there is no non-zero linguistic data covered by the branch. In this case, we obtain no information from the database so that equal probabilities are assigned to each class. $P(C_k|B) = \frac{1}{M}$ for $k = 1, \cdots, M$. In the case that a data element appears beyond the range of training data set, we then assign the appropriateness degrees of the minimum or maximum values of the universe to the data element depending on which side of the range it appears.

4.1 Linguistic ID3 Algorithm

Linguistic ID3 (LID3) is the learning algorithm we propose for building the linguistic decision tree based on a given linguistic database. Similar to the ID3 algorithm [22], search is guided by an information based heuristic, but the information measurements of a LDT are modified in accordance with label semantics. The measure of information defined for a branch B and can be viewed as an extension of the entropy measure used in ID3.

Definition 9 (*Branch Entropy*) *The entropy of branch B given a set of classes $C = \{C_1, \ldots, C_{|C|}\}$ is*

$$E(B) = -\sum_{t=1}^{|C|} P(C_t|B) \log_2 P(C_t|B) \tag{12}$$

Now, given a particular branch B suppose we want to expand it with the attribute x_j. The evaluation of this attribute will be given based on the *Expected Entropy* defined as follows:

Definition 10 (*Expected Entropy*)

$$EE(B, x_j) = \sum_{F_j \in \mathcal{F}_j} E(B \cup F_j) \cdot P(F_j|B) \tag{13}$$

where $B \cup F_j$ represents the new branch obtained by appending the focal element F_j to the end of branch B. The probability of F_j given B can be calculated as follows:

$$P(F_j|B) = \frac{\sum_{i \in D} P(B \cup F_j|x_i)}{\sum_{i \in D} P(B|x_i)} \tag{14}$$

We can now define the Information Gain (IG) *obtained by expanding branch B with attribute x_j as:*

$$IG(B, x_j) = E(B) - EE(B, x_j) \tag{15}$$

The goal of tree-structured learning models is to make subregions partitioned by branches be less "impure", in terms of the mixture of class labels, than the unpartitioned dataset. For a particular branch, the most suitable free attribute for further expanding (or partitioning), is the one by which the "pureness" is maximally increased with expanding. That corresponds to selecting the attribute with maximum information gain. As with ID3 learning, the most informative attribute will form the root of a linguistic decision tree, and the tree will expand into branches associated with all possible focal elements of this attribute. For each branch, the free attribute with maximum information gain will be the next node, from level to level, until the tree reaches the maximum specified depth or the maximum class probability reaches the given threshold probability.

4.2 Forward Merging Algorithm

From the empirical studies on UCI data mining repository [4], we showed that LID3 performs at least as well as and often better than three well-known classification algorithms across a range of datasets (see [19]). However, even with only 2 fuzzy sets for discretization, the number of branches increases exponentially with the depth of the tree. Unfortunately, the transparency of the LDT decreases with the increasing number of branches. To help to maintain transparency by generating more compact trees, a forward merging algorithm based on the LDT model is proposed in this section and experimental results are given to support the validity of our approach.

In a full developed linguistic decision tree, if any of two adjacent branches have sufficiently similar class probabilities according to some criteria, so that these two branches may give similar classification results and therefore can then be merged into one branch in order to obtain a more compact tree. We employ a *merging threshold* to determine whether or not two adjacent branches can be merged.

Definition 11 (*Merging Threshold*) *In a linguistic decision tree, if the maximum difference between class probabilities of two adjacent branches B_1 and B_2 is less than or equal to a given merging threshold T_m, then the two branches can be merged into one branch. Formally, if*

$$T_m \geq \max_{c \in C} |P(c|B_1) - P(c|B_2)| \qquad (16)$$

where $C = \{C_1, \cdots, C_{|C|}\}$ is the set of classes, then B_1 and B_2 can be merged into one branch MB.

Definition 12 (*Merged Branch*) *A merged branch MB nodes is defined as*

$$MB = \langle \mathcal{M}_{j_1}, \cdots, \mathcal{M}_{j_{|MB|}} \rangle$$

$|MB|$ are the number of nodes for the branch MB where the node is defined as:

$$\mathcal{M}_j = \{F_j^1, \cdots, F_j^{|\mathcal{M}_j|}\}$$

Each node is a set of focal elements such that F_j^i is adjacent to F_j^{i+1} for $i = 1, \cdots, |\mathcal{M}_j| - 1$. If $|\mathcal{M}| > 1$, it is called compound node, which means it is a compound of more than one focal elements because of merging. The associate mass for $\mathcal{M}_j$ is given by

$$m_x(\mathcal{M}_j) = \sum_{i=1}^{|\mathcal{M}_j|} m_x(F_j^i) \tag{17}$$

where w is the number of merged adjacent focal elements for attribute j.

Based on (9), we can obtain:

$$P(C_t|\mathbf{x}) = \prod_{r=1}^{|MB|} m_{x_r}(\mathcal{M}_r) \tag{18}$$

Therefore, based on (10) and (17) we use the following formula to calculate the class probabilities given a merged branch.

$$P(C_t|MB) = \frac{\sum_{i \in \mathcal{D}_t} P(C_t|\mathbf{x}_i)}{\sum_{i \in \mathcal{D}} P(C_t|\mathbf{x}_i)} \tag{19}$$

For example, Fig. 2 shows the change in test accuracy and the number of leaves (or the number of rules interpreted from a LDT) for different T_m on the wis-cancer

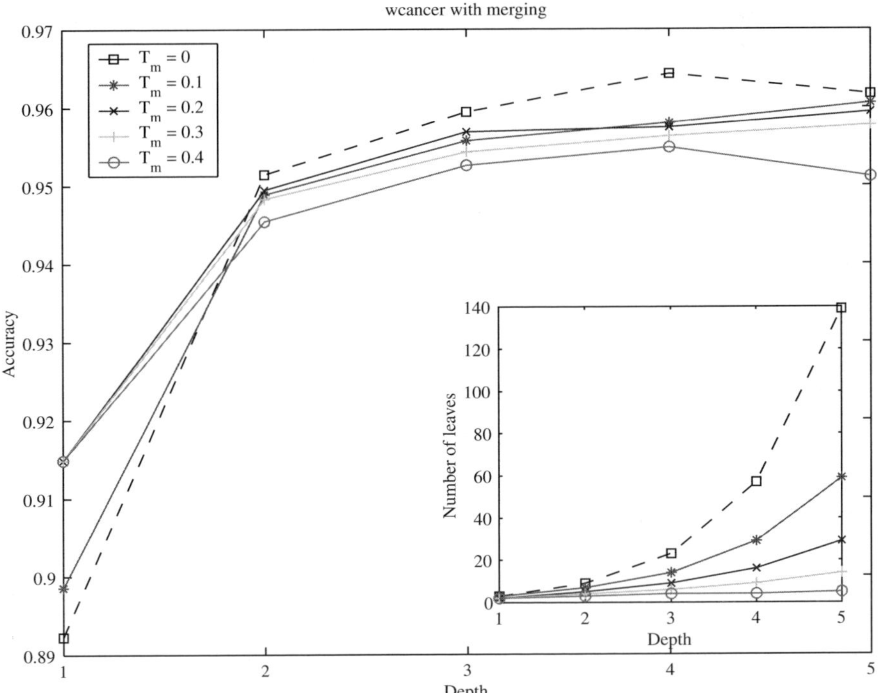

Fig. 2 The change in accuracy and number of leaves as T_m varies on the wis-cancer dataset with $N_F = 2$. While the dot trial $T_m = 0$ is with $N_F = 2$

dataset. It shows that the accuracy is not greatly influenced by merging, but the number of branches is greatly reduced. This is especially true for the curve marked by '+' corresponding to $T_m = 0.3$ where applying forward merging, the best accuracy (at the depth 4) is only reduced by approximately 1%, whereas, the number of branches is reduced by roughly 84%.

4.3 Linguistic Constraints

Here we assume that the linguistic constraints take the form of $\theta = \langle x_1$ is $\theta_1, \ldots, x_n$ is $\theta_n \rangle$, where θ_j represents a label expression based on $\mathcal{L}_j : j = 1, \ldots, n$. Consider the vector of linguistic constraint $\vec{\theta} = \langle \theta_1, \cdots, \theta_n \rangle$, where θ_j is the linguistic constraints on attribute j. We can evaluate a probability value for class C_t conditional on this information using a given linguistic decision tree as follows. The mass assignment given a linguistic constraint θ is evaluated by

$$\forall F_j \in \mathcal{F}_j \quad m_{\theta_j}(F_j) = \begin{cases} \dfrac{pm(F_j)}{\sum_{F_j \in \lambda(\theta_j)} pm(F_j)} & if : F_j \in \lambda(\theta_j) \\ 0 & otherwise \end{cases} \tag{20}$$

where $pm(F_j)$ is the prior mass for focal elements $F_j \in \mathcal{F}_j$ derived from the prior distribution $p(x_j)$ on Ω_j as follows:

$$pm(F_j) = \int_{\Omega_j} m_x(F_j) p(x_j) dx_j \tag{21}$$

Usually, we assume that $p(x_j)$ is the uniform distribution over Ω_j so that

$$pm(F_j) \propto \int_{\Omega_j} m_x(F_j) dx_j \tag{22}$$

For branch B with s nodes, the probability of B given $\vec{\theta}$ is evaluated by

$$P(B|\vec{\theta}) = \prod_{r=1}^{|B|} m_{\theta_{j_r}}(F_{j_r}) \tag{23}$$

and therefore, by Jeffrey's rule [13]

$$P(C_t|\vec{\theta}) = \sum_{v=1}^{|LDT|} P(C_t|B_v) P(B_v|\vec{\theta}) \tag{24}$$

The methodology for classification under linguistic constraints allows us to fuse the background knowledge in linguistic form into classification. This is one of the advantages of using high-level knowledge representation language models such as label semantics.

4.4 Linguistic Decision Trees for Predictions

Consider a database for prediction $\mathcal{D} = \{\langle x_1(i), \cdots, x_n(i), x_t(i)\rangle \,|\, i = 1, \cdots, |\mathcal{D}|\}$ where $x_1, \cdots, x_n$ are potential explanatory attributes and x_t is the continuous target attribute. Unless otherwise stated, we use trapezoidal fuzzy sets with 50% overlap to discretized each continuous attribute individually (x_t) universe and assume the focal sets are $\mathcal{F}_1, \cdots, \mathcal{F}_n$ and $\mathcal{F}_t$. For the target attribute x_t: $\mathcal{F}_t = \{F_t^1, \cdots, F_t^{|\mathcal{F}_t|}\}$. For other attributes: x_j: $\mathcal{F}_j = \{F_j^1, \ldots, F_j^{|\mathcal{F}_j|}\}$. The inventive step is, to regard the focal elements for the target attribute as class labels. Hence, the LDT[4] model for prediction has the following form: A linguistic decision tree for prediction is a set of branches with associated probability distribution on the target focal elements of the following form:

$$LDT = \{\langle B_1, P(F_t^1|B_1), \cdots, P(F_t^{|\mathcal{F}_t|}|B_1)\rangle, \cdots,$$
$$\langle B_{|LDT|}, P(F_t^1|B_{|LDT|}), \cdots, P(F_t^{|\mathcal{F}_t|})|B_{|LDT|})\rangle\}$$

where $F_t^1, \cdots, F_t^{|\mathcal{F}_t|}$ are the target focal elements (i.e. the focal elements for the target attribute or the output attribute).

$$P(F_t^j|\mathrm{x}) = \sum_{v=1}^{|LDT|} P(F_t^j|B_v)P(B_v|\mathrm{x}) \tag{25}$$

Given value $\mathrm{x} = \langle x_1, \cdots, x_n\rangle$ we need to estimate the target value $\widehat{x_t}$ (i.e. $\mathrm{x}_i \to \widehat{x_t}$). This is achieved by initially evaluating the probabilities on target focal elements: $P(F_t^1|\mathrm{x}), \cdots, P(F_t^{|\mathcal{F}_t|}|\mathrm{x})$ as described above. We then take the estimate of x_t, denoted $\widehat{x_t}$, to be the expected value:

$$\widehat{x_t} = \int_{\Omega_t} x_t \, p(x_t|\mathrm{x}) \, dx_t \tag{26}$$

where:

$$p(x_t|\mathrm{x}) = \sum_{j=1}^{|\mathcal{F}_t|} p(x_t|F_t^j) \, P(F_t^j|\mathrm{x}) \tag{27}$$

---

[4] We will use the same name 'LDT' for representing both linguistic decision trees (for classification) and linguistic prediction trees.

                                                    Z. Qin, J. Lawry

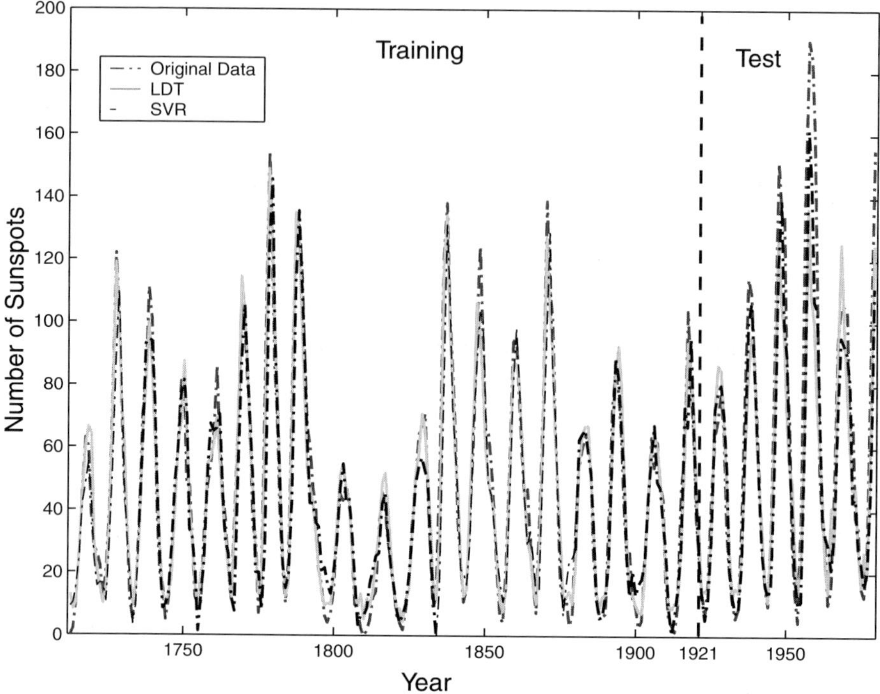

Fig. 3 The prediction results obtained from SVR and LID3 without merging, where the data on the left (1712-1921) are for training and the right (1921-1079) are for test

and

$$p(x_t|F_t^j) = \frac{m_{x_t}(F_t^j)}{\int_{\Omega_t} m_{x_t}(F_t^j)\,dx_t} \tag{28}$$

so that, we can obtain:

$$\widehat{x}_t = \sum_j P(F_t^j|\mathbf{x})\,E(x_t|F_t^j) \tag{29}$$

where:

$$E(x_t|F_t^j) = \int_{\Omega_t} x_t\,p(x_t|F_t^j)\,dx_t = \frac{\int_{\Omega_t} x_t\,m_{x_t}(F_t^j)\,dx_t}{\int_{\Omega_t} m_{x_t}(F_t^j)\,dx_t} \tag{30}$$

We tested our model with a real-world problem taken from the Time Series Data Library [8] and contains data of sunspot numbers between the years 1700-1979. The input attributes are x_{T-12} to x_{T-1} (the data for previous 12 years) and the output (target) attribute is x_T, i.e. one-year-ahead. The experimental results for LID3 and ε-SVR [6] are compared in Fig. 3. We can see the results are quite comparable. More details are available in [20].

5 Bayesian Estimation Tree Based on Label Semantics

Bayesian reasoning provides a probabilistic approach to inference based on the Bayesian theorem. Given a test instance, the learner is asked to predict its class according to the evidence provided by the training data. The classification of unknown example x by Bayesian estimation is on the basis of the following probability,

$$P(C_k|\mathrm{x}) = \frac{P(\mathrm{x}|C_k)P(C_k)}{P(\mathrm{x})} \tag{31}$$

Since the denominator in (31) is invariant across classes, we can consider it as a normalization parameter. So, we obtain:

$$P(C_k|\mathrm{x}) \propto P(\mathrm{x}|C_k)P(C_k) \tag{32}$$

Now suppose we assume for each variable x_j that its outcome is independent of the outcome of all other variables given class C_k. In this case we can obtain the so-called naive Bayes classifier as follows:

$$P(C_k|\mathrm{x}) \propto \prod_{j=1}^{n} P(x_j|C_k)P(C_k) \tag{33}$$

where $P(x_j|C_k)$ is often called the likelihood of the data x_j given C_k. For a qualitative attribute, it can be estimated from corresponding frequencies. For a quantitative attribute, either probability density estimation or discretization can be employed to estimate its probabilities.

5.1 Fuzzy Naive Bayes

In label semantics framework, suppose we are given focal set $\mathcal{F}_j$ for each attribute j. Assuming that attribute x_j is numeric with universe Ω_j, then the likelihood of x_j given C_k can be represented by a density function $p(x_j|C_k)$ determine from the database $\mathcal{D}_k$ and prior density according to Jeffrey's rule [13].

$$p(x_j|C_k) = \sum_{F \in \mathcal{F}_j} p(x_j|F)P(F|C_k) \tag{34}$$

From Bayes theorem, we can obtain:

$$p(x_j|F) = \frac{P(F|x_j)p(x_j)}{P(F)} = \frac{m_{x_j}(F)p(x_j)}{pm(F)} \tag{35}$$

where,

$$pm(F) = \int_{\Omega_j} P(F|x_j)p(x_j)dx_j = \frac{\sum_{x \in D} m_{x_j}(F)}{|D|} \qquad (36)$$

Substituting 35 in 34 and re-arranging gives

$$p(x_j|C_k) = p(x_j) \sum_{F \in \mathcal{F}_j} m_{x_j}(F) \frac{P(F|C_k)}{pm(F)} \qquad (37)$$

where $P(F|C_k)$ can be derived from $\mathcal{D}_k$ according to

$$P(F|C_k) = \frac{\sum_{x \in \mathcal{D}_k} m_{x_j}(F)}{|\mathcal{D}_k|} \qquad (38)$$

This model is called fuzzy Naive Bayes (FNB). If we weaken the independence assumption, we can obtain a fuzzy semi-Naive Bayes (FSNB). More details of FNB and FSNB can be found in [25].

5.2 Hybrid Bayesian Estimation Tree

Based on previous two linguistic models, a hybrid model was proposed in [18]. Given a decision tree T is learnt from a training database $\mathcal{D}$. According to the Bayesian theorem: A data element $x = \langle x_1, \ldots, x_n \rangle$ can be classified by:

$$P(C_k|x, T) \propto P(x|C_k, T)P(C_k|T) \qquad (39)$$

We can then divide the attributes into 2 disjoint groups denoted by $x_T = \{x_1, \cdots, x_m\}$ and $x_B = \{x_{m+1}, \cdots, x_n\}$, respectively. x_T is the vector of the variables that are contained in the given tree T and the remaining variables are contained in x_B. Assuming conditional independence between x_T and x_B we obtain:

$$P(x|C_k, T) = P(x_T|C_k, T)P(x_B|C_k, T) \qquad (40)$$

Because x_B is independent of the given decision tree T and if we assume the variables in x_B are independent of each other given a particular class, we can obtain:

$$P(x_B|C_k, T) = P(x_B|C_k) = \prod_{j \in x_B} P(x_j|C_k) \qquad (41)$$

Now consider x_T. According to Bayes theorem,

$$P(x_T|C_k, T) = \frac{P(C_k|x_T, T)P(x_T|T)}{P(C_k|T)} \tag{42}$$

Combining (40), (41) and (42):

$$P(x|C_k, T) = \frac{P(C_k|x_T, T)P(x_T|T)}{P(C_k|T)} \prod_{j \in x_B} P(x_l|C_k) \tag{43}$$

Combining (39) and (43)

$$P(C_k|x, T) \propto P(C_k|x_T, T)P(x_T|T) \prod_{j \in x_B} P(x_j|C_k) \tag{44}$$

Further, since $P(x_T|T)$ is independent from C_k, we have that:

$$P(C_k|x, T) \propto P(C_k|x_T, T) \prod_{j \in x_B} P(x_j|C_k) \tag{45}$$

where $P(x_j|C_k)$ is evaluated according to 37 and $P(C_k|x_T, T)$ is just the class probabilities evaluated from the decision tree T according to 11.

We tested this new model with a set of UCI [4] data sets. Figure 4 is a simple result. More results are available in [18]. From Figs. 4, we can see that the BLDT model generally performs better at shallow depths than LDT model. However, with the increasing of the tree depth, the performance of the BLDT model remains constant or decreases, while the accuracy curves for LDT increase. The basic idea of using Bayesian estimation given a LDT is to use the LDT as one estimator and the rest of the attributes as other independent estimators. Consider the two extreme cases for 45. If all the attributes are used in building the tree (i.e. $x_T = x$), the probability estimations are from the tree only, that is:

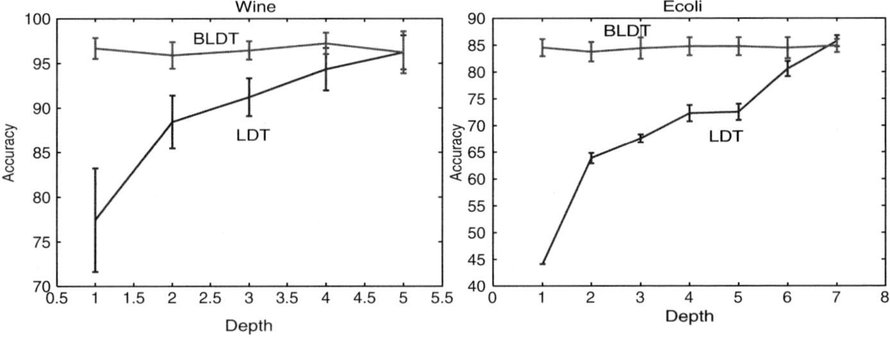

Fig. 4 Results for single LDT with Bayesian estimation: average accuracy with standard deviation on each dataset against the depth of the tree

$$P(C_k|\mathbf{x}, T) \propto P(C_k|\mathbf{x}_T, T)$$

If none of the attributes are used in developing the tree (i.e. $\mathbf{x} = \mathbf{x}_B$), the probability estimation will become:

$$P(C_k|\mathbf{x}, T) \propto \prod_{j \in \mathbf{x}_B} P(x_j|C_k)$$

which is simply a Naive Bayes classifier.

6 Linguistic Rule Induction

The use of high-level knowledge representation in data modelling allows for enhanced transparency in the sense that the inferred models can be understood by practioners who are not necessarily experts in the formal representation framework employed. Rule based systems inherently tend to be more transparent than other models such as neural networks. A set of concise understandable rules can provide a better understanding of how the classification or prediction is made. Generally, there are two general types of algorithms for rule induction, *top down* and *bottom up* algorithms. Top-down approaches start from the most general rule and specialize it gradually. Bottom-up methods star from a basic fact given in training database and generalize it. In this paper we will focus on a top-down model for generating linguistic rules based on Quinlan's *First-Order Inductive Learning* (FOIL) Algorithm [23].

The FOIL algorithm is based on classical binary logic where typically attributes are assumed to be discrete. Numerical variables are usually discretized by partitioning the numerical domain into a finite number of intervals. However, because of the uncertainty involved in most real-world problems, sharp boundaries between intervals often lead to a loss of robustness and generality. Fuzzy logic has been used to solve the problem of sharp transitions between two intervals. Fuzzy rule induction research has been popular in both fuzzy and machine learning communities as a means to learning robust transparent models. Many algorithms have been proposed including simple fuzzy logic rule induction [3], fuzzy association rule mining [27] and first-order fuzzy rule induction based on FOIL [5, 16]. In this paper, we will focus on an extension to the FOIL algorithm based on label semantics.

6.1 Linguistic Rules in Label Semantics

In Sect. 2 and 3, a basic introduction of label semantics is given and how it can be used for data modelling is discussed. In this section, we will describe a linguistic rule induction model based on label semantics. Now, we begin by clarifying the definition of a linguistic rule. Based on def. 5, a linguistic rule is a rule can be represented as a multi-dimensional logical expressions of fuzzy labels.

Definition 13 (*Multi-dimensional Logical Expressions of Labels*) $MLE^{(n)}$ *is the set of all multi-dimensional label expressions that can be generated from the logical label expression LE_j: $j = 1, \ldots, n$ and is defined recursively by:*

(i) If $\theta \in LE_j$ for $j = 1, \ldots, n$ then $\theta \in MLE^{(n)}$
(ii) If $\theta, \varphi \in MLE^{(n)}$ then $\neg\theta, \theta \wedge \varphi, \theta \vee \varphi, \theta \to \varphi \in MLE^{(n)}$

Any n-dimensional logical expression θ identifies a subset of $2^{\mathcal{L}_1} \times \ldots \times 2^{\mathcal{L}_n}$, denoted $\lambda^{(n)}(\theta)$ (see example 3), constraining the cross product of logical descriptions on each variable: $D_{x_1} \times \ldots \times D_{x_1}$. In such a way the imprecise constraint θ on n variables can be interpret as the precise constraint $D_{x_1} \times \ldots \times D_{x_1} \in \lambda^{(n)}(\theta)$

Given a particular data, how can we evaluated if a linguistic rule is appropriate for describing it? Based on the one-dimensional case, we now extend the concepts of appropriateness degrees to the multi-dimensional case as follows:

Definition 14 (*Multi-dimensional Appropriateness Degrees*) *Given a set of n-dimensional label expressions $MLE^{(n)}$:*

$$\forall\, \theta \in MLE^{(n)}, \forall x_j \in \Omega_j : j = 1, \cdots, n$$

$$\mu_\theta^n(\mathbf{x}) = \mu_\theta^n(x_1, \cdots, x_n) = \sum_{\langle F_1, \cdots, F_n \rangle \in \lambda^{(n)}(\theta)} (F_1, \cdots, F_n)$$

$$= \sum_{\langle F_1, \cdots, F_n \rangle \in \lambda^{(n)}(\theta)} \prod_{j=1}^{n} m_{x_j}(F_j)$$

The appropriateness degrees in one-dimension are for evaluating a single label for describing a single data element, while in multi-dimensional cases they are for evaluating a linguistic rule for describing a data vector.

Example 3. Consider a modelling problem with two variables x_1 and x_2 for which $\mathcal{L}_1 = \{small\ (s),\ medium\ (med),\ large(lg)\}$ and $\mathcal{L}_2 = \{low(lo),\ moderate\ (mod), high(h)\}$. Also suppose the focal elements for $\mathcal{L}_1$ and $\mathcal{L}_2$ are:

$$\mathcal{F}_1 = \{\{s\}, \{s, med\}, \{med\}, \{med, lg\}, \{lg\}\}$$

$$\mathcal{F}_2 = \{\{lo\}, \{lo, mod\}, \{mod\}, \{mod, h\}, \{h\}\}$$

According to the multi-dimensional generalization of definition 6 we have that

$$\lambda^{(2)}((med \wedge \neg s) \wedge \neg lo) = \lambda^{(2)}(med \wedge \neg s) \cap \lambda^{(2)}(\neg lo)$$

$$= \lambda(med \wedge \neg s) \times \lambda(\neg lo)$$

Now, the set of possible label sets is obtained according to the λ-function:

$$\lambda(med \wedge \neg s) = \{\{med\}, \{med, lg\}\}$$

$$\lambda(\neg lo) = \{\{mod\}, \{mod, h\}, \{h\}\}$$

Hence, based on def. 6 we can obtain:

$$\lambda^{(2)}((med \wedge \neg s) \wedge \neg lo) = \{\langle\{med\}, \{mod\}\rangle, \langle\{med\}, \{mod, h\}\rangle,$$

$$\langle\{med\}, \{h\}\rangle, \langle\{med, lg\}, \{mod\}\rangle, \langle\{med, lg\}, \{mod, h\}\rangle, \langle\{med, lg\}, \{h\}\rangle\}$$

The above calculation on random set interpretation of the given rule based on λ-function is illustrated in Fig. 5: given focal set $\mathcal{F}_1$ and $\mathcal{F}_2$, we can construct a 2-dimensional space where the focal elements have corresponding focal cells. Representation of the multi-dimensional λ-function of the logical expression of the given rule are represented by grey cells.

Given x $= \langle x_1, x_2 \rangle = \langle x_1 = \{med\} : 0.6, \{med, lg\} : 0.4\rangle, \langle x_2 = \{lo, mod\} : 0.8, \{mod\} : 0.2\rangle$, we obtain:

$$\mu_\theta(x) = (m(\{med\}) + m(\{med, lg\})) \times (m(\{mod\}) + m(\{mod, h\}) + m(\{h\}))$$

$$= (0.6 + 0.4) \times (0.2 + 0 + 0) = 0.2$$

And according to def. 6:

$$\mu^n_{\neg\theta}(x) = 1 - \mu_\theta(x) = 0.8$$

In another words, we can say that the linguistic expression θ covers the data x to degree 0.2 and θ can be considered as a linguistic rule. This interpretation of appropriateness is highlighted in next section on rule induction.

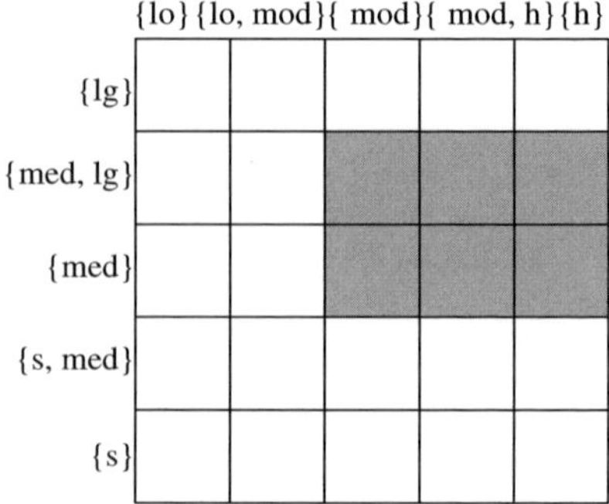

Fig. 5 Representation of the multi-dimensional λ-function of the logical expression $\theta = (med \wedge \neg s) \wedge \neg lo$ showing the focal cells $\mathcal{F}_1 \times \mathcal{F}_2$

6.2 Information Heuristics for LFOIL

In the last section, we have shown how to evaluate the appropriateness of using a linguistic rule to describe a data vector. In this section, a new algorithm for learning a set of linguistic rules is proposed based on the FOIL algorithm [23], it is referred to as *Linguistic FOIL* (LFOIL). Generally, the heuristics for a rule learning model are for assessing the usefulness of a literal as the next component of the rule. The heuristics used for LFOIL are similar but modified from the FOIL algorithm [23] so as to incorporate linguistic expressions based on labels semantics. Consider a classification rule of the form:

$$R_i = \theta \to C_k \ \ where \ \ \theta \in MLE^{(n)}$$

Given a data set $\mathcal{D}$ and a particular class C_k, the data belonging to class C_k are referred to as *positive examples* and the rest of them are *negative examples*. For the given rule R_i, the coverage of positive data is evaluated by

$$T_i^+ = \sum_{l \in \mathcal{D}_k} \mu_\theta(x_l) \tag{46}$$

and the coverage of negative examples is given by

$$T_i^- = \sum_{l \in (\mathcal{D} - \mathcal{D}_k)} \mu_\theta(x_l) \tag{47}$$

where $\mathcal{D}_k$ is the subset of the database which is consisted by the data belonging to class C_k. The information for the original rule R_i can by evaluated by

$$I(R_i) = -\log_2 \left(\frac{T_i^+}{T_i^+ + T_i^-} \right) \tag{48}$$

Suppose we then propose to another label expression φ to the body of R_i to generate a new rule

$$R_{i+1} = \varphi \wedge \theta \to C_k$$

where $\varphi, \theta \in MLE^{(n)}$. By adding the new literal φ, the positive and negative coverage becomes:

$$T_{i+1}^+ = \sum_{l \in \mathcal{D}_k} \mu_{\theta \wedge \varphi}(x_l) \tag{49}$$

$$T_{i+1}^- = \sum_{l \in (\mathcal{D} - \mathcal{D}_k)} \mu_{\theta \wedge \varphi}(x_l) \tag{50}$$

Therefore, the information becomes,

$$I(R_{i+1}) = -\log_2 \left(\frac{T_{i+1}^+}{T_{i+1}^+ + T_{i+1}^-} \right) \tag{51}$$

Then we can evaluate the information gain from adding expression φ by:

$$G(\varphi) = T_{i+1}^+ (I(R_i) - I(R_{i+1})) \tag{52}$$

We can see that the measure of information gain consists of two components. T_{i+1}^+ is the coverage of positive data by the new rule R_{i+1} and $(I(R_i) - I(R_{i+1}))$ is the increase of information. The probability of C_k given a linguistic rule R_i is evaluated by:

$$P(C_k|R_i) = \frac{\sum_{l \in \mathcal{D}_k} \mu_\theta(x_l)}{\sum_{l \in \mathcal{D}} \mu_\theta(x_l)} = \frac{T_i^+}{T_i^+ + T_i^-} \tag{53}$$

when $P(C_k|R_{i+1}) > P(C_k|R_i)$ (i.e., by appending a new literal, more positive examples are covered), we can obtain that $(I(R_i) - I(R_{i+1})) > 0$. By choosing a literal φ with maximum G value, we can form the new rule which covers more positive examples and thus increasing the accuracy of the rule.

6.3 Linguistic FOIL

We define a prior knowledge base $KB \subseteq MLE^{(n)}$ and a probability threshold $PT \in [0, 1]$. KB consists of fuzzy label expressions based on labels defined on each attribute. For example, given fuzzy labels $\{small_1 \ large_1\}$ to describe attribute 1 and $\{small_2 \ large_2\}$ to describe attribute 2. A possible knowledge base for the given two variables is: $KB = \{small_1, \neg small_1, large_1, \neg large_1, small_2, \neg small_2, large_2, \neg large_2\}$.

The idea for FOIL is as follows: from a general rule, we specify it by adding new literals in order to cover more positive and less negative examples according to the heuristics introduced in last section. After developing one rule, the positive examples covered by this rule are deleted from the original database. We then need to find a new rule based on this reduced database until all positive examples are covered. In this paper, because of the fuzzy linguistic nature of the expressions employed, typically data will be only partially covered by a given rule. For this reason we need a probability threshold PT as part of the decision process concerning rule coverage.

A pseudo-code of LFOIL are consists of two parts which are described follows:

Generating a Rule

- Let rule $R_i = \theta_1 \wedge \cdots \wedge \theta_d \rightarrow C_k$ be the rule at step i, we then find the next literal $\theta_{d+1} \in KB - \{\theta_1, \cdots, \theta_d\}$ for which $G(\theta_{d+1})$ is maximal.

- Replace rule R_i with $R_{i+1} = \theta_1 \wedge \cdots \wedge \theta_d \wedge \theta_{d+1} \rightarrow C_k$

- If $P(C_k|\theta_1 \wedge \cdots \wedge \theta_{i+1}) \geq PT$ then terminate else repeat.

Generating a Rule Base

Let $\Delta_i = \{\varphi_1 \rightarrow C_k, \cdots, \varphi_t \rightarrow C_k\}$ be the rule base at step i where $\varphi \in MLE$. We evaluate the coverage of Δ_i as follows:

$$CV(\Delta_i) = \frac{\sum_{l \in \mathcal{D}_k} \mu_{\varphi_1 \vee \cdots \vee \varphi_t}(x_l)}{|\mathcal{D}_k|} \tag{54}$$

We define a coverage function $\delta : \Omega_1 \times \cdots \times \Omega_n \rightarrow [0, 1]$ according to:

$$\delta(x|\Delta_i) = \mu_{\neg \Delta_i}(x) = \mu_{\neg(\varphi_1 \vee \cdots \vee \varphi_t)}(x) \tag{55}$$

$$= 1 - \mu_{(\varphi_1 \vee \cdots \vee \varphi_t)}(x) = 1 - \sum_{w=1}^{t} \mu_{R_w}(x)$$

where $\delta(x|\Delta_i)$ represents the degree to which x is *not* covered by a given rule base Δ_i. If CV is less than a predefined coverage threshold $CT \in [0, 1]$:

$$CV(\Delta_i) < CT$$

then we generate a new rule for class C_k according to the above rule generation algorithm to form a new rule base Δ_{i+1} but where the entropy calculations are amended such that for a rule $R = \theta \rightarrow C_k$,

$$T^{+} = \sum_{l \in \mathcal{D}_k} \mu_\theta(x_l) \times \delta(x_l|\Delta_i) \tag{56}$$

$$T^{-} = \sum_{l \in (\mathcal{D} - \mathcal{D}_k)} \mu_\theta(x_l) \tag{57}$$

The algorithm terminates when $CV(RB_{i+1}) \geq CT$ or $CV(RB_{i+1}) - CV(RB_i) < \epsilon$ where $\epsilon \in [0, 1]$ is a very small value, i.e., if there are no improvements in covering positive examples, we will stop the algorithm to avoid an infinite-loop calculation.

Given a rule base $\Delta_i = \{\varphi_1 \rightarrow C_k, \cdots, \varphi_t \rightarrow C_k\}$ and an unclassified data x, we can estimate the probability of C_k, $P(C_k|x)$, as follows: Firstly, we determine the rule $R_{max} = \varphi_j \rightarrow C_k$ for which $\mu_{\varphi_j}(x)$ is maximal:

$$\varphi_j = \arg\max_{k \in \Delta_i} \mu_{\varphi_k} \tag{58}$$

Therefore, given the unclassified data x, rule R_{max} is the most appropriate rule from the rule base we learned. For the rule $R_{max} \to C_k$ we evaluate two probabilities p_{max} and q_{max} where:

$$p_{max} = P(C_k|\varphi_j) \tag{59}$$

$$q_{max} = P(C_k|\neg\varphi_j) \tag{60}$$

We then use Jeffrey's rule [13] to evaluate the class probability by:

$$P(C_k|\mathrm{x}) = p_{max} \times \mu_{\varphi_j}(\mathrm{x}) + q_{max} \times (1 - \mu_{\varphi_j}(\mathrm{x})) \tag{61}$$

We tested this rule learning algorithms with some toy problems and some real-world problems. Although it does not give us very good accuracy but we obtained some comparable performance to decision tree but with much better transparency. More details are available in [21].

7 Conclusions and Discussions

In this chapter, label semantics, a higher level knowledge representation language, was used for modeling imprecise concepts and building intelligent data mining systems. In particular, several linguistic models have been proposed including: Linguistic Decision Trees (LDT) (for both classification and prediction), Bayesian estimation trees and Linguistic FOIL (LFOIL).

Through previous empirical studies, we have shown that in terms of accuracy the linguistic decision tree model tends to perform significantly better than both C4.5 and Naive Bayes and has equivalent performance to that of the Back-Propagation neural networks. However, it is also the case that this model has much better transparency than other algorithms. Linguistic decision trees are suitable for both classification and prediction. Some benchmark prediction problems have been tested with the LDT model and we found that it has comparable performance to a number of state-of-art prediction algorithms such as support vector regression systems. Furthermore, a methodology for classification with linguistic constraints has been proposed within the label semantics framework.

In order to reduce complexity and enhance transparency, a forward merging algorithm has been proposed to merge the branches which give sufficiently similar probability estimations. With merging, the partitioning of the data space is reconstructed and more appropriate granules can be obtained. Experimental studies show that merging reduces the tree size significantly without a significant loss of accuracy. In order to obtain a better estimation, a new hybrid model combining the LDT model and Fuzzy Naive Bayes has been investigated. The experimental studies show that this hybrid model has comparable performance to LID3 but with much smaller trees. Finally, a FOIL based rule learning system has been introduced within label semantics framework. In this approach, the appropriateness of using a rule to describe a data element is represented by multi-dimensional appropriateness measures. Based on the FOIL algorithm, we proposed a new linguistic rule induction algorithm according to which we can obtain concise linguistic rules reflecting the underlying nature of the system.

It is widely recognized that most natural concepts have non-sharp boundaries. These concepts are vague or fuzzy, and one will usually only be willing to agree to a certain degree that an object belongs to a concept. Likewise, in machine learning and data mining, the patterns we are interested in are often vague and imprecise. To model this, in this chapter, we have discretized numerical attributes with fuzzy labels by which we can describe real values. Hence, we can use linguistic models to study the underlying relationships hidden in the data. The linguistic models proposed in this chapter have advantages in the following respects[5]:

Interpretability

A primary motivation for the development of linguistic modeling is to provide an interface between numerical scales and a symbolic scale which is usually composed of linguistic terms. Transparency for a model is hard to define. In this chapter, we employ an intuitive way of judging the transparency for decision trees - the number of branches. By forward merging, the number of branches or the size of the tree is reduced, so that we may conclude the transparency of trees is enhanced. We also provide a methodology by which random sets of labels can be interpreted as logical expressions and vice versa.

Robustness

It is often claimed that fuzzy or other 'soft' approaches are more robust than discrete or 'crisp' approaches. In machine learning and data mining problems, the robustness can be considered as the insensitivity of predictive performance of models to small variations in the training data. In decision tree learning, soft boundaries are less sensitive to small changes than sharp boundaries. Hence, the performance of the linguistic models tends to be better than the corresponding discrete models because of the inherent robustness of these soft boundaries.

Information Fusion

One of the distinctive advantages of linguistic models is that they allow for information fusion. In this chapter, we discussed methods for classification with linguistic

---

[5] Hüllermeier [7] argues that these aspects are the potential contributions of fuzzy set theory to machine learning and data mining research.

constraints and linguistic queries based on linguistic decision trees. Other information fusion methods are discussed in [12]. How to efficiently use background knowledge is an important challenge in machine learning. For example, Wang [26] argues that Bayesian learning has limitations in combining the prior knowledge and new evidence. We also need to consider the inconsistency between the background knowledge and new evidence. We believe that it will become a popular research topic in approximate reasoning.

Acknowledgments This research was conduced when the first author was with the AI Group, Department of Engineering Mathematics of Bristol University, UK. The first author thanks Prof Lotfi Zadeh for some insightful comments on this research. He also thanks Drs Masoud Nikravesh, Marcus Thint and Ben Azvine for their interests in this research and support. At last, we thank BT/BISC fellowship for funding the publication of this chapter.

References

1. J.F. Baldwin, T.P. Martin and B.W. Pilsworth (1995) *Fril-Fuzzy and Evidential Reasoning in Artificial Intelligence*. John Wiley & Sons Inc.
2. J. F. Baldwin, J. Lawry and T.P. Martin (1997) Mass assignment fuzzy ID3 with applications. *Proceedings of the Unicom Workshop on Fuzzy Logic: Applications and Future Directions*, London pp. 278–294.
3. J. F. Baldwin and D. Xie (2004), Simple fuzzy logic rules based on fuzzy decision tree for classification and prediction problem, *Intelligent Information Processing II*, Z. Shi and Q. He (Ed.), Springer.
4. C. Blake and C.J. Merz. UCI machine learning repository. *http://www.ics.uci.edu/~mlearn/MLRepository.html*
5. M. Drobics, U. Bodenhofer and E. P. Klement (2003), FS-FOIL: an inductive learning method for extracting interpretable fuzzy descriptions, *International Journal of Approximate Reasoning*, 32: pp. 131–152.
6. S. R. Gunn (1998), Support vector machines for classification and regression. Technical Report of Dept. of Electronics and Computer Science, University of Southampton. *http://www.isis.ecs.soton.ac.uk/resources/svminfo*
7. E. Hullermeier (2005), Fuzzy methods in machine learning and data mining: status and prospects, to appear in *Fuzzy Sets and Systems*.
8. R. Hyndman and M Akram. Time series Data Library. Monash University. *http://www-personal.buseco.monash.edu.au/ffyndman/TSDL/index.htm*
9. C. Z. Janikow (1998), Fuzzy decision trees: issues and methods. *IEEE Trans. on Systems, Man, and Cybernetics-Part B: Cybernetics*, Vol. 28, No. 1.
10. J. Lawry (2001), Label semantics: A formal framework for modelling with words. *Symbolic and Quantitative Approaches to Reasoning with Uncertainty*, LNAI 2143: pp. 374–384, Springer-Verlag.
11. J. Lawry, J. Shanahan, and A. Ralescu (2003), *Modelling with Words: Learning, fusion, and reasoning within a formal linguistic representation framework*. LNAI 2873, Springer-Verlag.
12. J. Lawry (2004), A framework for linguistic modelling, *Artificial Intelligence*, 155: pp. 1–39.
13. R. C. Jeffrey (1965), *The Logic of Decision*, Gordon & Breach Inc., New York.
14. C. Olaru and L. Wehenkel (2003), A complete fuzzy decision tree technique. *Fuzzy Sets and Systems*. 138: pp.221–254.
15. Y. Peng, P. A. Flach (2001), Soft discretization to enhance the continuous decision trees. *ECML/PKDD Workshop: IDDM*.

16. H. Prade, G. Richard, and M. Serrurier (2003), Enriching relational learning with fuzzy predicates, *Proceedings of PKDD*, LNAI 2838, pp. 399–410.
17. Z. Qin and J. Lawry (2004), A tree-structured model classification model based on label semantics, *Proceedings of the 10th International Conference on Information Processing and Management of Uncertainty in Knowledge-based Systems (IPMU-04)*, pp. 261–268, Perugia, Italy.
18. Z. Qin and J. Lawry (2005), Hybrid Bayesian estimation trees based on label semantics, L. Godo (Ed.), *Proceedings of Eighth European Conference on Symbolic and Quantitative Approaches to Reasoning with Uncertainty*, Lecture Notes in Artificial Intelligence 3571, pp. 896–907, Springer.
19. Z. Qin and J. Lawry (2005), Decision tree learning with fuzzy labels, *Information Sciences*, Vol. 172/1–2: pp. 91–129.
20. Z. Qin and J. Lawry (2005), Prediction trees using linguistic modelling, *the Proceedings of International Fuzzy Association World Congress-05*, September 2005, Beijing, China.
21. Z. Qin and J. Lawry (2005), Linguistic rule induction based on a random set semantics, *the Proceedings of International Fuzzy Association World Congress-05*, September 2005, Beijing, China.
22. J. R. Quinlan (1986), Induction of decision trees, *Machine Learning*, Vol 1: pp. 81–106.
23. J. R. Quinlan (1990), Learning logical definitions from relations, *Machine Learning*, 5: 239–266.
24. J. R. Quinlan (1993), *C4.5: Programs for Machine Learning*, San Mateo: Morgan Kaufmann.
25. N. J. Randon and J. Lawry (2006), Classification and query evaluation using modelling with words, *Information Sciences, Special Issue - Computing with Words: Models and Applications*.
26. Pei Wang (2004), The limitation of Bayesianism, *Artificial Intelligence* 158(1): pp. 97–106.
27. D. Xie (2005), Fuzzy associated rules discovered on effective reduced database algorithm, To appear in the *Proceedings of IEEE-FUZZ*, Reno, USA.
28. L. A. Zadeh (1965), Fuzzy sets, *Information and Control*, Vol 8: pp. 338–353.
29. L. A. Zadeh (1996), Fuzzy logic = computing with words, *IEEE Transaction on Fuzzy Systems*. Vol. 4, No. 2: pp. 103–111.
30. L.A. Zadeh (2003), Foreword for modelling with words, *Modelling with Words*, LNAI 2873, Ed., J. Lawry, J. Shanahan, and A.Ralescu, Springer.
31. L.A. Zadeh (2005), Toward a generalized theory of uncertainty (GTU)– an outline, *Information Sciences*, Vol. 172/1–2, pp. 1–40.

Aggregating Subjective and Objective Measures of Web Search Quality using Modified Shimura Technique

Rashid Ali and M. M. Sufyan Beg

Abstract Web Searching is perhaps the second most popular activity on the Internet. But, as there are a number of search engines available, there must be some procedure to evaluate them. So, in this chapter, we propose an evaluation system for web search results. We are taking into consideration the "satisfaction" a user gets when presented with search results. The subjective evaluation based on the user feedback is augmented with the objective evaluation. The feedback of the user is inferred from watching the actions of the user on the search results presented before him in response to his query, rather than by a form filling method. This gives an implicit ranking of documents by the user. Then, the classical vector space model is used for computing the similarity of the documents selected by the user to that of the query. Also, the Boolean similarity measure is used to compute the similarity of the documents selected by the user to that of the query and thus another ranking of the documents based on this similarity measure is obtained. All the three rankings obtained in the process are then aggregated using the Modified Shimura Technique of Rank aggregation. The aggregated ranking is then compared with the original ranking given by the search engine. The correlation coefficient thus obtained is averaged for a set of queries. The averaged correlation coefficient is thus a measure of web search quality of the search engine. We show our experimental results pertaining to seven public search engines and fifteen queries.

1 Introduction

Internet has been very popular since its inception. Everyday, a number of Internet users search the web for some data and information using some query. A number of public search engines are available for this purpose. In an Internet search, the user writes some query to describe the nature of documents, and the search engine responds by returning a number of web pages that match the description. The results are ranked by the search engines and returned in the order of ranks. Since different search engines use different search algorithms and indexing techniques, they return different web pages in response to same query. Also, the same web pages are ranked differently by different search engines and returned at different positions in the list of search results. Then the question arises, which search engine one should use for web searching? A casual user prefers to look only first or second page of search results and in them only the top few. Therefore, a search engine, which gives relevance results in response to a query and returns them in proper order of relevance too, can be said to be a better one. To investigate this, search results need to be evaluated.

The evaluation procedure may be subjective or objective. In the present work, we propose a web search evaluation system, which combines both the subjective as well as objective techniques. For subjective evaluation, the users' vote is to be counted. For objective evaluation, different similarity measures based approaches such as Boolean similarity measures based; vector space model based approaches are used.

How are the users rating the results of a search engine should be taken into account to evaluate that search engine subjectively. Thus, it becomes imperative to obtain the feedback from the users. This feedback may either be explicit or implicit. The explicit feedback is the one in which the user is asked to fill up a feedback form after he has finished searching. This form is easy to analyze as the user may be asked directly to rank the documents as per the relevance according to his evaluation. But the problem is to obtain a correct feedback. The problem with the form-based approach is that it is too demanding from the user. In this approach, there is a lot of work for a casual user who might either fill it carelessly or not fill it at all. We, therefore, felt a need to devise a method to obtain the implicit feedback from the users. We watch the actions of the user on the search results presented before him in response to his query, and infer the feedback of the user there from.

We augment the subjective evaluation technique based on implicit user feedback as mentioned in the preceding paragraph with objective evaluation based on Vector Space Model and Boolean similarity measures. . Now, here, let the reason of using these objective evaluation techniques to be clarified. For the subjective evaluation, we assume the user to be an expert, sincere and honest. With these assumptions, our subjective technique solely serves the purpose in the best way. But then, it is so much difficult to get such users. A user usually has some personal effect on his actions that in turn will affect his feedback. Hence, the quality of the evaluation will be affected. So, we fill to include objective techniques in the process to check the shortcomings of the subjective evaluation. But, selecting only one objective technique may favor a particular search engine and may be harsh to the rest. So, we use here two objective techniques with the subjective evaluation to have an overall balanced judgment. These techniques re-rank the search results. These new ranks are then combined using Modified Shimura technique of Rank aggregation to get a "consensus" ranking. This "consensus" ranking when compared with the actual ranking provided by the search engine, gives a quantitative measure of search quality. We can then order the search engines on the basis of their quantitative measure.

1.1 Related Work

In the past, some efforts have been made to evaluate search results from different search engines. In most of the cases, a uniform sample of the web pages is collected by carrying out random walks on the web. The size of indices, which indirectly estimates the performance of a search engine, is then measured using this uniform sample. A search engine having larger index size has higher probability to give good search results. In (Henzinger et al 1999), (Henzinger et al 2000) and (Bar-Yossef et al 2000), some attempts involving this are available. In (Bharat and Broder 1998) also, the relative size and overlap of search engines is found but by using random queries, which are generated from a lexicon of about 400,000 words, built from a broad crawl of roughly 300,000 documents in the Yahoo hierarchy. In (Lawrence and Giles 1998) and (Lawrence and Giles 1999), the search engines are compared using a standard query log like that of NEC research institute. In (Hawking 1999), a frozen 18.5 million page snapshots of part of the web is created for proper evaluation of web search systems. In (Li and Shang 2000), for two different sets of ad-hoc queries, the results from *AltaVista*, *Google* and *InfoSeek* are obtained. These results are automatically evaluated for relevance on the basis of vector space model. These results are found to agree with the manual evaluation of relevance based on precision. Precision scores are given as 0, 1 or 2. But then this precision evaluation is similar to the form-filling exercise, already discussed for its demerits in Sect. 1. Precision evaluation of search engines is reported in (Shang and Li 2002). But then, "precision" being just the ratio of retrieved documents that are judged relevant, it doesn't say anything about the ranking of the relevant documents in the search results. Upon just the precision evaluation, other important aspects of web search evaluation such as recall, coverage, response time and web coverage etc. are also missed out. Also, there is no discussion on the satisfaction a user gets when presented with the search results.

An attempt has been made in (Beg and Ali 2005) to augment the implicit user feedback based subjective evaluation with vector space model based objective evaluation. In (Ali and Beg 2005), the subjective evaluation is augmented with Boolean similarity measure based objective evaluation. The subjective evaluation technique used in both (Beg and Ali 2005) and (Ali and Beg 2005) is implicit in nature. So, is free from the demerits of form filling method. But then in both of them, subjective and objective evaluations are performed separately and just averaged. Also, both consider just one objective technique to supplement the implicit feedback based subjective technique. So, it is likely that the evaluation is biased in favor of some of the search engines and is harsh to the rest.

In the present effort, we wish to combine both the objective techniques with user feedback based evaluation. But instead of making these evaluations separately and then taking their average, we make use of rank aggregation to get a consensus ranking from all. We use modified *Shimura* technique of rank aggregation for the purpose. The "consensus" ranking is then compared with original ranking by the search engine. The correlation coefficient thus obtained is a quantitative measure of web search quality of the search engine. The measure thus becomes the basis for evaluation of the search engine.

1.2 Useful Definitions

Here we have some definitions that are useful while evaluating search results.

Definition 1. Given a universe U and $S \subseteq U$, an *ordered list* (or simply, a *list*) l with respect to U is given as $l = [e_1, e_2, \ldots, e_{|s|}]$, with each $e_i \in S$, and $e_1 \succ e_2 \succ$

$\ldots \succ e_{|s|}$, where "$\succ$" is some ordering relation on S. Also, for $j \in U \wedge j \in l$, let $l(j)$ denote the position or rank of j, with a higher rank having a lower numbered position in the list. We may assign a unique identifier to each element in U and thus, without loss of generality we may get $U = \{1, 2, \ldots, |U|\}$.

Definition 2. If a list contains all the elements in U, then it is said to be a *full list*.

Example 1. A full list l_f given as $[e, a, d, c, b]$ has the ordering relation $e \succ a \succ d \succ c \succ b$. The Universe U may be taken as $\{1,2,3,4,5\}$ with say, $a \equiv 1, b \equiv 2, c \equiv 3, d \equiv 4, e \equiv 5$. With such an assumption, we have $l_f = [5, 1, 4, 3, 2]$. Here $l_f(5) \equiv l_f(e) = 1, l_f(1) \equiv l_f(a) = 2, l_f(4) \equiv l_f(d) = 3, l_f(c) \equiv l_f(c) = 4, l_f(2) \equiv l_f(b) = 5$.

Definition 3. The *Kendall Tau distance* between two full lists l_1 and l_2, each of cardinality $|l|$, is given as follows.

$$K(l_1, l_2) = \frac{|\{(i, j) \,|\, \forall l_1(i) < l_1(j), l_2(i) > l_2(j)\}|}{(1/2)\,|l|\,(|l| - 1)} \tag{1}$$

Definition 4. The *Spearman footrule distance* (SFD) between two full lists l_1 and l_2, each of cardinality $|l|$, is given as follows.

$$F(l_1, l_2) = \frac{\sum\limits_{\forall i} |l_1(i) - l_2(i)|}{\left\lfloor (1/2)\,|l|^2 \right\rfloor} \tag{2}$$

Definition 5. Given a set of k full lists as $L = \{l_1, l_2, \ldots, l_k\}$, the *normalized aggregated Kendall distance* of a full list l to the set of full lists L is given as $K(l, L) = \dfrac{\sum\limits_{i=1}^{k} K(l, l_i)}{k}$, while the *normalized aggregated footrule distance* of l to L is given as $F(l, L) = \dfrac{\sum\limits_{i=1}^{k} F(l, l_i)}{k}$.

Definition 6. Given a set of lists $L = \{l_1, l_2, \ldots, l_k\}$, *Rank Aggregation* is the task of coming up with a list l such that either $K(l, L)$ or $F(l, L)$ is minimized.

Definition 7. A list l containing elements, which are a strict subset of U, is called a *Partial List*. We have a strict inequality $|l| < |U|$.

Definition 8. Let the full lists $[u_1, u_2, \ldots, u_n]$ and $[v_1, \ldots, v_2, v_n]$ be the two rankings. The *Spearman Rank Order Correlation Coefficient* (r_s) between these two rankings is defined as follows.

$$r_s = 1 - \frac{6 \sum\limits_{i=1}^{n} \left[l_f(u_i) - l_f(v_i)\right]^2}{n\,(n^2 - 1)} \tag{3}$$

The Spearman Rank Order Correlation Coefficient (r_s) is a measure of closeness of two rankings. The coefficient r_s ranges between -1 and 1. When the two rankings are identical $rs = 1$, and when one of the rankings is the inverse of the other, then $r_s = -1$.

Definition 9. Without loss of generality, assume that full list be given as $[1, 2, \ldots, n]$. Let the partial list be given as $[v_1, v_2, \ldots \ldots, v_m]$. The *Modified Spearman Rank Order Correlation Coefficient* $(r_s{}')$ between these two rankings is defined as follows.

$$r_s' = 1 - \frac{\sum_{i=1}^{m} (i - v_i)^2}{m \left(\left[\max_{j=1}^{m} \{v_j\} \right]^2 - 1 \right)} \tag{4}$$

Example 2. For $|U| = 5$, let the full list be $l = \{1, 2, 3, 4, 5\}$ and the partial list l_j with $|l_j| = 3$ be $l_j = \{21, 11, 14\}$.

$$r_s' = 1 - \frac{(1 - 21)^2 + (2 - 11)^2 + (3 - 14)^2}{3 \times \left([\max \{21, 11, 14\}]^2 - 1 \right)} = 0.544.$$

2 Web Search Evaluation Using the User Feedback Vector Model

The underlying principle of our approach (Beg 2002) of subjective evaluation of search engines is to measure the "satisfaction" a user gets when presented with the search results. For this, we need to monitor the response of the user to the search results presented before him.

2.1 User Feedback Vector

We characterize the feedback of the user by a vector (V, T, P, S, B, E, C), which consists of the following.

(a) The sequence V in which the user visits the documents, $V = (v_1, v_2, \ldots, v_N)$. If document i is the kth document visited by the user, then we set $v_i = k$.
(b) The time t_i that a user spends examining the document i. We denote the vector $(t_1, t_2, \ldots, t_N)$ by T. For a document that is not visited, the corresponding entry in the array T is 0.
(c) Whether or not the user prints the document i. This is denoted by the Boolean p_i. We denote the vector $(p_1, p_2, \ldots, p_N)$ by P.
(d) Whether or not the user saves the document i. This is denoted by the Boolean s_i. We denote the vector $(s_1, s_2, \ldots, s_N)$ by S.
(e) Whether or not the user book-marked the document i. This is denoted by the Boolean b_i We denote the vector $(b_1, b_2, \ldots, b_N)$ by B.

(f) Whether or not the user e-mailed the document i to someone. This is denoted by the Boolean e_i. We denote the vector $(e_1, e_2, \ldots, e_N)$ by E.

(g) The number of words that the user copied and pasted elsewhere. We denote the vector $(c_1, c_2, \ldots, c_N)$ by C.

The motivation behind collecting this feedback is the belief that a well-educated user is likely to select the more appropriate documents early in the resource discovery process. Similarly, the time that a user spends examining a document, and whether or not he prints, saves, bookmarks, e-mails it to someone else or copies & pastes a portion of the document, indicate the level of importance that document holds for the specified query.

2.2 Search Quality Measure using User Feedback Vector

When feedback recovery is complete, we compute the following weighted sum σ_j for each document j selected by the user.

$$
\sigma_j = \left(w_V \frac{1}{2^{(v_j-1)}} + w_T \frac{t_j}{T_j^{\max}} + w_P p_j + w_S s_j + w_B b_j + w_E e_j + w_C \frac{c_j}{C_j^{total}} \right)
\tag{5}
$$

where $T_j^{\max}$ represents the maximum time a user is expected to spend in examining the document j, and C_j^{total} is the total number of words in the document j. Here, $w_V, w_T, w_P, w_S, w_B, w_E$ and w_C, all lying between 0 and 1, give the respective weights we want to give to each of the seven components of the feedback vector. The sum σ_j represents the importance of document j.

The intuition behind this formulation is as follows. The importance of the document should decrease monotonically with the postponement being afforded by the user in picking it up. More the time spent by the user in glancing through the document, more important that must be for him. If the user is printing the document, or saving it, or book-marking it, or e-mailing it to someone else, or copying and pasting a portion of the document, it must be having some importance in the eyes of the user. A combination of the above seven factors by simply taking their weighted sum gives the overall importance the document holds in the eyes of the user.

As regards the maximum time a user is expected to spend in examining the document j, we clarify that this is taken to be directly proportional to the size of the document. We assume that an average user reads at a speed of about 10 bytes per second. This includes the pages containing text as well as images. So a document of size 1 kB is expected to take a minute and 40 seconds to go through. The above mentioned default reading speed of 10 bytes per second may be set differently by the user, if he wishes so.

It may be noted that depending on his preferences and practice, the user would set the importance of the different components of the feedback vector. For instance, if a user does not have a printer at his disposal, then there is no sense in setting up the importance weight (w_P) corresponding to the printing feedback component (P). Similarly, if a user has a dial-up network connection, and so he is in a habit of saving the relevant documents rather than spending time on it while online, it would be better to give a higher value to w_S, and a lower value to w_T. In such a case, lower values may also be given to w_P, w_E and w_C, as he would not usually be printing or e-mailing or copying and pasting a document at a stretch while online. So, after explaining the modalities to him, the user is to be requested to modify the otherwise default values of 1 for all these weights. It may, however, be noted that the component of the feedback vector corresponding to the sequence of clicking, always remains to be the prime one and so w_V must always be 1.

Now, sorting the documents on the descending values of their weighted sum will yield a sequence $\Re_A$, which is in fact, a ranking of documents based on user feedback.

3 Web Search Evaluation Using Vector Space Model and Boolean Similarity Measures

Before we proceed, we must have a look at the text pre-processing operations, which are a pre-requisite for the application of any objective evaluation procedure.

3.1 Text Pre-Processing

First of all, we need to remove the stop-words, the words that have very little semantic meaning and are frequent homogeneously across the whole collection of documents. They are generally prepositions or articles like the, an, for etc. Over a period of time people have come up with a list of stop-words pertaining to a general domain. However, it may be argued that a stop-word is very much context dependant. A word like web may be treated as a stop-word in a collection of web-related articles, but not so in a set of literary documents.

The text pre-processing also includes an important step of *word stemming*, wherein all the words with same root are reduced to a single form. This is achieved by stripping each word of suffix, prefix or infix. This is to say that all words are reduced to their canonical form. For instance, the words like *drive*, *driver* and *driving*; all would be reduced to the stem word *drive*. This way the burden of being very specific while forming the query, is taken off from the shoulders of the user. A well- known algorithm for carrying out word stemming is *Porter Stemmer Algorithm* (Porter 1980).

It may be noted that the text pre-processing techniques are very much dependant on the language of the document. For instance, just the removal of suffixes may usually suffice as the stemming technique in the case of English language, but not necessarily so with other languages.

3.2 Vector Space Model

An n-dimensional vector space is taken with one dimension for each possible word or term. Therefore, n would be the number of words in a natural language. Each document or query is represented as a *term vector* in this vector space. Each component of *term vector* is either 0 or 1, depending on whether the term corresponding to that axis is absent or present in the text. Alternatively, each component of the term vector is taken as a function that increases with the frequency of the term (corresponding to that axis) with in the document and decreases with the number of documents in the collection that contain this term. This is to take into account the TF-IDF factor.

The frequency of a word in a given document is taken as a good measure of importance of that word in the given document (Salton and McGill 1983). This, of course, holds true only after the text pre-processing has been carried out.

An improved version of the term frequency is the *term frequency-inverse document frequency* (TF-IDF). In this, uniqueness of a keyword in a document, and hence the relevance, is measured by the fact that the word should be frequent in the current document but not so frequent in the rest of the documents of the given collection. The TF-IDF value for a given keyword w is given as

$$f_{TF-IDF} = \frac{f_w}{f_{w_{\max}}} \log \frac{\rho}{\rho_w} \tag{6}$$

where, f_w is the frequency of the keyword w in the document, $f_{w_{\max}}$ is the maximum frequency of any word in the document, ρ_w is the number of documents in which this keyword occurs, and ρ is the total number of documents in the collection.

As web is very dynamic with respect to nature of its content, the vector space model cannot be used directly in the performance evaluation of search engines (Li and Shang 2000). The value of log (ρ/ρ_w) is not available because we don't know the total number ρ of documents and number of documents ρ_w containing the given keyword. So, we use a simplified version of vector space model. Here, we assume log (ρ/ρ_w) to be constant with the argument that all keywords in our queries are technical terms, which appear approximately the same number of times. This simplified version may favour long documents and give documents with many appearances of same keywords a higher score. The term vectors of documents are normalized to one to compensate for different document lengths. As far as many occurrences of a keyword in a document are concerned; this in itself is an indication of the relevance of the document to the query.

Once the term vectors are obtained the similarity between a document and a query is obtained by computing the dot product of their term vectors. Larger the dot product, the greater would be the similarity. So, the document with larger dot product is more relevant to query.

Now, sorting the documents with the decreasing values of their respective dot products with that of the query will yield a sequence $\Re_B$, which is a ranking of documents based on vector space model.

3.3 Boolean Similarity Measures

There are a number of *Boolean similarity measures* (Li and Danzig 1994, Li and Danzig 1997) that can be used to compute the similarities of one document to another and documents to queries. Some of such well-known similarity measures are Dice's Coefficient, Jaccard's Coefficient, Cosine coefficient and Overlap Coefficient. In order to use these measures, documents and queries are to be represented as sets of keywords.

Radecki proposed two similarity measures, S and S*, based on Jaccard's coefficient .We assume that each query is transformed to a Boolean expression and denote $\psi(Q)$ and $\psi(C)$ as sets of documents in the response to query Q and in the cluster of documents represented by C.

The similarity value S between Q and C is defined as the ratio of common documents to total number of documents in $\psi(Q)$ and $\psi(C)$.

$$S(Q, C) = \frac{|\psi(Q) \cap \psi(C)|}{|\psi(Q) \cup \psi(C)|} \tag{7}$$

But, since all the documents in response to query belong to the cluster represented by C (i.e. $\psi(Q) \subseteq \psi(C)$), we can have

$$S(Q, C) = \frac{|\psi(Q)|}{|\psi(C)|} \tag{8}$$

This measure requires the actual results to the query and is mainly useful as index of comparison.

In similarity measure S^*, Boolean expression Q is transformed into its reduced disjunctive normal form (RDNF), denoted as $\tilde{Q}$, which is disjunction of a list of reduced atomic descriptors. If set T is the union of all the descriptors that appear in the to-be-compared Boolean expression pair, then a reduced atomic descriptor is defined as a conjunction of all the elements in T in either their true or negated form. Let T_Q and T_C be the set of descriptors that appear in Q and C respectively. Suppose $T_Q \cup T_C = \{t_1, t_2, \ldots, t_n)$, where n is the set size of $T_Q \cup T_C$, then the RDNF of Q and C are:

$$\left(\tilde{Q}\right)_{T_Q \cup T_c} = \left(\tilde{q}_{1,1} \wedge \tilde{q}_{1,2} \wedge \ldots \wedge \tilde{q}_{1,n}\right) \vee \left(\tilde{q}_{2,1} \wedge \tilde{q}_{2,2} \wedge \ldots \wedge \tilde{q}_{2,n}\right) \vee$$
$$\ldots \vee \left(\tilde{q}_{1,1} \wedge \tilde{q}_{1,2} \wedge .. \wedge \tilde{q}_{1,n}\right) \tag{9}$$

$$\left(\tilde{C}\right)_{T_Q \cup T_c} = \left(\tilde{c}_{1,1} \wedge \tilde{c}_{1,2} \wedge \ldots \wedge \tilde{c}_{1,n}\right) \vee \left(\tilde{c}_{2,1} \wedge \tilde{c}_{2,2} \wedge \ldots \wedge \tilde{c}_{2,n}\right) \vee$$
$$\ldots \vee \left(\tilde{c}_{m,1} \wedge \tilde{c}_{m,2} \wedge .. \wedge \tilde{c}_{m,n}\right) \tag{10}$$

where l and m are the number of reduced atomic descriptors in $\left(\widetilde{Q}\right)_{T_Q \cup T_c}$ and $\left(\widetilde{C}\right)_{T_Q \cup T_c}$ respectively and,

$$\widetilde{q}_{i,j} = \begin{cases} t_j & \text{true} \\ \neg t_j & \text{negated} \end{cases} \quad \dots 1 \leq i \leq 1, 1 \leq j \leq n \tag{11}$$

$$\widetilde{c}_{i,j} = \begin{cases} t_j & \text{true} \\ \neg t_j & \text{negated} \end{cases} \quad \dots 1 \leq i \leq m, 1 \leq j \leq n \tag{12}$$

where, $\neg$ is the NOT operator.

The similarity value S^* between the Boolean expressions (Q and C) is defined as the ratio of the number of common reduced atomic descriptors in $\widetilde{Q}$ and $\widetilde{C}$ to the total number of reduced atomic descriptors in them,

$$S^*(Q, C) = \frac{\left|\left(\widetilde{Q}\right)_{T_Q \cup T_c} \cap \left(\widetilde{C}\right)_{T_Q \cup T_c}\right|}{\left|\left(\widetilde{Q}\right)_{T_Q \cup T_c} \cup \left(\widetilde{C}\right)_{T_Q \cup T_c}\right|} \tag{13}$$

A new similarity measure $S^{\oplus}$ based on Radecki similarity measure S^*, was proposed by Li and Danzig. For this, a Boolean expression Q is transformed to its compact disjunctive normal form (CDNF) denoted as $\hat{Q}$, which is a disjunction of compact atomic descriptors. Each compact atomic descriptor itself is in turn a conjuction of subsets of descriptors present in its own Boolean expression. The CDNFs of Q and C are

$$\hat{Q} = \left(\hat{q}_{1,1} \wedge \hat{q}_{1,2} \wedge \dots \wedge \hat{q}_{1,x1}\right) \vee \left(\hat{q}_{2,1} \wedge \hat{q}_{2,2} \wedge \dots \wedge \hat{q}_{2,x2}\right) \vee$$
$$\dots \vee \left(\hat{q}_{l,1} \wedge \hat{q}_{l,2} \wedge \dots \wedge \hat{q}_{l,xl}\right) \tag{14}$$

$$\hat{C} = \left(\hat{c}_{1,1} \wedge \hat{c}_{1,2} \wedge \dots \wedge \hat{c}_{1,y1}\right) \vee \left(\hat{c}_{2,1} \wedge \hat{c}_{2,2} \wedge \dots \wedge \hat{c}_{2,y2}\right) \vee$$
$$\dots \vee \left(\hat{c}_{m,1} \wedge \hat{c}_{m,2} \wedge \dots \wedge \hat{c}_{m,xm}\right) \tag{15}$$

where, l and m are the numbers of compact atomic descriptors in $\hat{Q}$ and $\hat{C}$, x_i is the number of descriptors in the ith ($1 \leq i \leq l$) compact atomic descriptor of $\hat{Q}$, and y_j is the number of descriptors in the jth ($1 \leq j \leq m$) compact atomic descriptor of $\hat{C}$.

Each $\hat{q}_{i,u}$ and $\hat{c}_{j,v}$ in the CDNFs represents a descriptor in T_Q and T_C respectively. Specifically, we have $\hat{q}_{i,u} \; \varepsilon \; T_Q$, where $1 \leq i \leq l$ and $1 \leq u \leq x_i$ and $\hat{c}_{j,v} \; \varepsilon \; T_C$, where $1 \leq j \leq m$ and $1 \leq v \leq y_j$.

The individual similarity measure is defined as

$$s^{\oplus}\left(\hat{Q}^i, \hat{C}^j\right) \begin{cases} 0 \ldots \ldots . if\, T_Q^i \cap T_C^j = 0. or. \exists t \in T_Q^i, \neg t \in T_C^j \\ \dfrac{1}{2^{\left|T_C^j - T_Q^i\right|} + 2^{\left|T_Q^i - T_C^j\right|} - 1} \ldots \ldots \ldots \ldots . otherwise \end{cases} \qquad (16)$$

where, $\hat{Q}^i$ indicates the ith atomic descriptors of CDNF $\hat{Q}$, $\hat{C}^j$ indicates the jth compact atomic descriptor of CDNF $\hat{C}$. T_Q^i and T_C^j are the set of descriptors in $\hat{Q}^i$ and $\hat{C}^j$ respectively. The similarity of two expressions, $S^{\oplus}$ defined as the average value of the individual similarity measures ($s^{\oplus}$) between each atomic descriptor is given by

$$S^{\oplus}(Q, C) = \frac{\displaystyle\sum_{i=1}^{\left|\hat{Q}\right|} \sum_{j=1}^{\left|\hat{C}\right|} s^{\oplus}\left(\hat{Q}^i, \hat{C}^j\right)}{\left|\hat{Q}\right| \times \left|\hat{C}\right|} \qquad (17)$$

Example 3. Suppose Q be the query represented by the Boolean Expression $Q = (t_1 \vee t_2) \wedge t_3$ and the three to-be-compared documents or servers descriptors, say C_1, C_2 and C_3, be represented by the Boolean expressions $C_1 = t_1 \wedge t_2 \wedge t_4 \wedge t_5$, $C_2 = (t_1 \vee t_3) \wedge t_4$ and $C_3 = t_2 \wedge t_3 \wedge t_5$, respectively.

Let set T_Z be the union of all the descriptors in Boolean expression Z (Z = Q, C_1, C_2 or C_3). We have $T_Q = \{t_1, t_2, t_3\}$, $T_{C1} = \{t_1, t_2, t_4, t_5\}$, $T_{C2} = \{t_1, t_3, t_4\}$, $T_{C3} = \{t_2, t_3, t_5\}$.

The CDNF of Q, C_1, C_2 and C_3 are $Q = (t_1 \wedge t_3) \vee (t_2 \wedge t_3)$, $C_1 = t_1 \wedge t_2 \wedge t_4 \wedge t_5$, $C_2 = (t_1 \wedge t_4) \vee (t_3 \wedge t_4)$ and $C_3 = t_2 \wedge t_3 \wedge t_5$, respectively.

From the above, it is clear that

(i) CDNF of Q contains two compact atomic descriptors $\hat{Q}^1 = t_1 \wedge t_3$ and $\hat{Q}^2 = t_2 \wedge t_3$, the set of descriptors in them being $T_Q^1 = \{t_1, t_3\}$ and $T_Q^2 = \{t_2, t_3\}$ respectively. Similarly,

(ii) CDNF of C_1 contains only one compact atomic descriptors $\hat{C}_1^1 = t_1 \wedge t_2 \wedge t_4 \wedge t_5$, the set of descriptor being $T_{C_1}^1 = \{t_1, t_2, t_4, t_5\}$.

(iii) CDNF of C_2 contains two compact atomic descriptors $\hat{C}_2^1 = t_1 \wedge t_4$ and $\hat{C}_2^2 = t_3 \wedge t_4$, the set of descriptor in them being $T_{C_2}^1 = \{t_1, t_4\}$ and $T_{C_2}^2 = \{t_3, t_4\}$ respectively.

(iv) CDNF of C_3 contains only one compact atomic descriptors $\overset{\wedge}{C_3}{}^1 = t_2 \wedge t_3 \wedge t_5$, the set of descriptor being $T_{C_3}^1 = \{t_2, t_3, t_5\}$

$$\text{Thus, } s^{\oplus}\left(\hat{Q}{}^1, \hat{C}{}_1^1\right) = \frac{1}{2^{\left|T_{C1}^1 - T_Q^1\right|} + 2^{\left|T_Q^1 - T_{C1}^1\right|} - 1} = \frac{1}{2^3 + 2^1 - 1} = 0.1111$$

$$\text{and, } s^{\oplus}\left(\hat{Q}{}^2, \hat{C}{}_1^1\right) = \frac{1}{2^{\left|T_{C1}^1 - T_Q^2\right|} + 2^{\left|T_Q^2 - T_{C1}^1\right|} - 1} = \frac{1}{2^3 + 2^1 - 1} = 0.1111$$

$$\text{Hence, } S^{\oplus}(Q, C_1) = \frac{s^{\oplus}\left(\hat{Q}{}^1, \hat{C}{}_1^1\right) + s^{\oplus}\left(\hat{Q}{}^2, \hat{C}{}_1^1\right)}{2} = 0.1111$$

$$\text{Also, } s^{\oplus}\left(\hat{Q}{}^1, \hat{C}{}_2^1\right) = \frac{1}{2^{\left|T_{C2}^1 - T_Q^1\right|} + 2^{\left|T_Q^1 - T_{C2}^1\right|} - 1} = \frac{1}{2^1 + 2^1 - 1} = 0.3333,$$

$$s^{\oplus}\left(\hat{Q}{}^1, \hat{C}{}_2^2\right) = \frac{1}{2^{\left|T_{C2}^2 - T_Q^1\right|} + 2^{\left|T_Q^1 - T_{C2}^2\right|} - 1} = \frac{1}{2^1 + 2^1 - 1} = 0.3333$$

$$s^{\oplus}\left(\hat{Q}{}^2, \hat{C}{}_2^1\right) = 0 \qquad\qquad \text{(because } T_Q^2 \cap T_{C_2}^1 = \phi)$$

$$s^{\oplus}\left(\hat{Q}{}^2, \hat{C}{}_2^2\right) = \frac{1}{2^{\left|T_{C2}^2 - T_Q^2\right|} + 2^{\left|T_Q^2 - T_{C2}^2\right|} - 1} = \frac{1}{2^{\left|T_{C2}^2 - T_Q^2\right|} + 2^{\left|T_Q^2 - T_{C2}^2\right|} - 1}$$

$$= 0.3333$$

Hence, $S^{\oplus}(Q, C_2)$

$$= \frac{s^{\oplus}\left(\hat{Q}{}^1, \hat{C}{}_2^1\right) + s^{\oplus}\left(\hat{Q}{}^1, \hat{C}{}_2^2\right) + s^{\oplus}\left(\hat{Q}{}^2, \hat{C}{}_2^1\right) + s^{\oplus}\left(\hat{Q}{}^2, \hat{C}{}_2^2\right)}{4} = 0.250.$$

$$\text{Similarly, } s^{\oplus}\left(\hat{Q}{}^1, \hat{C}{}_3^1\right) = \frac{1}{2^{\left|T_{C3}^1 - T_Q^1\right|} + 2^{\left|T_Q^1 - T_{C3}^1\right|} - 1} = \frac{1}{2^2 + 2^1 - 1} = 0.20$$

$$\text{and } s^{\oplus}\left(\hat{Q}{}^2, \hat{C}{}_3^1\right) = \frac{1}{2^{\left|T_{C3}^1 - T_Q^2\right|} + 2^{\left|T_Q^2 - T_{C3}^1\right|} - 1} = \frac{1}{2^1 + 2^0 - 1} = 0.50$$

$$\text{Hence, } S^{\oplus}(Q, C_3) = \frac{s^{\oplus}\left(\hat{Q}{}^1, \hat{C}{}_3^1\right) + s^{\oplus}\left(\hat{Q}{}^2, \hat{C}{}_3^1\right)}{2} = 0.350.$$

Now, sorting the documents in a decreasing order of their Li and Danzig similarity measure values, we get a ranking $C_3 \succ C_2 \succ C_1$, where '$\succ$' indicates 'is more relevant to query than'.

3.4 Simplified Boolean Similarity Measure

It has been proved in (Li and Danzig 1994) that the Li and Danzig similarity measure $(S^\oplus)$ is equivalent to the Radecki similarity measure(S^*), which is based on the Jaccard's Coefficient. At the same time, $S^\oplus$ reduces time and space complexity from exponential to polynomial in the number of Boolean terms. But, as we are interested in the relevance ranking of the documents rather than their individual similarity measures with the query, we feel that the Li and Danzig measure $S^\oplus$ can be further simplified. We propose a simplified Boolean similarity measure $S^\otimes$ based on the Li and Danzig measure $S^\oplus$.

If $\hat{Q}$ and $\hat{C}$ be the CDNFs of the Boolean expressions Q and C as described above in (12) and (13), the simplified individual similarity measure is then defined as follows.

$$s^\otimes\left(\hat{Q}^i, \hat{C}^j\right) = \begin{cases} 0 \ldots .if\, T_Q^i \cap T_C^j = 0.or.\exists t \in T_Q^i, \neg t \in T_C^j \\ \dfrac{1}{\left|T_C^j - T_Q^i\right| + \left|T_Q^i - T_C^j\right| + 1} \ldots \ldots .otherwise \end{cases} \tag{18}$$

where $\hat{Q}^i$ indicates the ith atomic descriptors of CDNF $\hat{Q}$, $\hat{C}^j$ indicates the jth compact atomic descriptor of CDNF $\hat{C}$. T_Q^i and T_C^j are the set of descriptors in $\hat{Q}^i$ and $\hat{C}^j$ respectively.

The similarity of the two expressions, $S^\otimes$ is defined as the average value of the individual similarity measures ($s^\otimes$) between each atomic descriptor;

$$S^\otimes(Q, C) = \frac{\sum\limits_{i=1}^{\left|\hat{Q}\right|} \sum\limits_{j=1}^{\left|\hat{C}\right|} s^\otimes\left(\hat{Q}^i, \hat{C}^j\right)}{\left|\hat{Q}\right| \times \left|\hat{C}\right|} \tag{19}$$

The proposed simplified version reduces the computational effort substantially. Moreover, if we assume that Boolean expressions of the query Q and documents to be compared $(C_1, C_2, \ldots, C_n)$, just contain only AND terms, i.e their CDNF contain only a single compact descriptor and also, for each pair of the to-be-compared documents C_k and C_l, one of the following holds true,

$$\left| T^{j}_{C_k} - T^{i}_{Q} \right| = \left| T^{j}_{C_l} - T^{i}_{Q} \right| \text{ or } \left| T^{j}_{C_k} - T^{i}_{Q} \right| = \left| T^{i}_{Q} - T^{j}_{C_l} \right|$$

or

$$\left| T^{i}_{Q} - T^{j}_{C_k} \right| = \left| T^{i}_{Q} - T^{j}_{C_l} \right| \text{ or } \left| T^{i}_{Q} - T^{j}_{C_k} \right| = \left| T^{j}_{C_l} - T^{i}_{Q} \right|$$

then the relative rankings of the documents found using the simplified version, remains same with that found using Li and Danzig measure $S^{\oplus}$, as it is illustrated in following example.

Example 4. Suppose Q be the query represented by Boolean Expression $Q = t_1 \wedge t_2 \wedge t_3$ and the three to be compared documents or servers descriptions say C_1, C_2 and C_3 be represented by Boolean expressions $C_1 = t_1 \wedge t_2 \wedge t_3 \wedge t_4 \wedge t_5, C_2 = t_1 \wedge t_4 \wedge t_5$ and $C_3 = t_1 \wedge t_3 \wedge t_4 \wedge t_5$, respectively.
The CDNF of Q, C_1, C_2 and C_3 are
$Q = t_1 \wedge t_2 \wedge t_3, C_1 = t_1 \wedge t_2 \wedge t_3 \wedge t_4 \wedge t_5, C_2 = t_1 \wedge t_4 \wedge t_5$, and
$C_3 = t_1 \wedge t_3 \wedge t_4 \wedge t5$, respectively.
From the above, it is clear that

(i) CDNF of Q contains one compact atomic descriptor $\hat{Q}^1 = t_1 \wedge t_2 \wedge t_3$, the set of descriptors being $T^1_Q = \{t_1, t_2, t_3\}$. Similarly,

(ii) CDNF of C_1 contains one compact atomic descriptor $\hat{C}^1_1 = t_1 \wedge t_2 \wedge t_3 \wedge t_4 \wedge t_5$, the set of descriptor being $T^1_{C_1} = \{t_1, t_2, t_3, t_4, t_5\}$.

(iii) CDNF of C_2 contains one compact atomic descriptor $\hat{C}^1_2 = t_1 \wedge t_4 \wedge t_5$, the set of descriptor being $T^1_{C_2} = \{t_1, t_4, t_5\}$.

(iv) CDNF of C_3 contains one compact atomic descriptor $\hat{C}^1_3 = t_1 \wedge t_3 \wedge t_4 \wedge t_5$, the set of descriptor being $T^1_{C_3} = \{t_1, t_3, t_4, t_5\}$.

$$\text{Thus, } s^{\oplus}\left(\hat{Q}^1, \hat{C}^1_1 \right) = \frac{1}{2^{\left| T^1_{C1} - T^1_{Q} \right|} + 2^{\left| T^1_{Q} - T^1_{C1} \right|} - 1} = \frac{1}{2^2 + 2^0 - 1} = 0.250$$

$$\text{Hence, } S^{\oplus}(Q, C_1) = s^{\oplus}\left(\hat{Q}^1, \hat{C}^1_1 \right) = 0.250.$$

$$\text{Also, } s^{\oplus}\left(\hat{Q}^1, \hat{C}^1_2 \right) = \frac{1}{2^{\left| T^1_{C2} - T^1_{Q} \right|} + 2^{\left| T^1_{Q} - T^1_{C2} \right|} - 1} = \frac{1}{2^2 + 2^2 - 1} = 0.1429$$

$$\text{Hence, } S^{\oplus}(Q, C_2) = s^{\oplus}\left(\hat{Q}^1, \hat{C}^1_2 \right) = 0.1429$$

$$\text{Similarly, } s^{\oplus}\left(\hat{Q}^1, \hat{C}^1_3 \right) = \frac{1}{2^{\left| T^1_{C3} - T^1_{Q} \right|} + 2^{\left| T^1_{Q} - T^1_{C3} \right|} - 1} = \frac{1}{2^2 + 2^1 - 1} = 0.2000$$

$$\text{Therefore, } S^{\oplus}(Q, C_3) = s^{\oplus}\left(\hat{Q}^1, \hat{C}^1_3 \right) = 0.2000.$$

Now, sorting the documents in a decreasing order of their Li and Danzig similarity measure value, we get a ranking $C_1 \succ C_3 \succ C_2$ where '$\succ$' indicates 'is more relevant to query than'.

Now, in a similar way, we compute the simplified Similarity measures $S^\otimes$ for the same expressions.

Thus, $s^\otimes\left(\hat{Q}^1, \hat{C}_1^1\right) = \dfrac{1}{\left|T_{C1}^1 - T_Q^1\right| + \left|T_Q^1 - T_{C1}^1\right| + 1} = \dfrac{1}{2+0+1} = 0.3333$

Hence, $S^\otimes(Q, C_1) = s^\otimes\left(\hat{Q}^1, \hat{C}_1^1\right) = 0.3333$

Also, $s^\otimes\left(\hat{Q}^1, \hat{C}_2^1\right) = \dfrac{1}{\left|T_{C2}^1 - T_Q^1\right| + \left|T_Q^1 - T_{C2}^1\right| + 1} = \dfrac{1}{2+2+1} = 0.2000$

Hence, $S^\otimes(Q, C_2) = I^\otimes\left(\hat{Q}^1, \hat{C}_2^1\right) = 0.2000$

Similarly, $s^\otimes\left(\hat{Q}^1, \hat{C}_3^1\right) = \dfrac{1}{\left|T_{C3}^1 - T_Q^1\right| + \left|T_Q^1 - T_{C3}^1\right| + 1} = \dfrac{1}{2+1+1} = 0.2500$

Hence, $S^\otimes(Q, C_3) = s^\otimes\left(\hat{Q}^1, \hat{C}_3^1\right) = 0.2500.$

Now, sorting the documents in decreasing order of their Simplified similarity measure value, we get same ranking $C_1 \succ C_3 \succ C_2$ where '$\succ$' indicates 'is more relevant to query than'.

This may be noted that even without any constraints, $S^\otimes$ may give the same relative ranking as given by $S^\oplus$ in most of the cases. For example, if we compute the simplified Similarity measures $S^\otimes$ for expressions given in example 3, we have

$$s^\otimes\left(\hat{Q}^1, \hat{C}_1^1\right) = \dfrac{1}{\left|T_{C1}^1 - T_Q^1\right| + \left|T_Q^1 - T_{C1}^1\right| + 1} = \dfrac{1}{3+1+1} = 0.200,$$

$$s^\otimes\left(\hat{Q}^2, \hat{C}_1^1\right) = \dfrac{1}{\left|T_{C1}^1 - T_Q^2\right| + \left|T_Q^2 - T_{C1}^1\right| + 1} = \dfrac{1}{3+1+1} = 0.200.$$

Hence, $S^\otimes(Q, C_1) = \dfrac{s^\otimes\left(\hat{Q}^1, \hat{C}_1^1\right) + s^\otimes\left(\hat{Q}^2, \hat{C}_1^1\right)}{2} = 0.200.$

Also, $s^\otimes\left(\hat{Q}^1, \hat{C}_2^1\right) = \dfrac{1}{\left|T_{C2}^1 - T_Q^1\right| + \left|T_Q^1 - T_{C2}^1\right| + 1} = \dfrac{1}{1+1+1} = 0.333,$

$$s^\otimes\left(\hat{Q}^1, \hat{C}_2^2\right) = \dfrac{1}{\left|T_{C2}^2 - T_Q^1\right| + \left|T_Q^1 - T_{C2}^2\right| + 1} = \dfrac{1}{1+1+1} = 0.333,$$

$$s^{\otimes}\left(\hat{Q}^2, \hat{C}_2^1\right) = 0 \qquad\qquad \text{(because } T_Q^2 \cap T_{c_2}^1 = \phi\text{)},$$

$$s^{\otimes}\left(\hat{Q}^2, \hat{C}_2^2\right) = \frac{1}{\left|T_{C2}^2 - T_Q^2\right| + \left|T_Q^2 - T_{C2}^2\right| + 1} = \frac{1}{1+1+1} = 0.333,$$

$$\text{Hence, } S^{\otimes}\left(Q, C_2\right) = \frac{s^{\otimes}\left(\hat{Q}^1, \hat{C}_2^1\right) + s^{\otimes}\left(\hat{Q}^1, \hat{C}_2^2\right) + s^{\otimes}\left(\hat{Q}^2, \hat{C}_2^1\right) + s^{\otimes}\left(\hat{Q}^2, \hat{C}_2^2\right)}{4} = 0.250.$$

$$\text{Similarly, } s^{\otimes}\left(\hat{Q}^1, \hat{C}_3^1\right) = \frac{1}{\left|T_{C3}^1 - T_Q^1\right| + \left|T_Q^1 - T_{C3}^1\right| + 1} = \frac{1}{2+1+1} = 0.25,$$

$$\text{and } s^{\otimes}\left(\hat{Q}^2, \hat{C}_3^1\right) = \frac{1}{\left|T_{C3}^1 - T_Q^2\right| + \left|T_Q^2 - T_{C3}^1\right| + 1} = \frac{1}{1+0+1} = 0.50.$$

$$\text{Hence, } S^{\otimes}\left(Q, C_3\right) = \frac{s^{\otimes}\left(\hat{Q}^1, \hat{C}_3^1\right) + s^{\otimes}\left(\hat{Q}^2, \hat{C}_3^1\right)}{2} = 0.375.$$

Now, sorting the documents in decreasing order of their Simplified similarity measure value, we get the same ranking $C_3 \succ C_2 \succ C_1$, where '$\succ$' indicates 'is more relevant to query than'.

This way, we obtain the ranking $\Re_C$ of the documents by sorting them in the decreasing order of their Boolean similarity measures with the query.

4 Rank Aggregation Using Modified Shimura Technique

Rank aggregation is the problem of generating a "consensus" ranking for a given set of rankings. We begin with the Shimura technique of fuzzy ordering (Shimura 1973) because it is well suited for non-transitive rankings, as is the case here.

4.1 Shimura Technique of Fuzzy Ordering

For variables x_i and x_j defined on universe X, a relativity function $f(x_i|x_j)$ is taken to be the membership of preferring x_i over x_j. This function is given as

$$f\left(x_i \middle| x_j\right) = \frac{f_{x_j}(x_i)}{\max\left(f_{x_j}(x_i), f_{x_i}(x_j)\right)} \tag{20}$$

where, $f_{x_j}(x_i)$ is the membership function of x_i with respect to x_j, and $f_{x_i}(x_j)$ is the membership function of x_j with respect to x_i. For $X = [x_1, x_2, \ldots, x_n]$, $f_{x_i}(x_i) = 1$. $C_i = \min_{j=1}^{n} f\left(x_i \middle| x_j\right)$ is the membership ranking value for the ith variable. Now if a descending sort on C_i ($i = 1$ to n) is carried out, the sequence of $i's$ thus obtained would constitute the aggregated rank. For the lists $l_1, l_2, \ldots, l_N$

from the N participating search engines, we can have

$$f_{x_j}(x_i) = \frac{\left| k \in [1, N] \vee l_k(x_i) < l_k(x_j) \right|}{N} \tag{21}$$

For our case, N = 3, as we are considering the aggregation of the results from three different techniques described in the previous section.

4.2 Modified Shimura Technique

It is observed that classical Shimura technique gives worse performance in comparison to other rank aggregation techniques. We feel that the poor performance coming from the Shimura technique, is primarily due to the employment of "min" function in finding $C_i = \min_{j=1}^{n} f(x_i \mid x_j)$. The "min" function results in many ties, when a descending order sort is applied on C_i. There is no method suggested by Shimura to resolve these ties. So when resolved arbitrarily, these ties result in deterioration of the aggregated result. We, therefore replace this "min" function by an OWA operator (Yager 1988). The OWA operators, in fact, provide a parameterised family of aggregation operators, which include many of the well-known operators such as the maximum, the minimum, the k-order statistics, the median and the arithmetic mean.

We will be using the relative fuzzy linguistic quantifier "at least half" with the pair ($a = 0.0$, $b = 0.5$) for the purpose of finding the vector C_i as follows.

$$C_i = \sum_j w_j \cdot z_j \tag{22}$$

where z_j is the jth largest element in the ith row of the matrix $f(x_i \mid x_j)$.

Here, w_j is the weight of OWA based aggregation and is computed from the membership function Q describing the quantifier. In the case of a relative quantifier, with m criteria we have:

$$w_j = Q(j/m) - Q((j-1)/m), \ j = 0, 1, 2, \ldots, m \text{ with } Q(0) = 0.$$

The membership function Q of relative quantifier can be represented as

$$Q(r) = \begin{cases} 0 & if \ r < a \\ \frac{r-a}{b-a} & if \ b \leq r \leq a \\ 1 & if \ r > b \end{cases} \tag{23}$$

Now, as with the *Shimura* technique, if a descending sort on C_i(i=1 to m) is carried out, the sequence of i's thus obtained would constitute the aggregated rank $\Re_{COMP}$.

We will be using this modified *Shimura* technique for aggregating the three rankings $\mathfrak{R}_A$, $\mathfrak{R}_B$ and $\mathfrak{R}_C$ obtained from the three different evaluation procedures as described in preceding sections. Let us denote the aggregated ranking as $\mathfrak{R}_{COMP}$.

Let the full list $\mathfrak{R}_{SE}$ be the sequence in which the documents were initially listed by a search engine. Without loss of generality, it could be assumed that $\mathfrak{R}_{SE} = (1, 2, 3, \ldots, N_R)$, where N_R is the total number of documents listed in the result. We compare the sequences ($\mathfrak{R}_{COMP}$ and $\mathfrak{R}_{SE}$) and find Modified Spearman Rank Order Correlation Coefficient (r_s's). We repeat this procedure for a representative set of queries and take the average of r_s'. The resulting average value of r_s' is the required measure of the search quality (SQM). The overall procedure is illustrated in Fig. 1.

It may be noted that it is very common practice that a user views only those documents whose snippet displayed before him by the search engine he finds to be worth viewing. *Modified Spearman Rank Order Correlation Coefficient* is a better choice than *Spearman Rank Order Correlation Coefficient* to measure the closeness of the two rankings. Since, it is capable of working on a full list and a partial list and the sequence $\sum$ of documents viewed by a user is almost always a partial list, which in turn is used in getting the rankings $\mathfrak{R}_A$, $\mathfrak{R}_B$ and $\mathfrak{R}_C$ and hence aggregated ranking $\mathfrak{R}_{COMP}$ is also a partial list. Use of *Modified Spearman rank-order correlation coefficient* (r_s') saves computational efforts both in conversion of partial list to full list and also in the computation of r_s' for truncated lists.

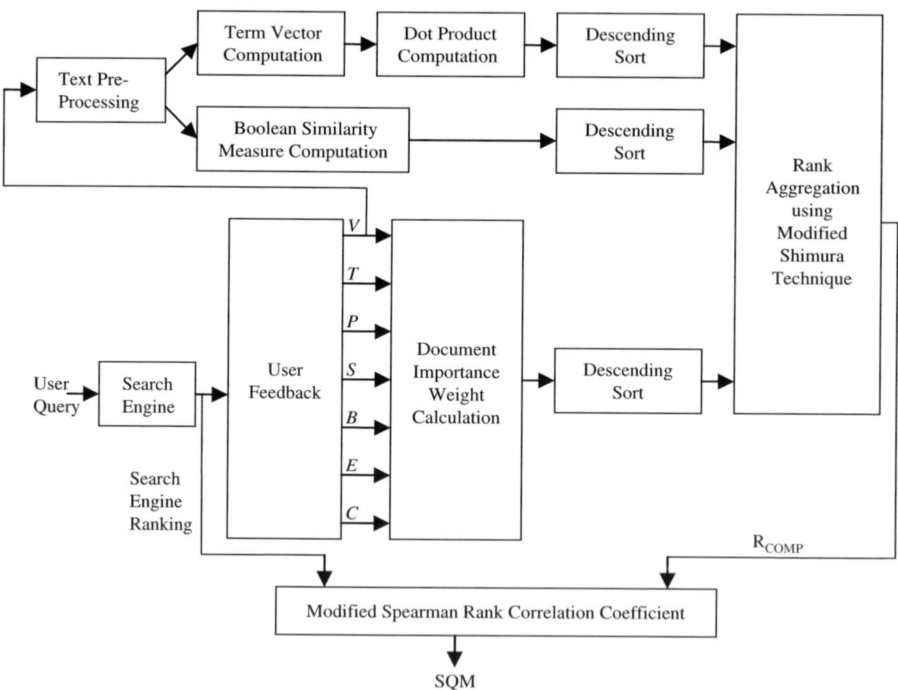

Fig. 1 Comprehensive Search Quality Evaluation

5 Experiments and Results

We experimented with a few queries on seven popular search engines, namely, *AltaVista, DirectHit, Excite, Google, HotBot, Lycos and Yahoo*. For the sake of simplicity, we obtain all our results with the weights in (3) being $w_V=1$, $w_T=1$, $w_P=1$, $w_S=1$, $w_B=1$, $w_E=1$ and $w_C=1$. For example, the observation corresponding to the query *similarity measure for resource discovery* is given in Table 1.

Table 1 shows that from the results of *AltaVista*, the document listed first was the document picked up first by the user, the document was read by the user for 20% of the time required to read it completely. It was not printed but saved and book marked. It was neither e-mailed to anyone nor was any of its portions copied and pasted elsewhere. The user then picked the second document listed by *AltaVista* and spent on it 20% of the time required actually to read it completely. It was not printed but was saved, book marked and e-mailed. None of its portion was copied and pasted. This gives an importance weight (σ_j) of 3.200 and 3.700 to the first and second documents, respectively. So the implicit ranking given by the user is document 2 $\succ$ document 1, where "$\succ$" indicates "more relevant than". .i.e $\Re_A =$ (2, 1) for *AltaVista* for the query. This way the value of $\Re_A$ is found for rest of the search engines.

In the vector space model, each query and document is represented by a term vector. Once the text pre-processing such as removal of stop-words, stemming is performed on query as well as on documents picked by user, normalized term vectors for the query and the documents are obtained. For example, the normalized term vector for the query *similarity measure for resource discovery* is (0.500,0.500,0.500,0.500). The vector contains only four components because there are only four terms in the query namely *"similarity", "measure", "resource" and*

Table 1 User Feedback model results for the Query: similarity measure for resource discovery

Search Engine	User Feedback (V,T,P,S,B,E,C)	Document Weight (σ_j)
AltaVista	(1,0.2,0,1,1,0,0.0)	3.200
	(2,0.2,0,1,1,1,0.0)	3.700
DirectHit	(1.0.2,0,1,1,0,0.0)	3.200
	(5,0.2,0,1,1,1,0.0)	3.700
Excite	(6,0.2,0,1,1,0,0.0)	3.200
	(4,0.2,0,1,1,1,0.0)	3.700
	(9,0.2,0,1,1,0,0.0)	2.490
Google	(1,0.4,0,0,1,0,0.0)	2.400
	(3,0.3,0,0,1,0,0.0)	1.800
	(5,0.3,1,0,0,0,0.0)	1.550
Hotbot	(2,0.2,0,1,1,0,0.0)	3.200
	(2,0.2,0,1,1,0,0.0)	3.200
Lycos	(3,0.5,1,0,1,0,0.0)	3.000
	(7,0.5,1,0,0,0,0.0)	1.750
Yahoo	(2,0.2,0,1,1,0,0.0)	3.200
	(4,0.2,0,0,1,0,0.0)	1.700
	(9,0.3,0,0,0,0,0.0)	0.550

"discovery". The fifth word in the query was *"for"* which is a stop-word and hence it is removed in pre-text processing. Each component has the same value (0.500) because all the four term appear in the query same number of times (only once). The vector is normalized to one since the magnitude of vector is 1. Similarly, normalized term vectors for the documents picked up from the results of the query are obtained. The observation corresponding to the query *similarity measure for resource discovery* is given in Table 2.

From the results of *AltaVista*, first and second documents were picked up as first and second document respectively. Table 2 shows that the term vector for first document is (0.500,0.500,0.500,0.500) which is same as that of the query. That means the first documents contains all the four terms present in the query and also each term appear in the document same number of times. The dot product of this term vector with that of query is 1.000. The term vector for the second document, on the other hand, is (0.378,0.882,0.252,0.126). Here, each component has different value since the four different terms appear different number of times in the document. The dot product of this with query term vector is 0.819. So the implicit ranking given by the vector space model is document 1 > document 2, where ">" indicates "more relevant than". We are not computing the term vectors for the rest of the documents listed by *AltaVista* as they were not clicked by user, thereby assuming that none of them contain relevant information. Thus, $\Re_B = (1, 2)$ for *AltaVista* for the query. This way the value of $\Re_B$ would be found for rest of the search engines.

In Boolean Similarity based model, each query and document is represented by a Boolean expression. We assume that Boolean expressions of the query Q and documents to be compared $(C_1, C_2, \ldots, C_n)$, just contain only AND terms i.e their CDNF contain only a single compact descriptor. Once the text pre-processing such

Table 2 Vector Space Model Results for the Query: similarity measure for resource discovery

Search Engine	Document Picked	Normalized Term Vector $c1,c2,c3, c4$	Dot product $\cos\theta$
AltaVista	1	(0.500,0.500,0.500,0.500)	1.000
	2	(0.378,0.882,0.252,0.126)	0.819
DirectHit	1	(0.500,0.500,0.500,0.500)	1.000
	5	(0.378,0.882,0.252,0.126)	0.819
	6	(0.500,0.500,0.500,0.500)	1.000
Excite	4	(0.378,0.882,0.252,0.126)	0.819
	9	(0.500,0.500,0.500,0.500)	1.000
	1	(0.436,0.655,0.436,0.436)	0.982
Google	3	(0.500,0.500,0.500,0.500)	1.000
	5	(0.378,0.882,0.252,0.126)	0.819
Hotbot	2	(0.500,0.500,0.500,0.500)	1.000
	2	(0.500,0.500,0.500,0.500)	1.000
Lycos	3	(0.706,0,706,0.314,0.314)	0.738
	7	(0.064,0.032,0.993,0.096)	0.592
	2	(0.500,0.500,0.500,0.500)	1.000
Yahoo	4	(0.436,0.655,0.436,0.436)	0.982
	9	(0.400,0.400,0.200,0.800)	0.900

as removal of stop-words, stemming is performed on query as well as on documents picked by user, set of descriptors in the compact atomic descriptors of the query and the documents are obtained. For example, set of descriptors in the compact atomic descriptor of the query *similarity measure for resource discovery* is (similarity, measure, resource, discovery). The set contains only four terms because there are only four AND terms in the Boolean expression of the query namely *"similarity"*, *"measure"*, *"resource"* and *"discovery"*. The fifth word in the query was *"for"* which is a stop-word and hence it is removed in pre-text processing. Similarly, sets of descriptors in the compact atomic descriptors of all the documents picked up by user from the results of the query are obtained. Once we have obtained set of descriptors present in the compact atomic descriptors of the query and documents, we can easily compute the simplified Boolean similarity measure using (19). The observation corresponding to the query *similarity measure for resource discovery* is given in Table 3.

From the results of *AltaVista*, first and second documents were picked up as first and second document respectively. Table 3 shows that the values of $\left| T^j_{C_k} - T^i_Q \right|$ and $\left| T^i_Q - T^j_{C_k} \right|$ to be used in (18) for the first document are 56 and 0 respectively. The Boolean similarity measure for this document, with that of query is 0.017857. The values of $\left| T^j_{C_k} - T^i_Q \right|$ and $\left| T^i_Q - T^j_{C_k} \right|$ for the second document, on the other hand, are 94 and 0 respectively. The Boolean similarity measure for this with query is 0.010638. So the implicit ranking given by the Boolean similarity based model is document $1 \succ$ document 2, where "$\succ$" indicates "more relevant than". We are not computing the Boolean similarity measures for the rest of the

Table 3 Boolean Similarity Model Results for the Query: similarity measure for resource discovery

Search Engine	Picked Document	$\left\| T^j_{C_k} - T^i_Q \right\|$	$\left\| T^i_Q - T^j_{C_k} \right\|$	$S^\otimes(Q, C_k)$
AltaVista	1	56	0	0.017857
	2	94	0	0.010638
DirectHit	1	56	0	0.017857
	5	94	0	0.010638
	6	56	0	0.017857
Excite	4	94	0	0.010638
	9	56	0	0.017857
	1	171	0	0.005848
Google	3	56	0	0.017857
	5	94	0	0.010638
Hotbot	2	56	0	0.017857
	2	56	0	0.017857
Lycos	3	450	0	0.002222
	7	496	0	0.002016
	2	56	0	0.017857
Yahoo	4	171	0	0.005848
	9	448	0	0.002232

Table 4 Aggregated Ranking ($\Re_{Comp}$) obtained using the Modified Shimura Technique and the Modified Spearman Rank Order Correlation Coefficient ($r_s{}'$) for the Query: similarity measure for resource discovery

Search Engine	Picked Document	$\Re_A$	$\Re_B$	$\Re_C$	Aggregated Ranking ($\Re_{Comp}$)	Correlation Coefficient ($r_s{}'$)
Altavista	1	2	1	1	1	1.000000
	2	1	2	2	2	
DirectHit	1	5	1	1	1	0.812500
	5	1	5	5	5	
Excite	6	4	6	6	6	0.687500
	4	6	9	9	9	
	9	9	4	4	4	
Google	1	1	3	3	1	0.930556
	3	3	1	5	3	
	5	5	5	1	5	
Hotbot	2	2	2	2	2	0.666667
	2	2	2	2	2	0.875000
Lycos	3	3	3	3	3	
	7	7	7	7	7	
	2	2	2	2	2	0.829167
Yahoo	4	4	4	4	4	
	9	9	9	9	9	

documents listed by *AltaVista,* as user did not click them, thereby assuming that none of them contain relevant information. Thus, $\Re_C = (1, 2)$ for the *AltaVista* for the query. This way the value of $\Re_C$ would be found for rest of the search engines.

All these three rankings $\Re_A$, $\Re_B$ and $\Re_C$ are then aggregated using *Modified Shimura Technique*. The Aggregated Ranking $\Re_{Comp}$ thus obtained is then compared with original ranking $\Re_{SE}$ to get the *Modified Spearman Rank Order Correlation Coefficient*. Thus, *AltaVista* gets $\Re_{Comp} = (1, 2)$ for the query. This would be compared with $\Re_{SE} = (1, 2)$ to give the *Modified Spearman Rank Order Cor-*

Table 5 List of Test Queries

1	*measuring search quality*
2	*mining access patterns from web logs*
3	*pattern discovery from web transactions*
4	*distributed associations rule mining*
5	*Document categorization query generation*
6	*term vector database*
7	*client -directory-server-model*
8	*Similarity measure for resource discovery*
9	*hypertextual web search*
10	*IP routing in satellite networks*
11	*focussed web crawling*
12	*concept based relevance feedback for information retrieval*
13	*parallel sorting neural network*
14	*Spearman rank order correlation coefficient*
15	*web search query benchmark*

Table 6 Modified Spearman Rank Order Correlation Coefficient (r_s') Obtained Using Aggregated Ranking ($\Re_{Comp}$) for the Queries given in Table 5

Query #	Altavista	DirectHit	Excite	Google	Hotbot	Lycos	Yahoo
1	0.800000	0.835417	0.840067	0.98125	0.977778	0.750000	0.877551
2	0.741667	0.875000	0.887500	0.90000	0.850505	0.888889	1.000000
3	0.729167	1.000000	0.777778	0.86250	0.765152	0.866667	0.947917
4	0.791667	0.757576	0.706349	0.83333	0.937500	0.400000	0.771429
5	0.645833	0.222222	0.875000	0.93750	0.833333	0.666667	0.791667
6	1.000000	1.000000	0.797619	1.00000	0.876190	1.000000	0.795833
7	0.930556	1.000000	0.800000	0.93333	0.793651	0.250000	0.906250
8	1.000000	0.812500	0.687500	0.93056	0.666667	0.875000	0.829167
9	0.685714	0.788360	0.848958	0.88889	0.854167	0.845714	0.807500
10	0.550505	1.000000	0.200000	1.00000	0.181818	0.671717	0.790476
11	0.888889	0.905000	0.733333	0.93056	0.882540	0.977778	0.913131
12	0.859375	0.861111	0.947917	0.91667	0.285714	1.000000	0.930556
13	1.000000	0.515873	0.222222	1.00000	0.181818	0.181818	0.666667
14	1.000000	0.914286	0.944444	0.97778	0.937500	1.000000	0.762500
15	0.181818	0.181818	0.181818	0.50000	0.285714	0.181818	0.181818
Average	0.787013	0.777944	0 .696700	0.90616	0.687337	0.703738	0.798164

relation Coefficient ($r_s' = 1.000$) for *AltaVista*. This way the value of r_s' would be found for rest of the search engines. All the three rankings $\Re_A$, $\Re_B$ and $\Re_C$ and their corresponding $\Re_{Comp}$ and r_s' are listed in Table 4 for the query *similarity measure for resource discovery*. We experimented with 15 queries in all. These queries are listed in Table 5 and their *modified spearman rank order correlation coefficient* (r_s') thus obtained using aggregated ranking ($\Re_{Comp}$) is given in Table 6.

The results of Table 6 are pictorially represented in Fig. 2. From Table 6 and Fig. 2, we observe that *Google* gives the best performance, followed by *Yahoo, AltaVista, DirectHit, Lycos, Excite,* and *Hotbot,* in that order.

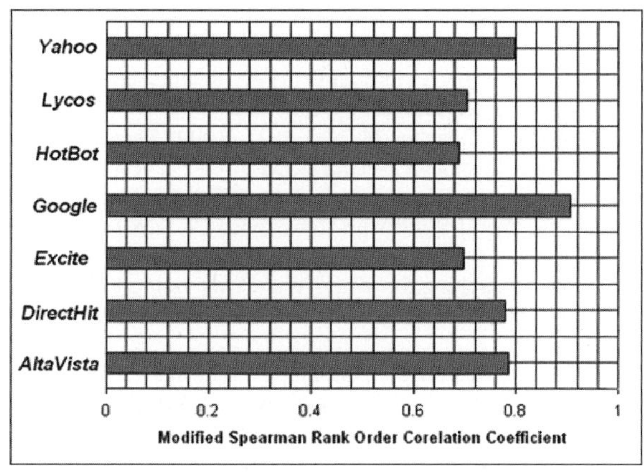

Fig. 2 Performance of Search Engines based on Aggregated Model

6 Conclusions

We have tried to augment the user feedback based subjective evaluation with objective evaluation for the public web search engines. For the subjective measure, we used a method that monitor the actions of the users on the search results presented before him in response to his query and then infer his preferences there from. For the objective measure, we have used Vector Space Model and Boolean similarity measures. We used the simplified version of Li Danzig Boolean similarity measure for computing the similarity between the query and the documents returned by the search engines. We are aggregating the ranking of documents obtained from these three evaluation processes using Modified Shimura Technique. The aggregated ranking is compared with the original ranking by the search engine and correlation coefficient is computed. Thus, The correlation coefficient is a quantitative measure of web search quality of the search engine. Hence, different search engines are evaluated on the basis of the measure. Our results for 15 queries and 7 public web search engines show that *Google* gives the best performance, followed by *Yahoo, AltaVista, DirectHit, Lycos, Excite,* and *Hotbot,* in that order.

References

Ali R, Beg M M S (2005) Aggregating Subjective measure of Web Search quality with Boolean Similarity measures. In: Proc. 2nd World Congress on Lateral computing (WCLC-05), December, Bangalore, India.

Bar-Yossef Z, Berg A, Chien S, Fakcharoenphol J, Weitz D (2000) Approximating Aggregate Queries about Web Pages via Random Walks. In: Proc. 26th VLDB Conference, Cairo, Egypt.

Beg M M S (2002) On Measurement and Enhancement of Web Search Quality. Ph.D. Thesis, Department of Electrical Engineering, I. I. T. Delhi, India.

Beg M M S, Ali R (2005) A Subjective cum Objective Measure of Web Search Quality. In: Proc. 2005 International Conference on Intelligent Sensing and Information Processing (ICISIP-05), Chennai, India, pp. 365–370.

Bharat K, Broder A (1998) A Technique for Measuring the Relative Size and Overlap of Public Web Search Engines. In: Proc. 7th International World Wide Web Conference (WWW9), April, pp. 379–388.

Hawking D, Craswell N, Thistlewaite P, Harman D (1999) Results and Challenges in Web Search Evaluation. Toronto '99, Elsevier Science, pp. 243–252.

Henzinger MR, Heydon A, Mitzenmacher M, Najork M (1999) Measuring Index Quality Using Random Walks on the Web. Computer Networks, 31, pp. 1291–1303.

Henzinger MR, Heydon A, Mitzenmacher M, Najork M (2000) On Near Uniform URL Sampling. In: Proc. 9th International World Wide Web Conference (WWW9), Amsterdam, Netherlands, May.

Lawrence S, Giles CL (1998) Searching the World Wide Web. Science, 5360(280):98.

Lawrence S, Giles CL (1999) Accessibility of Information on the Web. Nature, vol. 400, pp.107–109.

Li L, Shang Y (2000) A New Method for Automatic Performance Comparison of Search Engines. World Wide Web: Internet and Web Information Systems, Kluwer Academic, 3(4), pp. 241–247.

Li S H, Danzig P B (1994) Boolean similarity measures for resource discovery. Technical Report: USC-CS-94-579, Computer Science Department, University of Southern California, Los Angeles.

Li S H, Danzig P B (1997) Boolean Similarity Measures for Resource Discovery. IEEE Trans. Knowledge and Data Engineering, 9(6), pp. 863–876.

Porter M (1980) An Algorithm for Suffix Stripping. Program: Automated Library and Information Systems, 14(3).

Salton G, McGill MJ (1983) Introduction to Modern Information Retrieval. McGraw Hill.

Shang Y, Li L (2002) Precision Evaluation of Search Engines. World Wide Web: Internet and Web Information Systems, Kluwer Academic, 5(2), pp. 159–173.

Shimura M (1973) Fuzzy Sets Concept in RankOrdering Objects. J. Math. Anal. Appl., 43, pp. 717–733.

Yager R R (1988) On Ordered Weighted Averaging Aggregation Operators in Multicriteria Decision Making. IEEE Trans. Systems, Man and Cybernetics, 18(1), pp. 183–190.

Interpolative Realization of Boolean Algebra as a Consistent Frame for Gradation and/or Fuzziness

Dragan Radojević

Abstract L. Zadeh has ingeniously recognized importance and necessity of gradation in relations generally (theory of sets – fuzzy sets, logic – fuzzy logic, relations – fuzzy relations) for real applications. Common for all known approaches for treatment gradation is the fact that either they are not complete (from the logical point of view) or they are not in the Boolean frame. Here is given Interpolative Boolean algebra (IBA) as a consistent MV realization of finite (atomic) Boolean algebra. Since, axioms and lows of Boolean algebra are actually meter of value independent structure of IBA elements, all axioms and all laws of Boolean algebra are preserved in any type of value realization (two-valued, three-valued, . . ., [0, 1]). To every element of IBA corresponds generalized Boolean polynomial with ability to process all values of primary variables from real unit interval [0, 1]. The possibility of new approach is illustrated on two examples: generalized preference structure and interpolative sets as consistent realization of idea of fuzzy sets.

1 Introduction

According to J. Boole[1], objects of interest for Boolean algebra are relations: null-ary (for example: propositions in logic), unary (properties in theory of sets), binary (theory of graphs, preferential relations etc.), . . ., n-ary relations.

Classical (Aristotelian) two-valued realization of Boolean algebra is based on two-element Boolean algebra as its homomorphism. As a consequence, on the calculus level all laws of Boolean algebra are preserved.

In many real applications the classical two-valued ("black and white") realization is not adequate. L. Zadeh, after his famous and distinguished contribution in the modern control theory, has ingeniously recognized the necessity of gradation in relations generally (theory of sets – fuzzy sets , logic – fuzzy logic , relations – fuzzy relations).

Conventional fuzzy approaches rely on the same principle as many-valued (MV) logics . MV-logics are similar to classical logic because they accept *the principle*

---

[1] "The business of Logic is with the relations of classes and with the modes in which the mind contemplates those relations", The Calculus of Logic, 1848.

of truth-functionality . A logic is truth functional if the truth value of a compound sentence depends only on the truth values of the constituent atomic sentences, not on their meaning or structure. The consequences of this direction are in the best way described by Lukasiewicz, the innovator of MV-logic: *"Logic* (truth functional) *changes from its very foundations if we assume that in addition to truth and falsehood there is also some third logical value or several such values, . . ."* . Either simple MV-logics based on truth functional principle are not in the Boolean frame and/or they are not realization of Boolean algebra. According to fuzzy logic is based on truth functionality, since: *"This is very common and technically useful assumption"*. A contrary example: "Our world is a flat plate" was also a very common and technically useful assumption in the Middle Ages!?

One "argument" for destroying the Boolean frame in treating gradation (MV-case) can be the definition of Boolean axioms of contradiction and excluded middle according to Aristotle: *The same thing cannot at the same time both belong and not belong to the same object and in the same respect* (Contradiction) . . . *Of any object, one thing must be either asserted or denied* (Excluded middle).

If the goal is mathematics for gradation then it seems "reasonable" to leave these axioms as inadequate and accept the principle of truth functionality with all consequences or to go to the very source of Boolean algebra idea.

It is interesting that in his seminal paper G. Boole has said: *". . . **the symbols of the** (logic) **calculus do not depend for their interpretation upon the idea of quantity . . ."*** and only *"**in their particular application . . ., conduct us to the quantitative conditions of inference**"*. So, according to G. Boole, the principle of truth functionality is not a fundamental principle (and as a consequence this principle can't be the basis of any generalization).

A very important question is: ***Can fuzziness and/or gradation be realized in a Boolean frame as a realization of Boolean algebra?*** We have obtained a ***positive answer*** to this question as an unexpected result, during solving the problem of fuzzy measure (or capacity) meaning in decision making by theory of capacity .

The new approach to treating gradation in logic, theory of sets, relations etc., is based on interpolative realization of finite Boolean algebra (IBA) .

IBA has two levels: (a) *Symbolic or qualitative* – is a matter of finite Boolean algebra and (b) *Semantic or valued* – is in a general case, a matter of interpolation.

Structure of any IBA element and/or *principle of structural functionality* are a bridge between two levels and they are the basis of generalization, since they are value independent. The structure of analyzed element determines which *atom* (from a finite set of IBA atoms) is included in it and/or which is not. The principle of structural functionality: *The structure of any IBA combined element can be directly calculated on the basis of structures of its constituents,* using homomrphic mapping IBA on two-element Boolean algebra. Similarly, according to the principle of value (truth in the case of logic) functionality value of combined element of BA can be directly calculated on the basis of values of its components, in a two-valued case. So, the set of structures as a value independent homomrphic image of IBA elements is isomorphism to the set of values of IBA elements in a two-valued case.

In the new approach we keep still all Boolean axioms and identities which follow from these axioms. Now Contradiction is given by the following definition: *the same atomic property cannot at the same time both belong and not belong to the analyzed property*... and Excluded middle: *For any property one atomic property must be either asserted or denied.* From new generalized definitions of contradictions and excluded middle follow Aristotle's definitions as a special classical two-valued case. Value realization of these two axioms in a general case is: (as a consequence of the fact that the sum of the intensity values of atomic property is identical to 1 for any analyzed object) the intensity of realization of analyzed property for any object is the complement of its non-realization (or the realization of its negation as a property).

IBA has a crucially different approach to gradation compared to fuzzy approaches. Technically, as any element of finite Boolean algebra can be represent in a canonical disjunctive form it can also be represented in the form of a corresponding generalized Boolean polynomial. A generalized Boolean polynomial can process values from a real unit interval [0, 1]. So, all laws of Boolean algebra are preserved in the case of gradation.

The main characteristics of the new approach will be illustrated on the following interesting and illustrative examples:

1. A generalized preference structure, as a direct generalization of classical result ;
2. Interpolative sets as a consistent realization of the idea of fuzzy sets.

2 Finite Boolean Algebra

A finite Boolean algebra (and/or Boolean lattice) is a very well known algebraic structure . It is a partially ordered set of a special type. It is a distributive lattice with the most complex element "1", the unit of the Boolean algebra, and the simplest element "0", the zero of the Boolean algebra, that contains together with each element also its complement. The operations join and meet are usually denoted by the symbols $\cup$ and $\cap$. The complement of an element in a Boolean algebra is unique. A Boolean algebra can also be defined as

$$\mathrm{BA} = \langle BA, \cup, \cap, C \rangle,$$

where: BA is a non-empty set and $\cup$, $\cap$, C the operations, which for $\varphi, \psi, \xi \in BA$, satisfy the following *axioms*:

Commutativity

$$\varphi \cap \psi = \psi \cap \varphi, \quad \varphi \cup \psi = \psi \cup \varphi.$$

Associativity

$$\varphi \cap (\psi \cap \xi) = (\varphi \cap \psi) \cap \xi, \quad \varphi \cup (\psi \cup \xi) = (\varphi \vee \psi) \cup \xi.$$

Absorption

$$\varphi \cup (\varphi \cap \psi) = \varphi, \quad \varphi \cap (\varphi \cup \psi) = \varphi.$$

Distributivity

$$\varphi \cup (\psi \cap \xi) = (\varphi \cup \psi) \cap (\psi \cup \xi), \quad \varphi \cap (\psi \cup \xi) = (\varphi \cap \psi) \cup (\varphi \cap \xi).$$

Complements

$$\varphi \cup C\varphi = 1, \quad \varphi \cap C\varphi = 0.$$

From these axioms, the following *identities* also follow:
Idempotency

$$\varphi \cup \varphi = \varphi, \quad \varphi \cap \varphi = \varphi.$$

Boundedness

$$\begin{aligned} \varphi \cup 0 &= \varphi & \varphi \cap 1 &= \varphi \\ \varphi \cup 1 &= 1 & \varphi \cap 0 &= 0 \end{aligned}.$$

0 and 1 are complements

$$C0 = 1 \quad C1 = 0.$$

De Morgan's laws

$$C(\varphi \cup \psi) = (C\varphi) \cap (C\psi), \quad C(\varphi \cap \psi) = (C\varphi) \cup (C\psi).$$

Involution

$$C(C\varphi) = \varphi.$$

2.1 A Classical Two-valued Realization of BA

A two-valued realization of **BA** is adequate for those problems for which the following information is sufficient: an object has the analyzed property and/or a *BA* element is realized or the object hasn't the analyzed property and/or a *BA* element is not realized. This special case is very important since all laws of **BA** can be treated only on the basis of values actually on the basis of two values. This is the consequence of the fact that the mapping of any elements $\varphi \in BA$ to its values $\|\varphi\|$ in a two-valued case $\|\varphi\| : X^n \rightarrow \{0, 1\}$, is homomorphism:

$$\begin{aligned}
\big\| \varphi \cap \psi \big\| &= \big\| \varphi \big\| \wedge \big\| \psi \big\|, \\
\big\| \varphi \cup \psi \big\| &= \big\| \varphi \big\| \vee \big\| \psi \big\|, \\
\big\| C\varphi \big\| &= \neg \big\| \varphi \big\|,
\end{aligned}$$

where: $\varphi, \psi \in BA$ and

$$
\begin{array}{c|cc}
\wedge & 0 & 1 \\
\hline
0 & 0 & 0 \\
1 & 0 & 1
\end{array}
\qquad
\begin{array}{c|cc}
\vee & 0 & 1 \\
\hline
0 & 0 & 1 \\
1 & 1 & 1
\end{array}
\qquad
\begin{array}{c|c}
\neg & \\
\hline
0 & 1 \\
1 & 0
\end{array}
$$

So this mapping preserves all laws immanent to BA.

One very important consequence of this homomorphism is the realization of truth tables. This is a very common and technically useful tool, which actually preserves all properties of Boolean algebra, but only in a classical (two-valued) case.

Up to now only two-value realizations have been analyzed. The main constraint of two value realizations is the following characteristic: two different objects can't be discriminated on the basis of the same property in the case when both have this property and/or do not have this property. So, only in the case when one object has and the other doesn't have the analyzed property discrimination is possible.

3 Interpolative Boolean Algebra

Interpolative Boolean algebra (IBA) is devoted to the treatment of generalized valued realization . Generalization means that elements of IBA can have more than two values (as in classical case) including general case the hole real unit interval [0, 1]. Interpolative Boolean algebra has a finite number of elements. Finite number constraint should be seen in accordance with the fact that gradation gives the incomparably lager "power" to any element of an algebra. IBA attains much more in the sense of valued realization effects (adequate descriptions in a general case) with a smaller number of elements (so, with lower complexity). IBA is an atomic algebra (as a consequence of the finite number of elements). Atoms as the simplest elements of algebra have a fundamental role in the new approach.

All laws of Boolean algebra are value indifferent and they are the objects of *the symbolic level* of IBA. One of the basic notions from symbolic level is *the structure* of IBA element, which determines the atoms of which the analyzed element is built up and/or is not built up. On IBA *valued level* the elements of IBA are value realized so that all laws are preserved from symbolic level. The structure is value independent and it is the key of preserving the Boolean laws from a symbolic level on a valued level.

3.1 Symbolic Level

A symbolic or qualitative level is value independent and as a consequence, it is the same for all realizations on valued level: classical (two-valued), three-valued, . . . and a generalized MV-case. A symbolic or qualitative level is independent of the rank, or type, of the analyzed relation on value level: null-ary (in logic), unary (in the theory of sets), binary (in the graph theory, preference structures etc.) until general n-ary relations. On a symbolic or qualitative level: the main notion is a *finite set of elements with corresponding Boolean operators – atomic Boolean algebra*. This set is generated by a set of *primary elements – context* or *set generator* of analyzed Boolean algebra. No primary element can be realized as a Boolean function of the remaining primary elements. An *order relation* on this level is based only on the operator of *inclusion*. The *atomic element* of Boolean algebra is the simplest in the sense that it doesn't include in itself any other element except itself and a trivial zero constant. Meet (conjunction, intersection) of any two atomic elements is equal to a zero constant. Any element from analyzed Boolean algebra can be represented by a disjunctive canonical form: join (disjunction, union) of relevant atoms. The *structure* of analyzed element determines which atom is relevant (or included in it) and/or which is not relevant (or not included in it). The *structure function* of analyzed element is actually the characteristic function of the subset of its relevant atoms from the set of all atoms. Calculus of structure is based on *the principle of structural functionality*. This value independent principle is formally similar to the principle of "truth functionality", but fundamentally different since truth functionality is value dependent (the matter of valued level) and is actually valid only for two-valued case. The principle of structural functionality as value independent is the fundamental characteristic. The principle of *truth functionality* on a value level is only isomorphism of the principle of structural functionality and valid only for a classical (two valued) case.

The axioms and identities of IBA are the same as in Boolean algebra, given in paragraph 2. The axioms and identities of IBA are qualitatively (independently of IBA realization nature) and value irrelevant (independently of the number of possible values in valued realizations).

IBA is on a symbolic level identical to Boolean algebra with a finite number of elements.

$$\langle BA, \cap, \cup, C \rangle$$

Algebraic structure with: BA is a set with a finite number of elements, two binary operators $\cap$ and $\cup$, and one unary operator C, for which all axioms and identities from 2. are valid.

The element of IBA set in a symbolic way represents everything what on a valued level can characterize, image, determine, assert . . . the analyzed object in a qualitative sense. IBA element on a symbolic level is treated independently of its potential realization both in a qualitative (property, characteristic, relation, . . .) and quantitative sense (intensity of value realization). So two IBA-s are the same, on this level, if they have same number of elements. IBA elements are mutually different in their complexity and/or structure on which is based the partial order of IBA elements, based on the relation of inclusion.

The Basic operations on IBA elements are the basic Boolean operations. Result of any operations on IBA elements is IBA element.

Join (union, conjunction) of φ and ψ two different IBA elements generate $(\varphi \cup \psi)$ a new IBA element, with the following characteristic:

$$(\varphi \cup \psi) \Rightarrow \varphi \subseteq (\varphi \cup \psi), \psi \subseteq (\varphi \cup \psi).$$

Meet (intersection, disjunction) of φ and ψ two different IBA elements generates a new IBA element $(\varphi \cap \psi)$, with the following characteristic:

$$(\varphi \cap \psi) \Rightarrow \varphi \subseteq (\varphi \cap \psi), \psi \subseteq (\varphi \cap \psi).$$

Complement (negation) of any IBA element φ is a new IBA element $(C\varphi)$ uniquely determined with the following characteristic: $(C\varphi)$ doesn't contain anything what contains φ and contains everything else from analyzed Boolean algebra.

As a consequence the elements of Boolean algebra and/or IBA are partially ordered (Boolean lattice) on a symbolic level. The example of Boolean lattice, a partially ordered set generated by two primary elements $\Omega = \{a, b\}$ has sixteen elements and it is given in the Fig. 1. (Where: $a \Leftrightarrow b = (a \cap b) \cup (Ca \cap Cb)$ and

$$a \veebar b = (a \cap Cb) \cup (Ca \cap b)).$$

Primary IBA elements have the property that any of them can't be expressed only on the basis of the remaining primary IBA elements applying Boolean operation and they form a set $\Omega \subset BA$ of primary IBA elements. Set Ω generates a set of IBA elements, which is denoted by $BA(\Omega)$ as a consequence. If $n = |\Omega|$ then $2^{2^n} = |BA(\Omega)|$.

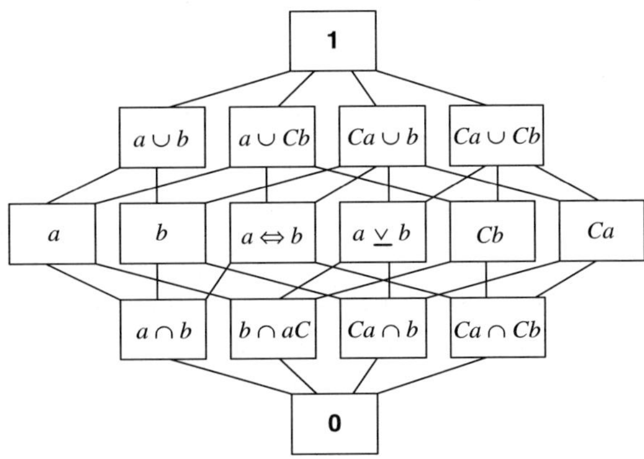

Fig. 1 Boolean lattice generated by primary relations $\Omega = \{a, b\}$

Example 1. If $\Omega = \{a, b\}$ is the set of primary elements – set generator of Boolean algebra, then the elements of Boolean algebra are given in the following table:

	$\varphi \in \mathrm{BA}\,(\{a, b\})$
1.	0
2.	$Ca \cap Cb$
3.	$a \cap Cb$
4.	$Ca \cap b$
5.	$a \cap b$
6.	Cb
7.	Ca
8.	$(a \cap b) \cup (Ca \cap Cb)$
9.	$(a \cap Cb) \cup (Ca \cap b)$
10.	a
11.	b
12.	$Ca \cup Cb$
13.	$a \cup Cb$
14.	$Cb \cup b$
15.	$a \cup b$
16.	$(a \cap b) \cup (a \cap Cb) \cup (Ca \cap b) \cup (Ca \cap Cb)$

Atomic IBA element is a total meet of primary elements $a_i \in \Omega$ and/or their complements so that every element occurs only once as a primary element or as its complement:

$$\alpha\,(S)\,(a_1, \ldots, a_n) = \bigcap_{a_i \in \Omega} \pi_S\,(a_i),$$

$$\pi_S\,(a_i) = \begin{cases} a_i, & a_i \in S \\ Ca_i, & a_i \notin S \end{cases}, \quad S \in \mathrm{P}\,(\Omega).$$

Example 2. The atomic IBA elements in the case $\Omega = \{a, b\}$ are given by the following expressions:

$$\begin{aligned}
\alpha\,(\emptyset)\,(a, b) &= Ca \cap Cb, \\
\alpha\,(\{a\})\,(a, b) &= a \cap Cb, \\
\alpha\,(\{b\})\,(a, b) &= Ca \cap b, \\
\alpha\,(\{a, b\})\,(a, b) &= a \cap b.
\end{aligned}$$

To any element $S \in \mathrm{P}\,(\Omega)$ of a power set of a primary set (subset of primary set), there corresponds one atomic element $\alpha\,(S)\,(a_1, \ldots, a_n)$. Atomic elements are the simplest elements of IBA because they don't include in themselves anything except their selves and a trivial IBA element "0". As a consequence intersection of two different atoms is identically equal to 0:

$$\alpha\,(S_i)\,(a_1, \ldots, a_n) \cap \alpha\,(S_j)\,(a_1, \ldots, a_n) = 0, i \neq j.$$

Universe of atomic elements $U_\alpha(\Omega)$ is a set of all atomic elements (generated by primary elements).

$$U_\alpha(\Omega) = \{\alpha(S)(a_1, \ldots, a_n) \,|\, S \in P(\Omega), a_i \in \Omega\}.$$

Based on the definition, a cardinal number of the universe of atomic elements is $2^{|\Omega|}$ in the case when $|\Omega|$ is the number of primary elements.

Example 3. In the case $\Omega = \{a, b\}$ universe of atomic elements is:

$$U_\alpha(\Omega) = \{(Ca \cap Cb), (a \cap Cb), (Ca \cap b), (a \cap b)\}.$$

Structure of any IBA element $\varphi \in BA(\Omega)$ is a set $\sigma'(\varphi)$ of atomic elements contained in analyzed IBA element φ. To every element of $\sigma'(\varphi)$ corresponds an element of power set $\sigma(\varphi) \in P(P(\Omega))$.

Example 4. Structures of IBA elements generated by $\Omega = \{a, b\}$ is given in the following table:

	φ	$\sigma(\varphi)$	$\sigma'(\varphi)$
1.	0	$\varnothing$	$\varnothing$
2.	$Ca \cap Cb$	$\{\varnothing\}$	$\{(Ca \cap Cb)\}$
3.	$a \cap Cb$	$\{\{a\}\}$	$\{(a \cap Cb)\}$
4.	$Ca \cap b$	$\{\{b\}\}$	$\{(Ca \cap b)\}$
5.	$a \cap b$	$\{\{a, b\}\}$	$\{(a \cap b)\}$
6.	Cb	$\{\varnothing, \{a\}\}$	$\{(Ca \cap Cb), (a \cap Cb)\}$
7.	Ca	$\{\varnothing, \{b\}\}$	$\{(Ca \cap Cb), (Ca \cap b)\}$
8.	$a \Leftrightarrow b$	$\{\varnothing, \{a, b\}\}$	$\{(Ca \cap Cb), (a \cap b)\}$
9.	$a \underline{\vee} b$	$\{\{a\}, \{b\}\}$	$\{(a \cap Cb), (Ca \cap b)\}$
10.	a	$\{\{a\}, \{a, b\}\}$	$\{(a \cap Cb), (a \cap b)\}$
11.	b	$\{\{b\}, \{a, b\}\}$	$\{(Ca \cap b), (a \cap b)\}$
12.	$Ca \cup Cb$	$\{\varnothing, \{a\}, \{b\}\}$	$\{(Ca \cap Cb), (a \cap Cb), (Ca \cap b)\}$
13.	$a \cup Cb$	$\{\varnothing, \{a\}, \{a, b\}\}$	$\{(Ca \cap Cb), (a \cap Cb), (a \cap b)\}$
14.	$Ca \cup b$	$\{\varnothing, \{b\}, \{a, b\}\}$	$\{(Ca \cap Cb), (Ca \cap b), (a \cap b)\}$
15.	$a \cup b$	$\{\{a\}, \{b\}, \{a, b\}\}$	$\{(a \cap Cb), (Ca \cap b), (a \cap b)\}$
16.	1	$\{\varnothing, \{a\}, \{b\}, \{a, b\}\}$	$\{(Ca \cap Cb), (a \cap Cb), (Ca \cap b), (a \cap b)\}$

Characteristic function of structure $\chi_{\sigma(\varphi)}$ or shortly ***structural function*** of any IBA element $\varphi \in BA(\Omega)$ is given by the following expression:

$$\chi_{\sigma(\varphi)}(S) = \begin{cases} 1, & S \in \sigma(\varphi) \\ 0, & S \notin \sigma(\varphi) \end{cases} ; \quad S \in P(\Omega),$$

where: Ω is a set of IBA primary elements and $P(\Omega)$ is a power set of Ω.

Example 5. Structural function of IBA elements BA $(\{a, b\})$

φ	$\chi_{\sigma(\varphi)}(\emptyset)$	$\chi_{\sigma(\varphi)}(\{a\})$	$\chi_{\sigma(\varphi)}(\{b\})$	$\chi_{\sigma(\varphi)}(\{a, b\})$
0	0	0	0	0
$Ca \cap Cb$	1	0	0	0
$a \cap Cb$	0	1	0	0
$Ca \cap b$	0	0	1	0
$a \cap b$	0	0	0	1
Cb	1	1	0	0
Ca	1	0	1	0
$a \Leftrightarrow b$	1	0	0	1
$a \underline{\vee} b$	0	1	1	0
a	0	1	0	1
b	0	0	1	1
$Ca \cup Cb$	1	1	1	0
$a \cup Cb$	1	1	0	1
$Ca \cup b$	1	0	1	1
$a \cup b$	0	1	1	1
1	1	1	1	1

Structural function $\chi_{\sigma(\varphi)}$ of IBA element φ determines the inclusion (value 1) and/or non inclusion (value 0) of analyzed atomic element in it. It is clear that the following equality holds:

$$\chi_{\sigma(\varphi)}(S) = \left\| \varphi\left(\chi_S(a_1), \ldots, \chi_S(a_n)\right) \right\|$$

where: $\left\| \varphi\left(\chi_S(a_1), \ldots, \chi_S(a_n)\right) \right\| \in \{0, 1\}$ is value of $\varphi \in \mathrm{BA}(\Omega)$ for

$$\chi_S(a_i) = \begin{cases} 1, & a_i \in S \\ 0, & a_i \notin S \end{cases} ; \ S \in \mathrm{P}(\Omega), a_i \in \Omega.$$

Comment: It is interesting that J. Boole in seminal paper "Calculus of logic" used an expression analogous to expression $\left\| \varphi\left(\chi_S(a_1), \ldots, \chi_S(a_n)\right) \right\|$ and called it "mod".

The value of structural function is determined on a symbolic level and of course it doesn't depend on the values of IBA elements.

The characteristic elements of IBA are "0" and "1".

IBA element "0" is trivial in the sense that it doesn't include in itself any atomic element.

$$\chi_{\sigma(0)}(S) = 0, \quad S \in \mathrm{P}(\Omega).$$

IBA element "1" is the most complex in the sense that its structure is equal to the universe of atomic elements.

$$\chi_{\sigma(1)}(S) = 1, \quad S \in \mathrm{P}(\Omega).$$

So, "1" includes in themselves (itself) all atomic elements from Ω and as a consequence it includes any IBA element too.

The structural function is ***homomrphic mapping*** of IBA elements:

$$\chi_{\sigma(\varphi \cap \psi)}\,(S) = \chi_{\sigma(\varphi)}\,(S) \wedge \chi_{\sigma(\psi)}\,(S)\,,$$
$$\chi_{\sigma(\varphi \cup \psi)}\,(S) = \chi_{\sigma(\varphi)}\,(S) \vee \chi_{\sigma(\psi)}\,(S)\,,$$
$$\chi_{\sigma(C\varphi)}\,(S) = \neg\chi_{\sigma(\varphi)}\,(S)\,,$$
$$\forall S \in P(\Omega)\,.$$

Since the structure is a result of homomrphic mapping of IBA elements, ***all axioms and identities of Boolean algebra are valid for the structure*** of IBA elements:

Commutativity

$$\chi_{\sigma(\varphi \cap \psi)}\,(S) = \chi_{\sigma(\psi \cap \varphi)}\,(S)\,, \quad \chi_{\sigma(\varphi \cup \psi)}\,(S) = \chi_{\sigma(\psi \cup \varphi)}\,(S)\,,$$
$$\forall S \in P(\Omega)\,.$$

Associativity

$$\chi_{\sigma(\varphi \cap (\psi \cap \phi))}\,(S) = \chi_{\sigma(\varphi)}\,(S) \wedge \left(\chi_{\sigma(\psi)}\,(S) \wedge \chi_{\sigma(\phi)}\,(S)\right)\,,$$
$$= \left(\chi_{\sigma(\varphi)}\,(S) \wedge \chi_{\sigma(\psi)}\,(S)\right) \wedge \chi_{\sigma(\phi)}\,(S)\,,$$
$$= \chi_{\sigma((\varphi \cap \psi) \cap \phi)}\,(S)\,.$$
$$\chi_{\sigma(\varphi \cup (\psi \cup \phi))}\,(S) = \chi_{\sigma((\varphi \cup \psi) \cup \phi)}\,(S)\,.$$
$$\forall S \in P(\Omega)\,.$$

Absorption

$$\chi_{\sigma(\varphi \cup (\varphi \cap \psi))}\,(S) = \chi_{\sigma(\varphi)}\,(S)\,,$$
$$\chi_{\sigma(\varphi \cap (\varphi \cup \psi))}\,(S) = \chi_{\sigma(\varphi)}\,(S)\,,$$
$$\forall S \in P(\Omega)\,.$$

Distributivity

$$\chi_{\sigma(\varphi \cup (\psi \cap \phi))}\,(S) = \left(\chi_{\sigma(\varphi)}\,(S) \vee \chi_{\sigma(\psi)}\,(S)\right) \wedge \left(\chi_{\sigma(\varphi)}\,(S) \vee \chi_{\sigma(\phi)}\,(S)\right)\,,$$
$$= \chi_{\sigma((\varphi \cup \psi) \cap (\varphi \cup \phi))}\,(S)\,.$$
$$\chi_{\sigma(\varphi \cap (\psi \cup \phi))}\,(S) = \chi_{\sigma((\varphi \cap \psi) \cup (\varphi \cap \phi))}\,(S)\,.$$
$$\forall S \in P(\Omega)\,.$$

Complements (Excluded middle and Contradiction)

$$\chi_{\sigma(\varphi \cup C\varphi)}\,(S) = 1,$$
$$\chi_{\sigma(\varphi \cap C\varphi)}\,(S) = 0,$$
$$\forall S \in \mathrm{P}\,(\Omega)\,.$$

From these axioms, the following *identities* also follow:

Idempotency

$$\chi_{\sigma(\varphi \cup \varphi)}\,(S) = \chi_{\sigma(\varphi)}\,(S)\,,$$
$$\chi_{\sigma(\varphi \cap \varphi)}\,(S) = \chi_{\sigma(\varphi)}\,(S)\,,$$
$$\forall S \in \mathrm{P}\,(\Omega)\,.$$

Boundedness

$$\chi_{\sigma(\varphi \cup 0)}\,(S) = \chi_{\sigma(\varphi)}\,(S)\,,$$
$$\chi_{\sigma(\varphi \cap 1)}\,(S) = \chi_{\sigma(\varphi)}\,(S)\,,$$
$$\chi_{\sigma(\varphi \cup 1)}\,(S) = \chi_{\sigma(1)}\,(S)\,,$$
$$\chi_{\sigma(\varphi \cap 0)}\,(S) = \chi_{\sigma(0)}\,(S)\,.$$
$$\forall S \in \mathrm{P}\,(\Omega)$$

0 and 1 are complements

$$\chi_{\sigma(C0)}\,(S) = 1\,, \quad \chi_{\sigma(C1)}\,(S) = 0\,,$$
$$\forall S \in \mathrm{P}\,(\Omega)\,.$$

De Morgan's laws

$$\chi_{\sigma(C(\varphi \cup \psi))}\,(S) = \chi_{\sigma(C\varphi)}\,(S) \wedge \chi_{\sigma(C\psi)}\,(S)\,,$$
$$= \chi_{\sigma(C\varphi \cap C\psi)}\,(S)\,.$$
$$\chi_{\sigma(C(\varphi \cap \psi))}\,(S) = \chi_{\sigma(C\varphi \cup C\psi)}\,(S)\,.$$
$$\forall S \in \mathrm{P}\,(\Omega)\,.$$

Involution

$$\chi_{\sigma(CC\varphi)}\,(S) = \chi_{\sigma(\varphi)}\,(S)\,, \quad \forall S \in \mathrm{P}\,(\Omega)\,.$$

Any IBA element $\varphi \in$ BA (Ω) can be expressed in **a disjunctive canonical form** as a union of relevant IBA atomic elements, determined by its structure:

$$\varphi(a_1, \ldots, a_n) = \bigcup_{S \in P(\Omega)} \chi_{\sigma(\varphi)}(S) \, \alpha(S)(a_1, \ldots, a_n)$$

where: Ω is a set of primary IBA elements and $P(\Omega)$ a power set of Ω.

Example 6. Disjunctive canonical form of IBA elements in the case $\Omega = \{a, b\}$

	Disjunctive canonical form
1.	0
2.	$Ca \cap Cb = Ca \cap Cb$
3.	$a \cap Cb = a \cap Cb$
4.	$Ca \cap b = Ca \cap b$
5.	$b \cap a = a \cap b$
6.	$Cb = (Ca \cap Cb) \cup (a \cap Cb)$
7.	$Ca = (Ca \cap Cb) \cup (Ca \cap b)$
8.	$a \Leftrightarrow b = (a \cap b) \cup (Ca \cap Cb)$
9.	$a \underline{\vee} b = (a \cap Cb) \cup (Ca \cap b)$
10.	$a = (a \cap Cb) \cup (a \cap b)$
11.	$b = (a \cap b) \cup (Ca \cap b)$
12.	$Ca \cup Cb = (Ca \cap Cb) \cup (a \cap Cb) \cup (Ca \cap b)$
13.	$a \cup Cb = (Ca \cap Cb) \cup (a \cap Cb) \cup (a \cap b)$
14.	$Ca \cup b = (Ca \cap Cb) \cup (Ca \cap b) \cup (a \cap b)$
15.	$a \cup b = (a \cap Cb) \cup (Ca \cap b) \cup (a \cap b)$
16.	$1 = (Ca \cap Cb) \cup (a \cap Cb) \cup (Ca \cap b) \cup (a \cap b)$

Structure of atomic element $\alpha(S)(a_1, \ldots, a_n)$ has only one element; it contains only itself, and is given by the following expression:

$$\chi_{\sigma(\alpha(S))}(SS) = \begin{cases} 1, & S = SS \\ 0, & S \neq SS \end{cases} ; \quad S, \; SS \in P(\Omega).$$

Example 7. Structure of atomic elements for $\Omega = \{a, b\}$ is given in the following table:

S	$\alpha(S)(a, b)$	$\chi_{\sigma(\alpha(S))}(\varnothing)$	$\chi_{\sigma(\alpha(S))}(\{a\})$	$\chi_{\sigma(\alpha(S))}(\{b\})$	$\chi_{\sigma(\alpha(S))}(\{a, b\})$
$\varnothing$	$Ca \cap Cb$	1	0	0	0
$\{a\}$	$a \cap Cb$	0	1	0	0
$\{b\}$	$Ca \cap b$	0	0	1	0
$\{a, b\}$	$a \cap b$	0	0	0	1

Structure of primary element $a_i \in \Omega$ contains all atomic elements in which the analyzed primary element figures affirmatively (not as a complement) and is given by the following expression:

$$\chi_{\sigma(a_i)}(S) = \begin{cases} 1, & a_i \in S \\ 0, & a_i \notin S \end{cases} ; \; S \in P(\Omega).$$

Example 8. Structure of primary elements $\Omega = \{a_1, a_2\}$ *are given in the following table:*

a_i	$\chi_{\sigma(a_i)}(\emptyset)$	$\chi_{\sigma(a_i)}(\{a_1\})$	$\chi_{\sigma(a_i)}(\{a_2\})$	$\chi_{\sigma(a_i)}(\{a_1, a_2\})$
a_1	0	1	0	1
a_2	0	0	1	1

Principle of structural functionality: Structure of any IBA elements can be directly calculated on the basis of structures of its components. This principle is a fundamental principle contrary to the famous truth functional principle.

The structural table of IBA elements represents the dependence in the sense of inclusion of any IBA element from IBA atomic elements.

The truth table is a matter of values, so it is even not possible on a symbolic level. The structural table is value independent and as a consequence the same for all value realizations. So, the structural table is a fundamental notion contrary to the truth table which is correct from the point of view of BA, only for a two-valued case. ***The structure, as value indifferent, preserves all laws from a symbolic level on a valued level.***

On a valued level the IBA is treated from the aspect of its value realization.

3.2 Value Level

On a valued level the IBA is value realized or, for short, realized. To elements of Boolean algebra from a symbolic level on value level can correspond: null-ary relation (truth), unary relations (properties), binary relations and n-ary relations, etc. On a value level a result from a symbolic level is concretized in the sense of value. In a classical case there are only two-valued but in a more general case there are three and more values up to the case of all values from a real unit interval [0, 1]. An element from a symbolic level on a value level preserves all its characteristics (described by Boolean axioms and laws) throughout corresponding values. For example, to the order, which is determined by inclusion on a symbolic level, corresponds the order on the valued level, determined by relation "less or equal". Any element from a symbolic level has its value realization on elements of analyzed universe. The value of analyzed element from a symbolic level on any element of universe is obtained by superposition of value realizations of its relevant atomic elements for this element of universe. The value of atomic element for the analyzed element of universe is a function – *Generalized Boolean polynomial,* of the values of primary elements realization for this element and a chosen operator of *generalized product.* The value realization of atomic elements from a symbolic level for any element of universe is non-negative and their sum is equal to 1. All tautologies and contradictions from a symbolic level are tautologies and contradictions, respectively, on the valued level.

3.2.1 Basic notions of valued level

A value realization of any IBA element $\varphi \in$ BA (Ω) is the *value* or *intensity* $\varphi^v (x)$ of owning a property which this element represents from the analyzed object x.

The universe of IBA realization is a finite or infinite set X of members on which IBA elements are realized on members (as object) of X themselves (unary relation – property)

$$\varphi^v : X \to [0, 1], \ \varphi \in \text{BA} (\Omega),$$

or on their ordered n-tuples (nary relation):

$$\varphi^v : X^n \to [0, 1], \ \varphi \in \text{BA} (\Omega), \ n \geq 2.$$

In the case of null-ary relation (propositional calculus for example) an object is the IBA element itself.

Generalized value realization means that any IBA element φ on a valued level obtains the value – intensity of realization φ^v. Intensity or gradation here means that the set of possible values can have more than two members including the most general case when the set of possible values is a real unit interval [0, 1].

Gradation in the value realization of IBA elements offers a much more efficient description (determination, characterization, specification, etc.) of analyzed objects. With a small number of properties (IBA elements) equipped with gradation on a valued level one can do much more than with a large number of two-valued properties.

The possibility that with fewer properties (IBA elements) one can do more (in the sense of expressivity) is the basic motive for introduction and treatment of MV in Boolean algebra and/or for development of IBA. So, gradation is the basis for balance between simplicity and efficiency in real problems.

A disjunctive canonical form for any IBA element $\varphi \in$ BA (Ω), (from a symbolic level)

$$\varphi (a_1, \ldots, a_n) = \bigcup_{S \in P(\Omega)} \chi_{\sigma(\varphi)} (S) \, \alpha (S) (a_1, \ldots, a_n)$$

has its value interpretation: *Generalized Boolean polynomial.* Boolean polynomials are defined in by J. Boole.

Generalized Boolean polynomial

The value or intensity φ^v of any IBA element $\varphi \in$ BA (Ω) on the analyzed object is equal to the sum of values of relevant atomic IBA elements contained in this element for the analyzed object:

$$\varphi^\otimes (a_1, \ldots, a_n) = \sum_{S \in P(\Omega)} \chi_{\sigma(\varphi)} (S) \, \alpha^\otimes (S) (\|a_1\|, \ldots, \|a_n\|)$$

where: $\alpha^\otimes (S) (\|a_1\|, \ldots, \|a_n\|)$ is a value realization of atomic function $\alpha (S) (a_1, \ldots, a_n)$.

The structure function is the same on both levels (symbolic and valued).

realization of atomic function $\alpha^{\otimes}(S)(\|a_1\|, \ldots, \|a_n\|)$ determines the intensity of atomic function $\alpha(S)(a_1, \ldots, a_n)$ on the basis of the values of primary IBA elements $\|a_i\| \in [0, 1]$, $a_i \in \Omega$, and in the most general case:

$$\alpha^{\otimes}(S) : [0, 1]^{|\Omega|} \to [0, 1], \quad S \in P(\Omega)..$$

The expression for value realization of atomic function is:

$$\alpha^{\otimes}(S)(\|a_1\|, \ldots, \|a_n\|) = \sum_{C \in P(\Omega \setminus S)} (-1)^{|C|} \bigotimes_{a_i \in S \cup C} \|a_i\|$$

where: $\otimes$ is a ***generalized product operator***.

Example 9. For the case when the set of primary IBA elements is $\Omega = \{a, b\}$, valued atomic functions in a general case are given by the following expressions:

$$\alpha^{\otimes}(\emptyset)(\|a\|, \|b\|) = 1 - \|a\| - \|b\| + \|a\| \otimes \|b\|,$$
$$\alpha^{\otimes}(\{a\})(\|a\|, \|b\|) = \|a\| - \|a\| \otimes \|b\|,$$
$$\alpha^{\otimes}(\{b\})(\|a\|, \|b\|) = \|b\| - \|a\| \otimes \|b\|,$$
$$\alpha^{\otimes}(\{a, b\})(\|a\|, \|b\|) = \|a\| \otimes \|b\|.$$

where: $\|a\|$ and $\|b\|$ are value realizations $\|a(x_1, \ldots, x_n)\|$ and $\|b(x_1, \ldots, x_n)\|$ ($\|a\|, \|b$ $X^n \to [0, 1]$, in general case) of primary elements from $\Omega = \{a, b\}$.

Generalized product $\otimes$ is any function $\otimes : [0, 1] \times [0, 1] \to [0, 1]$ that satisfie all four conditions of ***T-norms*** :

Commutativity

$$\|a\| \otimes \|b\| = \|b\| \otimes \|a\|.$$

Associativity

$$\|a\| \otimes (\|b\| \otimes \|c\|) = (\|a\| \otimes \|b\|) \otimes \|c\|.$$

Monotonicity

$$\|a\| \otimes \|b\| \le \|c\| \otimes \|d\| \quad if \quad \|a\| \le \|c\|, \quad \|b\| \le \|d\|.$$

1 as identity

$$\|a\| \otimes 1 = \|a\|.$$

And plus one additional condition:

Non-negativity condition:

$$\sum_{C \in P(\Omega \backslash S)} (-1)^{|C|} \bigotimes_{a_i \in S \cup C} \|a_i\| \geq 0,$$

where: $\Omega = \{a_1, \ldots, a_n\}$, $S \in P(\Omega)$.

The role of additional axiom "non-negativity" is to ensure non-negative values of valued atomic functions.

Comment: *In the case when the set of primary IBA elements is $\Omega = \{a, b\}$ the constraint on non-negativity is satisfied when the operator of generalized product satisfied the following non equality:*

$$max \left(\|a\| + \|b\| - 1, 0\right) \leq \|a\| \otimes \|b\| \leq min \left(\|a\|, \|b\|\right).$$

It is easy to show that in the case of three primary IBA elements the Lukasiewicz T-norm is not a candidate for the operator of generalized product $\otimes$.

Normalized value of valued atomic functions: is a very important characteristic (condition) of IBA value atomic functions. This condition is defined by the following expression:

$$\sum_{S \in P(\Omega)} \alpha^{\otimes} (S) \left(\|a_1\|, \ldots, \|a_n\|\right) = 1.$$

This condition in the two-valued case is trivial since only one atomic element is realized (equal to 1), and all others are not realized (equal to 0) for any analyzed object.

So, in a general case (contrary to a two-valued case) all atoms can be simultaneously realized for the analyzed object or all of them can simultaneously have the values of valued atomic functions greater than zero, but so that their sum is equal to 1.

So, value of any Boolean variable (relation) as a function of primary variables is given by the following expression:

$$\varphi^{\otimes} \left(\|a_1\|, \ldots, \|a_n\|\right) = \sum_{S \in P(\Omega)} \chi_{\sigma(\varphi)} (S) \sum_{C \in P(\Omega \backslash S)} (-1)^{|C|} \bigotimes_{a_i \in S \cup C} \|a_i\|.$$

Example 10. Boolean polynomials for Boolean elements in the case $\Omega = \{a, b\}$ are:

0

$$(Ca \cap Cb)^{\otimes} = 1 - \|a\| - \|b\| + \|a\| \otimes \|b\|$$

$$(a \cap Cb)^{\otimes} = \|a\| - \|a\| \otimes \|b\| \quad (Ca \cap b)^{\otimes} = \|b\| - \|a\| \otimes \|b\| \quad (a \cap b)^{\otimes} = \|a\| \otimes \|b\|$$

$$(Cb)^{\otimes} = 1 - \|b\|$$

$$(Ca)^{\otimes} = 1 - \|a\|$$

$$(a \Leftrightarrow b)^{\otimes} = 1 - \|a\| - \|b\| + 2\|a\| \otimes \|b\|$$

$$(a \underline{\vee} b)^{\otimes} = \|a\| + \|b\| - 2\|a\| \otimes \|b\|$$

$$(a)^{\otimes} = \|a\|$$

$$(b)^{\otimes} = \|b\|$$

$$(Ca \cup Cb)^{\otimes} = 1 - \|a\| \otimes \|b\|$$

$$(a \cup Cb)^{\otimes} = 1 - \|b\| + \|a\| \otimes \|b\|$$

$$(Ca \cup b)^{\otimes} = 1 - \|a\| + \|a\| \otimes \|b\|$$

$$(a \cup b)^{\otimes} = \|a\| + \|b\| - \|a\| \otimes \|b\|$$

1

Same result as value realization of lattice from Fig. 1. is given on the Fig. 2.

Properties of generalized valued realizations

From the fact that the structure of IBA element "1" includes all IBA atomic elements and on the basis of normalized value of valued atomic functions condition,

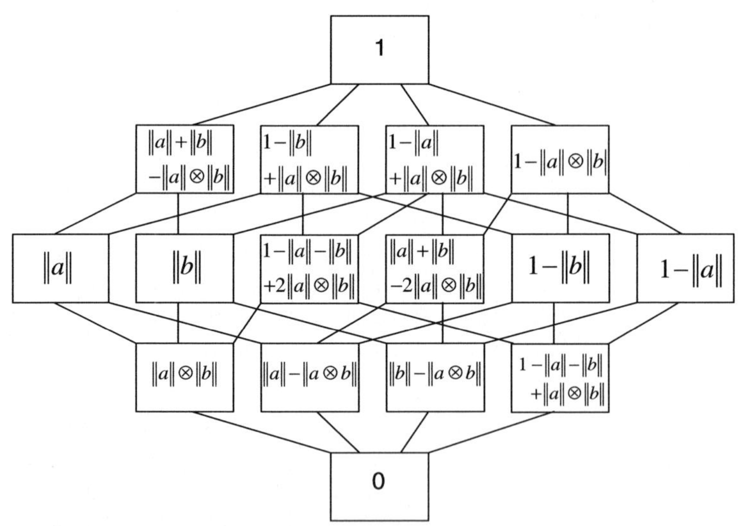

Fig. 2 Value realization of lattice generated by $\Omega = \{a, b\}$

it follows that: a value realization of IBA element "1" is equal to 1 for any analyzed object.

Because the structure of IBA element "0" does not include any IBA atom it follows that: a value realization of IBA element "0" is equal to 0 for any analyzed object.

Since the laws of **Commutativity, Associativity, Absorption and Distributivity** are valid for the structure of IBA elements and on the basis of valued representation of disjunctive canonical form and from the property that two Boolean expressions are equal if they have the same structure, it follows that all these laws (Commutativity, Associativity, Absorption and Distributivity) *are valid in the case of a generalized valued realization too*.

The laws of **Excluded middle** and **Contradiction (Complements)** are valid for the structure of IBA elements (symbolic level). The structure of the union of any IBA element and its complement is equal to the structure of IBA element "1". The structure of intersection of any IBA element and its complement is equal to the structure of IBA element "0". As a consequence these laws *are valid for a generalized value realization (valued level) too*.

Comment: Simultaneous appearance of two IBA elements (properties) in the same object in a general case doesn't mean automatically their intersection. So, one object can have some property (IBA element) with some intensity, simultaneously this object has a property which is a complement of the analyzed property with intensity as a complement value, but the value of their intersection is always equal to 0. So simultaneousness and intersections in a general case are not synonyms (as in a classical case).

Since the laws: **Idempotency, De Morgan's laws, Boundedness, Involution** hold for structures of IBA elements the same laws are valid on the value level for generalized valued realization of IBA elements too. So, since the structures of IBA elements are obtained by homomrphic mapping of IBA elements, it follows that the structures preserve all laws (Boolean laws) on which IBA are based. Since the laws of value realization are direct consequences of laws which are valid for the structures of IBA elements, it follows that all laws from symbolic level (Boolean laws) are preserved on a valued level in all possible valued realizations including the most general for which the set of possible values is a real unit interval [0,1].

4 Example of IBA Application

4.1 The Generalization of Preference Structures

Interpolative Boolean algebra as a MV algebra can be illustrated on the generalization of preference structures. A preference structure is the basic concept of preference modeling. Consider a set of alternatives A (objects, actions etc.) and suppose that a decision maker (DM) wants to judge them by pairwise comparison. Given two alternatives, the DM can act in one of the following three ways, : DM prefers one to the other - *strict preference relations* $(>)$ or $(<)$; two alternatives are indefferent to DM- *indifference relation* $(=)$; DM is unable to compare the two alternatives – *incomparability relation* $(<>)$

For any $(a, b) \in A^2$, we classify:

$(a, b) \in (>) \Leftrightarrow a > b$ DM prefers a to b;

$(a, b) \in (=) \Leftrightarrow a = b$ a to b is indifferent to DM

$(a, b) \in (<>) \Leftrightarrow a <> b$ DM is unable to compare a and b. A preference structure on A is a triplet $\{(>), (=), (<>)\}$

The binary relation $(\geq) = ((>) \vee (=))$ is called a *large preference relation* of a given preference structure $\{(>), (=), (<>)\}$.

The set of all possible binary relations generated by two primary relations $\Omega = \{(\leq), (\geq)\}$ (*large preference relations*) is a Boolean alagebra and/or a Boolean lattice given in the Fig. 3.

Atomic Interpolative relations as functions of primary relations

$$(=) (a, b) = ((\leq) \cap (\geq)) (a, b),$$
$$(<) (a, b) = ((\leq) \cap (C \geq)) (a, b),$$
$$(>) (a, b) = ((C \leq) \cap (\geq)) (a, b),$$
$$(<>) (a, b) = ((C \leq) \cap (C \geq)) (a, b), \quad a, b \in A.$$

Values (intensity) of atomic Interpolative relations as functions of intensity of primary relations

$$(=)^{\otimes} (a, b) = \|(\leq) (a, b)\| \otimes \|(\geq) (a, b)\|,$$
$$(<)^{\otimes} (a, b) = \|(\leq) (a, b)\| - \|(\leq) (a, b)\| \otimes \|(\geq) (a, b)\|,$$

$$(=)^{\otimes} (a, b) = \|(\leq) (a, b)\| \otimes \|(\geq) (a, b)\|,$$
$$(<)^{\otimes} (a, b) = \|(\leq) (a, b)\| - \|(\leq) (a, b)\| \otimes \|(\geq) (a, b)\|,$$

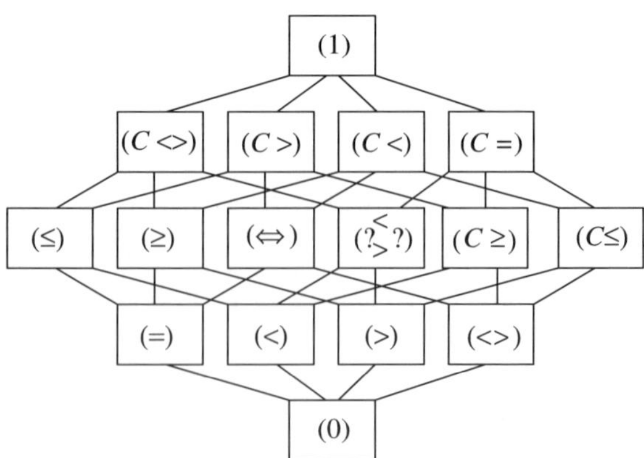

Fig. 3 Symbolic level: Boolean lattice generated by primary relations: $\Omega = \{(\leq), (\geq)\}$

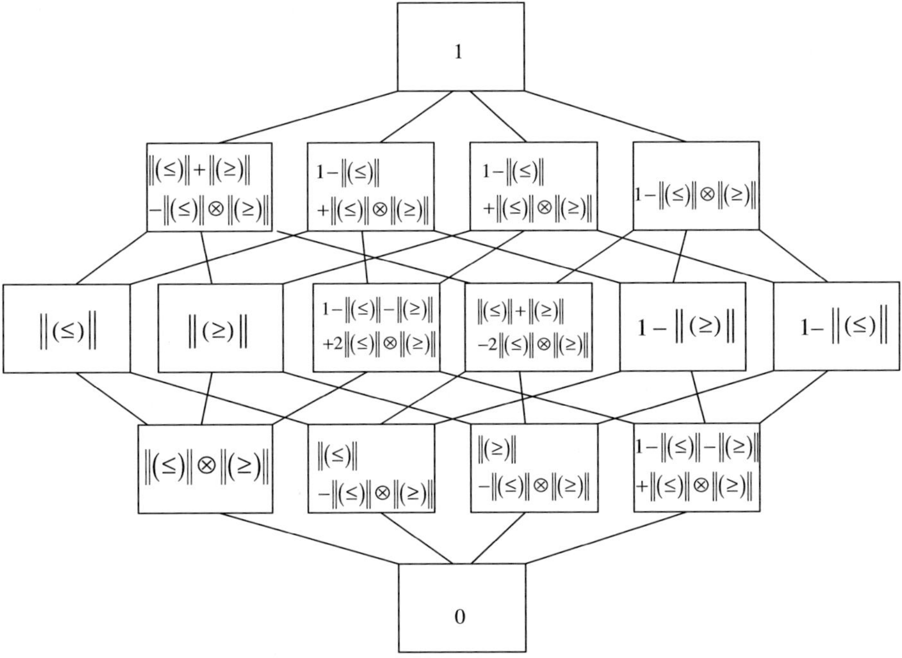

Fig. 4 Valued level: Boolean polynomials based on $\|(\leq)\|$ and $\|(\geq)\|$

$$(>)^{\otimes}(a,b) = \|(\geq)(a,b)\| - \|(\leq)(a,b)\| \otimes \|(\geq)(a,b)\|,$$
$$(<>)^{\otimes}(a,b) = 1 - \|(\leq)(a,b)\| - \|(\geq)(a,b)\| + \|(\leq)(a,b)\| \otimes \|(\geq)(a,b)\|,$$
$$a,b \in A,$$

All other Boolean polynomials as a function of primary relations are given in the Fig. 4.

4.2 Interpolative Sets

Theory of Interpolative sets (I-sets) is the realization of fuzzy sets idea in Boolean frame. The main characteristics of application of IBA algebra in the theory of I-sets is illustrated on the example of two I-sets A and B. The Boolean polynomials of corresponding set expressions are given in the following table.

Realization of the all possible set function for the given I-sets A and B, in the case when the generalized product is given as *min* function is represented on the Fig. 5.

It is clear that all properties of the classical set algebra are preserved in the case of I-set algebra and/or by using I-set approach (realization of IBA algebra) one can treat gradation in the Boolean frame, contrary to all fuzzy sets approaches.

1.	$(A \cap B)^{\otimes}(x) = A(x) \otimes B(x),$
2.	$(A \cap B^c)^{\otimes}(x) = A(x) - A(x) \otimes B(x),$
3.	$(A^c \cap B)^{\otimes}(x) = B(x) - A(x) \otimes B(x),$
4.	$(A^c \cap B^c)^{\otimes}(x) = 1 - A(x) - B(x) + A(x) \otimes B(x),$
5.	$A(x) = A(x),$
6.	$B(x) = B(x),$
7.	$((A \cap B) \cup (A^c \cap B^c))^{\otimes}(x) = 1 - A(x) - B(x) + 2A(x) \otimes B(x),$
8.	$((A \cap B^c) \cup (A^c \cap B))^{\otimes}(x) = A(x) + B(x) - 2A(x) \otimes B(x),$
9.	$(B^c)^{\otimes}(x) = 1 - B(x)$
10.	$(A^c)^{\otimes}(x) = 1 - A(x)$
11.	$(A \cup B)^{\otimes}(x) = A(x) + B(x) - A(x) \otimes B(x),$
12.	$(A^c \cup B)^{\otimes}(x) = 1 - A(x) + A(x) \otimes B(x),$
13.	$(A \cup B^c)^{\otimes}(x) = 1 - B(x) + A(x) \otimes B(x),$
14.	$(A^c \cup B^c)^{\otimes}(x) = 1 - A(x) \otimes B(x),$
15.	$(A \cap A^c)^{\otimes}(x) = 0, \quad (B \cap B^c)^{\otimes}(x) = 0,$
16.	$(A \cup A^c)^{\otimes}(x) = 1, \quad (B \cup B^c)^{\otimes}(x) = 1.$

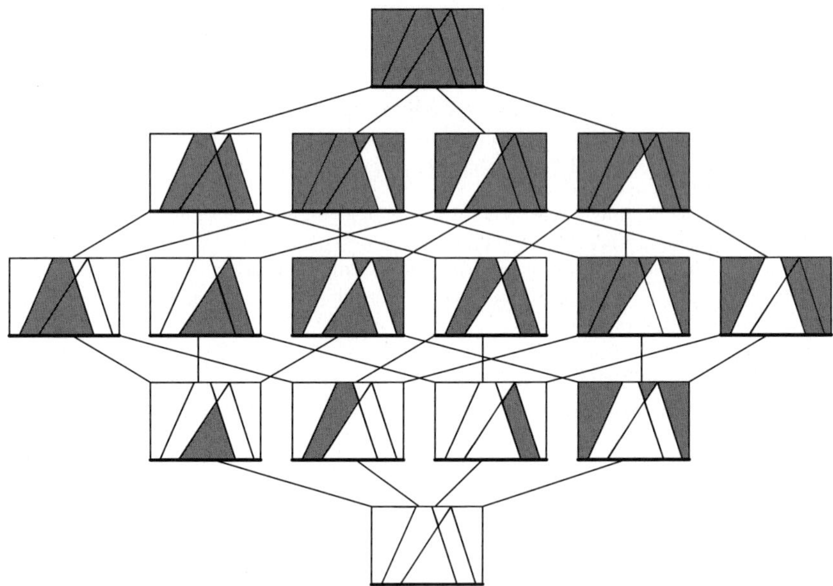

Fig. 5 Partially ordered I-sets generated by two I-sets $\Omega = \{A, B\}$

5 Conclusions

Interpolative Boolean algebra (IBA) is a consistent MV realization of finite (atomic) Boolean algebra. In new approach, based on IBA, to every element of any finite Boolean algebra, corresponds generalized Boolean polynomial with ability to process all values of primary variables from real unit interval [0, 1]. Since, axioms and lows of Boolean algebra are actually meter of value independent structure of IBA elements, all axioms and all laws of Boolean algebra are preserved in any type of value realization (two-valued, three-valued,..., [0, 1]). Possibility of approaches based on IBA are illustrated (a) on generalized preference structure – as straightaway generalization of classical result and (b) on interpolative sets as consistent realization of idea of fuzzy sets – all laws of set algebra are preserved in general case, contrary to conventionally fuzzy approaches.

References

G. Boole: The Calculus of Logic, *Cambridge and Dublin Mathematical Journal*, Vol. III, pp. 183–198, 1848.

L. Zadeh: Fuzzy Sets, *Information and Control*, no. 8, pages 338–353. 1965

L. Zadeh: Bellman R.E., Local and fuzzy logics, *Modern Uses of Multiple-Valued Logic*, J.M. Dunn and G. Epstein (eds.), 103–165. Dordrecht: D. Reidel, 1977.

L. Zadeh: Man and Computer, Bordeaux, France, 130-165, 1972. *Outline of a new approach to the analysis of complex systems and decision processes,* IEEE

S. Gottwald: *A Treats on Many-Valued Logics,* volume 9 of Studies in Logic and Computation. Research Studies Press, Bladock, 2000.

J. Lukasiewicz,: *Selected Works.* (ed.: L. Borkowski), North-Holland Publ. Comp., Amsterdam and PWN, Warsaw, (1970)

P. Hajek: *Metamathematics of Fuzzy Logic, Trends in Logica* – Studia logica library, Kluwer Academic Publishers, Dodrecth /Boston/London, 1998.

Radojevic D.: New [0,1]-valued logic: A natural generalization of Boolean logic, *Yugoslav Journal of Operational Research – YUJOR*, Belgrade, Vol. 10, No 2, 185–216, 2000

Radojevic D.: Interpolative relations and interpolative preference structures, *Yugoslav Journal of Operational Research – YUJOR* , Belgrade, Vol. 15, No 2, 2005

Radojevic D.: Logical measure - structure of logical formula, in *Technolopgies for Constructing Intelligent Systems* 2: Tools, Springer, pp 417–430, 2002.

Arrow K.J.: *Social Choice and Individual Values*, Wiley, New York, 1951

Sikorski R.: *Boolean Algebras,* Springer-Verlag, Berlin, New York, 1964.

Klement E. P, Mesiar R., Pap E.: *Triangular Norms*, Kluwer Academic Publ, Dordrecht, 2000

Systematic Design of a Stable Type-2 Fuzzy Logic Controller

Oscar Castillo, Luís Aguilar, Nohé Cázarez and Patricia Melin

Abstract Stability is one of the more important aspects in the traditional knowledge of Automatic Control. Type-2 Fuzzy Logic is an emerging and promising area for achieving Intelligent Control (in this case, Fuzzy Control). In this work, we use the Fuzzy Lyapunov Synthesis, as proposed by Margaliot [11], to build a Lyapunov Stable Type-1 Fuzzy Logic Control System. Then we make an extension from a Type-1 to a Type-2 Fuzzy Logic Control System, ensuring the stability on the control system and proving the robustness of the corresponding fuzzy controller.

1 Introduction

Fuzzy logic controllers (FLC's) are one of the most useful control schemes for plants in which we have difficulties in deriving mathematical models or having performance limitations with conventional linear control schemes. Error e and change of error $\dot{e}$ are the most used fuzzy input variables in most fuzzy control works, regardless of the complexity of controlled plants. Also, either control input u (PD-type) or incremental control input Δu (PI-type) is typically used as a fuzzy output variable representing the rule consequent ("then" part of a rule) [6].

Stability has been one of the central issues concerning fuzzy control since Mamdani's pioneer work [9], [10]. Most of the critical comments to fuzzy control are due to the lack of a general method for its stability analysis.

But as Zadeh often points out, fuzzy control has been accepted by the fact that it is task-oriented control, while conventional control is characterized as setpoint-oriented control, and hence do not need a mathematical analysis of stability. And as Sugeno says, in general, in most industrial applications, the stability of control is not fully guaranteed and the reliability of a control hardware system is considered to be more important than the stability [15].

The success of fuzzy control, however, does not imply that we do not need a stability theory for it. Perhaps the main drawback of the lack of stability analysis would be that we cannot take a model-based approach to fuzzy control design. In conventional control theory, a feedback controller can be primarily designed so that a close-loop system becomes stable [13], [14]. This approach of course restricts us

M. Nikravesh et al. (eds.), *Forging the New Frontiers: Fuzzy Pioneers II.*
© Springer-Verlag Berlin Heidelberg 2008

to setpoint-oriented control, but stability theory will certainly give us a wider view on the future development of fuzzy control.

Therefore, many researchers have worked to improve the performance of the FLC's and ensure their stability. Li and Gatland in [7] and [8] proposed a more systematic design method for PD and PI-type FLC's. Choi, Kwak and Kim [4] present a single-input FLC ensuring stability. Ying [18] presents a practical design method for nonlinear fuzzy controllers, and many other researchers have results on the matter of the stability of FLC's, in [1] Castillo et. al, and Cázarez et. al [2] presents an extension of the Margaliot work [11] to build stable type-2 fuzzy logic controllers in Lyapunov sense.

This work is based on Margaliot et. al. [11] work and in Castillo et. al. [1] and Cázarez et. al. [2] results, we use the Fuzzy Lyapunov Synthesis [11] to built an Stable Type-2 Fuzzy Logic Controller for a 1DOF manipulator robot, first without gravity effect to probe stability, and then with gravity effect to probe the robustness of the controller. The same criterion can be used for any number of DOF manipulator robots, linear or nonlinear, and any kind of plants.

This work if organized as follows: In Sect. II we presents an introductory explanation of type-1 and type-2 FLC's. In Sect. III we extend the Margaliot result to built a general rule base for any type (1 or 2) of FLC's. Experimental results are presented in Sect. IV and the concluding remarks are collected in Sect. V.

2 Fuzzy Logic Controllers

Type-1 FLCs are both intuitive and numerical systems that map crisp inputs to a crisp output. Every FLC is associated with a set of rules with meaningful linguistic interpretations, such as

$$R^l : \text{If } x_1 \text{ is } F_1^l \text{ and } x_2 \text{ is } F_2^l \text{ and } \ldots \text{ and } x_n \text{ is } F_n^l \text{ Then } w \text{ is } G^l$$

which can be obtained either from numerical data, or experts familiar with the problem at hand. Based on this kind of statement, actions are combined with rules in an antecedent/consequent format, and then aggregated according to approximate reasoning theory, to produce a nonlinear mapping from input space $U = U_1 x U_2 x \ldots U_n$ to the output space W, where $F_k^l \subset U_k, k = 1, 2, \ldots, n$, are the antecedent type-1 membership functions, and $G^1 \subset W$ is the consequent type-1 membership function. The input linguistic variables are denoted by $u_k, k = 1, 2, \ldots, n$, and the output linguistic variable is denoted by w.

A Fuzzy Logic System (FLS), as the kernel of a FLC, consist of four basic elements (Fig. 1): the type-1 fuzzyfier, the fuzzy rule-base, the inference engine, and the type-1 defuzzyfier. The fuzzy rule-base is a collection of rules in the form of R^l, which are combined in the inference engine, to produce a fuzzy output. The type-1 fuzzyfier maps the crisp input into type-1 fuzzy sets, which are subsequently used as inputs to the inference engine, whereas the type-1 defuzzyfier maps the type-1 fuzzy sets produced by the inference engine into crisp numbers.

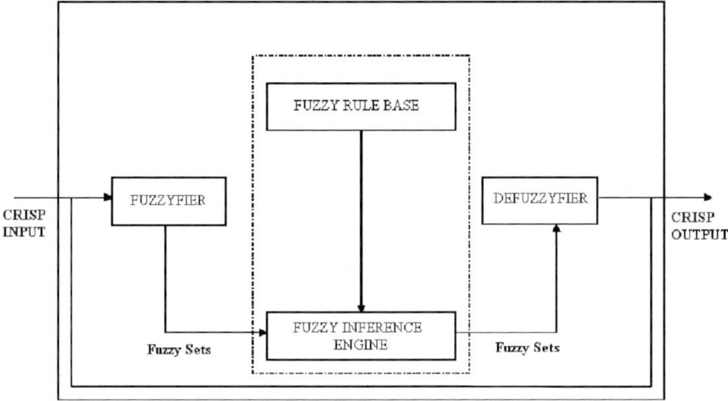

Fig. 1 Structure of type-1 fuzzy logic system

Fuzzy sets can be interpreted as membership functions u_X that associate with each element x of the universe of discourse, U, a number $u_X(x)$ in the interval [0,1]:

$$u_X : U \to [0, 1] \tag{1}$$

For more detail of Type-1 FLS see [17], [5], [3].

As the type-1 fuzzy set, the concept of type-2 fuzzy set was introduced by Zadeh [19] as an extension of the concept of an ordinary fuzzy set.

A FLS described using at least one type-2 fuzzy set is called a type-2 FLS. Type-1 FLSs are unable to directly handle rule uncertainties, because they use type-1 fuzzy sets that are certain. On the other hand, type-2 FLSs, are very useful in circumstances where it is difficult to determine an exact, and measurement uncertainties [12].

It is known that type-2 fuzzy set let us to model and to minimize the effects of uncertainties in rule-based FLS. Unfortunately, type-2 fuzzy sets are more difficult to use and understand than type-1 fuzzy sets; hence, their use is not widespread yet.

Similar to a type-1 FLS, a type-2 FLS includes a type-2 fuzzyfier, rule-base, inference engine and substitutes the defuzzifier by the output processor. The output processor includes a type-reducer and a type-2 defuzzyfier; it generates a type-1 fuzzy set output (from the type reducer) or a crisp number (from the defuzzyfier). A type-2 FLS is again characterized by IF-THEN rules, but its antecedent or consequent sets are now type-2. Type-2 FLSs, can be used when the circumstances are too uncertain to determine exact membership grades. A model of a type-2 FLS is shown in Fig. 2.

In the case of the implementation of the type-2 FLCs, we have the same characteristics as in type-1 FLC, but we used type-2 fuzzy sets as membership functions for the inputs and for the outputs. Fig. 3 shows the structure of a control loop with a FLC.

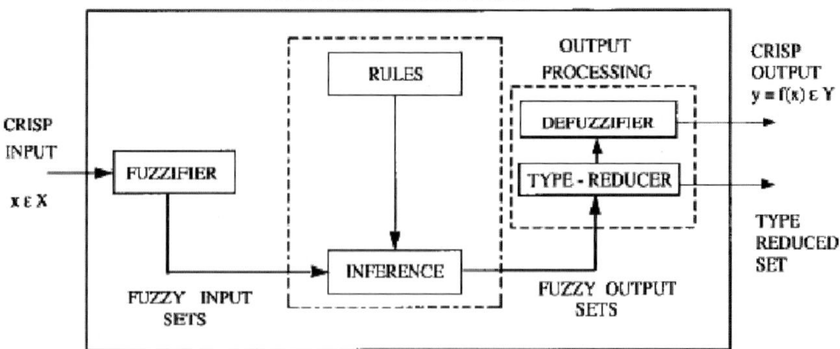

Fig. 2 Structure of type-2 fuzzy logic system

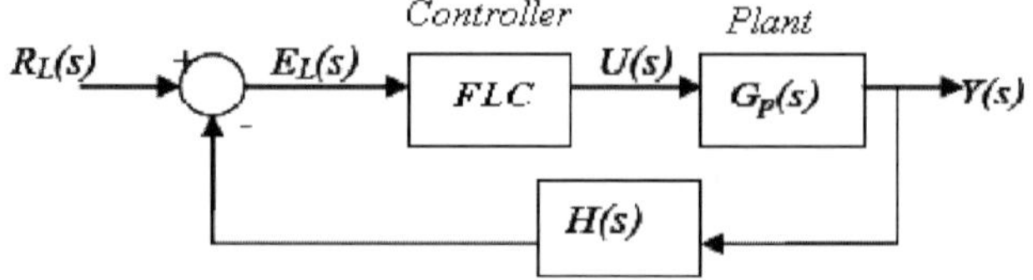

Fig. 3 Fuzzy control loop

3 Systematic Design of a Stable Fuzzy Controller

For our description we consider the problem of designing a stabilizing controller for a 1DOF manipulator robot system depicted in Fig. 4. The state-variables are $x_1 = \theta$ - the robot arm angle, and $x_2 = \dot{\theta}$ - its angular velocity. The system's actual dynamical equation, which we will assume unknown, is as the shows in (2) [14]:

$$M(q)\ddot{q} + C(q, \dot{q})\dot{q} + g(q) = \tau \tag{2}$$

To apply the fuzzy Lyapunov synthesis method, we assume that the exact equations are unknown and that we have only the following partial knowledge about the plant (see Fig. 4):

1. The system may have really two degrees of freedom θ and $\dot{\theta}$, referred to as x_1 and x_2, respectively. Hence, $\dot{x}_1 = x_2$.
2. $\dot{x}_2$ is proportional to u, that is, when u increases (decreases) $\dot{x}_2$ increases (decreases).

To facility our control design we are going to suppose no gravity effect in our model, see (3).

$$ml^2\ddot{q} = \tau \tag{3}$$

Fig. 4 1DOF Manipulator robot

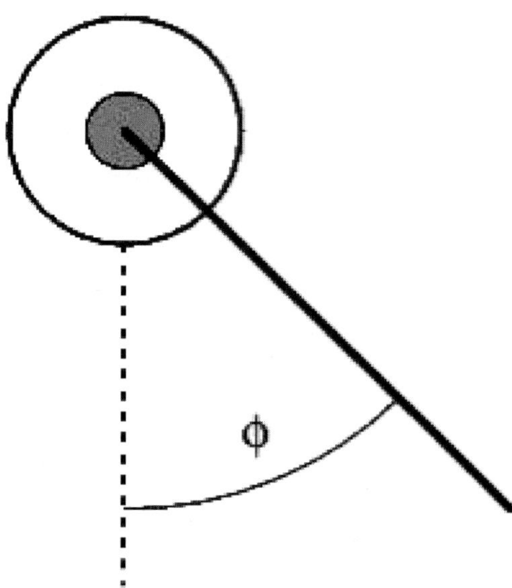

Our objective is to design the rule-base of a fuzzy controller that will carry the robot arm to a desired position $x_1 = \theta d$. We choose (4) as our Lyapunov function candidate. Clearly, V is positive-definite.

$$V(x_1, x_2) = \frac{1}{2}(x_1^2 + x_2^2) \tag{4}$$

Differentiating V, we have (5),

$$\dot{V} = x_1\dot{x}_1 + x_2\dot{x}_2 = x_1x_2 + x_2\dot{x}_2 \tag{5}$$

Hence, we require:

$$x_1x_2 + x_2\dot{x}_2 < 0 \tag{6}$$

We can now derive sufficient conditions so that condition (6) holds: If x_1 and x_2 have opposite signs, then $x_1x_2 < 0$ and (6) will hold if $\dot{x}_2 = 0$; if x_1 and x_2 are both positive, then (6) will hold if $\dot{x}_2 < -x_1$; and if x_1 and x_2 are both negative, then (6) will hold if $\dot{x}_2 > -x_1$.

We can translate these conditions into the following fuzzy rules:

- If x_1 is *positive* and x_2 is *positive* Then $\dot{x}_2$ must be *negative big*
- If x_1 is *negative* and x_2 is *negative* Then $\dot{x}_2$ must be *positive big*
- If x_1 is *positive* and x_2 is *negative* Then $\dot{x}_2$ must be *zero*
- If x_1 is *negative* and x_2 is *positive* Then $\dot{x}_2$ must be *zero*

However, using our knowledge that $\dot{x}_2$ is proportional to u, we can replace each $\dot{x}_2$ with u to obtain the fuzzy rule-base for the stabilizing controller:

- If x_1 is *positive* and x_2 is *positive* Then u must be *negative big*
- If x_1 is *negative* and x_2 is *negative* Then u must be *positive big*
- If x_1 is *positive* and x_2 is *negative* Then u must be *zero*
- If x_1 is *negative* and x_2 is *positive* Then u must be *zero*

It is interesting to note that the fuzzy partitions for x_1, x_2, and u follow elegantly from expression (5). Because $\dot{V} = x_2(x_1 + \dot{x}_2)$, and since we require that $\dot{V}$ be negative, it is natural to examine the signs of x_1 and x_2; hence, the obvious fuzzy partition is *positive, negative*. The partition for $\dot{x}_2$, namely *negative big, zero, positive big* is obtained similarly when we plug the linguistic values *positive, negative* for x_1 and x_2 in (5). To ensure that $\dot{x}_2 < -x_1$ ($\dot{x}_2 > -x_1$) is satisfied even though we do not know x_1's exact magnitude, only that it is *positive (negative)*, we must set $\dot{x}_2$ to *negative big (positive big)*. Obviously, it is also possible to start with a given, pre-defined, partition for the variables and then plug each value in the expression for $\dot{V}$ to find the rules. Nevertheless, regardless of what comes first, we see that fuzzy Lyapunov synthesis transforms classical Lyapunov synthesis from the world of exact mathematical quantities to the world of computing with words [20].

To complete the controller's design, we must model the linguistic terms in the rule-base using fuzzy membership functions and determine an inference method. Following [16], we characterize the linguistic terms *positive, negative, negative big, zero* and *positive big* by the type-1 membership functions shows in Fig. 5 for a Type-1 Fuzzy Logic Controller, and by the type-2 membership functions shows in Fig. 6 for a Type-2 Fuzzy Logic Controller. Note that the type-2 membership functions are extended type-1 membership functions.

To this end, we had systematically developed a FLC rule-base that follows the Lyapunov Stability criterion. At Sect. IV we present some experimental results using our fuzzy rule-base to build a Type-2 Fuzzy Logic Controller.

4 Experimental Results

In Sect. III we had systematically develop a stable FLC rule-base, now we are going to show some experimental results using our stable rule-base to built Type-2 FLC. The plant description used in the experiments is the same shown in Sect. III.

Our experiments were done with Type-1 Fuzzy Sets and Interval Type-2 Fuzzy Sets. In the Type-2 Fuzzy Sets the membership grade of every domain point is a crisp set whose domain is some interval contained in [0,1] [12]. On Fig. 6 we show some Interval Type-2 Fuzzy Sets, for each fuzzy set, the grey area is known as the Footprint Of Uncertainty (FOU) [12], and this one is bounded by an upper and a lower membership function as shown in Fig. 7.

Fig. 5 Set of Type-1
membership functions:
a) positive, b) negative,
c) negative big, d) zero and
e) positive big

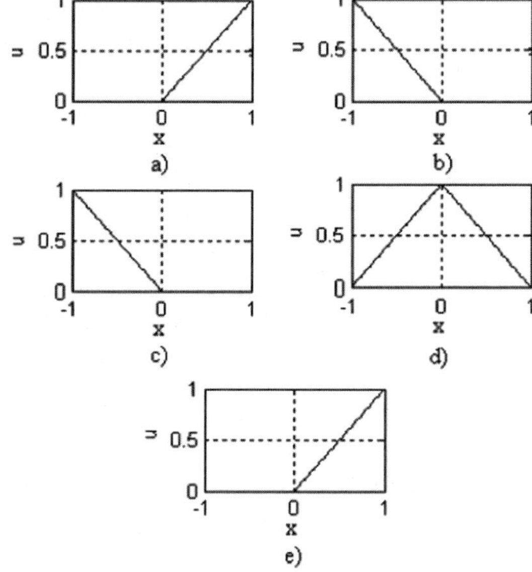

In our experiments we increase and decrease the value of ε to the left and to the right side having a εL and a εR values respectively to determine how much can be extended or perturbed the FOU with out loss of stability in the FLC.

We did make simulation with initial conditions θ having values in the whole circumference $[0, 2\pi]$, and the desired angle θd having values in the same range. The initial conditions considered in the experiments shown in this paper are an angle $\theta = 0\,rad$ and $\theta_d = 0.1\,rad$.

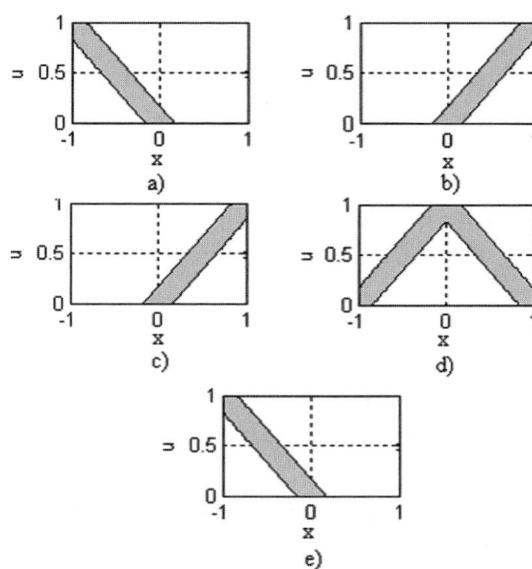

Fig. 6 Set of Type-2
membership functions:
a) negative, b) positive,
c) positive big, d) zero and
e) negative big

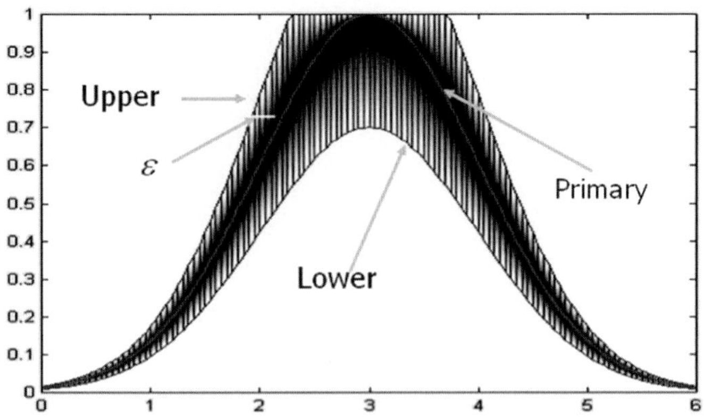

Fig. 7 Type-2 Fuzzy Set

In Fig. 8 we show a simulation of the plant made with a Type-1 FLC, as can be seen, the plant has been regulated in around 8seg, and in Fig. 9 we show the graph of (5) which is always negative defined and consequently the system is stable.

Figure 10 shows the simulation results of the plant made with the Type-2 FLC increasing and decreasing ε in the range of [0,1], as can be seen the plant has been regulated in the around of 10 seg, and the graph of (5) depicted at Fig. 11 is always negative defined and consequently the system is stable. As we can see, the time response is increasing about de value of ε is increasing.

With the variation of ε in the definition of the FOU, the control surface changes proportional to the change of ε, for that reason, the values of u for $\varepsilon \geq 1$ is practically zero, and the plant do not have physical response.

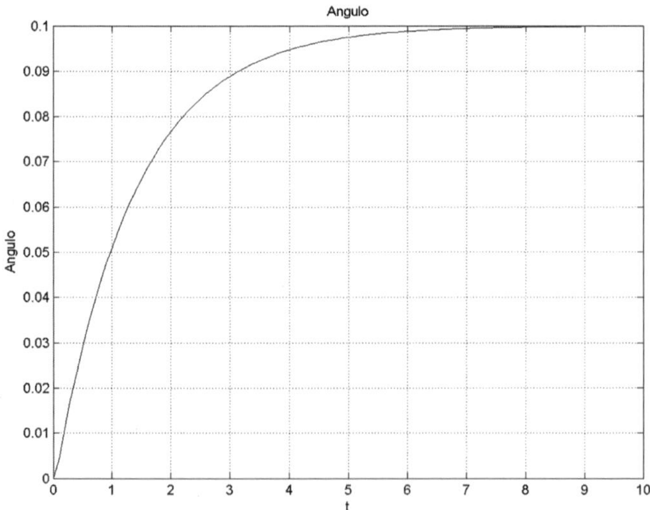

Fig. 8 Response for the type-1 FLC

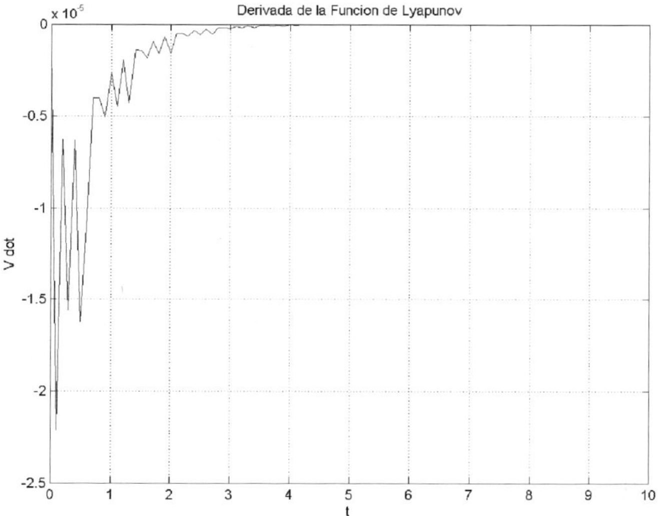

Fig. 9 $\dot{V}$ for the Type-1 FLC

To test the robustness of the built Fuzzy Controller, now we are going to use the same controller designed in Sect. III, but at this time, we are going to use it to control (2) considering the gravity effect as shows in (7).

$$ml^2\ddot{q} + gml \cos q = \tau \qquad (7)$$

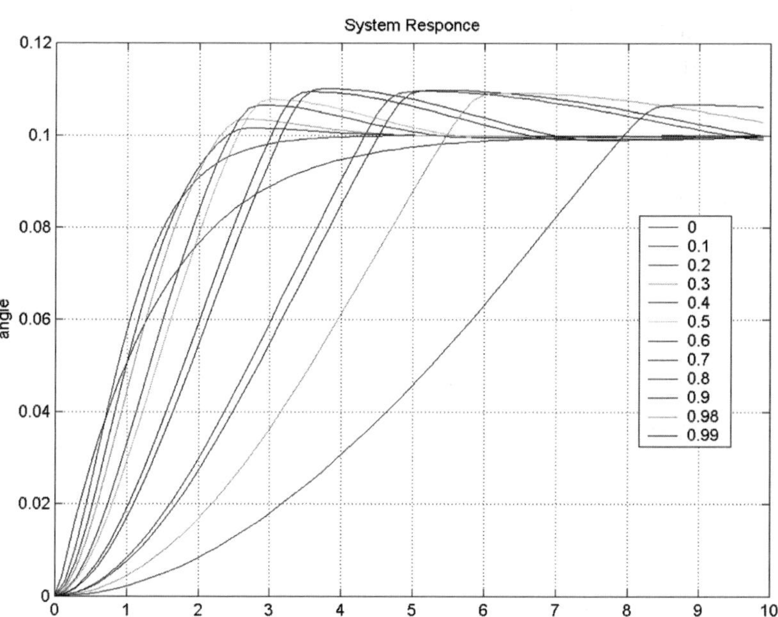

Fig. 10 Response for the Type-2 FLC ($\varepsilon \to [0, 1)$)

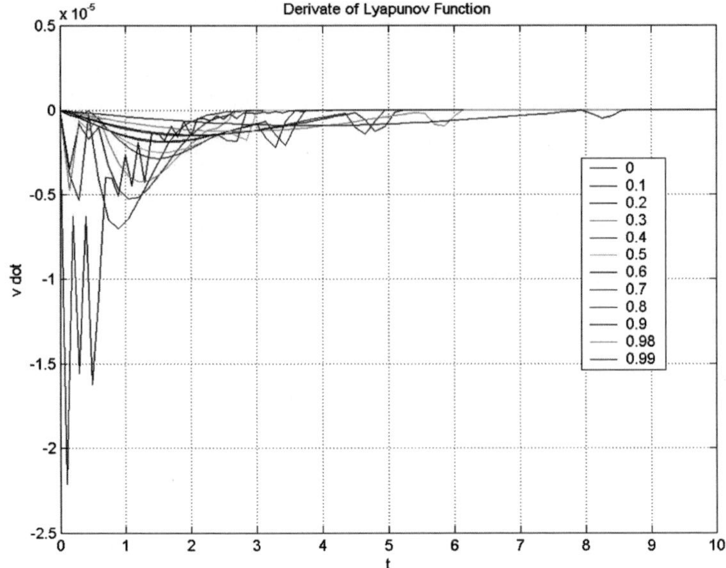

Fig. 11 $\dot{V}$ for the Type-2 FLC ($\varepsilon \rightarrow [0, 1]$)

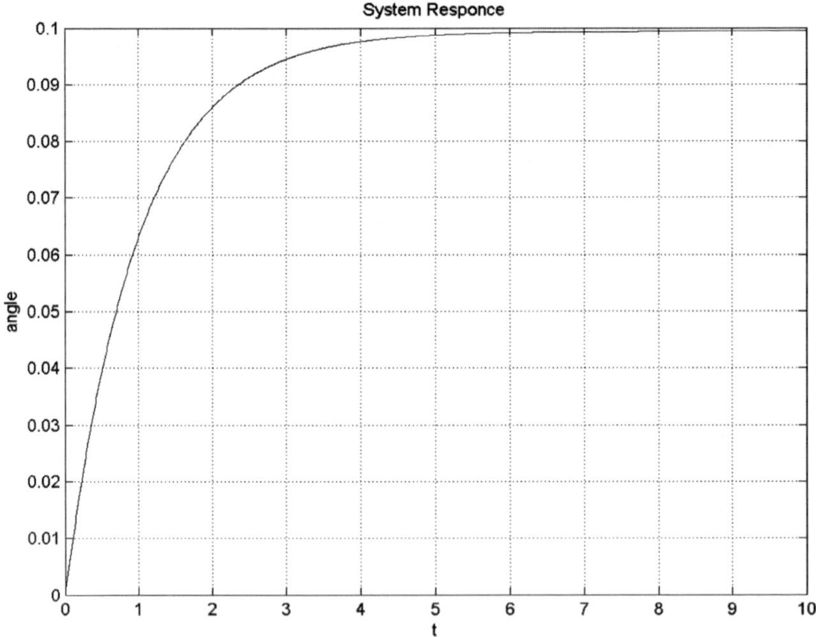

Fig. 12 Response for the Type-1 FLC

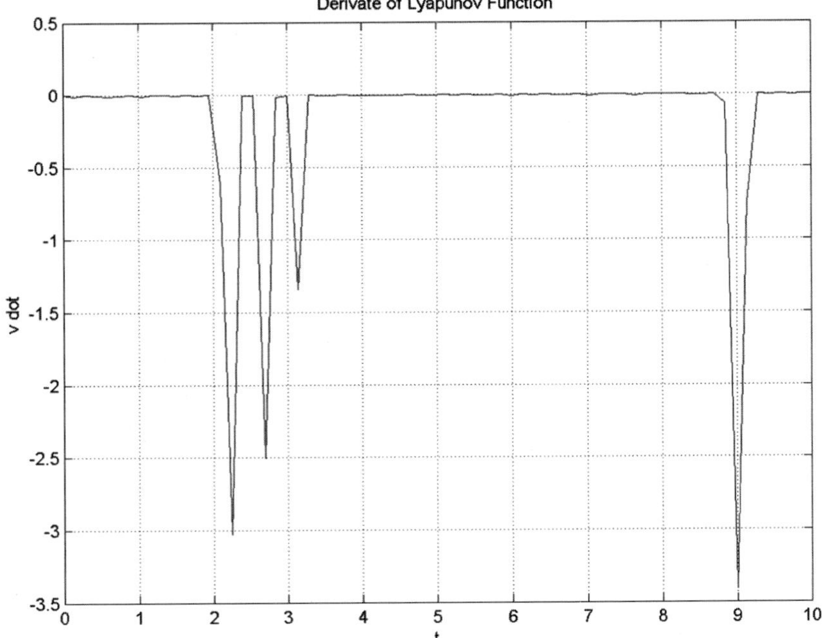

Fig. 13 $\dot{V}$ for the Type-1 FLC

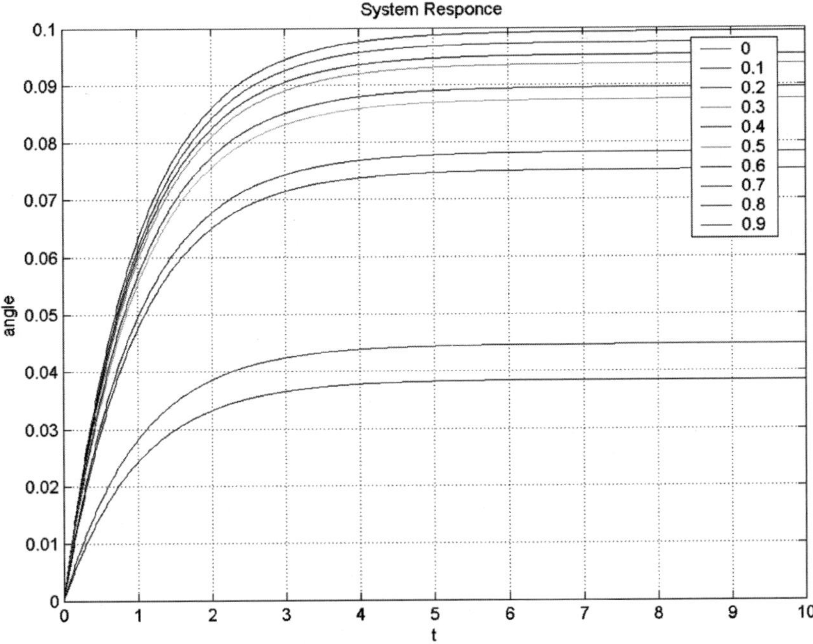

Fig. 14 Response for the Type-2 FLC ($\varepsilon \to [0, 1)$)

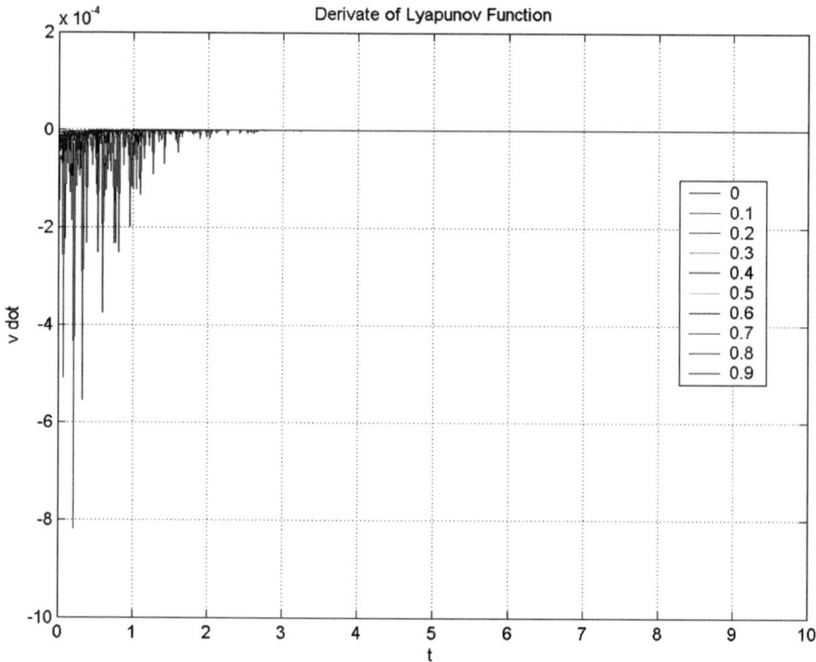

Fig. 15 $\dot{V}$ for the Type-2 FLC ($\varepsilon \to [0, 1]$)

At Fig. 12 we can see a simulation of the plant made with a Type-1 FLC, as can be seen, the plant has been regulated in around 8 seg, and Fig. 13 shows the graph of (5) which is always negative defined and consequently the system is stable.

Figure 14 shows the simulation results of the plant made with the Type-2 FLC increasing and decreasing ε in the range of [0,1], and the graph of (5) depicted at Fig. 15 is always negative defined and consequently the system is stable. As we can see, that if we use an adaptive gain like in [1] all the cases of ε can be regulated around 8 seg.

5 Conclusions

As in [1] and [2], the Margaliot approach for the design of FLC is now proved to be valid for both, Type-1 and Type-2 Fuzzy Logic Controllers.

On Type-2 FLC's membership functions, we can perturb or change the definition domain of the FOU without losing of stability of the controller; in the case seen at this paper, like in [1] we have to use an adaptive gain to regulate the plant in a desired time.

For our example of the 1DOF manipulator robot, the stability holds extending the FOU on the domain [0,1), this same was happened in [1] and [2]; we proved that a FLC designed following the Fuzzy Lyapunov Synthesis is stable and robust.

Acknowledgments The authors thank Tijuana Institute of Technology, DGEST, CITEDI-IPN, CONACYT and COSNET for supporting our research activities.

References

1. O. Castillo, L. Aguilar, N. Cázarez and D. Rico, "Intelligent Control of Dynamical Systems with Type-2 Fuzzy and Stability Study", Proceedings of the International Conference on Artificial Intelligence (IC-AI'2005, IMCCS&CE'2005), Las Vegas, USA, June 27-30, 2005 (to appear)
2. N. R. Cázarez, S. Cárdenas, L. Aguilar and O. Castillo, "Lyapunov Stability on Type-2 Fuzzy Logic Control", IEEE-CIS International Seminar on Computational Intelligence, México Distrito Federal, México, October 17-18 2005 (submitted)
3. G. Chen and T.T. Pham, "Introduction to fuzzy sets, fuzzy logic, and fuzzy control systems", CRC Press, USA, 2000.
4. B.J. Choi, S.W. Kwak, and B. K. Kim, "Design and Stability Analysis of Single-Input Fuzzy Logic Controller", IEEE Trans. Fuzzy Systems, vol. 30, pp. 303–309, 2000
5. J.-S.R. Jang, C.-T. Sun and E. Mizutani, "Neuro-Fuzzy and Soft Computing: a computational approach to learning and machine intelligence", Prentice Hall, USA, 1997.
6. J. Lee, "On methods for improving performance of PI-type fuzzy logic controllers", IEEE Trans. Fuzzy Systems, vol. 1, pp.298–301, 1993.
7. H.-X. Li and H.B. Gatland, "A new methodology for designing a fuzzy logic controller", IEEE Trans. Systems Man and Cybernetics, vol. 25, pp. 505–512, 1995.
8. H.-X. Li and H.B. Gatland, "Conventional fuzzy control and its enhancement", IEEE Trans. Systems Man and Cybernetics, vol. 25, pp. 791–797, 1996.
9. E. H. Mamdani and S. Assilian, "An experiment in linguistic systhesis with a fuzzy logic controller", Int. J. Man-Machine Studies, vol. 7, pp. 1–13, 1975.
10. E.H. Mamdani, "Advances in the linguistic synthesis of fuzzy controllers", Int. J. Man-Machine Studies, vol. 8, pp. 669–679, 1976.
11. M. Margaliot and G. Langholz, "New Approaches to Fuzzy Modeling and Control: Design and Analysis", World Scientific, Singapore, 2000.
12. J.M. Mendel, "Uncertain Rule-Based Fuzzy Logic: Introduction and new directions", Prentice Hall, USA, 2000.
13. K. Ogata, "Ingeniería de Control Moderna", 3ª Edición, Prentice Hall Hispanoamericana, Spain, 1998.
14. H. Paul and C. Yang, "Sistemas de control en ingeniería", Prentice Hall Iberia, Spain, 1999.
15. M. Sugeno, "On Stability of Fuzzy Systems Expressed by Fuzzy Rules with Singleton Consequents", IEEE Trans. Fuzzy Systems, vol.7, no. 2, 1999.
16. L.X. Wang, " A Course in Fuzzy Systems and Control", Prentice Hall, 1997.
17. J. Yen and R. Langari, "Fuzzy Logic: Intelligence, Control, and Information", Prentice Hall, New Jersey, USA, 1998.
18. H.Ying, "Practical Design of Nonlinear Fuzzy Controllers with Stability Analysis for Regulating Processes with Unknown Mathematical Models", Automatica, vol. 30, no. 7, pp. 1185–1195, 1994.
19. L.A. Zadeh, "The concept of a linguistic variable and its application to approximate reasoning", J. Information Sciences, vol.8, pp.43–80,1975.
20. L.A. Zadeh, "Fuzzy Logic = Computing With Words", IEEE Trans. Fuzzy Systems, vol. 4, no. 2, pp. 103–111, 1996.

Soft Computing for Intelligent Reservoir Characterization and Decision Analysis

Masoud Nikravesh

Abstract Reservoir characterization plays a crucial role in modern reservoir management. It helps to make sound reservoir decisions and improves the asset value of the oil and gas companies. It maximizes integration of multi-disciplinary data and knowledge and improves the reliability of the reservoir predictions. The ultimate product is a reservoir model with realistic tolerance for imprecision and uncertainty. Soft computing aims to exploit such a tolerance for solving practical problems. In reservoir characterization, these intelligent techniques can be used for uncertainty analysis, risk assessment, data fusion and data mining which are applicable to feature extraction from seismic attributes, well logging, reservoir mapping and engineering. The main goal is to integrate soft data such as geological data with hard data such as 3D seismic and production data to build a reservoir and stratigraphic model. While some individual methodologies (esp. neurocomputing) have gained much popularity during the past few years, the true benefit of soft computing lies on the integration of its constituent methodologies rather than use in isolation.

Key words: Pattern recognition · Petroleum · Soft Computing · Reservoir characterization · Fuzzy Logic · Neural Network

1 Introduction

With oil and gas companies presently recovering, on the average, less than a third of the oil in proven reservoirs, any means of improving yield effectively increases the world's energy reserves. Accurate reservoir characterization through data integration (such as seismic and well logs) is a key step in reservoir modeling & management and production optimization.

There are many techniques for increasing and optimizing production from oil and gas reservoirs:

- precisely characterizing the petroleum reservoir
- finding the bypassed oil and gas
- processing the huge databases such as seismic and wireline logging data,

M. Nikravesh et al. (eds.), *Forging the New Frontiers: Fuzzy Pioneers II.*
© Springer-Verlag Berlin Heidelberg 2008

- extracting knowledge from corporate databases,
- finding relationships between many data sources with different degrees of uncertainty,
- optimizing a large number of parameters,
- deriving physical models from the data
- Optimizing oil/gas production.

This paper address the key challenges associated with development of oil and gas reservoirs. Given the large amount of by-passed oil and gas and the low recovery factor in many reservoirs, it is clear that current techniques based on conventional methodologies are not adequate and/or efficient. We are proposing to develop the next generation of Intelligent Reservoir Characterization (IRESC) tool, based on Soft computing (as a foundation for computation with perception) which is an ensemble of intelligent computing methodologies using neuro computing, fuzzy reasoning, and evolutionary computing. We will also provide a list of recommendations for the future use of soft computing. This includes the hybrid of various methodologies (e.g. neural-fuzzy or neuro-fuzzy, neural-genetic, fuzzy-genetic and neural-fuzzy-genetic) and the latest tool of "computing with words" (CW) (Zadeh, 1996, 1999, Zadeh and Kacprzyk, 1999a and 1999b, and Zadeh and Nikravesh, 2002). CW provides a completely new insight into computing with imprecise, qualitative and linguistic phrases and is a potential tool for geological modeling which is based on words rather than exact numbers.

2 The Role of Soft Computing Techniques

Soft computing is bound to play a key role in the earth sciences. This is in part due to subject nature of the rules governing many physical phenomena in the earth sciences. The uncertainty associated with the data, the immense size of the data to deal with and the diversity of the data type and the associated scales are important factors to rely on unconventional mathematical tools such as soft computing. Many of these issues are addressed in a recent books, Nikravesh et al. (2003a, 2003b), Wong et al (2001), recent special issues, Nikravesh et al. (2001a and 2001b) and Wong and Nikravesh (2001) and recent papers by Nikravesh et al. (2001c) and Nikravesh and Aminzadeh (2001).

Intelligent techniques such as neural computing, fuzzy reasoning, and evolutionary computing for data analysis and interpretation are an increasingly powerful tool for making breakthroughs in the science and engineering fields by transforming the data into information and information into knowledge.

In the oil and gas industry, these intelligent techniques can be used for uncertainty analysis, risk assessment, data fusion and mining, data analysis and interpretation, and knowledge discovery, from diverse data such as 3-D seismic, geological data, well log, and production data. It is important to mention that during 1997, the US industry spent over \$3 billion on seismic acquisition, processing and interpretation. In addition, these techniques can be a key to cost effectively locating and producing our remaining oil and gas reserves. Techniques can be used as a tool for 1) Lowering Exploration Risk, 2) Reducing Exploration and Production cost, 3) Improving recovery through more efficient production, and 4) Extending the life of producing wells.

3 Hybrid Systems

So far we have seen the primary roles of neurocomputing, fuzzy logic and evolutionary computing. Their roles are in fact unique and complementary. Many hybrid systems can be built. For example, fuzzy logic can be used to combine results from several neural networks; GAs can be used to optimize the number of fuzzy rules; linguistic variables can be used to improve the performance of GAs; and extracting fuzzy rules from trained neural networks. Although some hybrid systems have been built, this topic has not yet reached maturity and certainly requires more field studies.

In order to make full use of soft computing for intelligent reservoir characterization, it is important to note that the design and implementation of the hybrid systems should aim to improve prediction and its reliability. At the same time, the improved systems should contain small number of sensitive user-definable model parameters and use less CPU time. The future development of hybrid systems should incorporate various disciplinary knowledge of reservoir geoscience and maximize the amount of useful information extracted between data types so that reliable extrapolation away from the wellbores could be obtained.

4 Intelligent Reservoir Characterization

Figure 1 shows techniques to be used for intelligent reservoir characterization (IRESC). The main goal is to integrate soft data such as geological data with hard data such as 3-D seismic, production data, etc. to build reservoir and stratigraphic models. In this case study, we analyzed 3-D seismic attributes to find similarity cubes and clusters using three different techniques: 1. k-means, 2. neural network (self-organizing map), and 3. fuzzy c-means. The clusters can be interpreted as lithofacies, homogeneous classes, or similar patterns that exist in the data. The relationship between each cluster and production-log data was recognized around the well bore and the results were used to reconstruct and extrapolate production-log data away from the well bore. The results from clustering were superimposed on the reconstructed production-log data and optimal locations to drill new wells were determined.

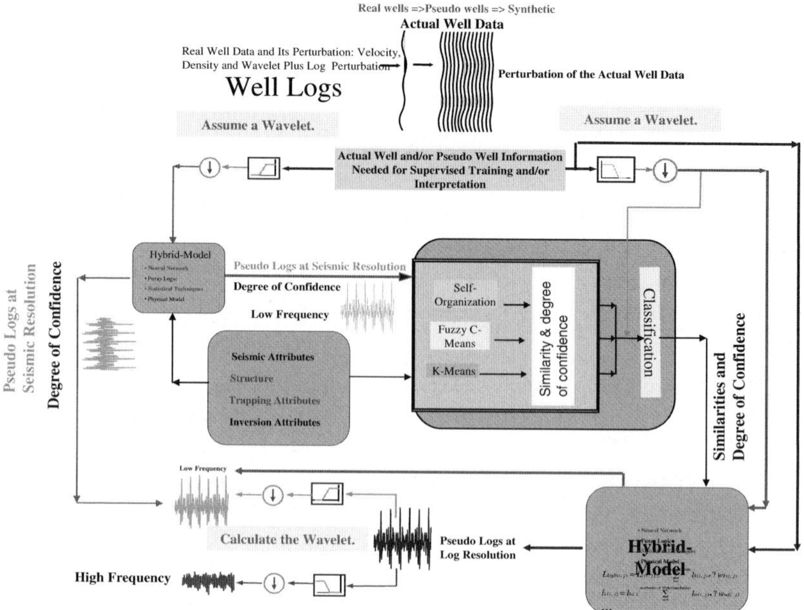

Fig. 1 Technique used in IRESC Software

5 Example

Our example is from a field that produces from the RedRiver Reservoir. A representative subset of the 3-D seismic cube, production log data, and an area of interest were selected in the training phase for clustering and mapping purposes. The subset (with each sample equal to 2 m sec of seismic data) was designed as a section passing through all the wells. However, only a subset of data points was selected for clustering purposes, representing the main Red River focus area. This subset covers the horizontal and vertical boreholes of producing wells. For clustering and mapping, there are two windows that must be optimized, the seismic window and the well log window. Optimal numbers of seismic attributes and clusters need to be determined, depending on the nature of the problem. Expert knowledge regarding geological parameters has also been used to constrain the maximum number of clusters to be selected. In this study, seventeen seismic attributes, five inversion attributes, six pseudo log attributes in seismic resolution and seven structure/trapping attributes, equaling a total of 35 attributes have been used (Table 1).

Clustering was based on three different techniques, k-means (statistical), neural network, and fuzzy c-means clustering. Different techniques recognized different cluster patterns and one can conclude that the neural network predicted a different structure and patterns than the other techniques. Finally, based on a qualitative and quantitative analysis given the prediction from high resolution data using the technique presented in Fig. 1, specific clusters that have the potential to include producing zones were selected. In this sub-cluster, the relationship between production-log

Table 1 List of the attributes calculated in this study

1. Amplitude envelope	1–17; Seismic Attributes
2. Amplitude weighted cosine phase	
3. Amplitude weighted frequency	Structure and Trapping Attributes.
4. Amplitude weighted phase	Six horizons and with four attributes out of seven attributes.
5. Apparent polarity	
6. Average frequency	Column A: line identifier
7. Cosine instantaneous phase	Column B: trace or cross-line identifier
8. Derivative	Column C: easting in feet
9. Derivative instantaneous amplitude	Column D: northing in feet
10. Dominant Frequency	1 Column E: horizon time in m secs
11. Instantaneous Frequency	2 Column F: time_resd, first order residual of horizon time, negative is high or above plane
12. Instantaneous Phase	
13. Integrated absolute amplitude	
14. Integrate	3 Column G: aspect, angle of updip direction at horizon (present day)
15. Raw seismic	
16. Second derivative instantaneous amplitude	Column H: next deeper horizon time (used for calculation of iso values)
17. Second derivative	4 Column I: iso, incremental time to next horizon
18. Acoustic Impedance	
19. Low Frequency of 18.	5 Column J: iso_resd, first order residual of iso time, negative is tinner (faster) than plane
20. Reflectivity Coefficients	
21. Velocity	
22. Density	6 Column K: iso_aspect of updip direction (at time of burial)
23. computed_Neutron_Porosity	
24. computed_Density_Porosity	7 Column L: cum_iso_resd, cumulative iso_resd from Winnipeg to this horizon
25. computed_Pwave	
26. computed_Density	
27. computed_True_Resistivity	18–22; Inversion Attributes
28. computed_Gamma_Ray	
	23–28; Pseudo Logs Attributes

data and clusters has been recognized and the production-log data has been reconstructed and extrapolated away from the wellbore. Finally, the production-log data and the cluster data were superimposed at each point in the 3-D seismic cube.

Figure 2 was generated using IRESC techniques (Fig. 1). Figure 3 shows both qualitative and quantitative analysis of the performance of the proposed technique. In this study, we have been able to predict the D1-Zone thickness whose its presence is very critical to production from D-Zone. D1-Zone thickness it is in the order of 14 feet or less and it is not possible to be recognized using seismic resolution information which is usually in the order of 20 feet and more in this area.

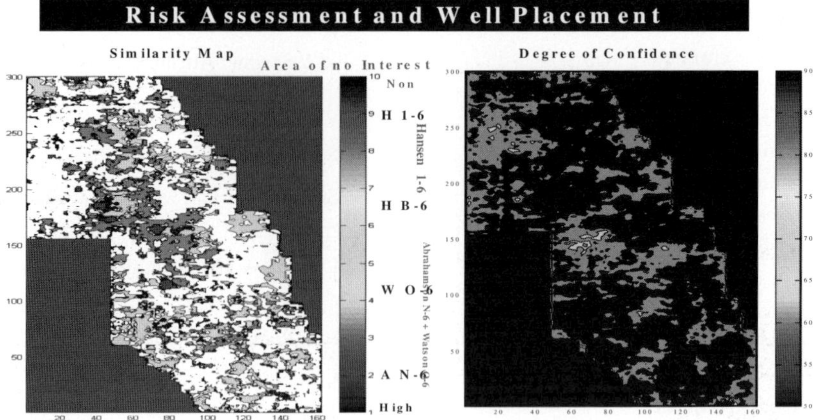

Fig. 2 Performance of IRESC technique for prediction of the high-potential and no-potential producing D-Zone based on virtual logs

Figures 4 through 6 show the performance of the IRESC technique for the prediction of classes (potential for production of high and no potential) and also the prediction of Phi*Dh which is a representative of the production zone in Red Reviver reservoirs. We have also been able to precisely predict not only the D-zone which is in the order of 50 feet, but both D1-zone which is in the order of 15 feet and D2-Zone which is in the order of 35 feet. The technique can be used for both risk assessment and analysis with high degree of confidence. To further use this information, we use three criteria to select potential locations for infill drilling or recompletion:

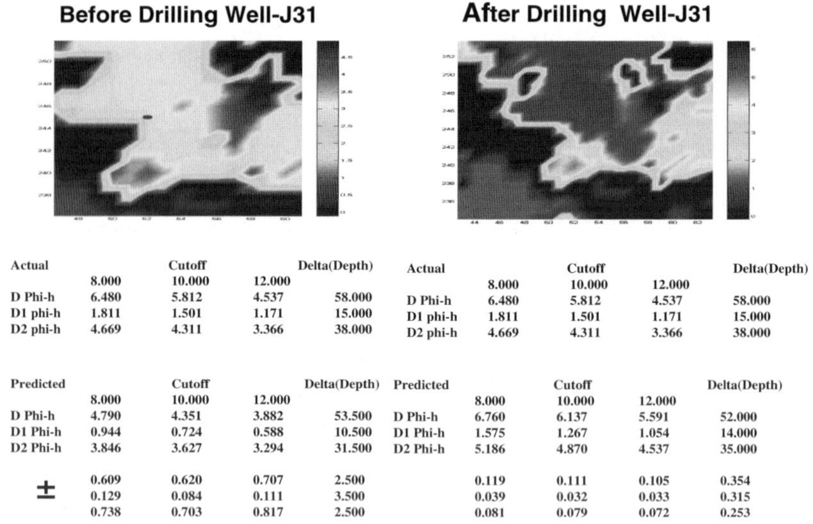

Actual		Cutoff		Delta(Depth)
	8.000	10.000	12.000	
D Phi-h	6.480	5.812	4.537	58.000
D1 phi-h	1.811	1.501	1.171	15.000
D2 phi-h	4.669	4.311	3.366	38.000

Actual		Cutoff		Delta(Depth)
	8.000	10.000	12.000	
D Phi-h	6.480	5.812	4.537	58.000
D1 phi-h	1.811	1.501	1.171	15.000
D2 phi-h	4.669	4.311	3.366	38.000

Predicted		Cutoff		Delta(Depth)
	8.000	10.000	12.000	
D Phi-h	4.790	4.351	3.882	53.500
D1 Phi-h	0.944	0.724	0.588	10.500
D2 Phi-h	3.846	3.627	3.294	31.500
$\pm$	0.609	0.620	0.707	2.500
	0.129	0.084	0.111	3.500
	0.738	0.703	0.817	2.500

Predicted		Cutoff		Delta(Depth)
	8.000	10.000	12.000	
D Phi-h	6.760	6.137	5.591	52.000
D1 Phi-h	1.575	1.267	1.054	14.000
D2 Phi-h	5.186	4.870	4.537	35.000
	0.119	0.111	0.105	0.354
	0.039	0.032	0.033	0.315
	0.081	0.079	0.072	0.253

Fig. 3 Qualitative and quantitative analysis and performance of IRESC technique for prediction of the high-potential and no-potential producing D-Zone

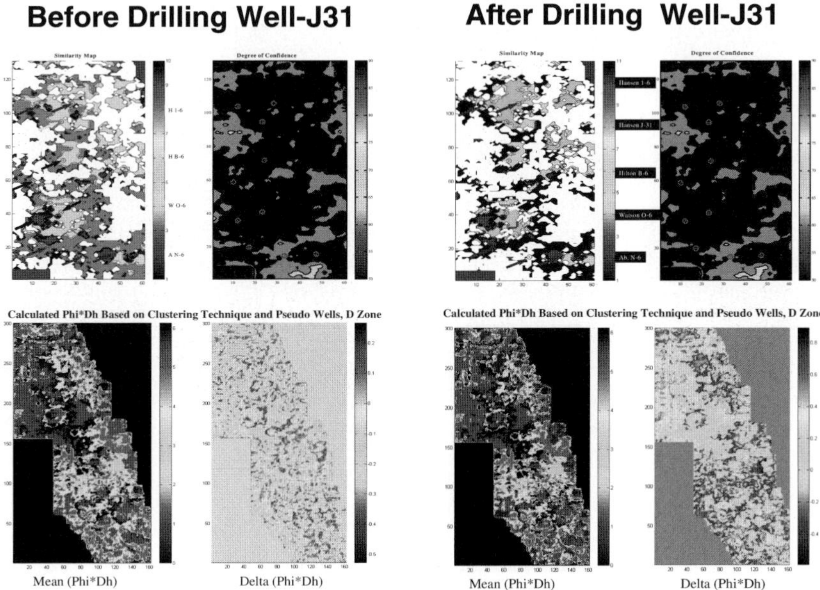

Fig. 4 Qualitative and quantitative analysis and performance of IRESC technique for prediction of D-Zone and Phi*Dh

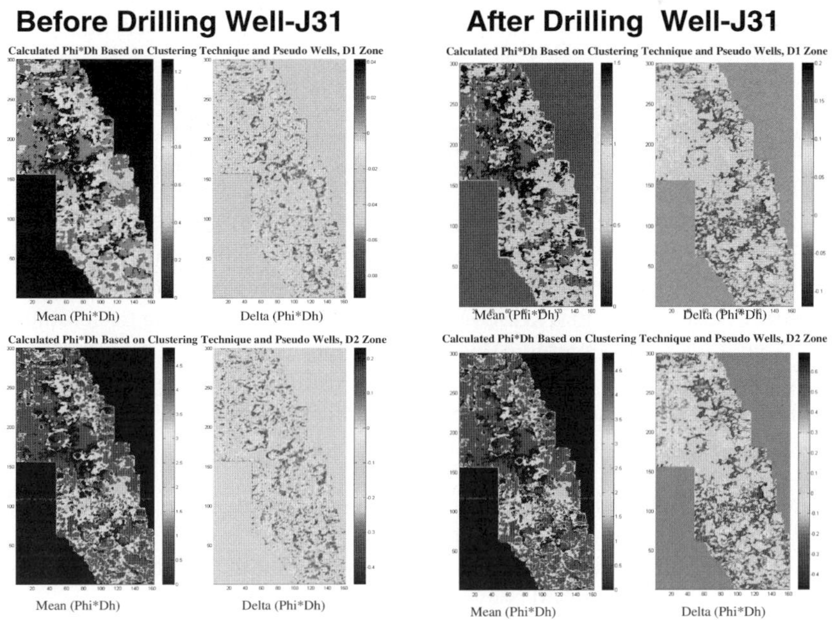

Fig. 5 Performance of IRESC technique for prediction of Phi*Dh for D1-Zone and D2-Zone

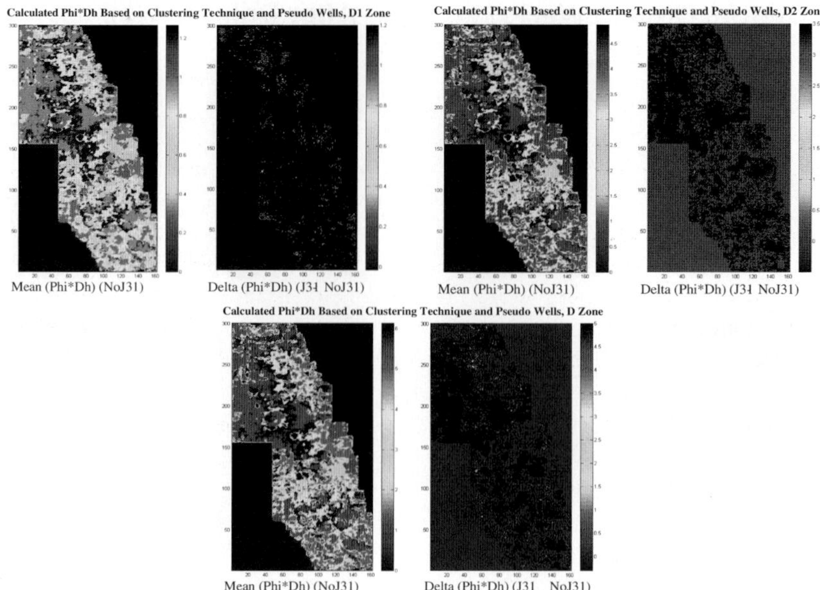

Fig. 6 Performance of IRESC technique for prediction of Phi*Dh for D-Zone, D1-Zone and D2-Zone and error bar at each point before and after drilling a new well

1. continuity of the selected cluster, 2. size and shape of the cluster, and 3. existence of high Production-Index values inside a selected cluster with high Cluster-Index values. Based on these criteria, locations of the new wells can be selected.

6 Future Trends and Conclusions

This paper addressed the key challenges associated with development of oil and gas reservoirs, given the large amount of by-passed oil and gas and the low recovery factor in many reservoirs. We are proposed the next generation of Intelligent Reservoir Characterization (IRESC) tool, based on Soft computing (as a foundation for computation with perception) which is an ensemble of intelligent computing methodologies using neuro computing, fuzzy reasoning, and evolutionary computing. The IRESC addresses the fundamental problems of current complex problems and its significant technical features are:

- Data Fusion: Integrating data from different sources
- Data Mining: Discovery of Knowledge
- Knowledge Engineering or Acquisition: Mapping the set of knowledge in a particular problem domain and converting it into a knowledge base
- Knowledge Management: Incorporating subjective information and knowledge
- Uncertainty Management: Quantifying and handling risk and uncertainty

- Scaling: Effective use of data orders of magnitude scale differences
- Economy: Time requirements to build models and update them

We have also discussed the main areas where soft computing can make a major impact in geophysical, geological and reservoir engineering applications in the oil industry. These areas include facilitation of automation in data editing and data mining. We also pointed out applications in non-linear signal (geophysical and log data) processing. And better parameterization of wave equations with random or fuzzy coefficients both in seismic and other geophysical wave propagation equations and those used in reservoir simulation. Of significant importance is their use in data integration and reservoir property estimation. Finally, quantification and reduction of uncertainty and confidence interval is possible by more comprehensive use of fuzzy logic and neural networks. The true benefit of soft computing, which is to use the intelligent techniques in combination (hybrid) rather than isolation, has not been demonstrated in a full extent. This section will address two particular areas for future research: hybrid systems and computing with words.

6.1 Computing with Word

One of the major difficulties in reservoir characterization is to devise a methodology to integrate qualitative geological description. One simple example is the core descriptions in standard core analysis. These descriptions provide useful and meaningful observations about the geological properties of core samples. They may serve to explain many geological phenomena in well logs, mud logs and petrophysical properties (porosity, permeability and fluid saturations).

Computing with words (CW) aims to perform computing with objects which are propositions drawn from a natural language or having the form of mental perceptions. In essence, it is inspired by remarkable human capability to manipulate words and perceptions and perform a wide variety of physical and mental tasks without any measurement and any computations. It is fundamentally different from the traditional expert systems which are simply tools to "realize" an intelligent system, but are not able to process natural language which is imprecise, uncertain and partially true. CW has gained much popularity in many engineering disciplines (Zadeh, 1996, 1999, Zadeh and Kacprzyk, 1999a and 1999b, and Zadeh and Nikravesh, 2002). In fact, CW plays a pivotal role in fuzzy logic and vice-versa. Another aspect of CW is that it also involves a fusion of natural languages and computation with fuzzy variables.

In reservoir geology, natural language has been playing a very crucial role for a long time. We are faced with many intelligent statements and questions on a daily basis. For example: "if the porosity is high then permeability is likely to be high"; "most seals are beneficial for hydrocarbon trapping, a seal is present in reservoir A, what is the probability that the seal in reservoir A is beneficial?"; and "high resolution log data is good, the new sonic log is of high resolution, what can be said about the goodness of the new sonic log?"

CW has much to offer in reservoir characterization because most available reservoir data and information are too imprecise. There is a strong need to exploit the tolerance for such imprecision, which is the prime motivation for CW. Future research in this direction will surely provide a significant contribution in bridging reservoir geology and reservoir engineering. Given the level of interest and the number of useful networks developed for the earth science applications and specially oil industry, it is expected soft computing techniques will play a key role in this field. Many commercial packages based on soft computing are emerging. The challenge is how to explain or "sell" the concepts and foundations of soft computing to the practicing explorationist and convince them of the value of the validity, relevance and reliability of results based on the intelligent systems using soft computing methods.

References

1. Nikravesh, M., F. Aminzadeh, and L. A. Zadeh (2003a), *Intelligent Data Analysis for Oil Exploration,* Developments in Petroleum Science, 51; ISBN: 0-444-50685-3, Elsevier (March 2003).
2. Nikravesh, M., L.A. Zadeh and V. Korotkikh (2003b), *Fuzzy Partial Differential Equations and Relational Equations: Reservoir Characterization and Modeling,* to be published in the Series Studies in Fuzziness and Soft Computing, Physica-Verlag, Springer (Expected, August 2003).
3. Nikravesh, M., F. Aminzadeh and L.A. Zadeh (2001a), Soft Computing and Earth Sciences (Part 2), Journal of Petroleum Science and Engineering, Volume 31, Issue 2-4, January 2001; Special Issue.
4. Nikravesh, M., F. Aminzadeh and L.A. Zadeh (2001b), Soft Computing and Earth Sciences, Journal of Petroleum Science and Engineering, Volume 29, Issue 3-4, May 2001; Special Issue, 2001b.
5. Nikravesh, M., R. D. Adams and R. A. Levey (2001c), Soft computing: tools for intelligent reservoir characterization (IRESC) and optimum well placement (OWP), *Journal of Petroleum Science and Engineering, Volume 29, Issues 3-4, May 2001, Pages 239–262.*
6. Nikravesh, M. and F. Aminzadeh (2001), Mining and fusion of petroleum data with fuzzy logic and neural network agents, *Journal of Petroleum Science and Engineering, Volume 29, Issues 3-4, May 2001, Pages 221–238.*
7. Wong, P. M., F Aminzadeh, and M. Nikravesh (2001), Soft *Computing for Reservoir Characterization,* in Studies in Fuzziness, Physica Verlag, Germany
8. P.M. Wong and M. Nikravesh (2001), A thematic issue on "Field Applications of Intelligent Computing Techniques," Journal of Petroleum Geology, 24(4), 379–476; Special Issue.
9. Zadeh, L. A. (1999), From Computing with Numbers to Computing with Words – From Manipulation of Measurements to Manipulation of Perceptions, IEEE Transactions on Circuits and Systems, 45, 105–119, 1999.
10. Zadeh, L. and Kacprzyk, J. (eds.) (1999a), *Computing With Words in Information/Intelligent Systems 1: Foundations,* Physica-Verlag, Germany (1999a).
11. Zadeh, L. and Kacprzyk, J. (eds.) (1999b), *Computing With Words in Information/Intelligent Systems 2: Applications,* Physica-Verlag, Germany (1999b).
12. Zadeh, L.A. (1996) Fuzzy Logic = Computing with Words, *IEEE Trans. on Fuzzy Systems* (1996) 4,103–111.
13. Zadeh, L. A. and M. Nikravesh (2002), Perception-Based Intelligent Decision Systems, AINS; ONR Summer 2002 Program Review, 30 July-1 August, UCLA.

Fuzzy-Based Nosocomial Infection Control

Klaus-Peter Adlassnig, Alexander Blacky and Walter Koller

Abstract Nosocomial, or hospital-acquired, infections (NIs) are a frequent complication affecting hospitalized patients. The growing availability of computerized patient records in hospitals allows automated identification and extended monitoring of the signs of NI for the purpose of reducing NI rates. A fuzzy- and knowledge-based system to identify and monitor NIs at intensive care units according to the European Surveillance System HELICS was developed. It was implemented into the information technology landscape of the Vienna General Hospital and is now in routine use.

1 Introduction

Nosocomial, or hospital-acquired, infections (NIs) are by far the most common complications affecting hospitalized patients. Currently, 5 to 10 percent of patients admitted to acute care hospitals acquire one or more infections. These adverse events affect approximately 2 million patients each year in the United States, result in some 90,000 deaths, and add an estimated $ 4.5 to $ 5.7 billion per year to the cost of patient care [1].

The growing availability of computerized patient records in hospitals allows extended monitoring of the signs of NIs for the purpose of reducing NI rates by early initiation of appropriate therapy. In addition, ward- and institution-based surveillance data of NIs analyzed by infection control personnel are used as a basis to implement preventive measures.

2 Methods

Based on the methodological and practical results obtained from the development and application of CADIAG-II/RHEUMA (computer-assisted diagnosis, version 2, applied to rheumatology), a fuzzy-based differential diagnostic consultation system for rheumatology [2, 3, 4, 5], the MONI system (cf., Fig. 1) was developed [6, 7]. MONI is a fuzzy- and knowledge-based system for the evaluation of definitions of

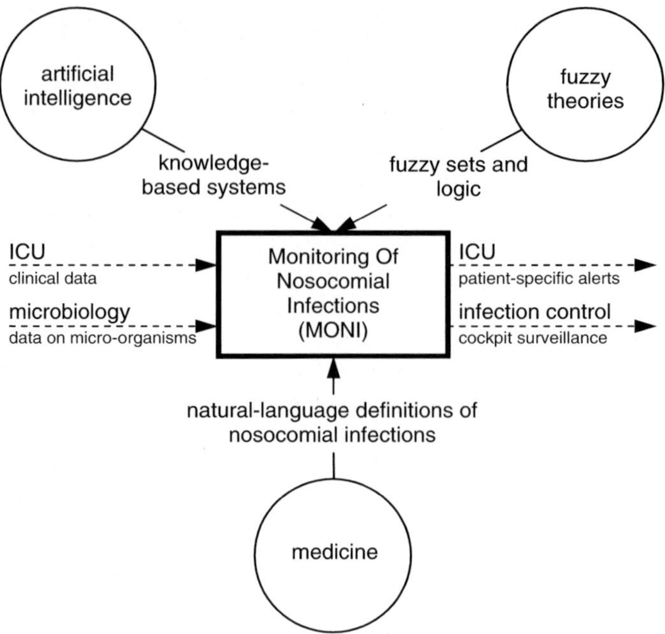

Fig. 1 Medical and formal areas that constitute the methodological basis of MONI

NIs according to the European Surveillance System HELICS [8]. The definitions are derived from those issued by the Centers for Disease Control and Prevention (CDC) in Atlanta, USA [9, 10]. They are expressed as natural language text (see Fig. 2) and are analyzed and transferred into a fuzzy-rule-based representation with a step-wise

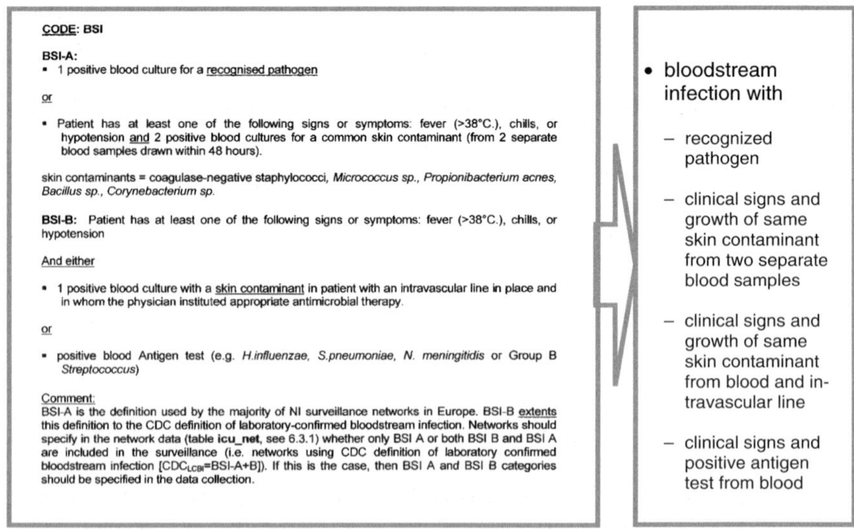

Fig. 2 Natural language definition for BSIs (left part) from which four different entities of BSIs are derived (right part); excerpt from [8]

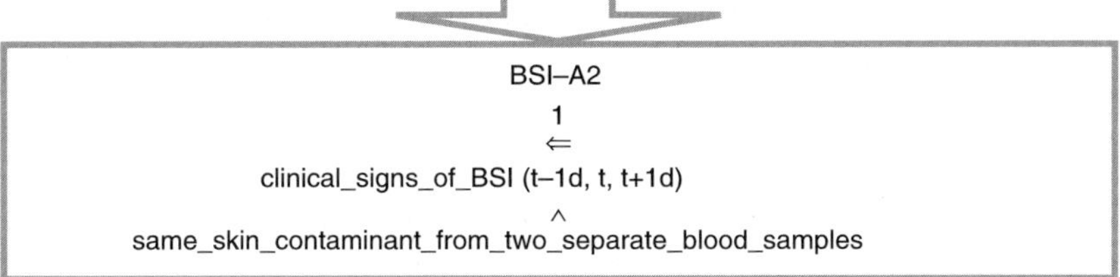

Fig. 3 Example of one BSI definition ($\wedge$ denotes fuzzy logical and, $\Leftarrow$ denotes full implication if the premise on the right side is definitely fulfilled); each of the two conjuncts will be decomposed to less aggregated entities

decomposition of major medical concepts (such as "clinical signs of bloodstream infection (BSI)") to subconcepts (such as "CRP (C-reactive protein) increased") until data items from the patient records in the hospital can be mapped into these definitions (see Fig. 3, Fig. 4, and Fig. 5). MONI applies fuzzy sets to formally represent all medical entities such as symptoms, signs, laboratory test results (see Fig. 5), clinical findings, intermediate medical concepts, and diseases. Fuzzy logic, then, is the method of knowledge processing in MONI.

The rules and the fuzzy sets were formed in a dialogue between a medical knowledge engineer and an expert of the infection control unit. The rules are based on the natural language definitions in [8]; the fuzzy sets allow for gradual transition from one medical concept to an adjacent concept such as "CRP normal" to "CRP increased". Common sense knowledge and clinical experience tell how "stretchable" the respective medical concepts are, i.e., how far the gradual transition area must extend in order to capture early and slight forms of NIs.

The rules and the fuzzy sets were formed in a dialogue between a medical knowledge engineer and an expert of the infection control unit. The rules are based on the natural language definitions in [8]; the fuzzy sets allow for gradual transition from one medical concept to an adjacent concept such as "CRP normal" to "CRP increased". Common sense knowledge and clinical experience tell how "stretchable" the respective medical concepts are, i.e., how far the gradual transition area must extend in order to capture early and slight forms of NIs.

3 Results

MONI is operated in 12 adult intensive care units (ICUs) accommodating a total of up to 96 beds at the Vienna General Hospital, a 2,200-bed university hospital and the main teaching hospital of the Medical University of Vienna. It is fully integrated into the information technology (IT) landscape of the university hospital. Twenty-four

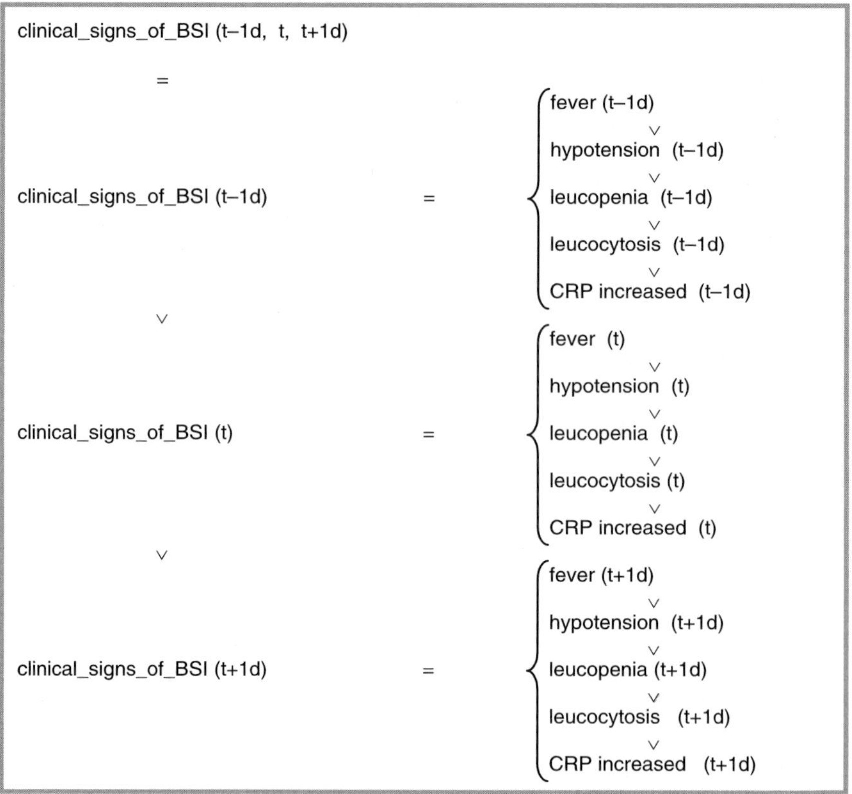

Fig. 4 Decomposition of the concept "clinical signs of BSI" ($\vee$ denotes fuzzy logical or; t-1d, t, and t+1d denote yesterday, today, and tomorrow, respectively); this concept can be used for both prospective and retrospective monitoring

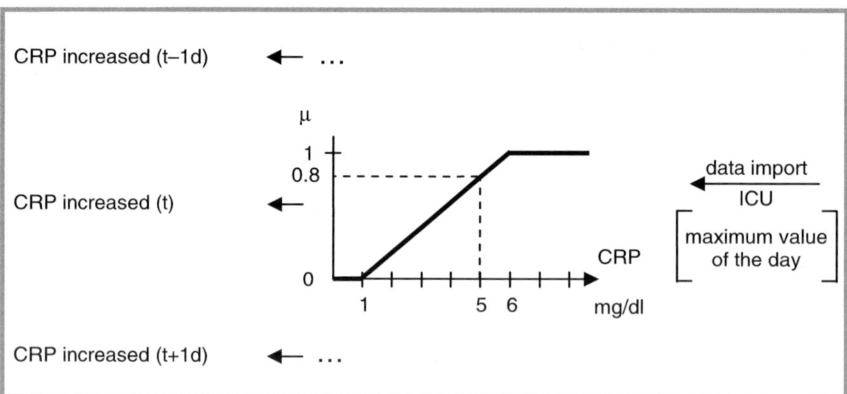

Fig. 5 Example of the definition of a fuzzy set; a measured CRP value of 5 mg/dl, for example, is compatible with the linguistic term "CRP increased" with a degree of 0.8, this degree of compatibility μ is then propagated through the rule-based inference network of MONI via fuzzy logic

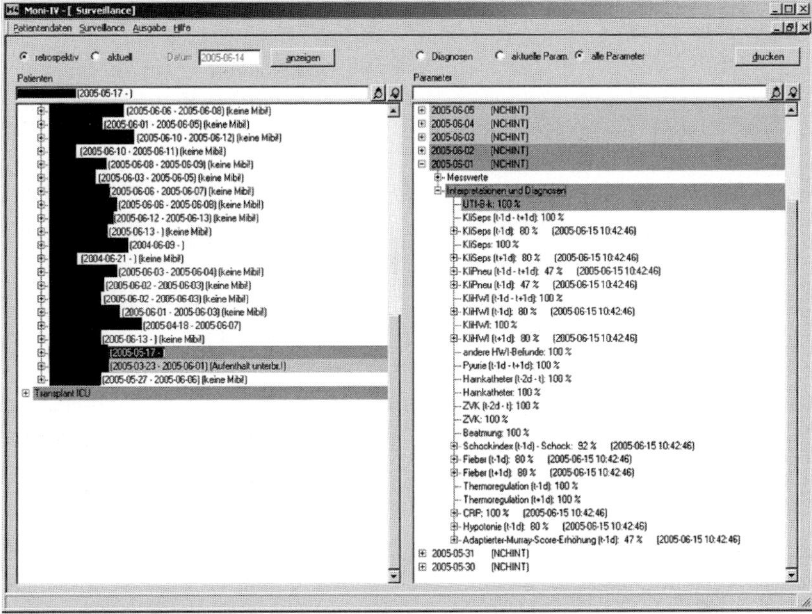

Fig. 6 In one patient, the definition of a catheter-associated symptomatic urinary tract infection (see above UTI-B-k) was completely fulfilled (100%): the underlying patient data and the interpreted symptoms derived from these data are shown

definitions of NIs were implemented in MONI. They cover BSIs, ICU-acquired pneumonias, urinary tract infections, and central venous catheter-related infections.

The recognition and monitoring of NIs according to the HELICS definitions for ICUs can be viewed in the following screenshots (see Figs. 6–8).

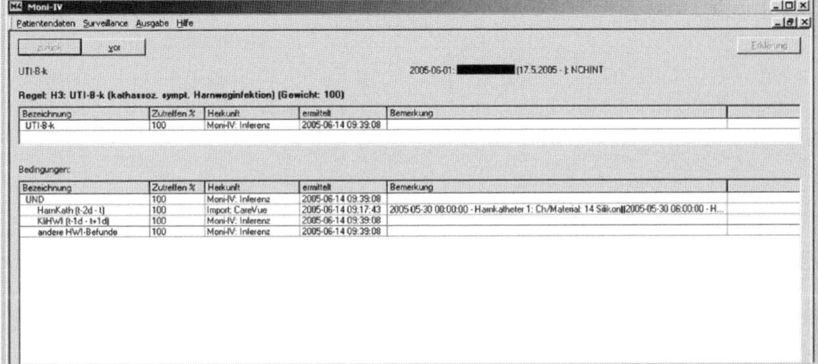

Fig. 7 Backtracking of the logical chain of reasoning shows that the patient has a urinary catheter; this data element was documented in the respective patient data management system (PDMS) and passed on to the MONI system through several intermediate steps of abstraction

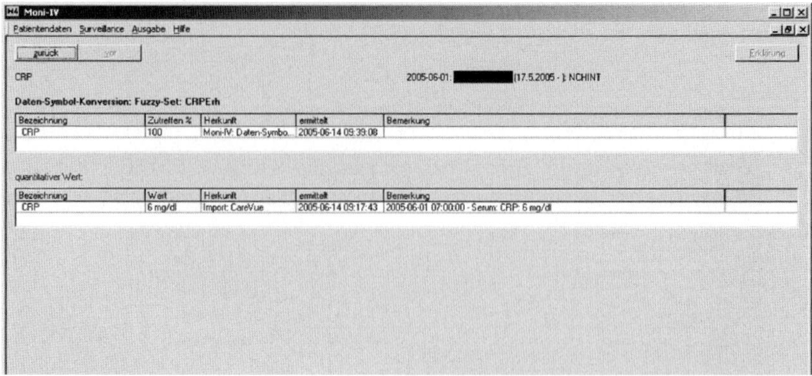

Fig. 8 A CRP value of 6 mg/dl signifies that an elevated CRP is present with a fuzzy compatibility of 100%; therefore, a clinical sign necessary to fulfill the definition of symptomatic urinary tract infection is given

At present, the system is being fine-tuned by physicians and infection control personnel. The preliminary results confirm the technical appropriateness of the system. The medical results are convincing, and a refined scientific study measuring the clinical correctness and potential impact on patient care and health care costs is under way. Besides, the medical results already allow a large number of NIs to be identified; these numbers of infection were considered impossible to identify in the past.

4 Conclusions

Medical knowledge is usually produced by human medical researchers for other human beings, who, as a rule, are practicing physicians. Natural language is used to document this knowledge, as is done with the definitions of NIs.

To automate and thus support the medical decision-making process, these texts have to be converted into a formal representation. The latter then allows application of this knowledge to medical data of specific patients through several layers of data abstraction. With the methodological framework of fuzzy theories, this is performed successfully and applied in clinical real-world settings.

References

1. Burke JP (2003) Infection Control—A Problem for Patient Safety. The New England Journal of Medicine 348(7):651–656.
2. Adlassnig KP (1980) A Fuzzy Logical Model of Computer-Assisted Medical Diagnosis. Methods of Information in Medicine 19:141–148.
3. Adlassnig KP, Kolarz G (1986) Representation and Semiautomatic Acquisition of Medical Knowledge in CADIAG-1 and CADIAG-2. Computers and Biomedical Research 19:63–79.

4. Adlassnig KP, Kolarz G, Scheithauer W, Grabner H (1986) Approach to a Hospital-Based Application of a Medical Expert System. Medical Informatics 11:205–223.
5. Leitich H, Adlassnig KP, Kolarz G (1996) Development and Evaluation of Fuzzy Criteria for the Diagnosis of Rheumatoid Arthritis. Methods of Information in Medicine 35:334–342.
6. Fabini B (2001) Monitoring of Infectious Risk Situations and Nosocomial Infections in the Hospital, PhD Thesis, Vienna University of Technology, Vienna, Austria.
7. Heisz H (2004) Praktisch orientierte Konzepte der Inferenz mit Fuzzy-Regeln auf Grundlage des nosokomialen Diagnosesystems MONI-IV, Dissertation, Technische Universität Wien, Wien, Österreich, 2004 (in German).
8. Hospital in Europe Link for Infection Control through Surveillance (HELICS) (2004) Surveillance of Nosocomial Infections in Intensive Care Units – Protocol Version 6.1, (Based on Version 5.0 including technical amendments), September 2004, Project commissioned by the EC / DG SANCO / F/ 4, Agreement Reference number: VS/1999/5235 (99CVF4-025), 2004, 1–51. http://helics.univ-lyon1.fr/protocols/icu_protocol.pdf (last accessed: 1 September 2006)
9. Garner JS, Jarvis WR, Emori TG, Horan TC, Hughes JM (1988) CDC Definitions for Nosocomial Infections, 1988. American Journal of Infection Control 16:128–140.
10. Garner JS, Jarvis WR, Emori TG, Horan TC, Hughes JM (1996) CDC Definitions of Nosocomial Infections. In Olmsted R.N. (Ed.), APIC Infection Control and Applied Epidemiology: Principles and Practice, Mosby, St. Louis, A-1–A-20.

Fuzzy Association Rules for Query Refinement in Web Retrieval

M. Delgado, M.J. Martín-Bautista, D.Sánchez, J.M. Serrano and M.A. Vila

Abstract In this paper, we present an application for helping users to find new query terms in web retrieval via fuzzy association rules. Once the user has made an initial query, a set of documents is retrieved from the web. Representing these documents as text transactions, each item in the transaction means the presence of the term in the document. From the set of transactions, fuzzy association rules are extracted. Based on the thresholds of support and certainty factor, a selection of rules is carried out and the terms in those rules are offered to the user to be added to the query and to improve the retrieval.

1 Introduction

The dissatisfaction of users with the answer of search robots in web retrieval is a very common problem. This problem is due, most of the times, to the terms used to query, which meet the search criteria, but do not reflects exactly what the user is really searching. The query refinement process, also called query expansion, is a possible solution to this problem. To solve this problem, the query can be modified by adding or removing terms to discard uninteresting retrieved documents and/or to retrieve interesting documents that were not retrieved by the query. This problem has been named as *query refinement* or *query expansion* in the field of Information Retrieval [14].

In this work, we propose the use of mining techniques to refine queries in web retrieval. For this purpose, we use fuzzy association rules to find dependence relations among the presence of terms in an initial set of retrieved documents. A group of selected terms from the extracted rules generates a vocabulary related to the search topic that helps the user to refine the query with the aim of improving the retrieval effectiveness. Data mining techniques have been applied successfully in the last decade in the field of Databases, but also to solve some classical Information Retrieval problems such as document classification [26] and query refinement [35].

This paper is organized as follows: The concepts of association rules, fuzzy association rules and fuzzy transactions are presented briefly in Sect. 2. A survey of query refinement solutions found in the literature is given in Sect. 3, while the process to refine queries via fuzzy association rules is explained in Sect. 4. The

M. Nikravesh et al. (eds.), *Forging the New Frontiers: Fuzzy Pioneers II.*

obtention of the document representation and the extraction of fuzzy association rules are given in Sects. 5 and 6, respectively. Finally, some experimental examples are shown in Sect. 7 and the conclusions and future work are presented in 8.

2 Association Rules and Fuzzy Association Rules

We use association rules and fuzzy association rules to find the terms to be added to the original query. In this section, we briefly review association rules and some useful extensions able to deal with weighted sets of items in a fuzzy framework.

2.1 Association Rules

Given a database of transactions, where each transaction is an itemset, we can extract association rules [1]. Formally, let T be a set of transactions containing items of a set of items I. Let us consider two sets of items $I_1, I_2 \subset I$, where $I_1, I_2 \neq \phi$ and $I_1 \cap I_2 = \phi$. An association rule [1] $I_1 \Rightarrow I_2$ is an implication rule meaning that the apparition of itemset I_1 in a transaction implies the apparition of itemset I_2 in the same transaction. The reciprocal does not have to happen necessarily [22]. I_1 and I_2 are called antecedent and consequent of the rule, respectively. The rules obtained with this process are called boolean association rules or, in general, association rules since they are generated from a set of boolean or crisp transactions.

2.2 Fuzzy Association Rules

Fuzzy association rules are defined as those rules extracted from a set of fuzzy transactions FT where the presence of an item in a transaction is given by a fuzzy value of membership [25, 3, 24, 19, 10]. Though most of these approaches have been introduced in the setting of relational databases, we think that most of the measures and algorithms proposed can be employed in a more general framework. A broad review, including references to papers on extensions to the case of quantitative attributes and hierarchies of items, can be found in [11].

In this paper we shall employ the model proposed in [10]. This model considers a general framework where data is in the form of fuzzy transactions, i.e., fuzzy subsets of items. A (crisp) set of fuzzy transactions is called a FT-set, and fuzzy association rules are defined as those rules extracted from a FT-set. Fuzzy relational databases can be seen as a particular case of FT-set. Other datasets, such as the description of a set of documents by means of fuzzy subsets of terms, are also particular cases of FT-sets but fall out of the relational database framework.

Given a FT-set $\widetilde{T}$ on a set of items I and a fuzzy transaction $\widetilde{\tau} \in \widetilde{T}$, we note $\widetilde{\tau}(i)$ the membership degree of i in τ $\forall i \in I$. We also define $\widetilde{\tau}(I_0) = \min_{i \in I_0} \widetilde{\tau}(i)$ for every itemset $I_0 \subseteq I$. With this scheme, we have a degree in [0,1] associated to each pair $\langle \widetilde{\tau}, I_0 \rangle$. Sometimes it is useful to see this information in a different way by means of what we call the *representation* of an itemset. The idea is to see an itemset as a fuzzy subset of transactions. The representation of an itemset $I_0 \subseteq I$ in a FT-set $\widetilde{T}$ is the fuzzy subset $\widetilde{\Gamma}_{I_0} \subseteq \widetilde{T}$ defined as

$$\widetilde{\Gamma}_{I_0} = \sum_{\widetilde{\tau} \in \widetilde{T}} \widetilde{\tau}\,(I_0)\big/\widetilde{\tau} \tag{1}$$

On this basis, a fuzzy association rule is an expression of the form $I_1 \Rightarrow I_2$ that holds in a FT-set T iff $\widetilde{\Gamma}_{I_1} \subseteq \widetilde{\Gamma}_{I_2}$. The only difference with the definition of crisp association rule is that the set of transactions is a FT-set, and the inclusion above is the usual between fuzzy sets.

2.3 *Measures for Association and Fuzzy Association Rules*

The assessment of association rules is usually based on the values of *support* and *confidence*. Support is the percentage of transactions containing an itemset, calculated by its probability, while confidence measures the strength of the rule calculated by the conditional probability of the consequent with respect to the antecedent of the rule. Only itemsets with a support greater than a threshold *minsupp* are considered, and from the resulting association rules, those ones with a confidence less than a threshold *minconf* are discarded. Both thresholds must be fixed by the user before starting the process.

To deal with the imprecision of fuzzy transactions, we need to obtain the support and the confidence values with alternative methods which can be found mainly in the framework of approximate reasoning. We have selected the the evaluation of quantified sentences presented in [40], calculated by means of method GD presented in [13]. Moreover, as an alternative to confidence, we propose the use of *certainty factors* to measure the accuracy of association rules, since they have been revealed as a good measure in knowledge discovery too [17].

3 Query Refinement with Fuzzy Association Rules

The query refinement process, also called query expansion, is a possible solution to the problem of dissatisfaction of the user with the answer of an information retrieval system, given a certain query. This can occur because the user does not know the vocabulary of the topic of the query, or the query terms do not come to user's mind at the query moment, or just because the vocabulary of the user does not match with the indexing words of the collection. This problem is even strong when the user is searching in the web, due to the amount of available information which makes that the user feels overwhelmed with the retrieved set of documents. The process of query refinement solves this problem by modifying the search terms so the system results are more adequate to user's needs.

3.1 Related Work

There are mainly two different approaches in query refinement regarding how the terms are added to the query. The first one is called *automatic query expansion* [7, 18] and consists of the augmentation of query terms to improve the retrieval process without the intervention of the user. The second one is called *semi-automatic query-expansion* [30, 37], where new terms are suggested to the user to be added to the original query in order to guide the search towards a more specific document space.

We can also distinguish different cases based on the source from which the terms are selected. By this way, terms can be obtained from the collection of documents [2, 39], from user profiles [23], from user behavior [21] or from other users' experience [16], among others. If a document collection is considered as a whole from which the terms are extracted to be added to the query, the technique is called *global analysis*, as in [39]. However, if the expansion of the query is performed based on the documents retrieved from the first query, the technique is denominated *local analysis*, and the set of documents is called *local set*.

Local analysis can also be classified into two types. On the one hand, *local feedback* adds common words from the top-ranked documents of the local set. These words are identified sometimes by clustering the document collection [2]. In this group we can include the relevance feedback process, since the user has to evaluate the top ranked documents from which the terms to be added to the query are selected. On the other hand, *local context analysis* [39], which combines global analysis and context local feedback to add words based on relationships of the top-ranked documents. The calculus of co-occurrences of terms is based on passages (text windows of fixed size), as in global analysis, instead of complete documents. The authors show that, in general, local analysis performs better than global one.

3.2 Using Fuzzy Association Rules

We consider fuzzy association rules (which generalize the crisp ones) as a way to find presence dependence relations among the terms of a document set. A group of selected terms from the extracted rules generate a vocabulary related to the search topic that helps the user to refine the query.

The process occurs as follows: before query refinement can be applied, we assume that a retrieval process is performed. The user's initial query generates a set of ranked documents. If the top-ranked documents do not satisfy user's needs, the query improvement process starts. Since we start from the initial set of documents retrieved from a first query, we are dealing with a *local analysis* technique. And, since we just considered the top-ranked documents, we can classify our technique as a *local feedback* one.

From the initial retrieved set of documents, called *local set*, association rules are found and additional terms are suggested to the user in order to refine the query. As we have explained in Sect. 2, there are two general approaches to query refinement:

automatic and semi-automatic. In our case, as we offer to the user a list of terms to add to the query, the system performs a semi-automatic process. Finally, the user selects from that list the terms to be added to the query, and the query process starts again. The whole process is summarized in the following:

Semi-automatic Query Refinement Process Using Association Rules

1. The user queries the system
2. A first set of documents is retrieved
3. From this set, the representation of documents is extracted and association rules are generated
4. Terms that appear in certain rules are shown to the user (subsection 6.1)
5. The user selects those terms more related to her/his needs
6. The selected terms are added to the query, which is used to query the system again

Once the first query is constructed, and the association rules are extracted, we make a selection of rules where the terms of the original query appear. However, the terms of the query can appear in the antecedent or in the consequent of the rule. If a query term appears in the antecedent of a rule, and we consider the terms appearing in the consequent of the rule to expand the query, a generalization of the query will be carried out. Therefore, a generalization of a query gives us a query on the same topic as the original one, but looking for more general information. However, if query terms appear in the consequent of the rule, and we reformulate the query by adding the terms appearing in the antecedent of the rule, then a specialization of the query will be performed, and the precision of the system should increase. The specialization of a query looks for more specific information than the original query but in the same topic. In order to obtain as much documents as possible, terms appearing in both sides of the rules can also be considered.

3.3 *Document Representation for Association Rule Extraction*

From the initial retrieved set of documents, a valid representation for extracting the rules is needed. Different representations of text for association rules extraction can be found in the literature: bag of words, indexing keywords, term taxonomy and multi-term text phrases [12]. In our case, we use automatic indexing techniques coming from Information Retrieval [34] to obtain *word items*, that is, single words appearing in a document where stop-list and/or stemming processes can be applied. Therefore, we represent each document by a set of terms where a weight meaning the presence of the term in the document can be calculated. There are several term weighting schemes to consider [33]. In this work, we study three different weighting schemes [22]:

- *Boolean weighting scheme:* It takes values $\{0, 1\}$ indicating the absence or presence of the word in the document, respectively.
- *Frequency weighting scheme:* It associates to each term a weight meaning the relative frequency of the term in the document. In a fuzzy framework, the normalization of this frequency can be carried out by dividing the number of occurrences of a term in a document by the number of occurrences of the most frequent term in that document [6].
- *TFIDF weighting scheme:* It is a combination of the within-document word frequency (*TF*) and the inverse document frequency (*IDF*). The expressions of these schemes can be found in [33]. We use this scheme in its normalized form in the interval [0, 1] according to [5]. In this scheme, a term that occurs frequently in a document but infrequently in the collection is assigned a high weight.

6 Extraction of Fuzzy Association Rules

In a text framework, we consider each document as a transaction. Let us consider $T_D = \{d_1, \ldots, d_n\}$ as the set of transactions from the collection of documents D, and $I = \{t_1, \ldots, t_m\}$ as the text items obtained from all the representation documents $d_i \in D$ with their membership to the transaction expressed by $W_i = (w_{i1}, \ldots, w_{im})$. On this set of transactions, we extract association rules, discarding those rules below threshold *minconf* (confidence threshold) or *mincf* (certainty factor threshold). We must note that in this process, we do not distinguish the crisp and the fuzzy case. The specific cases will be given by the item weighting scheme that we consider in each case.

We must point out that, as it has been explained in [15, 32], in the applications of mining techniques to text, documents are usually categorized, in the sense of documents which representation is a set of keywords, that is, terms that really describe the content of the document. This means that usually a full text is not considered and its description is not formed by all the words in the document, even without stop words, but also by keywords. The authors justify the use of keywords because of the appearing of useless rules. Some additional commentaries about this problem regarding the poor discriminatory power of frequent terms can be found in [30], where the authors comment the fact that the expanded query may have worst performance than the original one due to the poor discriminatory ability of the added terms.

Therefore, the problem of selecting good terms to be added to the query has two faces. On the one hand, if the terms are not good discriminators, the expansion of the query may not improve the result. But, on the other hand, in dynamic environments or systems where the response-time is important, the application of a pre-processing stage to select good discriminatory terms may not be suitable. In our case, since we are dealing with a problem of query refinement in Internet, information must be shown on-line to the user, so a time constraint is present.

Solutions for both problems can be given. In the first case, discriminatory schemes almost automatic can be used alternatively to a preprocessing stage for selecting the most discriminatory terms. This is the case of the *TFIDF* weighting scheme (see Sect. 5). In the second case, when we work in a dynamic environment, we have to remind that to calculate the term weights following the *TFIDF* scheme, we need to know the presence of a term in the whole collection, which limits in some way its use in dynamic collections, as usually occurs in Internet. Therefore, instead of improving document representation in this situation, we can improve the rule obtaining process. The use of alternative measures of importance and accuracy such as the ones presented in subsection 3.3 is considered in this work in order to avoid the problem of non appropriate rule generation.

6.1 The Selection of Terms for Query Refinement

The extraction of rules is usually guided by several parameters such as the minimum support *(minsupp)*, the minimum value of certainty factor *(mincf)*, and the number of terms in the antecedent and consequent of the rule. Rules with support and certainty factor over the respective thresholds are called *strong rules*. Strong rules identify dependence in the sense of nontrivial inclusion of the set of transactions where each itemset (set of terms in this case) appears. This information is very useful for us in order to refine the query. First, the minimum support restriction ensures that the rules apply to a significant set of documents. Second, the minimum accuracy restriction, though allowing for some exceptions, ensures that the inclusion holds to an important degree.

Once the strong association rules are extracted, the selection of useful terms for query refinement depends on the appearance in antecedent and/or consequent of the terms. Let us suppose that $qterm$ is a term that appears in the query and let $term \in S$, $S_0 \subseteq S$. Some possibilities are the following:

- Rules of the form $term \Rightarrow qterm$ such that $qterm \Rightarrow term$ has low accuracy. This means that the appearance of $term$ in a document "implies" the appearance of $qterm$, but the reciprocal does not hold significantly, i.e., $\Gamma_{term} \subseteq \Gamma_{qterm}$ to some extent. Hence, we could suggest the word $term$ to the user as a way to restrict the set of documents obtained with the new query.
- Rules of the form $S_0 \Rightarrow qterm$ with $S_0 \subseteq S$. We could suggest the set of terms S_0 to the user as a whole, i.e., to add S_0 to the query. This is again uninteresting if the reciprocal is a strong rule.
- Rules of the form $qterm \Rightarrow term$ with $term \in S$ and $term \Rightarrow qterm$ a not strong rule. We could suggest the user to replace $qterm$ by $term$ in order to obtain a set of documents that include the actual set (this is interesting if we are going to perform the query again in the web, since perhaps $qterm$ is more specific that the user intended).
- Strong rules of the form $S_0 \Rightarrow qterm$ or $term \Rightarrow qterm$ such that the reciprocal is also strong. This means co-occurrence of terms in documents. Replacing $qterm$ by S_0 (or $term$) can be useful in order to search for similar documents

where *qterm* does not appear. These rules can be interesting if we are going to perform the query again in Internet, since new documents not previously retrieved and interesting for the user can be obtained by replacing *qterm* by *term*.

7 Experimental Examples

The experiments have been carried out in the web with the search engine *Google* *(http://www.google.es)*. Three different queries have been submitted to the engine, with the search and results in English, namely: *networks*, *learning* and *genetic*.

The purpose of our system is to find additional terms that can modify the query but narrow the set of retrieved documents in most of the cases, and/or improve the retrieval effectiveness. Therefore, if the user has the intention of searching for documents about *genetic* with a Computer Science and an Artificial Intelligence meaning, but she/he does not know more vocabulary related to that concept, the resulting rules can suggest her/him some terms to add to the query. This new query can discard the documents related to other meanings (always that the additional terms are not in the vocabulary of the other meanings).

Once the queries have been submitted to the search engine for the first time, an initial set of documents is retrieved, from which we take the first 100 top-ranked documents. Since we start from the initial set of documents retrieved from a first query, we are dealing with a local analysis technique. And, since we just considered the top-ranked documents, we can classify our technique as a local feedback one. From this local set, a document representation is obtained as in classical information retrieval, and a transformation of this representation into a transactional one is carried out. These transactions are mined for each query to obtain a set of association rules so additional terms can be offered to the user to refine the query. The number of results in each query, the number of text transactions and the number of terms (items) can be seen in Table 1.

It must be remarked the difference in the length of the dimensions of the set of transactions obtained. In traditional data mining, the number of transactions is usually greater while the number of items is lower. In our case it is the opposite, although the goodness of the rules has not to be affected.

The weighted schemes considered are those proposed in subsection 3.3, that is, the boolean, the frequency and the TFIDF weighting scheme. We must point out that the first one is crisp, while the other two are fuzzy values. The threshold of support is established to 2% for the crisp and the frequency case, while for the TFIDF we decide to remove the threshold, since no rules appear with more than a 2% for all

Table 1 Queries with their number of results, transactions and terms

Query	N. Results	N. Transactions	N. Terms
networks	94.200.000	100	839
learning	158.000.000	100	832
genetic	17.500.000	100	756

Table 2 Number of rules for each query with different weighting schemes

Query	Boolean	Norm. Freq.	TFIDF
networks	1118	95	56
learning	296	73	10
genetic	233	77	10

the queries. For the obtention of the rules, we have established a level of the rule of 5, which implies that the number of components appearing in the rule (antecedent and consequent) can not be more than 5 adding both sides of the rule).

The number of rules obtained for each weighting scheme with these thresholds can be seen in Table 2. In this table, we can observe the main advantages of the fuzzy weighting schemes against the crisp case. We must remember that the boolean scheme assigns 0 if the term does not appear in the document, and 1 if the terms appears, no matter how many times. This implies that the importance of a term will be 1 either if the term appears 1 or 20 times in the same document, which does not reflect the real presence of a term in a document. From the point of view of rules, this generates a huge number of them which give not very realistic presence relations among the terms, so they are not very useful for the user.

In the case of the TFIDF case, this scheme assigns a low weight to those items appearing very frequently in the whole collection. When the TFIDF scheme is used, the term query, for instance, *networks* is assigned a weight of 0, since it appears in all the documents of the collection. This means that no rule with the term *networks* will appear in the set of extracted rules in this case. This effect is the same that is obtained with the selection of rules, where high frequent terms are not considered since they do not give new information. However, this lack of new information does not mean that the terms appearing in the same rule as the query term do not help to refine the query to decrease the number of retrieved documents and increase the satisfaction of the user.

The best scheme to analyze cases is the normalized frequency scheme. This scheme assigns a weight to a term meaning the normalized relative frequency of the term in the document, which is more realistic than the boolean scheme but less discriminatory than the TFIDF one. For instance, in the document set retrieved as the answer of query *genetic*, there are documents related to Biology and to Computer Science. If a novel user does not know the vocabulary of the topic, and the intention of the search is looking for *genetic* in the field of Computer Science, rules such as *programming*$\Rightarrow$*genetic*, can suggest to the user a new term, *programming*, in order to add it to the query so the results of the refined query are more suitable to user's needs. This case is of type *term*$\Rightarrow$*qterm*, where the rule *programming*$\Rightarrow$*genetic* holds with a certainty factor of 1 while the opposite rule *genetic*$\Rightarrow$*programming* holds with a certainty factor of 0.013.

Other example in this case is related to the query *learning*. Let us suppose that the user has the intention of searching about learning and the new technologies, but only use the query term *learning* so millions of documents are retrieved by the search engine. Some interesting rules obtained in this case related learning and new technologies are shown in Table 3, where the terms appearing in the antecedent of

Table 3 Confidence/ Certainty Factor values of some rules with the normalized frequency weighting scheme for the query learning

	learning	technology	web	online
learning	-	0.04/0.01	0.06/0.01	0.15/0.03
technology	0.94/0.94	-	-	-
web	0.8/0.79	-	-	-
online	0.79/0.76	-	-	-

the rules are shown in the left column and the terms appearing in the consequent of the rules are shown in the first row of the table.

We can also observe a case of substitution of terms when both $term{\Rightarrow}qterm$ and its reciprocal are strong rules. For instance, with the query of *networks*, the rule *rights$\Rightarrow$reserved* and its reciprocal *reserved$\Rightarrow$rights*, appears with a support of 2.3% and a certainty factor of 1. This means that these two terms are equivalent to be used as additional terms to refine the query.

Regarding the information retrieval effectiveness values, as we add terms to the query, in our experiments the precision increases while the recall decreases. For instance, let us suppose again the example of the user looking for documents related to *genetic* in the field of Computer Science. If the user submits the query with only the term *genetic*, the recall value is 1 while the precision value is of 0.16 in the top-ranked first 100 documents. As the rule *programming$\Rightarrow$genetic* has a support of 6% and a certainty factor of 1, it will be selected to show to the user the term *programming* to be added to the query. With the refined query, the recall decreases to 0.375, but the precision increases to 1.

8 Conclusions

A possible solution to the Information Retrieval problem of query refinement in the web by means of fuzzy association rules has been presented in this paper. The fuzzy framework allows us to represent documents by terms with an associated weight of presence. This representation improves the traditional ones based on binary presence/ausence of terms in the document, since it allows us to distinguish between terms appearing in a document with different frequencies. This representation of documents by weighted terms is transformed into a transactional one, so text rules can be extracted following a mining process.

From all the extracted rules, a selection process is carried out, so only the rules with a high support and certainty factor are chosen. The terms appearing in these rules are shown to the user, so a semi-automatic query refinement process is carried out. As it has been shown in the experimental examples, the refined queries reflect better the user's needs and the retrieval process is improved. The selection of rules and the chance to make the query more general, specific or to change the terms with the same meaning in order to improve the results lead us to consider this approach an useful tool for query refinement.

Acknowledgments This work is supported by the research project Fuzzy-KIM, CICYT TIC2002-04021-C02-02.

References

1. Agrawal, R., Imielinski, T. & Swami, A. "Mining Association Rules between Set of Items in Large Databases". In *Proc. of the 1993 ACM SIGMOD Conference*, 207–216 ,1993.
2. Attar, R. & Fraenkel, A.S. "Local Feedback in Full-Text Retrieval Systems". *Journal of the Association for Computing Machinery 24(3)*:397–417, 1977.
3. Au, W.H. & Chan, K.C.C. "An effective algorithm for discovering fuzzy rules in relational databases". In *Proc. Of IEEE International Conference on Fuzzy Systems*, vol II, 1314–1319, 1998.
4. Bodner, R.C. & Song, F. "Knowledge-based approaches to query expansion in Information Retrieval". In McCalla, G. (Ed.) *Advances in Artificial Intelligence*:146–158. New-York, USA: Springer Verlag, 1996.
5. Bordogna, G., Carrara, P. & Pasi, G. "Fuzzy Approaches to Extend Boolean Information Retrieval". In Bosc., Kacprzyk, J. *Fuzziness in Database Management Systems*, 231–274. Germany: Physica Verlag, 1995.
6. Bordogna, G. & Pasi, G. "A Fuzzy Linguistic Approach Generalizing Boolean Information Retrieval: A Model and Its Evaluation". *Journal of the American Society for Information Science 44(2)*:70–82, 1993.
7. Buckley, C., Salton. G., Allan, J. & Singhal, A. "Automatic Query Expansion using SMART: TREC 3". *Proc. of the 3rd Text Retrieval Conference*, Gaithersburg, Maryland, 1994.
8. Chen, H., Ng, T., Martinez, J. & Schatz, B.R. "A Concept Space Approach to Addressing the Vocabulary Problem in Scientific Information Retrieval: An Experiment on the Worm Community System". *Journal of the American Society for Information Science 48(1)*:17–31, 1997.
9. Croft, W.B. & Thompson, R.H. "I^3R: A new approach to the design of Document Retrieval Systems". *Journal of the American Society for Information Science 38(6)*, 389–404, 1987.
10. Delgado, M., Marín, N., Sánchez, D. & Vila, M.A., "Fuzzy Association Rules: General Model and Applications". *IEEE Transactions on Fuzzy Systems 11*:214–225, 2003a.
11. Delgado, M., Marín, N., Martín-Bautista, M.J., Sánchez, D. & Vila, M.A., "Mining Fuzzy Association Rules: An Overview". *2003 BISC International Workshop on Soft Computing for Internet and Bioinformatics"*, 2003b.
12. Delgado, M., Martín-Bautista, M.J., Sánchez, D. & Vila, M.A. "Mining Text Data: Special Features and Patterns". In *Proc. of EPS Exploratory Workshop on Pattern Detection and Discovery in Data Mining*, London, September 2002a.
13. Delgado, M., Sánchez, D. & Vila, M.A. "Fuzzy cardinality based evaluation of quantified sentences". *International Journal of Approximate Reasoning 23*: 23–66, 2000c.
14. Efthimiadis, E. "Query Expansion". *Annual Review of Information Systems and Technology 31*:121–187, 1996.
15. Feldman, R., Fresko, M., Kinar, Y., Lindell, Y., Liphstat, O., Rajman, M., Schler, Y. & Zamir, O. "Text Mining at the Term Level". In *Proc. of the 2nd European Symposium of Principles of Data Mining and Knowledge Discovery*, 65–73, 1998.
16. Freyne J, Smyth B (2005) Communities, collaboration and cooperation in personalized web search. In Proc. of the 3rd Workshop on Intelligent Techniques for Web Personalization (ITWP'05). Edinburgh, Scotland, UK
17. Fu, L.M. & Shortliffe, E.H. "The application of certainty factors to neural computing for rule discovery". *IEEE Transactions on Neural Networks 11(3)*:647–657, 2000.
18. Gauch, S. & Smith, J.B. "An Expert System for Automatic Query Reformulation". *Journal of the American Society for Information Science 44(3)*:124–136, 1993.

19. Hong, T.P., Kuo, C.S. & Chi, S.C., "Mining association rules from quantitative data." *Intelligent Data Analysis 3*:363–376, 1999.
20. Jiang, M.M., Tseng, S.S. & Tsai, C.J. "Intelligent query agent for structural document databases." *Expert Systems with Applications 17*:105–133, 1999.
21. Kanawati R, Jaczynski M, Trousse B, Andreoli JM (1999) Applying the Broadway recommendation computation approach for implementing a query refinement service in the CBKB meta search engine. In Proc. of the French Conference of CBR (RaPC99), Palaiseau, France
22. Kraft, D.H., Martín-Bautista, M.J., Chen, J. & Sánchez, D., "Rules and fuzzy rules in text: concept, extraction and usage". *International Journal of Approximate Reasoning 34*, 145–161, 2003.
23. Korfhage RR (1997) Information Storage and Retrieval. John Wiley & Sons, New York
24. Kuok, C.-M., Fu, A. & Wong, M. H., "Mining fuzzy association rules in databases," *SIGMOD Record 27(1)*:41–46, 1998.
25. Lee, J.H. & Kwang, H.L., "An extension of association rules using fuzzy sets". In *Proc. of IFSA'97*, Prague, Czech Republic, 1997.
26. Lin, S.H., Shih, C.S., Chen, M.C., Ho, J.M., Ko, M.T., Huang, Y.M. "Extracting Classification Knowledge of Internet Documents with Mining Term Associations: A Semantic Approach". In *Proc. of ACM/SIGIR'98*, 241–249. Melbourne, Australia, 1998.
27. Miller, G. "WordNet: An on-line lexical database". *International Journal of Lexicography 3(4)*:235–312, 1990.
28. Mitra, M., Singhal, A. & Buckley, C. "Improving Automatic Query Expansion". In *Proc. Of ACM SIGIR*, 206–214. Melbourne, Australia, 1998.
29. Moliniari, A. & Pasi, G. "A fuzzy representation of HTML documents for information retrieval system." *Proceedings of the fifth IEEE International Conference on Fuzzy Systems, vol. I*, pp. 107–112. New Orleans, EEUU, 1996.
30. Peat, H.P. & Willet, P. "The limitations of term co-occurrence data for query expansion in document retrieval systems". *Journal of the American Society for Information Science 42(5)*,378–383, 1991.
31. Qui, Y. & Frei, H.P. "Concept Based Query Expansion". In *Proc. Of the Sixteenth Annual International ACM-SIGIR'93 Conference on Research and Development in Information Retrieval*, 160–169, 1993.
32. Rajman, M. & Besançon, R. "Text Mining: Natural Language Techniques and Text Mining Applications". In *Proc. of the 3rd International Conference on Database Semantics (DS-7)*. Chapam & Hall IFIP Proceedings serie, 1997.
33. Salton, G. & Buckley, C. "Term weighting approaches in automatic text retrieval". *Information Processing and Management 24(5)*, 513–523, 1988.
34. Salton, G. & McGill, M.J. *Introduction to Modern Information Retrieval*. McGraw-Hill, 1983.
35. Srinivasan, P., Ruiz, M.E., Kraft, D.H. & Chen, J. "Vocabulary mining for information retrieval: rough sets and fuzzy sets". *Information Processing and Management 37*:15–38, 2001.
36. Van Rijsbergen, C.J., Harper, D.J. & Porter, M.F. "The selection of good search terms". *Information Processing and Management 17*:77–91, 1981.
37. Vélez, B., Weiss, R., Sheldon, M.A. & Gifford, D.K. "Fast and Effective Query Refinement". In *Proc. Of the 20th ACM Conference on Research and Development in Information Retrieval (SIGIR'97)*. Philadelphia, Pennsylvania, 1997.
38. Voorhees, E. "Query expansion using lexical-semantic relations. *Proc. of the 17th International Conference on Research and Development in Information Retrieval (SIGIR)*. Dublin, Ireland, July, 1994.
39. Xu, J. & Croft, W.B. "Query Expansion Using Local and Global Document Analysis". In *Proc. of the Nineteenth Annual International ACM SIGIR Conference on Research and Development in Information Retrieval*, 4–11, 1996.
40. Zadeh, L.A. "A computational approach to fuzzy quantifiers in natural languages". *Computing and Mathematics with Applications 9(1)*:149–184, 1983.

Fuzzy Logic in a Postmodern Era

Mory M. Ghomshei, John A. Meech and Reza Naderi

Abstract An event is a spatio-temporally localizable occurrence. Each event in our universe can be defined within a two-dimensional space in which one dimension is causality and the other is serendipity. Unfortunately, the majority of scientists in the Modern Era in their fascination with rules of causality and wanting to believe in a complete deterministic expression of the universe have banished all notions of serendipity to the realm of fiction, religion and/or, the occult. But the hegemony of Newtonian causality finally crumbled under the gravity of Heisenberg's Uncertainty Principle which demonstrated that an external observer can never acquire enough information to fully express the state of a system. This was a quantum physical expression of what was later eloquently put by Heidegger in his philosophical definition of Ereignis, to designate an unpredictable and often uncontrollable disruption of the spatio-temporal causal continuity. In the Postmodern Era, when events such as 9/11 occur beyond any assessable realm of causal relationships, we can no longer afford to discard the serendipity component of events if we wish to understand with clarity. Instead we must devise rules of conformity between the causal and non-causal fields of reality. Fuzzy Logic provides such a vigorous system of thinking that can lead us to this accord. This paper uses the tools of Fuzzy Logic to find pathways for events taking place within a causal-serendipity space. As a first approach, an event is defined on a hyperbolic path in which the degree of serendipity multiplied by the degree of causality is constant. This allows for the diminution of serendipity as scientific knowledge about a subject increases and the enhancement of serendipity to become dominant when data are scarce or measurements uncertain. The technique is applied to several different types of causality – direct, chain-like, parallel, and accumulation.

1 Introduction

Modernity (the 19th and most of the 20th century) has been overwhelmed with acquisition of knowledge that was defined as our only connection to the so-called "absolute" realities of the universe (as separate from the human mind). In other words, the mind of the scientist together with other scientific instruments is considered a non-intervening probe into the realm of reality. Post-Modern philosophers who have

M. Nikravesh et al. (eds.), *Forging the New Frontiers: Fuzzy Pioneers II.*

their roots of thinking in Kant and more recently, Heideger, and Kierkegard, have disengaged themselves from the concept of an external existing "absolute" reality and instead, have concerned themselves with knowledge as it exists in the human mind. Derrida, Lyotard, and Foucault are some of the most eminent Post-Modern philosophers who, though rejecting many of each other's views, are all skeptical of "absolute" or universal truth-claims. Modernity's dogmatic view on causality and deterministic exposition of the universe comes from this Aristotelian absolutism.

By the end of the Modern era, human interaction with the universe has became so complex that simplistic, deterministic models based on absolutism fail to provide satisfactory or even, satisfying results. The birth of Fuzzy Logic was a necessity of our time first sensed by engineers, rather than philosophers, since they were in immediate contact with evolving complexity in human need. Fuzzy Logic was first used in control systems, where the perception of human comfort interacts with external physical realities. The approximation of a response surface became a "good-enough" solution to provide adaptable control that out-performed exact mathematical functions when applied to a real-world problem. But even Zadeh, when he first proposed this logic believed that the largest impact of his ideas would be found in the field of the "soft" sciences – natural language programming and understanding, linguistics, automated translation, technology-transfer, and philosophy.

It has taken considerable time and effort for Fuzzy Logic to be accepted by the scientific world. For many years, Zadeh and his disciples were viewed by the mainstream scientific and engineering community as pariahs preaching folk-art. But as the number of followers has continued to grow, as the numbers of successful applications has steadily increased, and as so many other fields of science and technology have begun to grasp the truthfulness of fuzzy-thinking and its intrinsic ability to solve problems for which the "hard" sciences cannot offer global solutions, the nay-sayers have fallen by the way-side and Fuzzy Logic now sits at the pinnacle of all other forms of logic.

This paper is an attempt to look at Fuzzy Logic as an instrument to deal with practical philosophical issues in an era in which fuzzy-thinking is now the norm rather than an exception.

2 Philosophical Overview

What is the relationship between necessity on one hand and possibility and actuality on the other? The basis for causality is that one thing is necessary to make the coming to existence of another thing. That other thing that comes to existence based on a determinant cause must have been something that was possible all along. Of course, something impossible (e.g., a contradiction) cannot come to existence. Hence we are talking about a cause necessitating the coming to existence of something possible (i.e., making something an actuality that was previously possible). So, causality is, in fact, the transformation of possibility to actuality, and cause is something that changes a "possible" to an "actual".

Having established a framework to examine causality, we briefly review the relationship between cause and effect, first from a philosophical perspective and then, from a mathematical point of view. The first serious attempt to explain the causal relationship was offered by Hume. Hume's argument goes as follows: The idea of necessity so far as it is revealed to our mind is always ascribed to the relation between causes and effects. There are two things that can be perceived about two objects in a casual relationship, that they are *contiguous* in time and place, and that the object we call the "cause" *precedes* the other that we call the "effect" (Hume, 1987). Hume established that reason, as distinguished from experience, can not make us conclude that a cause is absolutely requisite to the beginning of the existence of a thing.

According to Hume (1987), by merely observing the "efficacy" of a relation between two objects, there is no way to discern cause from effect. It is only after we observe very many of the same instances in which the same objects are conjoined together that we begin to name one object the cause and the other the effect. However by witnessing them repeatedly in the past, the natural propensity of our imagination expects that we will see them repeated together in the future. Hence, necessity does not belong to the structure of the objective reality. Rather, it belongs to the human mind.

Through this interpretation, we only know passively about necessity and efficacy which struck Kant in two different ways. First, as this interpretation is related to metaphysics, it "interrupted Kant's dogmatic slumber". Secondly, as this interpretation relates to fields of scientific theory and morality, it was more like an alarming skeptical voice needing a response. Why was Hume's skeptic attack on necessity and efficacy alarming to Kant? The main difference between the worldview of all forms of empiricism (including skepticism) and that of Kant is that to empirical thought, the world is given, while for Kant, it is to be created.

Since Kant, we now have two explanations for the nature of a causal relationship, i.e. two attempts to understand, i.e., explain, how a cause is related to its effect–that offered by the empiricists and that by Kant. No explanation is offered by anyone else, and as stated above, their relation is utterly shrouded in mystery.

3 Cause, Effect, Change, and Necessity

Returning to the basic idea of causality as a process that brings the possible into existence, a more fundamental question arises: What changes take place when a possible thing becomes real (actual)? It is known since Aristotle that to rationalize change, we must speak of something that persists throughout that change. To say otherwise about change is nonsensical. The persistent element, the substance, or the essence of the thing that changes, remains unchanged. Hence coming into existence (or actuality) does not change the essence of a thing. Hence change is not in the essence of a thing, but rather is in its being (Kierkegaard, 1985). If "possible", by virtue of having a "cause", becomes necessary to "exist", what is the necessary change in the

Fig. 1 Linguistics in "cause and effect" philosophy

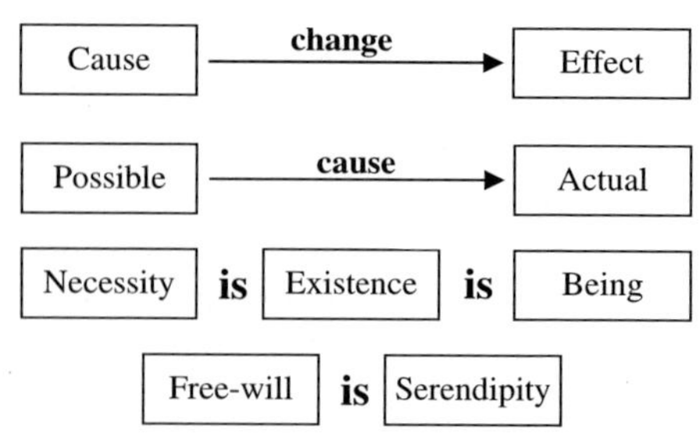

possible that brings it into existence? To examine this question, Kierkegaard asks: Can necessary come into existence?

Necessary cannot "not exist", or it is a contradiction. Necessary always *is*. So necessary cannot come into existence, because this would mean it was not necessary before. "Being" is the essence of necessary (i.e., necessary means necessary to be). We indicate that something that has come into existence (actual) and something that has the potential to come into existence (possible) are the same (i.e., they have the same essence). This essence cannot be "Being", or it would already exist and does not have to come into existence. Hence "necessary" is essentially different from "possible" and "actual", whereas "possible" and "actual" are "essentially" the same thing. What does this mean? If "possible" becomes "actual", it cannot be because actual becomes a necessity in the possible. Nothing can become necessary or it is a contradiction. Nothing comes to existence by way of necessity, and neither can necessity come into existence nor can something by coming into existence, be necessary. No event exists because it is necessary, but necessary exists simply because it is necessary.

Actual is no more necessary than Possible, for necessary is absolutely different from both (Kierkegaard, 1985). If coming to existence (i.e. actualization of the possible), is not necessary, then how does it occur? The simple answer is: "coming into existence occurs in freedom, not by way of necessity" (Kierkegaard, 1985). We can assume there is a degree of liberty (free-will) in any causal relation. From an observer's viewpoint, this degree may be insignificant in some cases (e.g. Newtonian mechanics), but very significant in others (e.g., human relations).

4 Causality: Transition from Modern to Post-Modern

The systemization of human knowledge has been an effort undertaken by philosophy from the early Greek. This systematization achieved significance as achievements in scientific theory and practice gained momentum at the dawn of industrialization in Europe. As exact sciences portrayed a clockwork vision of the world, the philosophical reaction was equally embodied in an effort to portray a model of human knowledge and understanding analogous to the precision of the exact sciences. The sum total of the vision attained by many great achievements in scientific theory and practice is a culture we have lived with through the era referred to as

Modernity - a culture marked by its absolutism and idealogism that Lyotard (1985) famously referred to as *grand narratives*. Before the so-called Post-Modern era, the rejection of grand narratives is an effort shared by many philosophers such as Nietzsche and Heidegger who helped shape the transition from Modernity to Post-Modernism.

Nietzsche's famous criticism of the sanctity of reason is most eloquently captured in his characterization of the modern philosophical lack of historical sense as a form of archaic "Egyptianism" (Nietzsche, 1968). With this metaphor Nietzsche referred to the mummifying tendency of modern philosophers to consider all things only in terms of abstract concepts taken out of real experience. These philosophers reduce the whole realm of change, transience, and coming into a fantasy of pure Being *in-itself*. Reason is the cause of our falsification of the evidence of the senses (Nietzsche, 1968). With this, Nietzsche achieved a reversal of the traditional Western philosophical valuation of reason over the senses. Western philosophical tradition has been largely based on the skeptical devaluation of the senses, and of the so-called *apparent world*. The task of reason was always to keep the prudent philosopher from falling prey to errors of perception. Thus the realm of philosophical inquiry set itself apart from that of theology, the former using reason as its main organ and the latter relying on faith. By debunking reason, Nietzsche reset faith or *passion* (to use less religious language) as the main tool in the search for truth. Kierkegaard (1992) achieved a similar end when he underlined the problems with Descartes' proof for the existence of God.

The relevance of faith (or passion) to our discussion is its intervention in our perceptions. The transition from computing with numbers to computing with perceptions, as put forward by Zadeh (2003), teaches us to understand how faith or passion affects our perceptions. The conventional AI community has mimicked the human mind by considering things as binary in nature, i.e., either true or false. This approach derived from the use of binary systems as the fundamental building block to program a computer. But the AI community is naïve if it assumes that a person's faith belongs only to their transcendental experience (such as God and religion) and has nothing to do with their mundane perceptions. In a Post-Modern era, we should not isolate the scientist from faith, especially when we wish to understand complex events perpetrated by one person on another. The true perception of a scientist is an integral of the entire universe of human experience from mundane to sublime and everything in between.

By replacing the traditional metaphysical opposition to subject and object with the unified phenomena of being-in-the-world, Heidegger places reason and the theoretical stance towards the world as a derivative mode of Being. What has been understood as Heidegger's so-called *pragmatism* is a testimony to the human being's (Dasein[1]) unity with the world. The human being is essentially cognitively ambivalent to the world. Knowledge, in the cognitive sense, is only acquired when the fundamental pragmatic relationship with the world is disrupted. Thus, for Heidegger,

---

[1] Heidegger's term for "being-there" the existence that self-conscious human beings uniquely possess.

reason, knowledge, and the scientific outlook on the world are essential derivatives of a primordially pragmatic relationship with the world.

So it can be said that for Heidegger, the reality of human practice in the context of everyday doing/making is a fundamental element of the reality of human consciousness. As a result, human actions (including "perceptions") precede theoretical speculation of which it is fundamentally independent. It was a mistake of the Modern mind to posit reason as the foundation of action. Heidegger put it eloquently when he said: "in the most hidden ground of his essence, man truly is, only when, in his way, he is like the rose – without why" (Heidegger, 1996).

The deconstruction of metaphysics begun by Heidegger influenced Derrida immensely. In 1983, he chose the term, *déconstruction*, as his attempt to translate and re-appropriate the Heideggerian terms *Destruktion* and *Abbau* using a word from the French language. Derrida's writings question the authority of philosophy. According to him, the essence of the metaphysical (or physical) discourse depends on the validity in establishing intent through the text upon which it is based. But can such intent be completely unambiguous in any discourse? This is the question that led Derrida (1967a) in his effort to rethink the correspondence between object and name in both phenomenological and structural contexts.

Derrida's deconstruction is grounded in the transition heralded by structuralism from the traditional Augustinian view of language. According to Augustine, language acquires meaning through a one-to-one relationship between a signifier (a word) and a signified (an object in the external world). This model of language took a significant fall with the advent of the structural linguistics developed by Saussure who stated that a given signifier acquires meaning with respect to its difference from other signifiers in the language. Thus - to the interest of the AI community - the word *orange* does not mean what is meant when referring to a specific visual phenomenon in the external world, but rather, by its difference and similarity to other signifiers such as *red, green* and *blue*[2].

Lacan (1997) radicalized this view by positing an impenetrable *wall* between "signifiers" and "signifieds". There is no interaction between the realm of signifiers and those objects in the external world to which they are assumed to refer. For Lacan, language sustains its integrity by virtue of the existence of a certain *privileged* signifier maintaining an intimate tie with the external world and thereby prevents other signifiers from falling into meaninglessness. This privileged signifier prevents a subject from becoming psychotic use of language where words become mere objects in themselves and lose all power of representation (Lacan, 1997). Derrida further extended Lacan's thesis about the fundamental separation of the realm of signifiers from signifieds by emphasizing the instrumental role of "context", beyond which nothing can be defined (e.g., there is nothing that is simply cure or simply poison). In other words, context refuses to settle down as the mere *appearance* of *true* knowledge (Derrida, 1967b).

---

[2] This is why we can speak of colours without having any guarantee whatsoever that what one subject means by "orange" is at all *actually* the same as what another subject understands by the term. What really matters is that the *relation* between one colour and another is the same for all.

Perhaps, the Post-Modern era can best be thought of as a time that has awakened to the recognition and appreciation of the richness and fullness of the degree of density of all aspects of the human experience (social, scientific, cultural, linguistic, religious, etc.). Embracing this richness by means that allow an unmitigated or distorted representation is an undertaking that can transform many (if not all) aspects of our lives. The Modern era was built on an understanding of the world through which the God-made and Man-made worlds were governed by absolute timeless universal laws, and the efforts of human beings in the Modern era has always focused on approximating the "laws of nature". The agenda in the Post-Modern era was mistakenly taken as debunking the hope of discovering new universal causal laws. That is incorrect. Rather, the new agenda is to debunk the belief that universal causal laws are the *only* dimension on which reality can be modeled, explained, predicted, and/or justified.

The idea that the source of events is not merely definable by a measurable and traceable chain of phenomena is not new to philosophy. Kant realized that philosophy must account for freedom of choice as another dimension for the origin of actions. But if freedom of choice itself is caused by a chain of measurable and traceable phenomena, then there is not much left for freedom of choice as they are completely consumed by the necessity imposed by the preceding chain of events. Kant approached this issue in his categorization of the third antinomy of pure reason. His solution was to project freedom into a separate realm from that in which causality operates. Kant's model maps causality as a fundamental rule of our understanding of the realm of appearances, or phenomena that are ordinary experiences, whereas freedom is mapped on the realm of things in themselves, or the so-called, noumena. World events can only be defined through such a model in which freedom has the power to act despite the causes that yield appearances within which such freedom must act. In such cases, freedom of choice causes a brand new chain of events unrelated to what preceded in the world of appearance. In a more romantic sense, our world is being constantly re-set by totally unpredictable events.

The realm in which freedom acts is referred to as the intelligible cause as opposed to causality by appearances. Thus the intelligible cause, with its causality, is outside the apparent chain of events; while its effects, on the contrary, are encountered in a series of empirical conditions (Kant, 1997). The intelligible cause, as a dimension of defining empirical events, is totally inaccessible by *understanding* and is beyond measurable and traceable chains of phenomena. Kant believes that a free agent stands totally outside the condition of time, for time is only a condition of appearance, but not of things in themselves.

Kant's argument was further radicalized by Kierkegaard (1985) in his treatment of possibility, actuality and necessity. In Kierkegaard's view, "necessity" which is the logical category under which efficacy falls, is totally unable to explain why a "possible" becomes an "actual". To Kierkegaard, any phenomenon of coming into existence is caused by a free act that emphasizes the importance of a coefficient of freedom in this regard: all types of "coming into existence" occur in "freedom", not because of "necessity".

Two major philosophical trends in the 20th Century were heavily influenced by Kant's model with some reservations with the dualist view latent in Kant's metaphysics–phenomenology and structuralism respectively. Phenomenology in Heidegger's view, took questioning of the definition of events to a deeper lever in which the question was replaced by that of "Being a human" in the world. In this view what we used to call a phenomenon is only a secondary and less significant question. A phenomenon is an external manifestation of what Heidegger refers to as *"Ereignis"*, "enowning or presencing", through which "Dasein" experience and practice being and acting. It is not a case of talking "about" something and representing something objective, but rather of being owned over into *"Ereignis"*. This amounts to a transformation of the human from "rational animal" to Da-sein (Heidegger, 1999). Heidegger's contribution to defining phenomena through causal relations is a pragmatic answer wherein, it depends on the state of a Dasein's Being. In this view, relations between cause and effect can never be taken out of the context.

Structuralism on the other hand, approached event interactions in a different way. In any significant human creation such as language, culture, and society, the best way to capture and represent the interrelationships and interdependencies of the elements is not explained by causality, but rather through a mutual effect exerted by substructures on each other. In this nexus, the efficacy relationship is minor and secondary, and has very local/tactical importance. Language is the ultimate model for a structural relationship and the analysis of language ultimately reveals how other structures such as society and culture hold up through mutual interactions of their substructures. Structures find meaning in an interdependency, wherein they lean upon each other as do cards in a house of cards.

Derrida single-handedly delivered a substantial criticism of structuralism and phenomenology, and so, causation found yet another philosophical interpretation. He did not agree with structuralism that language has an arch or phallus upon which an objective evaluation of meaning is rendered. Hence the house of cards is destructible in a moment's notice. He believed the only thing rendering meaning to signifiers is the context within which signifiers are placed. As such, meaning is completely relative to the context. There is only context and nothing else. This is the end of any Objectivism upon which a cause can be objectively and unambiguously related to an effect. A cause and effect relationship can be only represented, understood, and acted upon within a given context.

5 Non-linearity in a Cause and Effect System

Causality is one of the most important concepts that have concerned humanity from the beginning of our intellectual endeavours. Emphasizing the central role of causality in the human mind, Zadeh (2003) attempted to bring the subject from its traditional high-heavens of philosophical debate to the ground-zero of mathematical treatment. In his view, human knowledge occupies a wide spectrum beginning with a crisp end wherein concepts are arithmetically-definable to an amorphous end that passes through a fuzzy universe.

During the era of Modernity (now being overlapped by Post-Modernity) we were so fascinated by linear relations between cause and effect that we rejected or ridiculed domains of knowledge that did not conform to this assumption. Predictions of events were based on a linear causal relationship, a system input by a series of discernable causes and output by a series of measurable effects. A non-linear causal relation, on the contrary, is explained by a system that is inputted not only by discernable causes, but also by certain non-discernable factors. In mathematical terms, a linear cause and effect system would be:

$$E_i = \Sigma S * C_j \tag{1}$$

where "S" is the system, C is the causal input matrix to the system, and E is the output effect matrix.

A non-linear system can be expressed by:

$$E_i = \Sigma S * C_j + s \tag{2}$$

where "s" stands for all factors that may contribute to the effects but are beyond the perceived system of causality (S). We call this set of non-linearity elements "serendipity". Note that philosophers can refer to this as "free-will", alluding to a concept that there are elements of "absolute free-will" in the universal system that act beyond the rules of causality.

Some may link the "s" elements to trans-causal domains, while others may call it the "ignorance elements". We do not attempt to link these to any transcendental field, as this leads to a debate beyond the purpose of this paper. Rather, we acknowledge the existence of "s" factors (without being sarcastic) and try to incorporate these phenomena into our predictive assessment of reality by providing a context to express observed non-linearities in some cause-effect systems.

6 Context and Tools

First, we must provide some tools to assess the magnitude (or importance) of the "s" elements in a predictive model. This can be done empirically from a study of historical events (i.e., similar events that have happened in the past). As the nature of "s" is totally unknown, we can legitimately assume that it is independent of our perceived system of causality (S). This independence is not necessarily in the domain of "realities" that exist beyond human perception. Rather, it has meaning only in the context of our perceptions of reality. Note that we cannot make judgments beyond the context of our perceptions. In other words, "pure" reality is beyond us. This takes us to the Post-Modern idea that "there is no reality beyond the context" as frantically put by the late Post-Modern guru, Jacques Derrida.

Setting aside the debate on context and reality, independence of the "s" factors from the causal system provides a new context for communication. In mathematical terms, this can be defined as a two dimensional coordinate system in which the orthogonal axes represent causality C and serendipity s. Every event is identified by a point on this 2-D space. Events that lie close to the C axis are dominated by linear causal factors that can be identified and factored into a prediction model. Those events in the proximity of the s axis are serendipitous (i.e., not controlled by an identifiable set of causes).

The history of past events can be viewed within this context to provide us with pathways for future events. One of these paths would be related to empirically-accumulating knowledge about certain events. Increased knowledge will move the event along a hyperbolic curve, towards the C axis. In mathematical terms, our knowledge about an event moves along a curve in which causality times serendipity is constant (or relatively so):

$$C \cdot s = K \tag{3}$$

The graph of C versus s is an inverse relationship with its position dependent on the value of K. The higher the K value, the farther the curve lies from the origin. Different pathways can be defined for different types of events. Higher values of K mean that the event is more amorphous, while lower values of K lead to crisp events. Examples of crisp types of events are natural phenomena such as earthquakes (close to the serendipity axis and virtually unpredictable at meaningful time scales) or rain (close to the causality axis in which our ability to accurately predict such weather has grown enormously over the past century).

Zadeh (2003) believes causality is totally amorphous as

> *there does not exist a mathematically precise definition of causality such that, given any two events A and B, a definition can be used to answer questions such as: 'is there a causal connection between A and B?' and 'what is the strength of the causal relation?'*

In our model of events, we chose not to link amorphicity directly to causality (defined as an independent component of events). Rather, amorphicity comes from a 2-D space where one dimension defines the degree of causality, while the other defines the degree of serendipity. In our model, a causal relation can be identified, if it can be separated from its serendipitous components. The strength of a causal relationship can be mathematically defined once the other dimension (i.e., serendipity) is accounted for. The model provides a mathematical tool to characterize an amorphous event by segregating causality from serendipity. Note that in our model, amorphicity has a generalized 2-D definition and we do not attribute it totally to the serendipitous component alone. Rather, the model treats amorphicity as a **measure of complexity**. Those event categories that are more complex (i.e., more advanced in the causality-serendipity metric) are naturally more shapeless. Even with an event 100% determined by causality, as the degree of complexity advances, a degree of amorphicity appears.

The human mind makes a complex thing amorphous in order to develop a general feeling about the event. On the serendipity-causality domain, a "ghost" is perhaps, the ultimate amorphous picture because of the complexity of our mind's interaction with a ghost's reality (whatever that may be!). Shapelessness comes from a high degree of complexity where serendipity interacts with causality. If things are totally serendipitous or totally causal, they are not so shapeless and the model can

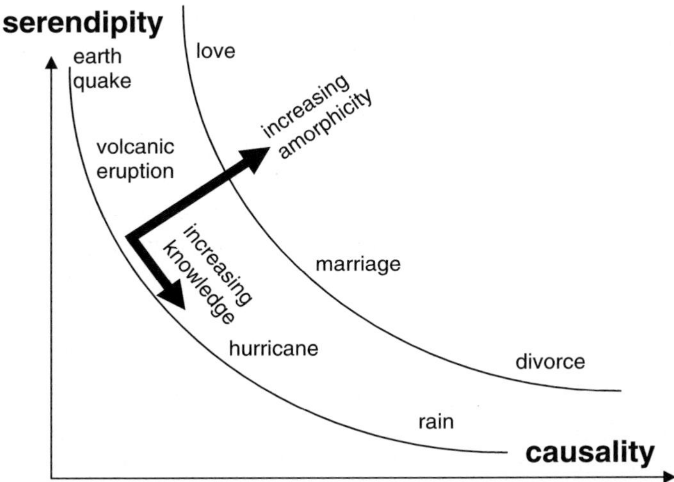

Fig. 2 Crisp events, such as those that occur in nature, are less amorphous. Fuzzy events, such as those of human relations, are more amorphous

mathematically deal with these situations. When causality and serendipity both increase, their interaction creates complexity that we call shapeless (or amorphous).

A totally serendipitous event with low amorphicity is "belief in other life-forms in the universe". Examples of totally serendipitous events with high amorphicity could be "belief in a God" or "belief in Intelligent Design or the Theory of Creativity".

A philosophical discussion of whether an event takes place by causality or free-will does not really enter this model. Events can be considered serendipitous because of lack of knowledge about the causal factors or because of a lack of known causal factors with the existent event. So if we cannot delineate the causal factors, we call the event serendipitous.

With amorphous events, causes cannot be identified with full clarity. In other words, as the subject of discussion or interest moves from the mind of science to that of humanity, events become more amorphous. While a probabilistic approach may apply for low values of K, with an amorphous (high K-value) event, a fuzzy logic approach will be more flexible in dealing with the extreme complexities of the event. Examples of amorphous (high K) events are human relations such as love, marriage and divorce (Fig. 2) in which the degree of complexity (or amorphicity) in all of these is high but the causal component becomes more relevant, as we move from love (which is mainly serendipitous) towards marriage and divorce.

7 Amorphicity of Fuzzy Logic

One of the important attributes of Fuzzy Logic is its intrinsic low level of information (or amorphicity) within a particular context. This provides a degree of liberty for other intervening parameters including free-will. One of the attributes of human language is its inherent elasticity and/or plasticity that provide a tolerating space for complex human inference in text interpretation. Modernity placed a great deal of importance on maximizing the information density in expressing a so-called reality. The Post-Modern trend gives more importance to the context rather than to the reality itself. Note that the context belongs to each of us as observers or participants, while the reality is typically beyond us. The most extreme side of the Post-Modern spectrum considers only the context as relevant; in Derrida's famous statement that "il n'y a pas de hors-texte" (there is nothing beyond the text). This emphasis on context instead of reality is necessary in human communication (both between each other and with reality). When we put too many information constraints on a phenomenon, we deprive it from the liberty needed to transition from possibility to reality. Heisenberg's Uncertainty Principle is compatible with this transition. In a 100% predictable system, nothing will happen (i.e., electrons cannot move) because of a lack of tolerance for unpredictable elements that are necessary in any phenomena. More liberty is naturally needed for human possibilities to become realities. Thus it is with good reason that human language is not crisp. Obviously, the degree of amorphicity increases with increasing human intervention in an event. The degree of amorphicity is further increased as we move from body to soul (the language of love being more amorphous that the language of economics).

8 Post-Modernity Appreciates Amorphicity

Modernity placed too much value on simple binary oppositions dominant in 19th Century and early 20th Century thinking to the degree that philosophers and scientists isolated knowledge from ignorance, or correct from incorrect, and presence from absence. Post-Modernity began with skepticism towards this duality. Note that this denial of a duality is not the same as the Relativism of the early 20th Century. While Relativism abandoned authority (or absolutism), the Post-Modern view denies duality thus creating an opening towards the meaning of authority or absolute. Modern philosophers saw the "cause" as the authority (or absolute). This Aristotelian confrontation with reality persisted throughout the Middle Ages, and into the Renaissance, eventually becoming central to the Modern Western scientific attitude.

9 Epilogue

Rapid changes that our Modern world has undergone during the last decades of the 20th Century are dramatically affecting all dimensions of human activity on earth. Societal norms are breaking, intellectual frameworks are deforming, and our relationship with nature is being completely re-defined. All these changes suggest that a second Renaissance is being shaped, or at least, is being felt all around us. This second Renaissance, has been given a name, before it has been defined. Nobody knows where Post-Modernity is leading us.

But it is clear that Modernity is running out of ammunition in its sacred war against uncertainty. So, our "Brave New World" is better off to concede and come to terms with uncertainty. To do so, we must respect the unknown and understand (or even celebrate) amorphicity. We must find new tools to probe the "unseen" and the "irrational". We must use all our faculties, even those that go beyond reason, to find solutions for complex problems of our time. Fuzzy Logic, although still in its infancy, has the modesty to be a beginning. In a philosophical scope, Fuzzy Logic should be seriously explored, and its abilities re-examined to find the new metric needed to challenge the complexities of our Post-Modern framework.

An important question to ponder:

Can freewill and ethics exist under the supreme control of causality?

If we believe in an ethical society then we must find a way to free ourselves from the chains of causality and let freewill shoulder part of the blame particularly when wrongful events occur. The demonic supremacy of causality is graphically demonstrated in **Othello:** Act V; Scene 2 when the Moor convinces himself to kill Desdemona despite his love for her:

It is the cause,
It is the cause, my soul
Let me not name it to you, you chaste stars!
It is the cause.

William Shakespeare

In Praise of Fuzzy Logic

War and conflict are commonly blamed on causality while Peace is the legitimate child of freewill and choice. The arithmetic of an ethical society is "Fuzzy Logic" which resembles geometry, since it provides space for freewill and choice to maneuver with causality within the progress of events.

My friend is busy doing regression of fools
He counts the peas in his empty plate
Peas and beans do not correlate this late
By night, forget this arithmetic of ghouls
Follow the geometry of the moon and integrate.

Mory Ghomshei

References

J. Derrida, 1967a. Structure, Sign and Play in Discourse of the Human Sciences, in Writing and Difference, University of Chicago Press, 1978.

J. Derrida, 1967b. Difference, in Speech and Phenomena, Northwest University Press, Illinois.

M. Heidegger, 1996. The Principle of Reason. Translated by Reginal dLily, Indiana University Press, p. 148.

M. Heidegger, 1999. Contributions to Philosophy–Vom Ereignis, Indiana University Press.

D. Hume, 1987. Treatise of Human Nature, Book One, edited by D.G.C. MacNabb, Fontana/Collins, 7th Ed., p. 283.

I. Kant, 1997. Critique of Pure Reason, Cambridge Press, p.785.

S. Kierkegaard, 1985. Philosophical Fragments, edited and translated by Howard V. Hong and Edna H. Hong with Introduction and Notes, Princeton University Press, p. 371.

S. Kierkegaard, 1992. Concluding Unscientific Postscript to Philosophical Fragments, edited and translated by Howard V. Hong and Edna H. Hong with Introduction and Notes, Princeton University Press, p. 371.

J. Lacan, 1997. The Function of Language in Psychoanalysis, Johns Hopkins University Press, Baltimore and London.

J.F. Lyotard, 1985. The Post-Modern Condition: A Report on Knowledge, University of Minnesota Press, p. 110.

F. Nietzsche, 1968. Twilight of the Idols, The Portable Nietzsche, edited and translated by Walter Kaufmann, Penguin Books Ltd., London, p. 692.

L. Zadeh, 2003. Causality is indefinable–toward a theory of hierarchical definability. in Intelligence in a Materials World: selected papers from IPMM-2001, CRC Press, NY, 237.

Implementation of Fuzzy Set Theory in the Field of Energy Processes Applied in Nuclear Power Technology

Traichel Anke, Kästner Wolfgang, Hampel and Rainer

Abstract The paper deals with application of a special Fuzzy Model with external dynamics for observation of nonlinear processes in Nuclear Power Plants (NPP) by example in the kind of an general overview. The application is the water level measurement in pressure vessels with water-steam mixture during accidental conditions, like pressure drops. Functionality of this fuzzy model will be demonstrated for a basic structure.

Key words: Fuzzy Set Theory · Model Based Measuring Methods · Fuzzy Model with External Dynamics · Pressure Vessel · Hydostatic Water Level Measurement

1 Introduction

The continuous monitoring and diagnosis of actual process state specially in the case of arising accidental conditions as well as during accidents in NPP's can be improved by application of intelligent methods and high performance algorithms of signal processing. Different kinds of such methods can be established as analytical redundancies (observer) like Model-based Methods and Knowledge-based Methods such as Fuzzy Set Theory. Figure 1 shows a general conception for observing process parameters by combined classical and fuzzy Model-based Measuring Methods (MMM) [1].

The features of safety-related processes in nuclear science like complexity, nonlinearity, and dynamics often require application of fuzzy models. The presented system for water level monitoring demonstrates that fuzzy models can be used advantageously in connection with classical algorithms, particularly as a higher quality resolution.

M. Nikravesh et al. (eds.), *Forging the New Frontiers: Fuzzy Pioneers II.*

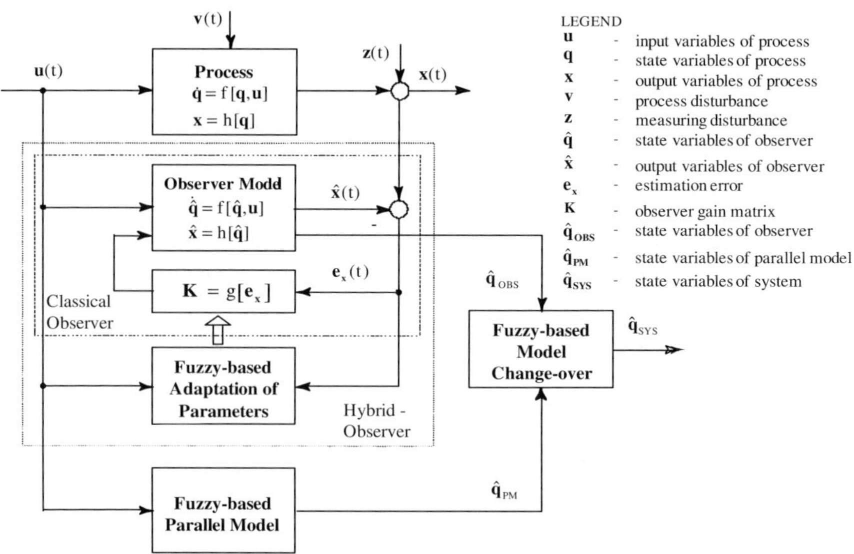

Fig. 1 Conception of a model- and rule-based system for continuous process monitoring by MMM

2 Determination of Narrow Range Water Level in a Pressure Vessel

The paper deals with application of knowledge-based algorithms (observer model, rule-based model) for modelling of the narrow range collapsed level (lcn) in a pressure vessel with water-steam mixture during accidental depressurizations (blow down). Therefore a Fuzzy-based Parallel-Model was developed as an observer to generate redundant process information to determine subsequently the mixture level (lm). Based on the analysis of experimental and simulation data the input and output variables as well as the structure of rule-based algorithms are defined. The narrow range collapsed level lcn characterizes the water inventory within a limited measuring range (Fig. 2) [2].

For reproduction of narrow range collapsed level the following effects have to be considered (Fig. 3) [1, 2]:

1. rapid increase of both levels as a result of the initiated evaporation (by opening of a leak),
2. gradual decrease of all water levels as a result of the boiling (by loss of mass beyond the leak),
3. rapid decrease of both water levels as a result of the collapse of water-steam mixture (closing of leak).

For calculation of the narrow range water level (a kind of water-steam-mixture level) (lcn) during depressurizations using Model Based Methods, it is necessary to consider the nonlinear dynamic behaviour. To build up dynamic processes and non-linearities, the time dependence has to be generated in a defined fuzzy structure.

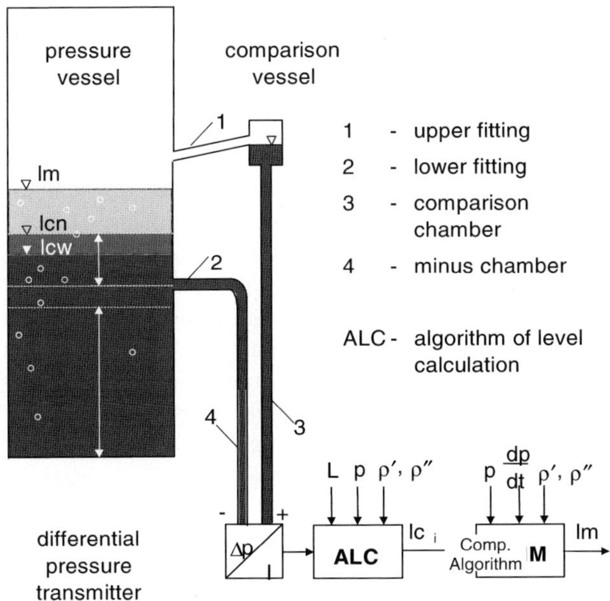

Fig. 2 Priciple scheme of Pressure vessel with a hydrostatic level measuring system

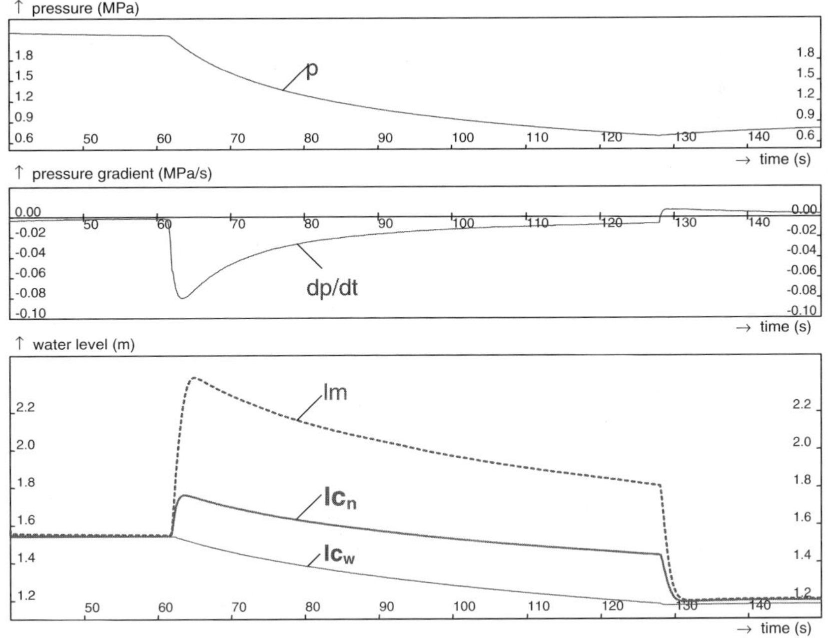

Fig. 3 Pressure, pressure gradient, and water levels (wide range collapsed level lcw, narrow range collapsed level lcn, mixture level lm for one blow down experiment

3 Fuzzy-Based Parallel-Model

With respect to the high process dynamics a Fuzzy Model with external dynamics (Fig. 4) was developed. The wide range collapsed level lcw is a global, accurate measurable process variable characterizing the actual process state. Furthermore it is not affected by a limitation of measuring range and therefore it is favoured as basic parameter for the determination of the narrow range collapsed level lcn. This leads to the idea to determine the spread value between narrow range collapsed level and wide range collapsed level in form of the ratio factor Kc. A definite classification and description of this ratio factor Kc was realized by using of significant process parameters and process signals like initial pressure. The direct output variable of the Fuzzy Model is the time derivation dKc /dt of the ratio factor. As second output variable the initial value of the ratio factor Kc|0 is determined. The absolute value of the ratio factor Kc which is the output variable of the Dynamic Fuzzy Model structure is fed back as additional input variable of the Fuzzy Model [2].

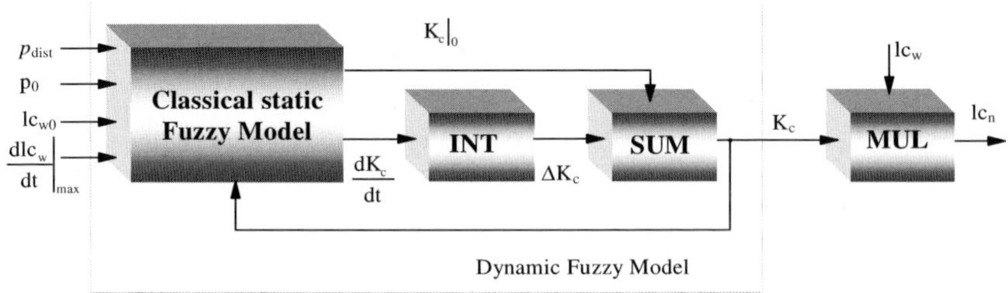

Fig. 4 Dynamic Fuzzy Model for the calculation of the narrow range collapsed level

4 Fuzzy Model with External Dynamics

The underlying basic design of the above-mentioned model is presented in Fig. 5a [1, 3].

Figure 5b. demonstrates the equivalent classical method (Signal Flow Diagram) by a differential equation. In automation technology the description is carried out by a first order lag element. The simplest transfer function is given by equations (1) and (2).

$$\frac{dx_a(t)}{dt} = \frac{K}{T}x_e(t) - \frac{1}{T}x_a(t) \tag{1}$$

$$x_a(t) = \int \left(\frac{dx_a(t)}{dt}\right) dt = \int y\,dt \tag{2}$$

As an advantage the fuzzy transfer function contains the whole non linearities of the process, a special knowledge about several transfer parameters, like time

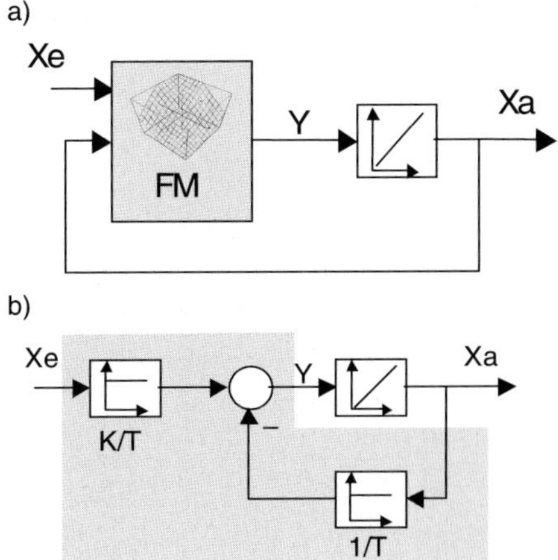

Fig. 5 (**a**) Structure of the Fuzzy Model with external dynamics (**b**) Comparison to a Classical Signal Flow Diagram (PT1 – first order lag element))

constants and gain, is not necessary. The input and output variables of a classical Fuzzy Model stay for the dynamic model. The proof of the stability of the Fuzzy Model with external dynamics will be deduced from the integral process knowledge. The dynamic Fuzzy Model enfolds the real fuzzy algorithm, the integration of the direct output and the feedback of this value (Fig. 5a). The direct output is the time derivation of the desired output and thereby interpretable. For comparison two kinds of Fuzzy Models with external dynamics with different distribution of the membership functions will be used. Therefore the simulation of transfer behaviour of PT1-classical model and fuzzy model with external dynamics will be carried out.

With the basic rule for the fuzzy model and the distribution of the Fuzzy Sets the rule basis will be designed [1, 3]:

IF x_e AND x_a THEN y

For demonstration, two variants of the characteristic field and the dynamic system behaviour are shown in Fig. 7. In Fig. 7a, a linear module (i.e. K and T are constants) is used. The result is a plane characteristic field. In Fig. 7b the parameters K and T are nonlinear functions with the highest density in the center of the deformed characteristic field. The represented transfer functions are the result of a jump of xe(t) (step response) with the starting point xe(t = 0) = 0.

Y		Xa				
		NG	NM	ZR	PM	PG
	NG	ZR	NK	NM	NG	NS
	NM	PK	ZR	NK	NM	NG
Xe	ZR	PM	PK	ZR	NK	NM
	PM	PG	PM	PK	ZR	NK
	PG	PS	PG	PM	PK	ZR

Fig. 6 Rule Basis of Dynamic Fuzzy Model

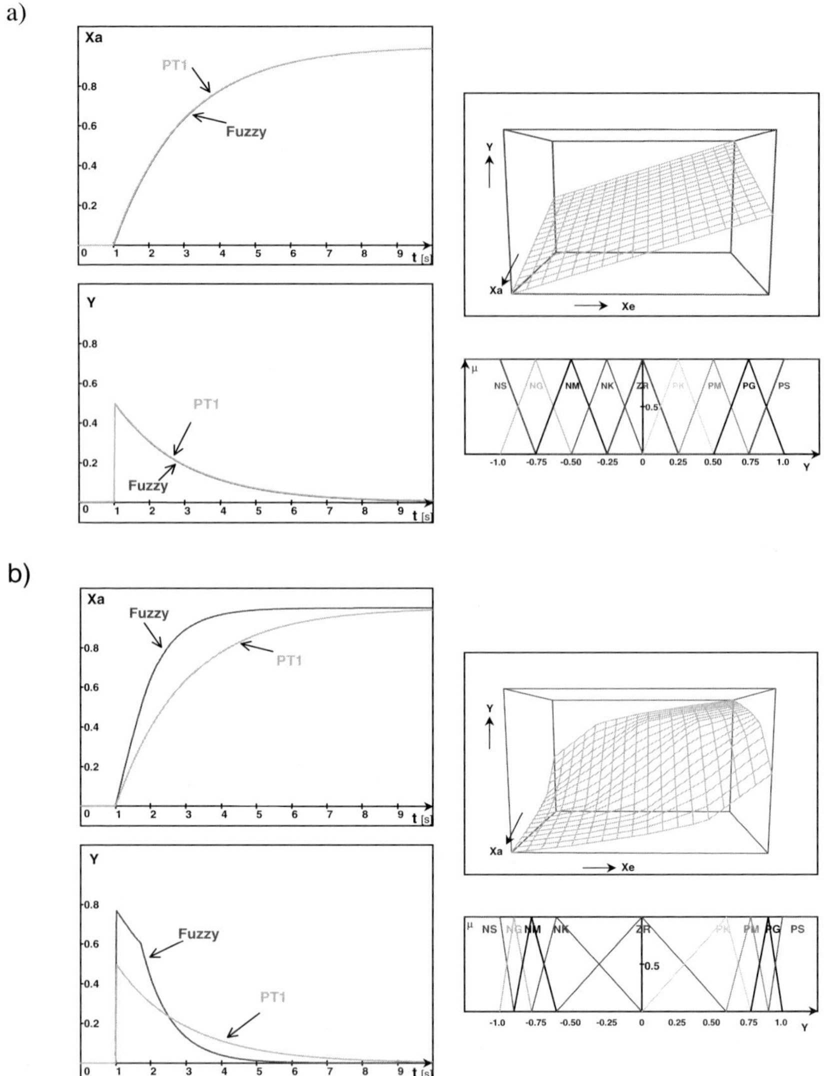

Fig. 7 (**a**) Step Response and Characteristic Field – Linear Module (**b**) Step Response and Characteristic Field – Nonlinear Module without Tolerance Area

The kind of deformation in this example leads to an influence of the time constant T, the stationary values will not be influenced. This can be reached generating a plateau in the center of the characteristic field by changed distribution of fuzzy sets. It is obvious that any characteristic field deformation by changing the Fuzzy Sets distribution of the output variable, nonlinear influences the static and dynamic system behaviour. Using this principle, it is possible to generate all kinds of nonlinear transfer function modules.

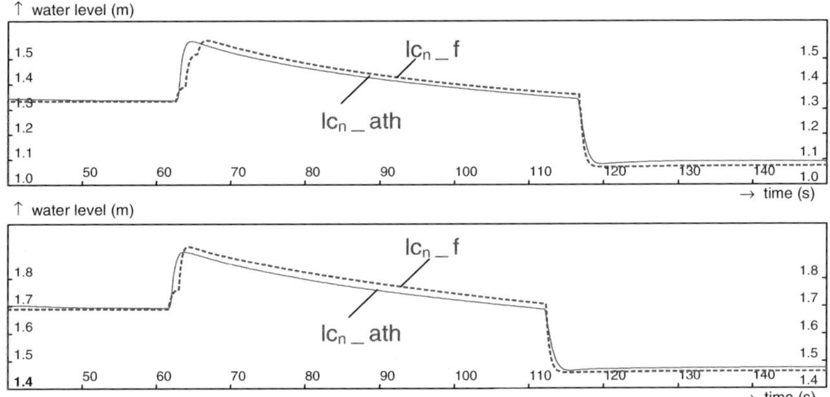

Fig. 8 Comparison between the fuzzy-based reproduced (f) and the real response characteristic (ath) of narrow range collapsed level for two blow down experiments

5 Results

To characterize the quality and capability regarding reproduction of narrow range collapsed level by the developed Fuzzy Model some results in conclusion are presented. For the above mentioned blow down experiment the reproduced water levels by the Fuzzy Model (_f) are compared with the real levels (_ath; simulated with the help of the complex Thermohydraulic Calculation Code ATHLET). The reproduction result for the narrow range collapsed level (lcn) is shown in Fig. 8. The relative error of the fuzzy-based reproduced level (lcn_f) to the real measured narrow range collapsed level is about 5% in maximum, what can be classified as a sufficient result [2].

As a result of the combination of model-based and fuzzy-based algorithms a new quality of methods for the monitoring and diagnosis was achieved with the following advantages: use of the analytical knowledge (models) and inclusion of knowledge which is indescribable by analytical algorithms (experiences), description of strong non-linear processes, and description of complex multi-variable processes and their dependencies.

The investigations were supported by the German Ministry for Economy and Labour (BMWA).

References

1. A. Traichel, "Neue Verfahren zur Modellierung nichtlinearer thermodynamischer Prozesse in einem Druckbehälter mit siedendem Wasser-Dampf Gemisch bei negativen Drucktransienten", Fakultät für Maschinenbau der Universität Karlsruhe, Dissertation, 2004

2. A. Traichel, W. Kästner, R. Hampel, "Fuzzy Modeling of Dynamic Non-Linear Processes - Applied for Water Level Measu¬rement", Eds: R. Hampel, M. Wagenknecht, N. Chaker, IPM, Series: Advances in Soft Computing, Physica Verlag, Heidelberg, 2000 (Editor in Chief: J. Kacprzyk)
3. R. Hampel, A. Traichel, S. Fleischer, W. Kästner, N. Chaker, "System Control And Diagnosis Using Soft Computing Methods. Control 2003, Moscow, MPEI

Methods of Possibilistic Optimization
with Applications

A.V. Yazenin

Abstract In the article the approach to optimization with fuzzy random data from the general methodological positions of uncertainty measure theory is developed. The corresponding principles and models of decision making are formulated. Applications of the approach to portfolio selection problem are considered.

Key words: Possibilistic-probabilistic optimization · fuzzy variable · fuzzy random variable · portfolio selection

1 Introduction

In the possibilistic optimization the theory and methods of extreme problems solution with fuzzy parameters are considered. The values of fuzzy parameters are characterized by possibility distributions. The development of the scientific basis began in 80th, when L.A. Zadeh [16] and S.Nahmias [7] formed in their works the fundamental of modern theory of possibility.

Naturally, possibilistic optimization begins on fuzzy optimization. The article of R.Bellman and L.A. Zadeh [1] where the "symmetric scheme" of decision making problems solution is considered can be called foundational. The category of fuzziness is referred in the approach to objectives – problem restrictions. Further the category of fuzziness is used to describe the problem parameters.

The analysis of the existent works of the field shows that the membership functions that model the value set of fuzzy parameters of the problem are interpreted as possibility distributions and the apparatus of problem analysis remains the same: computation of fuzziness based on generalization principle.

However, the bases of possibility theory and probability theory are monotone set functions, fuzzy measures (half-measure, uncertainty measure) [2, 10]. Under the approach there is an opportunity to research the optimization problem with fuzzy random data [10, 11] from the point of uncertainty measure, that defines its universality. That helps to show up the internal and external structure of the decision making problem, to make the connection between possibilistic, stochastic and interval models.

M. Nikravesh et al. (eds.), *Forging the New Frontiers: Fuzzy Pioneers II.*

In the work the mentioned approach is described, the calculation of possibility based on decision making problems solution is developed, the class of basic models of possibility optimization is introduced. The indirect methods on the problems are obtained. Within the developed approach the optimization problems with fuzzy random data are considered. For the purpose the calculation of fuzzy random data based on notion of fuzzy random variable is developed. The numeric characteristics calculation methods are obtained for different approaches [4]. As application of possibilistic-probabilistic optimization the generalization of portfolio analysis problems with fuzzy random data is considered. Fuzzy random variable in the context improves the expected value and risk of the portfolio. In the work the possibilistic-probabilistic models of portfolio analysis are developed. The obtained results expand the sphere of portfolio theory application and help to construct adequate decision making models.

2 Preliminaries

Let Γ be a set of elements denoted as $\gamma \in \Gamma$, $P(\Gamma)$ is a power set of Γ, E^n denotes the n – dimensional Euclidean space.

Following [8, 10] we introduce necessary definitions and notations.

Definition 1. *A possibility measure is a set function* $\pi : P(\Gamma) \to E^1$ *with properties:*

$$1.\ \pi\{\emptyset\} = 0,\ \pi\{\Gamma\} = 1; \quad 2.\ \pi\{\bigcup_{i \in I} A_i\} = \sup_{i \in I} \pi\{A_i\},$$

for any index set I and $A_i \in P(\Gamma)$.

Triplet $(\Gamma, P(\Gamma), \pi)$ is a possibilistic space.

Definition 2. *A possibilistic (fuzzy) variable is a mapping $Z : \Gamma \to E^1$. Distribution of possibilistic values of variable Z is function $\mu_Z : E^1 \to E^1$, defined as $\mu_Z(z) = \pi\{\gamma \in \Gamma : Z(\gamma) = z\}, \forall z \in E^1$.*

$\mu_Z(z)$ is a possibility that variable Z may accept value z.

Definition 3. *r-level set of a fuzzy variable Z is given by $Z_r = \{z \in E^1 / \mu_Z(z) \geq r\}, r \in (0, 1]$.*

A necessity measure v is a dual concept notion to a possibility measure and defined as

$v(A) = 1 - \pi(A^c)$, where "$\rfloor$" means the complement of a set $A \in P(\Gamma)$. Taking into consideration results [8, 10], give the definition of a fuzzy random variable.

Let (Ω, B, P) be a probability space.

Definition 4. *Fuzzy random variable X is a real function $X(\cdot, \cdot) : \Omega \times \Gamma \to E^1$, such that for any fixed $\gamma \in \Gamma$ $X_\gamma = X(\omega, \gamma)$ is a random variable on (Ω, B, P).*

Everything becomes clear when distribution $\mu_X(x, \omega)$ is defined as in case of fuzzy variable:

$$\mu_X(x, \omega) = \pi\{\gamma \in \Gamma : X(\omega, \gamma) = x\} \,\forall x \in E^1.$$

The expected value $E\{X(\omega, \gamma)\}$ of a random variable $X(\omega, \gamma)$ can be defined in different ways.

We define distribution of a random variable expected value according to [8] through averaged random variable:

$$\mu_{EX}(x) = \pi\{\gamma \in \Gamma : E\{X(\omega, \gamma)\} = x\} \,\forall x \in E^1.$$

It's easy to show an expected value of a fuzzy random variable defined this way has the basic properties of random variable expected value.

Solving applied problems we are interested not only in expected value but also in variance and covariance of fuzzy random variables. There exist at least two approaches to their definition. The characteristics are fuzzy within the **first** *approach [4] and nonfuzzy within the* **second** *one [3].*

In **first** *approach variance and covariance are defined by probability theory formulae. We obtain formulas for variance and covariance as function of fuzzy variable.*

In **second** *approach*

$$\mathrm{cov}(X, Y) = \frac{1}{2} \int_0^1 (\mathrm{cov}(X_\omega^-(r), Y_\omega^-(r)) + \mathrm{cov}(X_\omega^+(r), Y_\omega^+(r)))dr,$$

where $X_\omega^-(r)$, $Y_\omega^-(r)$, $X_\omega^+(r)$, $Y_\omega^+(r)$ are r–level set endpoints of fuzzy variables X_ω, Y_ω respectively.

It's obvious variance $D(X) = \mathrm{cov}(X, X)$ and moments of the second order are without fuzziness.

It's important that the definition methods of the second order moments in the **first** *and* **second** *approaches are different in principle. In the* **first** *approach we identify possibility distribution and in the* **second** *one we make numeric calculations.*

3 Possibility-probability Optimization Models and Decision Making

Within fuzzy random data functions that form goals and restrictions of decision making problem make sense of mapping $R_i(\cdot, \cdot, \cdot) : W \times \Omega \times \Gamma \to E^1, i = \overline{0, m}$, where W is a set of acceptable solutions, $W \subset E^n$. Thus, a set of acceptable outcomes can be obtained by combination of solution set elements with elements of random and fuzzy parameter sets. That's why any concrete solution can't be directly connected either with goal achievement degree no with restriction system execution degree.

Existence of two different types of uncertainty in efficiency function complexifys reasoning and formalization of solution selection optimality principles. However decision making procedure based on **expected possibility** principle is quite natural.

Its content is elimination of two types of uncertainty that is a realization of two types of decision making principles [8, 10, 11]:

averaging of fuzzy random data that allows to get to decision making problem with fuzzy data;

choice of optimal solution with more possible values of fuzzy parameters or with possibility not lower than preset level.

Adequate mean of suggested optimality principle formalization is a mathematical apparatus of fuzzy random variables.

Let τ be a possibility or necessity measure that is $\tau \in \{\pi, \nu\}$. Taking into consideration stated decision making principles we came to the following optimization problem settings within fuzzy random factors.

Problem of maximizing goal achievement measure with liner possibility (necessity) restrictions

$$\tau \left\{ E R_0 \left(w, \omega, \gamma\right) \Re_0 \, 0 \right\} \rightarrow \max,$$

$$\begin{cases} \tau \left\{ E R_i \left(w, \omega, \gamma\right) \Re_i \, 0 \right\} \geq a_i, i = \overline{1, m}, \\ w \in W. \end{cases}$$

Problem of level optimization with liner possibility (necessity) restrictions

$$k \rightarrow \max,$$

$$\begin{cases} \tau \left\{ E R_0 \left(w, \omega, \gamma\right) \Re_0 k \right\} \geq a_0, \\ \tau \left\{ E R_i \left(w, \omega, \gamma\right) \Re_i \, 0 \right\} \geq a_i, i = \overline{1, m}, \\ w \in W. \end{cases}$$

In the stated problems $\Re_0$, $\Re_i$ are binary relations, $\Re_0, \Re_i \in \{\leq, \geq, =\}, \alpha_i \in (0, 1]$, k is an additional (level) variable.

The represented models are the natural generalization of possibility optimization models on case of fuzzy random data [12, 13, 14].

4 Applications

As some researchers equitably denote the main drawback of Markowitz [6] approach to portfolio selection problem is an absence of statistic data which are used for model parameters estimation. However financial market is instable and changeable so investment decision making leans on both expert estimations which are tolerant and fuzzy and statistic information. In some instances profitabilities and prices of separate financial assets are characterized by tolerant time series. In this case a fuzzy random variable is an adequate model of profitability.

Let $R_i \left(\cdot, \cdot\right) : \Omega \times \Gamma \rightarrow E^1$ be a fuzzy random variable that represents profitability of an i-asset and Γ are elements of probability space (Ω, B, P) and possibilistic space $(\Gamma, P(\Gamma), \pi)$ respectively. Then expected value of portfolio is a fuzzy random variable

$$R_p\left(w, \omega, \gamma\right) = \sum_{i=1}^{n} w R_i\left(\omega, \gamma\right)$$

Here $w = (w_1, \ldots, w_n)$ is the vector representing the portfolio: $w \geq 0$, $\sum_{i=1}^{n} w_i = 1$.

Expected profit and risk of the portfolio under fixed w are presented by following functions:

$$\widehat{R}_p\left(w, \gamma\right) = E R_p\left(w, \omega, \gamma\right),\ \widehat{V}_p\left(w, \gamma\right) = E\left(R_p\left(w, \omega, \gamma\right) - \widehat{R}_p\left(\omega, \gamma\right)\right)^2 \text{ [4]},$$

$$\text{or } \hat{R}_p\left(w, \gamma\right),\ V_p\left(w\right) = \text{cov}\left(R_p\left(w, \omega, \gamma\right), R_p\left(w, \omega, \gamma\right)\right) \text{ [3]}.$$

Detailed analysis of possibilistic-probabilistic models and methods of portfolio analysis with the pair of criteria (profitability - risk) is made in [15].

5 Conclusion

In the present paper the approach to analysis of problems possibilistic-probabilistic optimization is described. In the frames of proposed schema of possibilistic-probabilistic optimization the models of portfolio analysis and optimization methods can be developed.

References

1. R.E.Bellman, L.A.Zadeh, Decision making in a fuzzy environment, Management Sci., 17(1970)141–162.
2. Dubois D., Prade H. Possibility theory: an approach to computerized processing of uncertainty, Plenum press, 1988.
3. Y.Feng, L.Hu, H.Shu. The variance and covariance of fuzzy random variables, Fuzzy sets and systems 120(2001) 487–497.
4. Yu.Khokhlov, A.V.Yazenin. The calculation of numerical characteristics of fuzzy random data, Vestnik TvGU, No.2. Series "Applied mathematics", 2003, pp.39–43
5. H.Kwakernaak. Fuzzy random variables – 1.Definitions and theorems, Inf. Sci.. 15(1978) 1–29.
6. H.Markowitz. Portfolio selection: efficient diversification of investments. Wiley. New York, 1959.
7. S.Nahmias. Fuzzy variables, Fuzzy sets and systems 1(1978) 97–110.
8. S.Nahmias. Fuzzy variables in random environment, In: Gupta M.M. et. al. (eds.). Advances in fuzzy sets Theory. NHCP. 1979.
9. M.D.Puri, D.Ralesky. Fuzzy random variables, J. Math. Anal. Appl. 114(1986) 409–422.
10. A.V.Yazenin, M.Wagenknecht, Possibilistic optimization. Brandenburgische Technische Universitat. Cottbus, Germany, 1996.
11. A.V.Yazenin. Linear programming with fuzzy random data, Izv. AN SSSR. Tekhn. kibernetika 3(1991) 52–58.

12. A.V.Yazenin, On the problem of maximization of attainment of fuzzy goal possibility, Izv. RAN, Teoriya i sistemy upravleniya 4(1999) pp.120 – 123.
13. A.V.Yazenin, Fuzzy and stochastic programming, Fuzzy sets and systems 22(1987) 171–180.
14. A.V.Yazenin, On the problem of possibilistic optimization, Fuzzy sets and systems 81(1996) 133–140.
15. A.V.Yazenin, Optimization with fuzzy random data and its application in financical analysis, Proceedings of International Conference on Fuzzy Sets and Soft Computing in Economics and Finance (June 17–20, Saint-Petersburg, Russia), 2004, V.1, p.16–31.
16. L.A.Zadeh. Fuzzy sets as a basis for a theory of possibility, Fuzzy sets and systems 1(1978) 3–28.

Using Nᴛʜ Dimensional Fuzzy Logic
for Complexity Analysis, Decision Analysis
and Autonomous Control Systems

Tony Nolan OAM JP and Emily Baker

Abstract Fuzzy Logic is the key element in our research to try and understand the interrelationships in the universe. We have been using fuzzy logic in conjunction with matrixes and topographical / geometrical frameworks to map and examine the elements of knowledge, decision support, complexity, autonomous control, and the cause & effect of different phenomena in general. Our research is mainly about measuring the differences between either solid objects or mental concepts over space / time observations and then to model and project those differences. For us the secret of understanding our universe is not the objects in general, but more the specific action and reaction of objects in conjunction with each other. Fuzzy logic is the perfect tool for this type of approach as it allows us to model different movements and differences between objects, either solid or conceptual. The paper is made up of two parts, the first is more about the ideas, concepts and back ground of our projects, then the other part is broken down into projects or activities, where each is described in more detail. Each of the projects uses common elements of the other projects, so it is not so easy to explain on paper; of course we are always happy to reply to questions.

Key words: Hyperpanometrics · Artificial Intelligence · Complexity Chains · Decision Support · Autonomous Control · Knowledge Nets · Symbiotic Relativity · Fuzzy Logic · Granulation · Intervals · Hyper-dimensions

1 Introduction

Over our history as an intelligent species, we have treated the discovery of our universe as though we were doing an unsolvable jigsaw puzzle. Through both observation and experimentation, we have gathered and treated each piece of the puzzle in isolation, rarely understanding how the different pieces connected together. When we had joined a few pieces together, we found that we had only completed a minor segment. Our philosophy and logic was at an infant stage, which limited our understanding of the complexity and interdependency of the pieces to each other and of the value of these connections to the total picture.

M. Nikravesh et al. (eds.), *Forging the New Frontiers: Fuzzy Pioneers II.*
© Springer-Verlag Berlin Heidelberg 2008

It was not until the creation of fuzzy logic, that we could begin to have a better understanding of how the pieces interact as a whole, and that no piece was totally independent, but that it was necessary to evaluate all the relevant pieces to get a better understanding of how to solve the puzzle. As our knowledge increased and the processes became more refined, constructs of our reality became clearer and the complexities became more readily understood.

Prior to 1965, I feel that we were approaching the point where our logic could no longer support our discoveries. As our understanding of the complex interactions were developing, we required a more fluid and flexible logic. In 1965, Lotfi Zadek developed the fuzzy logic system to aid us in the next step of our modeling and understanding. Yet, Fuzzy Logic was confined; not by its existence but by the environment which contains it. Fuzzy Logic needs a fresh environment to expand truly to its fullest potential as a mathematical system that is not limited by the inflexibilities of our calculations.

Fuzzy logic reaches its greatest potential when it transcends beyond the two dimensional, and three dimensional, and extends into the Nth dimensional. The ability of fuzzy logic creates a symbiotic and balanced environment, where the slightest variance in either value or variable, automatically realigns itself into a proportional complexity of the sum of all parts. This ability when combined within a framework, and with a mechanism of transposition between the real world and computational environments, allows for a greater degree of complexity analysis and modeling, not available in pre fuzzy logic times.

I have found that when you combine fuzzy logic with the aspects of finite mathematics, granulation, clustering, Euclidean distances, and interval mathematics, there are many different possibilities and opportunities open for the development and operation of autonomous control systems, artificial life forms, smart probes, control systems, decision-making, knowledge mapping and signposting.

The goal of our research is to gain an understanding of the universe and how it all fits together. We believe that there is a symbiotic relationship between fuzzy logic and frameworks that give it form and purpose.

Fuzzy Logic is the key to unlocking the door, behind which hides the secrets of the reactionary balance between cause and effect, and through which we gain an understanding of how our universe fits together in both artificial and natural systems. The framework therefore provides the fuzzy logic with an environment within which, using a bounded rationality approach, we can apply a greater degree of analysis and comparison and hence improve our understanding.

The following sections are descriptions of different projects that use fuzzy logic and frameworks to model, simulate and explore both the objective and subject.

Nolan's Matrix The matrix works on the theory of bounded rationality through finite mathematics, and fuzzy logic, contained within a quasi-fractal framework. The matrix uses fuzzy logic principles for establishing decision modeling, simulations, and a model for autonomous control systems. These outcomes can be tested in a 'What If' environment, and clustered by the degree of importance and the amount of effect each variable has against each other variable in regard to the total problem or scenario under investigation. By using Euclidean Distance Measures, we have the advantage of using Multi Dimensional data inputs that can be represented within a

3d environment. Thus, we have the ability to graph cognition, and the ability to manipulate the outputs to gain an understanding of decision-making processes. We also the ability to inter link these outputs to machine controllers and the inputs through sensors, to create a cognitive environment, where a machine can learn and control various functions.

With the use of relational databases, knowledge classification systems, and object oriented programming. It is possible to create Artificial Life (AL) forms, which have the basic abilities to gather, process, store and retrieve data. Combining knowledge referencing with distance measures allows the AL forms the beginning of intelligence decision-making, and helps with the analysis of human decision making. These systems, though basic at the moment, also have the ability to recommend behavior modifications and changes with work teams and collaborations. AL's are able to compare different decision making scenarios and prompt different solutions, from an extensive knowledge base, and through better Human Computer Interfaces contribute to problem solving.

Interval data through the matrix also gives possibilities to various mappings of the data, and graphical representations of outputs. This is of use in diagrammatical reasoning activities, and representation of clustered elements of data, and gives rise to sense making and bounded rationality explorations of decision-making and data manipulation.

SAHIS Sahis is an attempt to create a computer system that could mimic human thought. To be able to mimic human thought you need to be able to replicate certain processes of human thought. The computer should be able to recognize different paradigms, concepts, ideas, knowledge, and emotions, and then include them in the analysis. In effect, it is a relational knowledge database that can examine and recognize both patterns and interrelationships. SAHIS is to have a workable and understandable interface so any user can easily relate to the data being presented for human analysis.

Through using of a Finite Fuzzy Logic Granulated Clustering Hyperspace Euclidean Distance Matrix, there many different possibilities and opportunities open for the development and operations of Multi Dimensional Control Systems, Artificial Life Forms, Smart Probes, Decision Making, Signposting and Simulations. The matrix works on the theory of Bounded Rationality through Finite Mathematics, and Fuzzy Logic principles. An examination of previous works in philosophy and logic, this was expanded to include other disciplines such as management, information science, computer science, geometry, hyperspace mathematics, psychology, sociology, communications theory, learning, cognition, intelligence, knowledge, library classification systems, finite mathematics, critical thinking, neural networks, fuzzy logic, statistics, etc. This led to the observation, that the solution needs to be a cross discipline approach.

So far the benefits of this project have been the creation of the matrix. This has allowed the graphing of human decision-making, group thought and environment and criminal profiling. We have developed a better understand of human information behavior, the philosophy of decision making, logic, and fuzzy logic modeling and profiling. The artificial life form is designed to work in collaboration with humans, and not to be a slave / master relationship.

Future developments should be a-lifes that can help human beings with decision making, lateral thing, problem solving and retaining corporate knowledge. They would be able to explore hazardous environments and be able to learn / report valuable analysis of the environment. They should show us ways to develop computer systems that can better interact with human beings, for more efficient interactions.

Complexity Chains Complexity chains are a new way to plot Nth dimensional data, where the data point is plotted using a unique set of reference points in a sequenced chain, where each data point becomes the sum of all the previous values in the chain. This means that highly complex analysis can be undertaken on many variables and their interactions. By linking the hyper-data points into a data-path using time series analysis, and examining the convergence and divergence of the different data paths, it becomes possible to see patterns and interactions very quickly. The data points can be un-compacted to their original values.

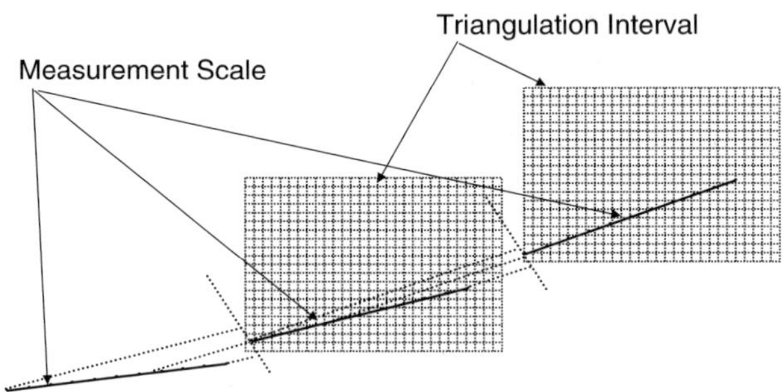

Hyperpanometrics When we examine our universe, we have sets of rules or laws, from which we gain an understanding, create measurements and reproduce in model form a scaled series of observations to explain a series of interactions between objects of interest. Our universe gives an environment from which to explore, navigate, experiment and study so that we can make sense of it, providing we remain within the boundaries that are established by nature.

However, there are limitations on how well we are able to use mathematics to create models of our universe, and to what degree we can represent the complexities of our environment. More often that not, it is impossible to accurately represent a greater number of variables or complexities, due to the mathematical limitations of being confined to the laws of our Reality. Through bounded rationality, it is however possible to limit some for the complexities, and entertain some types of exploration. Through the development and use of computer modelling, it is possible to extending the exploration of different scenarios, and then link them back to real work observations.

The theory of Hyperpanometrics is that, in effect we change the laws of nature, to be able to model the complexities of our universe. We establish alternative universes, where we have the power to establish our own laws of behaviour, and can adjust the degrees of measurement and inter-relativity. We are able to give data observations & variables physical properties, and define their behaviour the same way as we do in our universe. Through establishing a series of reference points that are common to both the real universe and the artificial universe, it is then possible to have observations that have meaning in both environments.

Hyperpanometrics is what you might call the measure's measurement, it is designed to work with the differences between various types of standardised measurements to then be able to map those differences and then plot them in an artificial universe. The concept is to convert any series of observations into interval data, and then to be able to plot these observations. The system requires a series of observations and that the measurements are consistent. The same way that we can use Newtons Law of Motion or Einstein's Theory of Relativity to explain movements in our universe can be used to explain a data track through a series of Variable Gateways within an alternative universe.

Alternative universes are the creation and exploration of a mathematically generated reality, where the laws of nature are defined and control by a different set of rules to the ones that exist in our reality. This allows for a greater number of possible connections, interactions and combinations of elements that are limited or impossible in our reality. Through bounded rationality it is possible to constrain and adjust the different levels of complexities, and set boundaries for definitions and constructions of the artificial elements. By using a index of reference points, that represent factors in both our reality and the artificial universe, it becomes possible to translate the variables, observations and results from one to the other, and thus achieve a linked correlation between the two.

Hyperpanometrics is the process of using a series of plotted data points on a map or graph, where their positions are determined through a calculative navigational triangulation within a geometric or topological framework. Each data-point is a derived constructed unique representation of a number of time/space restricted observations, from a number of variables. A waypoint [constructed data point] is then plotted in reference to its time/space defined gateway. Each gateway is a separate finite mini time/space based reality, that when combined with other gateways constitute the artificial universe. Desired waypoints are linked together to form a data-pathway. Thus it becomes possible to observe the behaviours of these data-pathways in relationship to each other. The examination and measurement of the bearing/degree of movement [divergence, parrel, convergence] of intra-variable and inter-variable movements within a hyperspacial environment, provides an insight into the dynamic relationships within complexity modelling.

Just like in Newtonian physics or newtons law, of motion. An object will go in a straight line unless acted on. When you combine this with Einstein's theory of relativity, where space and time limit an observational point. When you have parrel observation points, logic dictates that unless a data path is attracted towards an attraction point in some way, the data path will also change. Hyperpanometrics uses the adjustment of these attraction points, in accordance with construct and variable observations. These data paths will transverse through gateways from one parrel observation point to another, in a continuous motion, unless altered. The analysis of this system comes from the bearing and distance from one data path to another, and the movement between the data paths as they converge or diverge. The direction of that movement is dictated from the change in the values of the attraction points. The number and type of variables define the size and shape of the gateway. Gateways can be in either 2d or 3d. Target data path and baseline data path. Hyperpanometrics is the measurement, we aren't interested in the directed observation, but more the movement between data points. Hence its an between group or within group principle. Movement is refined into the logarithmic. It is possible to have difference geometric shapes as gateways, with a different configuration, as this can be used to observe the influences and impacts.

Hyperpanometrics is about establishing alternative universes or micro world simulations, where the laws of nature are rewritten to allow observations of how different variables react within a bounded rational and then transposed back to the real world. This allows for the observation and comparison of data the same way as we do in the real world, which can enhance the behaviors of data.

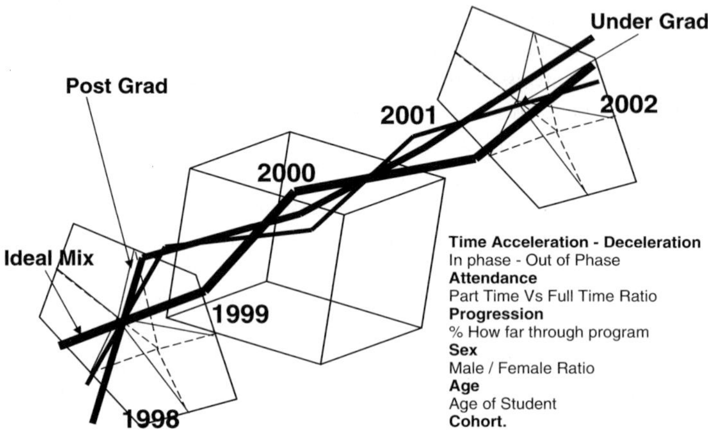

Instinct Circuits This is a circuit that has the ability to perform basic decision-making activities, because of a matrix of sensors and drivers. This allows a type of primitive autonomous control, for data collection and learning, and forms the basics of an artificial life form.

Knowledge Nets Knowledge nets are mainly the same as neural nets except that they use a geometric sphere, where a knowledge classification system is used for

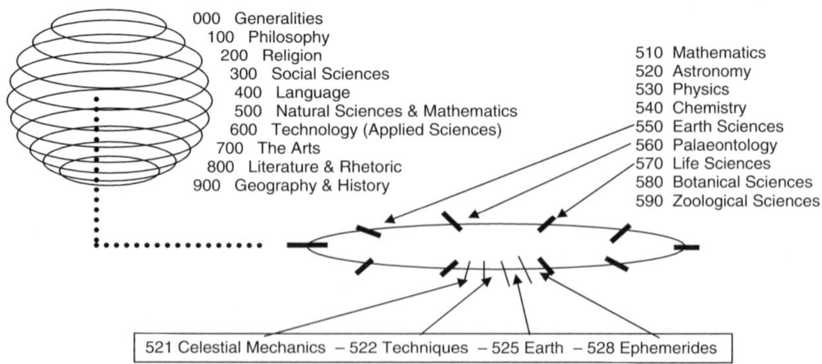

the mapping of decisions. By using a decision matrix with a knowledge classification framework and meta tagging of each contributing variable creates a distance measure to represent the importance of the variable relative to all the variables in the decision. This then provides a 3D decision tree within the sphere that represents each of the branches relative to each other, by subject and impact. Through using a knowledge-mapping standard for all decisions, and equal scaling on the impact of each factor, decision analysis becomes much easier due to the ease of comparison, especially in regards to 'WHAT IF' scenarios.

Signposting & Dendromatrix Nolan's matrix allows for snapshots of human decision-making or environmental factors, and allows for the profiling of the interactions between elements as a signature of behavior. Through using relational databases as well as time series observations, it is possible to signpost situations and develop responses. This process also allows us the use of 'WHAT IF' scenarios. It produces a decision tree that is interactive, and is a snap shot of human thought, or of a decision matrix from a computer or autonomous robot. This allows the elements to be clustered and examined in like areas of knowledge. Thus giving a common basis for analysis between non-related concepts.

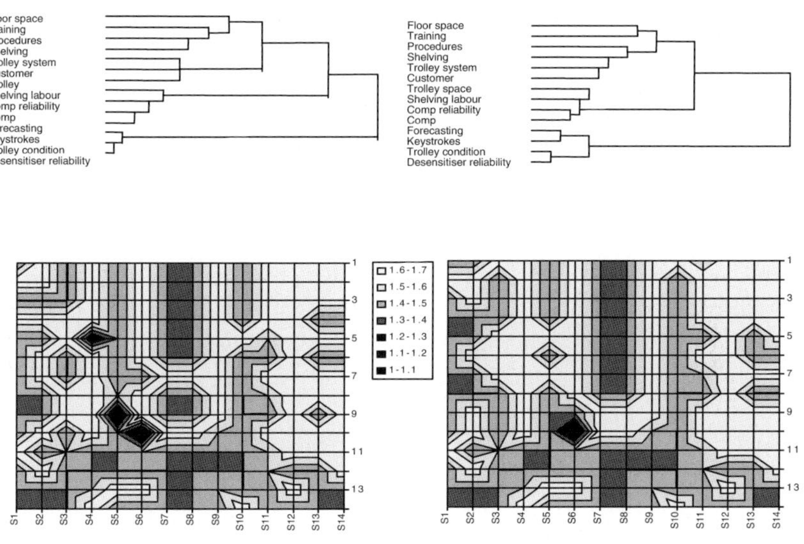

Symbiotic Relativity Einstein's theories of relativity specifically view time and space as central to a person's point of observation, which has been adopted in the information, knowledge and cognitive sciences for some time. However, Einstein's theories were written in a time and place where little was realised of just how dynamic and inter-relational the universe is.

With the observation of what had previously appeared as random events that happened under the previously observable threshold, we now realise that certain events are the result of cause and effect, as the threshold of understanding lowers.

In previous thinking, when two or more entities come closer, all effected entities become subjugated to a single dominant entity. However, now there is a growing realisation that action and reaction per mutate throughout all stratas of the universe. Symbiotic relativity is about recognising that every thing is interconnected, randomness is just action and reaction below the observational threshold, and that not all primary relational occurrences become subjugated to a single dominance, but there is interdependence with a shared interdependence in some cases.

It is in the factors of probability, causality, co-incidence, serendipity, randomness and interdependencies where there is evidence of a link or connection between two or more objects that we are more able to distinguish between what's able to be explained and what isn't.

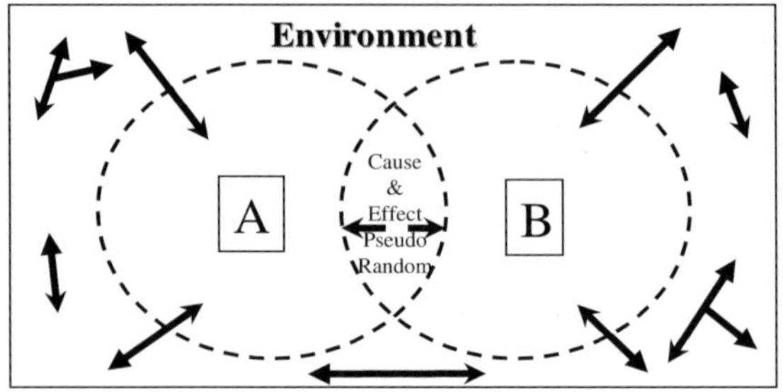

Pattern Trees: An Effective Machine Learning Approach

Zhiheng Huang, Masoud Nikravesh, Tamás D. Gedeon and Ben Azvine

Abstract Fuzzy classification is one of the most important applications of fuzzy logic. Its goal is to find a set of fuzzy rules which describe classification problems. Most of the existing fuzzy rule induction methods (e.g., the fuzzy decision trees induction method) focus on searching rules consisting of t-norms (i.e., AND) only, but not t-conorms (OR) explicitly. This may lead to the omission of generating important rules which involve t-conorms explicitly. This paper proposes a type of tree termed *pattern trees* which make use of different aggregations including both t-norms and t-conorms. Like decision trees, pattern trees are an effective machine learning tool for classification applications. This paper discusses the difference between decision trees and pattern trees, and also shows that the subsethood based method (SBM) and the weighted subsethood based method (WSBM) are two specific cases of pattern trees, with each having a fixed pattern tree structure.

A novel pattern tree induction method is proposed. The comparison to other classification methods including SBM, WSBM and fuzzy decision tree induction over datasets obtained from UCI dataset repository shows that pattern trees can obtain higher accuracy rates in classifications. In addition, pattern trees are capable of generating classifiers with good generality, while decision trees can easily fall into the trap of over-fitting. According to two different configurations, simple pattern trees and pattern trees have been distinguished. The former not only produce high prediction accuracy, but also preserve compact tree structures, while the latter can produce even better accuracy, but as a compromise produce more complex tree structures. Subject to the particular demands (comprehensibility or performance), simple pattern trees and pattern trees provide an effective methodology for real world applications.

Weighted pattern trees have been proposed in which certain weights are assigned to different trees, to reflect the nature that different trees may have different confidences. The experiments on British Telecom (BT) customer satisfaction dataset show that weighted pattern trees can slightly outperform pattern trees, and both are slightly better than fuzzy decision trees in terms of prediction accuracy. In addition, the experiments show that both weighted and unweighted pattern trees are robust to over-fitting. Finally, a limitation of pattern trees as revealed via BT dataset analysis is discussed and the research direction is outlined.

M. Nikravesh et al. (eds.), *Forging the New Frontiers: Fuzzy Pioneers II.*

Key words: Pattern trees · fuzzy model classification · fuzzy decision trees · machine learning · data mining

1 Introduction

The major advantage of using fuzzy rules for classification applications is to maintain transparency as well as a high accuracy rate. The difficulty in obtaining an optimal classification model is to balance the complexity of a rule base and the classification accuracy. Generally, fuzzy rule models can produce arbitrary accuracy if a unlimited number of rules are allowed [15]. However, this inevitably violates the original motivation of using fuzzy rules – transparency. To address this, many fuzzy rule induction methods have been developed. In Wang and Mendel [14] an algorithm for generating fuzzy rules by learning from examples has been presented. Yuan and Shaw [18] have proposed the induction of fuzzy decision trees for generating fuzzy classification rules (the extension of the classic decision tree induction method by Quinlan [10]). Chen, Lee and Lee [1] have presented a subsethood based method (SBM) for generating fuzzy rules. Rasmani and Shen [12] have proposed a weighted fuzzy subsethood based rule induction method (WSBM). To avoid the exponential growth of the size of the rule base when the number of input variables increases, Raju, Zhou and Kisner [11] have proposed hierarchical fuzzy systems. Recently, Kóczy, Vámos and Biró [6], and Wong, Gedeon and Kóczy [16] have presented fuzzy signatures which model the complex structure of the data points in a hierarchical manner.

Most of the existing fuzzy rule induction methods including fuzzy decision trees [18] focus on searching for rules which only use t-norm operators [13] such as the MIN and algebraic MIN. Disregarding of the t-conorms such as MAX and algebraic MAX is due to the fact that any rule using t-conorms can be represented by several rules which use t-norms only. This is certainly true and it is helpful to simplify the rule induction process by considering t-norms only. However, it may fail to generate important rules in which fuzzy terms are explicitly connected with t-conorms. This will be clearly shown in an artificial dataset in Sect. 4.6. Research has been conducted to resolve this problem. For example, Kóczy, Vámos and Biró [6] have proposed fuzzy signatures to model the complex structures of data points using different aggregation operators including MIN, MAX, and average etc. Mendis, Gedeon and Kóczy [7] have investigated different aggregations in fuzzy signatures. Nikravesh [9] has presented evolutionary computation (EC) based multiple aggregator fuzzy decision trees.

Huang and Gedeon [3] have first introduced the concept of pattern trees and proposed a novel pattern tree induction method by means of similarity measures and different aggregations. Like decision trees, pattern trees are an effective machine learning approach for classification applications. The experiments carried out on UCI datasets show that the pattern trees can obtain higher accuracy rates than the SBM, WSBM and the fuzzy decision trees in classifications. In addition, pattern trees perform more consistently than fuzzy decision trees. The former are capable of generating classifiers with good generality, while the latter can easily fall into the trap of over-fitting.

Simple pattern trees and pattern trees [4] are distinguished with two different configurations. The former not only produce high prediction accuracy, but also preserve compact tree structures, while the latter can produce even better accuracy, but as a compromise, produce more complex tree structures. Subject to the particular demands (comprehensibility or performance), simple pattern trees or pattern trees provide an effective methodology for real world applications.

Weighted pattern trees [5] have been proposed in which certain weights are assigned to different trees. As a result, it enhances the semantic meaning of pattern trees. The experiments on BT customer satisfaction dataset show that weighted pattern trees can slightly outperform pattern trees. In fact, both weighted and unweighted pattern trees with only two or three tree levels are good enough for most experiments carried out in this paper. This provides a very transparent way to model real world applications.

The rest of the paper is arranged as follows: Sect. 2 provides the definitions for similarity, aggregations and pattern trees. Section 3 proposes a novel pattern tree induction method, which is configured differently to build simple pattern trees and pattern trees. Section 4 shows that the SBM and WSBM are two specific cases of pattern trees. It also outlines the difference between decision trees and pattern trees, and the advantage of using trees. Section 5 presents the experimental results using Saturday Morning Problem dataset and datasets from UCI machine learning repository [8]. Section 6 suggests the concept of weighted pattern trees and shows how to use them for classification. Section 7 presents the experimental results applying both weighted and unweighted pattern trees induction to BT customer satisfaction dataset. Finally, Sect. 8 concludes the paper and points out some further research work.

2 Similarity, Aggregations and Pattern Trees

2.1 Similarity

Let A and B be two fuzzy sets [19] defined on the universe of discourse U. The commonly used fuzzy similarity measures can be defined in Table 1, where $\cap$ and $\cup$ denote a certain t-norm operator and a t-conorm respectively. Usually, the MIN ($\wedge$) and MAX ($\vee$) operators are used. According to the definition, $0 \leq S(A, B) \leq 1$. Without losing generality, the most commonly used similarity measure namely Jaccard is used to construct the pattern trees in Sect. 3. In practice, it is computed as

$$S(A, B) = \frac{\sum_{j=1}^{m}[\mu_A(x_j) \wedge \mu_B(x_j)]}{\sum_{j=1}^{m}[\mu_A(x_j) \vee \mu_B(x_j)]}, \tag{1}$$

Table 1 Similarity measures

Name	Definition
Simple matching	$A \cap B$
Jaccard	$\frac{A \cap B}{A \cup B}$
Dice	$2 \frac{A \cap B}{A + B}$

where x_j, $j = 1, \ldots, m$, are the crisp values discretized in the variable domain, and $\mu_A(x_j)$ and $\mu_B(x_j)$ are the fuzzy membership values of x_j for A and B.

An alternative similarity definition is proposed in this paper for pattern tree construction. Consider that the root mean square error (RMSE) of fuzzy sets A and B can be computed as

$$RMSE(A, B) = \sqrt{\frac{\sum_{j=1}^{m} (\mu_A(x_j) - \mu_B(x_j))^2}{m}}, \tag{2}$$

the RMSE based fuzzy set similarity can thus be defined as

$$S(A, B) = 1 - RMSE(A, B). \tag{3}$$

The large value $S(A, B)$ takes, the more similar A and B are. Of course, this alternative definition retains $0 \leq S(A, B) \leq 1$ given $\mu_A(x_j), \mu_B(x_j) \in [0, 1]$.

2.2 Fuzzy Aggregations

Fuzzy aggregations are logic operators applied to fuzzy membership values or fuzzy sets. They have three sub-categories, namely t-norm, t-conorm, and averaging operators such as weighted averaging (WA) and ordered weighted averaging (OWA) [17].

Triangular norms were introduced by Schweizer and Sklar [13] to model distances in probabilistic metric spaces. In fuzzy sets theory, triangular norms (t-norm) and triangular conorms (t-conorm) are extensively used to model logical operators *and* and *or*. The basic t-norm and t-conorm pairs which operate on two fuzzy membership values a and b, $a, b \in [0, 1]$ are shown in Table 2.

Table 2 Basic t-norms and t-conorms pairs

Name	t-norm	t-conorm
MIN/MAX	$min\{a, b\} = a \wedge b$	$max\{a, b\} = a \vee b$
Algebraic AND/OR	ab	$a + b - ab$
Šukasiewicz	$max\{a + b - 1, 0\}$	$min\{a + b, 1\}$
EINSTEIN	$\frac{ab}{2 - (a + b - ab)}$	$\frac{a + b}{1 + ab}$

Although the aggregations shown above only apply to a pair of fuzzy values, they can apply to multiple fuzzy values as they retain associativity.

Definition 1. A WA operator of dimension n is a mapping $E : \mathbb{R}^n \to \mathbb{R}$, that has an associated n-elements vector $w = (w_1, w_2, \ldots, w_n)^T$, $w_i \in [0, 1]$, $1 \leq i \leq n$, and $\sum_{i=1}^{n} w_i = 1$ so that

$$E(a_1, \ldots, a_n) = \sum_{j=1}^{n} w_j a_j. \tag{4}$$

Definition 2. An OWA operator [17] of dimension n is a mapping $F : \mathbb{R}^n \to \mathbb{R}$, that has an associated n-elements vector $w = (w_1, w_2, \ldots, w_n)^T$, $w_i \in [0, 1]$, $1 \leq i \leq n$, and $\sum_{i=1}^{n} w_i = 1$ so that

$$F(a_1, \ldots, a_n) = \sum_{j=1}^{n} w_j b_j, \tag{5}$$

where b_j is the jth largest element of the collection $\{a_1, \ldots, a_n\}$.

A fundamental difference of OWA from WA aggregation is that the former does not have a particular weight w_i associated for an element, rather a weight is associated with a particular ordered position of the element.

Example 1. Assume $w = (0.2, 0.3, 0.1, 0.4)^T$, then the WA operator on the vector of $(0.6, 1, 0.3, 0.5)$ is

$$E(0.6, 1, 0.3, 0.5) = 0.2 \times 0.6 + 0.3 \times 1 + 0.1 \times 0.3$$
$$+ 0.4 \times 0.5 = 0.65,$$

and the OWA operator on the vector is

$$F(0.6, 1, 0.3, 0.5) = 0.2 \times 1 + 0.3 \times 0.6 + 0.1 \times 0.5$$
$$+ 0.4 \times 0.3 = 0.55.$$

It is worth noting two special OWA operators are equal to MAX and MIN:

- If $w^* = (1, 0, \ldots, 0)$, then

$$F(a_1, \ldots, a_n) = max\{a_1, \ldots, a_n\}. \tag{6}$$

- If $w_* = (0, 0, \ldots, 1)$, then

$$F(a_1, \ldots, a_n) = min\{a_1, \ldots, a_n\}. \tag{7}$$

The main factor in determining which aggregation should be used is the relationship between the criteria involved. Compensation has the property that a higher degree of satisfaction of one of the criteria can compensate for a lower degree of satisfaction of another criterion. w^* means full compensation (*or*) and w_* means no compensation (*and*). Normally, an OWA operator lies in between these two extremes. An OWA operator with much of nonzero weights near the top will be more *or* than *and*.

2.3 Pattern Trees

A pattern tree is a tree which propagates fuzzy terms using different fuzzy aggregations. Each pattern tree represents a structure for an output class in the sense that how the fuzzy terms aggregate to predict such a class. The output class is located at the top as the root of this tree. The fuzzy terms of input variables are on different levels (except for the top) of the tree. They use fuzzy aggregations (as presented in Sect. 2.2) to aggregate from the bottom to the top (root).

Assume two fuzzy variables A and B each have two fuzzy linguistic terms A_i and B_i, $i = \{1, 2\}$, and the task is to classify the data samples to either class X or Y. The primitive pattern trees for class X are shown in Fig. 1. Each primitive tree consists of only one leaf node (fuzzy set) and it uses such a leaf node to predict the output class (X). Typically, primitive pattern trees do not lead to high prediction accuracy, and multi-level pattern trees are required to maintain satisfactory performance. In fact, multi-level trees are built via the aggregation of primitive trees. For example, the two level pattern tree for class X as shown in Fig. 2 is built using the primitive trees in Fig. 1. Let *candidate trees* be trees which begin as primitive trees and aggregate with other trees (termed *low level trees*) to generate more optimal pattern trees in terms of the similarity measure to the fuzzy values of output. A candidate tree $B_1 \Rightarrow X$ aggregates with a low level tree $A_2 \Rightarrow X$ using the *and* operator to lead to a new candidate tree $B_1 \wedge A_2 \Rightarrow X$. This new candidate tree then aggregates with another low level tree $A_1 \Rightarrow X$ using the *or* operator to generate $(B_1 \wedge A_2) \vee A_1 \Rightarrow X$ (as shown in Fig. 2). The low level trees are so called due to the fact that they always have less levels, and thus are shallower than the candidate trees. In this example, the low level trees are primitive trees only. However, it is not always the case as low level trees may be multi-level pattern trees as well (so long as they are shallower than the candidate trees).

For a classification application which involves several output classes, the worked model should have as many pattern trees as the number of output classes, with each pattern tree representing one class. When a new data sample is tested over a pattern tree, it traverses from the bottom to the top and finishes with a truth value, indicating the degree to which this data sample belongs to the output class of this pattern tree. The output class with the maximal truth value is chosen as the prediction class. For

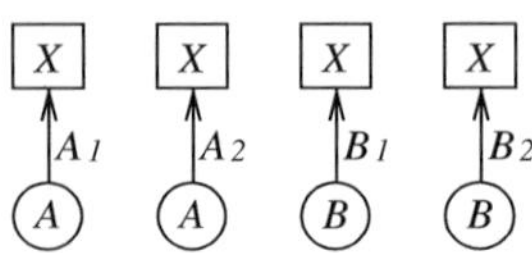

Fig. 1 Primitive pattern trees

Fig. 2 Two example pattern trees

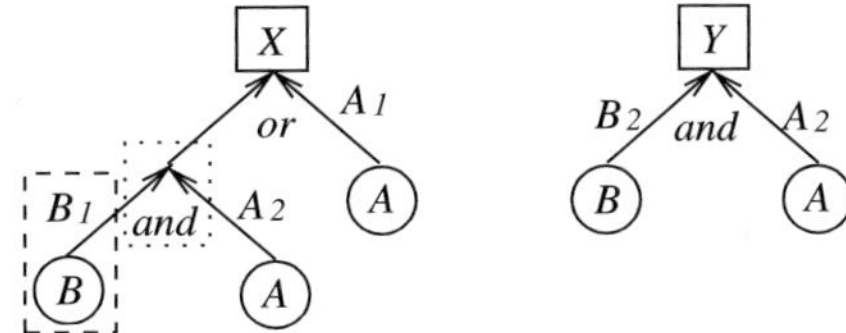

example, consider that a fuzzy data $A_1 = 0.8$, $A_2 = 0.2$, $B_1 = 0$, and $B_2 = 1$ is given for classification. As the truth values of this data over pattern trees for class X and Y are 0.8 and 0.2 respectively, X is chosen as the output class.

Conventional fuzzy rules can be extracted from pattern trees. For example, the following rules can be obtained from the pattern trees in Fig. 2.

$$Rule1 : IF\ A = A_1\ THEN\ class = X \tag{8}$$

$$Rule2 : IF\ A = A_2\ AND\ B = B_1\ THEN\ class = X \tag{9}$$

$$Rule3 : IF\ A = A_2\ AND\ B = B_2\ THEN\ class = Y \tag{10}$$

In addition to the conventionally used fuzzy aggregations MIN/MAX, pattern trees can use any aggregations as described in Sect. 2.2.

3 Proposed Pattern Tree Induction Method

Without losing generality, assume a dataset has n input variables A_i, $i = 1, 2, \ldots, n$ and one output variable B. For simplicity, further assume that both the input and output variables have m fuzzy linguistic terms denoted as A_{ij} and B_j, $i = 1, 2, \ldots, n$, and $j = 1, 2, \ldots, m$. That is, each data in the dataset is represented by a fuzzy membership value vector of dimension $(n + 1) \times m$. The task is to build m pattern trees for the m output classes (fuzzy terms). This section proposes a pattern tree induction method in pseudo code to build a pattern tree for class B_1. The pattern trees for other classes can be built following the same procedure.

Class TreeNode shows the structure of *TreeNode*, which represents leaf nodes (such as the dashed box) and internal nodes (the dotted box) in Fig. 2. *isLeaf* indicates if a node is a leaf node or an internal node. *aggreVals* stores the aggregated values on this node. Note that it stores the fuzzy values of the fuzzy set if the node is a leaf node. *similarity* is the similarity measure between *aggreVals* and the output class (B_1 in this case), and *parent* points to its parent node.

Class 1 TreeNode{

```
1:   boolean isLeaf
2:   double[] aggreVals
3:   double similarity
4:   TreeNode parent
5: }
```

Each node is further classified into either a leaf node or an internal node according to the status of *ifLeaf*. Leaf nodes and internal nodes are represented by two subclasses of *TreeNode*, namely the *LeafNode* and the *InternalNode*, as shown in classes 2 and 3 respectively. For leaf nodes, *usedAttribute* and *usedTerm* store the indices of the used fuzzy set. For instance, the use of $usedAttribute = 1$ and $usedTerm = 2$ stands for the use of fuzzy term A_{12}. For internal nodes, *aggreOpe* indicates which aggregation operator is used, *lambda* is the value associated with *aggreOpe* if *aggreOpe* = *WA* or *OWA*. *children* points to this node's child nodes.

Class 2 LeafNode extends TreeNode{

1:      int $usedAttribute$
2:      int $usedTerm$
3: }

Class 3 InternalNode extends TreeNode{

1:      int $aggreOpe$
2:      double $lambda$
3:      TreeNode[] $children$
4: }

Algorithm 4 is the main algorithm which constructs a pattern tree for a class. In particular, it finds and returns the root of the best pattern tree in terms of the similarity measure between the tree and the output class B_1. This algorithm takes two extra input arguments apart from the training dataset, namely the *numCandidateTrees* and *numLowLevelTrees*, indicating how many candidate trees and low level trees are used for generating optimal pattern trees. In particular, a vector consisting of *nm* primitive pattern trees is built by method *buildPrimitiveTrees()*. These primitive pattern trees are sorted in descending order according to the similarity measure to the fuzzy values of output class B_1. A copy of such vector *nodes* is made and is trimmed to size *numCandidateTrees*. In other words, the *numCandidateTrees* primitive trees which have the highest similarities are chosen to form a repository of candidate trees. $buildPatternTree()$ is then invoked to find the most optimal tree and the root node of this tree is returned.

Algorithm 5 shows how to build a pattern tree. It takes five input arguments: a vector of candidate tree, a vector of primitive trees, a vector of low level trees, the number of low level trees, and the level of the constructed tree so far. Candidate

---

Algorithm 4 main()

Input: A fuzzified dataset ds, int $numCandidateTrees$, and int $numLowLevelTrees$
Output: The root of the constructed pattern tree for class B_1
1: Vector $primitiveTrees$ =buildPrimitiveTrees(ds);
2: $primitiveTrees$.sort()
3: Vector $nodes$ = $primitiveTrees$.clone()
4: $nodes$.setSize($numCandidateTrees$)
5: TreeNode $root$ = buildPatternTree($nodes$, $primitiveTrees$, null, $numLowLevelTrees$, 0)
6: return $root$

Algorithm 5 buildPatternTree()

Input: Vector *nodes*, Vector *primitiveTrees*, Vector *lowLevelTrees*,
int *numLowLevelTrees*, and int *level*
Output: the root of the constructed pattern tree
 1: double *maxSim* = *nodes*[0]. similarity
 2:   Vector *candidateTrees* = new Vector()
 3: Vector *thisLowLevelTrees* = *lowLevelTrees*.clone()
 4: **for** $i = 0$ **to** *numCandiateTrees* **do**
 5:       Vector *aggregatedPrimitives* = aggregate(*nodes*[i], *primitiveTrees*)
 6:       Vector *aggregatedLowLevels* = aggregate(*nodes*[i], *thisLowLevelTrees*)
 7:       *lowLevelTrees*.addAll(*aggregatedPrimitives*)
 8:       *lowLevelTrees*.addAll(*aggregatedLowLevels*)
 9:       *candidateTrees*.addAll(*aggregatedPrimitives*)
10:       *candidateTrees*.addAll(*aggregatedLowLevels*)
11: **end for**
12: *lowLevelTrees*.sort()
13: *lowLevelTrees*.removeDuplicates()
14: *lowLevelTrees*.setSize(*numLowLevelTrees*)
15: *candidateTrees*.sort()
16: *candidateTrees*.removeDuplicates()
17: *candidateTrees*.setSize(*numCandidateTrees*)
18: double *newMaxSim* = *candiateTrees*[0].similarity
19: **if** *newMaxSim* $>$ *maxSim* **then**
20:      return buildPatternTree(*candidateTrees*, *primitiveTrees*,
         lowLevelTrees, *numLowLevelTrees*, *level* + 1)
21: **else**
22:      return *nodes*[0]
23: **end if**

---

trees are initially *numCandidateTrees* primitive trees which have the highest similarities (see Algorithm4). They are used to aggregate with the low level trees in a parallel manner. The vector of primitive trees remains the same in building the optimal pattern tree. The number of low level trees specifies how many low level trees are maintained to aggregate with the candidate trees, such that their aggregation may lead to higher similarities. The level of the tree indicates how many levels of the tree have been reached so far. The higher the level, the more complex the pattern tree. *buildPatternTree*() invokes itself when building the pattern tree. Each invocation results in the number of the level being increased by 1.

As candidate trees *nodes* are sorted in a descending order with respect to similarities, *maxSim* is the highest similarity among all the candidate trees. The candidate trees *candidateTrees* for the next level is initiated to be an empty vector. *thisLowLevelTrees* is cloned from *lowLevelTrees* to perform the aggregations with candidate trees in this invocation time. Within the **for** loop at lines 4 − 11, all current candidate pattern trees are aggregated with *primitiveTrees* and *lowLevelTrees* resulting in two aggregated vectors of trees (*aggregatedPrimitives* and *aggregatedLowLevels*) respectively. All these trees are added into vectors *lowLevelTrees* and *candidateTrees*. After the loop, both the *lowLevelTrees* and *candidateTrees* are sorted and the duplicate pattern trees are removed. They are then trimmed to the size of *numLowLevelTrees* and *numCandidateTrees* respectively.

Algorithm 6 aggregate()

Input: TreeNode $candidateTree$ **and Vector** $trees$
Output Aggregated pattern trees
1: Vector $optimalTrees$ = new Vector()
2: **for** $i = 0$ **to** $trees.size()$ **do**
3:   **if** (!isSubSet($trees[i]$, $candidateTree$)) **then**
4:       TreeNode $optimalTree$ = aggregateOptimal($candidateTree$, $trees[i]$)
5:       $optimalTrees$.add($optimalTree$)
6:   **end if**
7: **end for**
8: return $optimalTrees$

Now $newMaxSim$ is the highest similarity among all the current candidate trees. If it is greater than the previous one $maxSim$, the method buildTree() is kept on invoking to next level with updated candidate trees ($candidateTrees$) and low level trees ($lowLevelTrees$). Otherwise, the tree building process stops and the root of the pattern tree which has the highest similarity is returned.

The method $aggregate()$ is given in algorithm 3. It takes a pattern tree $candidate$ and a vector of pattern trees $trees$ as inputs and outputs a vector of aggregated pattern trees. In particular, for each pattern tree $trees[i], i = 0, \ldots, trees.size()$, in $trees$, if $trees[i]$ is not a subset of $candidateTree$, then it is used to aggregate with $candidate$ by invoking $aggregateOptimal()$ method. The condition of non-subset prevents repeated parts existing in the same pattern tree. Note that $trees[i]$ can be either a primitive tree or a multiple level tree. The aggregated results are added into vector $optimalTrees$ and the vector is returned. Note that the $optimalTrees$ remains the same size as $trees$.

The method $aggregateOptimal()$ is given in algorithm 3. It takes two pattern trees $candidate$ and $tree$ as inputs and outputs an optimal aggregated pattern tree. In particular, among all possible aggregation operators, this method chooses the optimal one to aggregate $candidate$ with $tree$. The optimal aggregated tree is constructed and returned.

The method $applyAggregation()$ is given in algorithm 3. It shows how to calculate the aggregated values when the aggregation operator and two vectors of fuzzy term values are given. This method additionally requires the use of the class fuzzy values if the aggregation operator is OWA or WA.

Method $calLambda()$ calculates a singleton value λ at lines 10 and 15 so that the vector of $aggregatedVals[i]$ at line 12 and 17 has the closest distance from $classVals$ in the sense of the mean square error.

Figure 3 summarizes the process of building a pattern tree. Each node T_i, $i = 0, 1, \ldots$, represents a pattern tree, with the horizontal axis indicating the trees's level, and the vertical axis indicating its similarity to the output. The level and similarity represent the tree's complexity and performance respectively. The goal is to find a pattern tree which has a high similarity with a low level. Assume initially three primitive trees T_0, T_1 and T_2 are available and they are denoted as level 0 trees. Let $numCandidateTrees = 2$ and $numLowLevelTrees = 3$, the nodes within the ellipses and rectangles represent candidate and low level trees respectively. In particular, T_0 and T_1 are candidate trees, and T_0, T_1 and T_2 are low level trees at

Fig. 3 Summary of building
a pattern tree

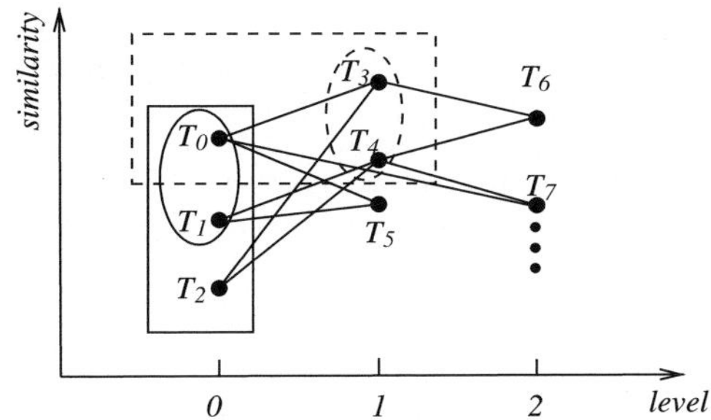

level 0. Each candidate tree aggregates with each low level tree to generate a level 1 tree: T_0 aggregates with T_1 to T_5, with T_2 to T_3; T_1 aggregates with T_0 to T_5, with T_2 to T_4. As T_3 and T_4 are level 1 trees and they have the highest similarities, they are chosen as candidate trees at level 1. The low level trees are updated to T_3, T_0 and T_4 which have the highest similarities.

The process is carried out until the highest similarity at one level decreases compared to that in the previous level (e.g., T_6's similarity is less than T_3), or no candidate tree is available for further process. As the size of used fuzzy terms increases exponentially with the growth of the level, a threshold tree level can be set in practice to limit the over-growth of the pattern tree. In particular, the pattern tree induction stops when the tree reaches the threshold tree level, or the highest similarity decreases from one level to the next, whichever happens first. Note that the value of threshold tree level depends on the number of input variables. If the variables are all relevant for the classification, the more input variables are, the higher value the threshold should be.

Generally, the higher values of $numCandidateTrees$ and $numLowLevelTrees$ are, the more thoroughly the search goes and the less likely the search is to be trapped in a local optimum. High values may lead to an exhaustive search in building a globally optimal pattern tree, which is unfortunately not practical for

---

Algorithm 7 aggregateOptimal()

Input: TreeNode $candidate$ and TreeNode $tree$
Output: An optimal aggregated pattern tree
1: double $maxSim$ = -1
2: double[] $candidateVals$ = $candidate$.aggregatedVals
3: double[] $treeVals$ = $tree$.aggregatedVals
4: **for** $i = 1$ **to** $numAggreOpes$ **do**
5:     $tempAggreVals$ = applyAggregation(i,
       $candidateVals$, $treeVals$, B_1.fuzzyVals)
6:     $tempSim$ = similarity($tempAggreVals$, B_1.fuzzyVals)
7:     **if** $maxSim < tempSim$ **then**
8:         $maxSim = tempSim$
9:         $aggregationOpe = i$
10:     **end if**
11: **end for**
12: TreeNode $parent$ = new InternalNode($aggregationOpe, candidate, tree$)
13: return $parent$

Algorithm 8 applyAggregation()

Input: Int $aggOpe$, **double[]** $aggreVals1$, **double[]** $aggreVals2$, **and double[]** $classVals$
    (required if the operator is WA or OWA)
Output: Aggregated values $aggregatedVals$
 1: **if** $aggOpe$!= **OWA or** $aggOpe$!= **OWA then**
 2:    **for** $i = 1$ **to sizeOf**($aggreVals1$) **do**
 3:      $aggregatedVals[i] =$
                   aggOpe($aggreVals1[i], aggreVals2[i]$)
 4:    **end for**
 5: **else if** $aggOpe ==$ **OWA then**
 6:    **for** $i = 1$ **to sizeOf**($aggreVals1$) **do**
 7:      $maxVals[i] = \max\{aggreVals1[i], aggreVals2[i]\}$
 8:      $minVals[i] = \min\{aggreVals1[i], aggreVals2[i]\}$
 9:    **end for**
10:    $\lambda =$ calLambda($maxVals, minVals, classVals$)
11:    **for** $i = 1$ **to sizeOf**($aggreVals1$) **do**
12:      $aggregatedVals[i] = \lambda * maxVals[i]+$

$$(1 - \lambda) * minVals[i]$$

13:    **end for**
14: **elss if** $aggOpe ==$ **WA then**
15:    $\lambda =$ calLambda($aggreVals1, aggreVals2, classVals$)
16:    **for** $i = 1$ **to sizeOf**($aggreVals1$) **do**
17:      $aggregatedVals[i] = \lambda * aggreVals1[i]+$
             $(1 - \lambda) * aggreVals2[i]$
18:    **end for**
19: **end if**

NP hard problems. In this paper, the setting of $numCandidateTrees = 2$ and $numLowLevelTrees = 3$ is used through out all examples and experiments to trade off the search effort and the capacity of escaping from local optima.

The configuration of $numCandidateTrees = 1$ and $numLowlevelTrees = 0$ forms specific pattern trees. In this configuration, only primitive trees (no low level trees) are considered to aggregate the candidate tree. The generated pattern trees have one and only one fuzzy set at each level except for level 0 (bottom level), forming snake-like trees such as in Fig. 4. They are denoted as *simple pattern trees* for later reference, in distinction with normal pattern trees. In this paper, two small examples are given to illustrate the pattern tree induction. The first one uses the simple pattern tree configuration with $numCandidateTrees = 1$ and $numLowlevelTrees = 0$, while the second uses the configuration of $numCandidateTrees = 2$ and $numLowlevelTrees = 3$.

Example 2. Assume an artificial dataset has two input variables A and B, with each having two fuzzy linguistic terms A_i and B_i, $i = 1, 2$. Also assume this dataset has two output classes X and Y. All the fuzzy membership values are shown in Table 3. To simplify the representation, only the construction of pattern tree for class X is presented and the MIN/MAX, OWA, and WA aggregations are considered here. The process of building the pattern tree is shown in Fig. 4. The pattern trees constructed in the process are denoted as T_i, $i = 0, \ldots, 9$. The are marked in the order of the time that the trees are constructed. That is, the trees with lower indices

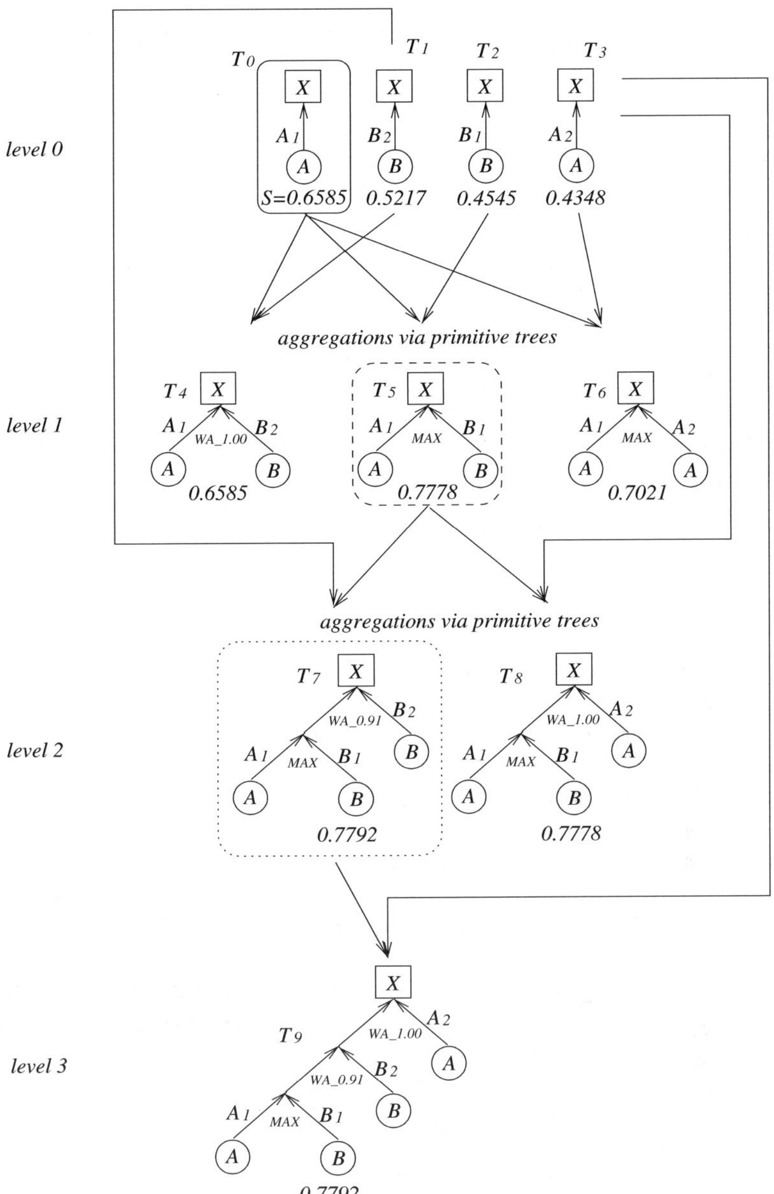

Fig. 4 Pattern tree construction for example 2

are constructed earlier than the ones with higher indices. The candidate trees at each level are surrounded by rectangular boxes with rounded corners. To distinguish them at different levels, solid, dashed, and dotted lines are used for level 0, 1, and 2 respectively. The number shown below each tree node is the similarity measure to

Table 3 An artificial dataset

A		B		Class	
A_1	A_2	B_1	B_2	X	Y
0.8	0.2	0.0	1.0	0.9	0.1
0.9	0.1	0.1	0.9	0.8	0.2
0.5	0.5	0.9	0.1	0.7	0.3
0.2	0.8	0.8	0.2	0.9	0.1
0.4	0.6	0.4	0.6	0.3	0.7
0.3	0.7	0.5	0.5	0.1	0.9

the fuzzy values of output class X. In particular, the process of building a pattern tree includes the following steps:

1. Build four primitive pattern trees T_0, T_1, T_2 and T_3, with the similarities calculated from (1) being 0.6585, 0.5217, 0.4545, and 0.4348 respectively.
2. Assign T_0 which has the highest similarity measure as the candidate tree at level 0.
3. Now consider the tree aggregation from level 0 to level 1. Try the rest of the primitive pattern trees T_1, T_2 and T_3 in turn to aggregate with T_0 using different aggregation operators MIN, MAX, OWA, and WA. The resulted similarity measures are shown in Table 4
4. Choose the aggregated tree T_5 as the new candidate tree at level 1. This is because the aggregation of T_0 and T_2 with MAX operator results in the highest similarity of 0.7778 which is greater than the similarity produced by T_0 alone. Also, the aggregated values of T_5, $(A_1\ MAX\ B_1) = \{0.8, 0.9, 0.9, 0.8, 0.4, 0.5\}$, are stored in the leaf node for further aggregation.
5. Now consider the tree aggregation from level 1 to level 2. Try the rest of the primitive trees T_1 and T_3 in turn to aggregate with T_5 using different aggregation operators. The resulting similarity measures are shown in Table 5.
6. Choose the aggregated tree T_7 as the new candidate tree at level 2, as this aggregation results in the highest similarity 0.7792 which is greater than the similarity produced by T_5. Note that a $\lambda = 0.91$ is stored in the node along with the WA operator. Also, the aggregated values of $T_7 = \{0.8185, 0.9000, 0.8259, 0.7444, 0.41856, 0.5\}$ are stored in this node for further aggregation.
7. Now consider the tree aggregation from level 2 to level 3. Try the rest of the primitive tree T_3 to aggregate with T_7 using different aggregation operators. None of the resulting similarity measures are greater than the current 0.7792. The pattern tree building process stops and the pattern tree T_7 is returned.

Table 4 Similarity measures of aggregations in step 3

	T_1	T_2	T_3
MIN	0.5610	0.3000	0.3500
MAX	0.6087	0.7778	0.7021
OWA	0.6065	0.7740	0.6815
WA	0.6585	0.6543	0.6356

Table 5 Similarity measures of aggregations in step 5

	T_1	T_3
MIN	0.5349	0.4762
MAX	0.7500	0.7143
OWA	0.7609	0.7109
WA	0.7792	0.7778

It is worth noting that each primitive tree is being checked to see if it is a subset tree of the candidate tree before the aggregation happens. For example, at level 1 for candidate tree T_5, only T_1 and T_3 among all primitive trees are allowed to aggregate. This is because other primitive trees have already appeared in T_5.

Example 3. Following the conditions given in the previous example, the construction of pattern tree using the configuration of *numCandidateTrees* $= 2$ and *numLowlevelTrees* $= 3$ is shown in Fig. 5. The candidate trees at each level are surrounded by rectangular boxes with rounded corners, while the low level trees are surrounded by rectangular boxes. Again, solid, dashed, and dotted lines are used for level 0, 1, and 2 respectively. Note that no low level trees are available at level 0. The building process is described by the following steps:

1. Build four primitive pattern trees T_0, T_1, T_2, and T_3, with the similarities calculated from (1) being 0.6585, 0.5217, 0.4545, and 0.4348 respectively.
2. At level 0, assign T_0 and T_1 which have the highest similarities as the candidate trees. There is no low level tree at this level.
3. Now consider the tree aggregation from level 0 to level 1. For the first candidate tree T_0, try the rest of the primitive pattern trees T_1, T_2, and T_3 in turn to aggre-

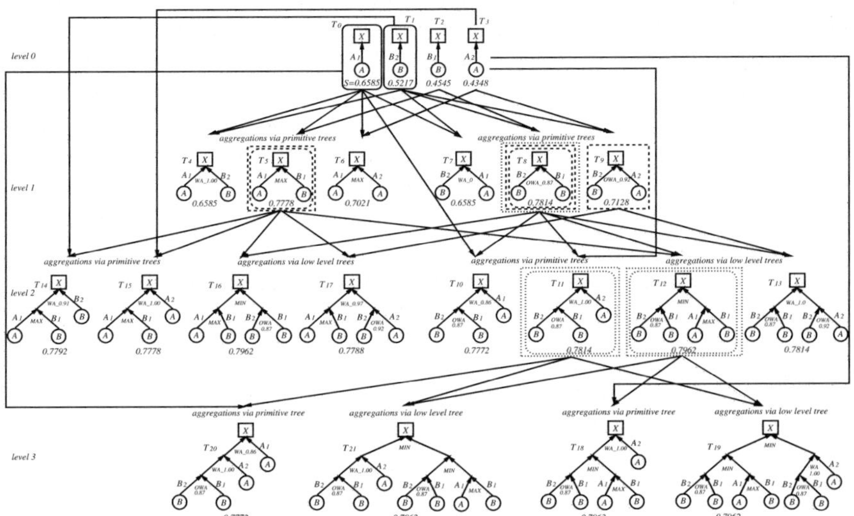

Fig. 5 Pattern tree construction for example 3

gate with, the best aggregated pattern trees are T_4, T_5 and T_6, with similarities being 0.6585, 0.7778, and 0.7021. As there is no low level trees available, no aggregations between the candidate tree and low level trees happen.

4. For the second candidate tree T_1, try the rest of the primitive pattern trees T_0, T_2 and T_3 in turn to aggregate with, the best aggregated pattern trees are T_7, T_8 and T_9, with similarities being 0.6585, 0.7814, and 0.7128. again, no aggregations between the candidate tree and low level trees happen.

5. At level 1, assign T_8 and T_5 which are level 1 trees and have the highest similarities as the candidate trees. Also, assign T_8, T_5 and T_9 which have the highest similarities as the low level trees.

6. Now consider the tree aggregation from level 1 to level 2. For the first candidate tree T_8, try the primitive pattern trees T_0 and T_3 which do not appear in the candidate tree. The resulted aggregated trees are T_{10} and T_{11}, with similarities being 0.7772 and 0.7814 respectively. Then the low level trees T_5 and T_9 are used to aggregate with T_8, resulting in T_{12} and T_{13} with similarities being 0.7962 and 0.7814.

7. Similarly for the second candidate tree T_5 at level 1, four aggregated trees T_{14}, T_{15}, T_{16}, and T_{17} are generated, with similarities being 0.7792, 0.7778, 0.7962, and 0.7788 respectively.

8. This process is carried out in the same manner from level 2 to level 3. T_{18}, T_{19}, T_{20}, and T_{21} are constructed. As none of them has a higher similarity than T_{12}, the building process stops and T_{12} is returned.

It is worth noting that the duplication removal is performed in selecting the candidate trees and low level trees. For example, at level 2, although T_{16} has a high similarity measure, it is not selected as a candidate tree or low level tree as tree T_{12} is identical to it, and also T_{12} is constructed earlier. It can be shown from Fig. 4 and Fig. 5 that, the search space for example 2 is a sub-space of that for example 3. Generally, the higher values of $numCandidateTrees$ and $numLowlevelTrees$, the more of the search space the building process reaches. On the other hand, the configuration of $numCandidateTrees = 1$ and $numLowlevelTrees = 0$ attempts to find an optimal simple tree with little computation effort.

4 Comparison to SBM, WSBM and Decision Trees

As the fuzzy subsethood based method (SBM) [1] and its extension, the weighted subsethood based method (WSBM) [12], are two specific cases of patterns trees (see below), they are chosen along with the well-known fuzzy decision tree induction [18] to compare with the proposed pattern tree induction method. For a clear representation of the comparison, the small artificial dataset as shown in Table 3 is used to generate fuzzy rules for SBM, WSBM, decision trees induction, and pattern trees induction.

4.1 SBM

The subsethood based method consists of three main steps:

1. Classifying training data into subgroups according to class values, with each group having the data which prefer voting for one class.
2. Calculating *fuzzy subsethood* values of a certain output class to each input fuzzy term. A fuzzy subsethood value of X with regard to A_1, $Sub(X, A_1) = \frac{X \cap A_1}{X}$, represents the degree to which X is a subset of A_1 (see [1, 12]), where $\cap$ is a t-norm operator (MIN is used).
3. Creating rules based on fuzzy subsethood values. In particular, for each input variable, the fuzzy term that has the highest subsethood value (must be greater than or equal to a pre-specified threshold value $\alpha \in [0, 1]$, 0.9 used in [1]) will be chosen as an antecedent for the resulting fuzzy rules.

Regardless of the subgrouping process and the use of *subsethood* rather than *similarity* measure, the SBM always generates a pattern tree which has only one level. In this tree, the fuzzy terms which have the greatest subsethood values (must be greater than or equal to α) per input variable are aggregated via a t-norm operator (MIN) to predict a class concerned.

For the artificial dataset, two groups are created. The first has the first four data points, which prefer voting for class X rather than Y, and the second has the remaining two data points. Assume class X is considered, the subsethood values of X to A_i, B_i, $i = 1, 2$, are calculated as 0.6970, 0.4848, 0.4848, and 0.6061 respectively. If α is set to 0.9 as in [1], no fuzzy rules (pattern trees) can be generated. However, for comparison purposes, α is set to 0.6 to generate the pattern tree for class X as shown in Fig. 6. This figure also shows the pattern tree for Y which is constructed in the same manner.

4.2 WSBM

WSBM is a weighted version of SBM. It uses a certain weighting strategy to represent fuzzy terms rather than choosing the one which has the greatest subsethood value per variable. In particular, the weight for fuzzy term A_i is defined as

$$w(X, A_i) = \frac{Sub(X, A_i)}{max_{j=1,2} Sub(X, A_j)}. \tag{11}$$

Fig. 6 Pattern trees generated by SBM using the artificial dataset

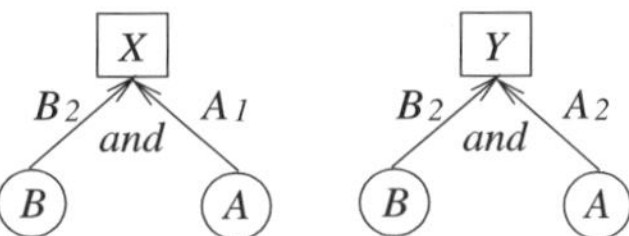

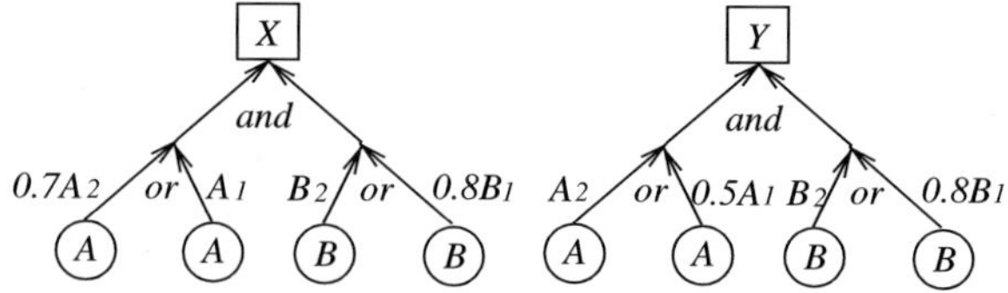

Fig. 7 Pattern trees generated by WSBM using the artificial dataset

Using this equation, the weights for A_1, A_2, B_1 and B_2 in the classification of X are calculated as 1, 0.7, 0.8, and 1 respectively. The fuzzy rules generated by WSBM form fixed structured pattern trees as shown in Fig. 7.

Unlike the proposed pattern tree induction method which can generate different structured patter trees, SBM and WSBM generate trees with fixed structures. In particular, SBM trees have only one level, on which the fuzzy terms which have the highest subsethood values per input variable aggregate via a t-norm operator. WSBM trees have two levels. On the bottom level, all fuzzy terms (with different weights) for each variable aggregate via a t-conorm operator, on the top level the aggregated values further aggregate via a t-norm operator. The fixed structures of SBM and WSBM may prevent them from obtaining high accuracy in classification applications (see Sect. 4.6).

4.3 Fuzzy Decision Tree Induction

The decision tree induction [10] is a classic machine learning method. It has been extended to induce fuzzy decision tree by Yuan and Shaw [18], where the fuzzy entropy is used to guide the search of most effective branches. A fuzzy decision tree as shown in Fig. 8 can be built using [18] over the artificial dataset. It has a root on the top and several leaf nodes on the bottom. When a new data sample needs to be classified by the fuzzy decision tree, it traverses from the root to the bottom. The output class of a leaf node in the bottom, which the data sample reaches with the highest truth value, is chosen as the prediction class.

For the same training dataset, fuzzy decision tree induction may generate different results with different numbers of minimal data points per leaf node, which are used as criteria to terminate the tree building process. Considering only six data samples are available, this can be set to 1 to 6. Among all these, the best result (as shown in Fig. 8, when the number of leaf nodes is set to 1 or 2) has been chosen for comparison in Sect. 4.6.

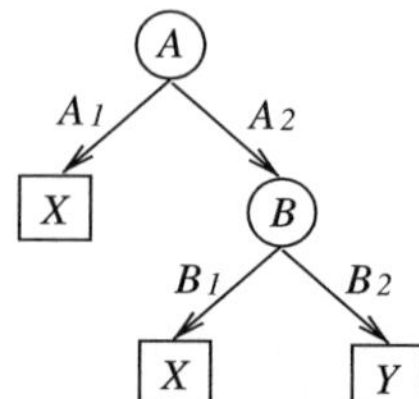

Fig. 8 A decision tree generated by fuzzy decision tree induction using the artificial dataset

4.4 Relation between Fuzzy Decision Trees and Pattern Trees

A fuzzy decision tree can be converted to a set of pattern trees whose size is equal to the number of output classes. In particular, the easiest way is to convert a fuzzy decision tree to a set of semi-binary pattern trees. That is, there are maximally two branches allowed for each node (except for the root) in the pattern tree. The conversion process is outlined as follows. For each fuzzy rule (a branch from the root to a leaf node) in a decision tree, the input fuzzy terms are connected by a t-norm operator (usually MIN) in different levels of the pattern tree. A fuzzy rule consisting of n fuzzy terms results in a $(n - 1)$-level binary pattern tree as the bottom level contains two fuzzy terms. The fuzzy rules which have the same classification class are connected by a t-conorm (usually MAX) at the top level (level n) to construct a pattern tree for this output class. The number of levels of the generated semi-binary pattern trees remains the same as the fuzzy decision tree, regardless whether the decision tree is binary or not. Fig. 2 shows the pattern trees which are equivalent to the decision tree as shown in Fig. 8. They both have the same rule base as listed in $(8) - (10)$.

The conversion from a decision tree to pattern trees is not unique. For example, an alternative conversion from Fig. 8 is shown in Fig. 9, which can be represented by two rules as following.

$$Rule1 : IF A = A_1 \ OR \ B = B_1 \ THEN \ class = X, \tag{12}$$

$$Rule2 : IF A = A_2 \ AND \ B = B_2 \ THEN \ class = Y. \tag{13}$$

Note that these two fuzzy rules are functionally equal to rules (8), (9) and (10). Such a conversion is closely related to Zadeh's work on compactification [20]. When the size of fuzzy terms for each variable increases, the conversion of a fuzzy decision tree to multi-branch pattern trees are desirable and the work on that is on-going.

On the other hand, pattern trees can be converted to a decision tree. The fuzzy decision tree shown in Fig. 8 can be converted from either Fig. 2 or Fig. 9.

It is worth noting that decision trees and pattern trees are different in terms of four aspects: 1) the former focus on separating data samples which have *different* output classes, while the latter focus on representing the structures of data samples which have the *same* output classes; 2) for each internal node, the former consider *all* fuzzy terms of the chosen input variable, whist the latter only consider *one*; 3) the former normally make use of MIN and MAX aggregations *only*, while the latter can use *any* aggregations as described in Sect. 2.2; and 4) the tree induction methods are completely different, with the former based on the heuristics of *entropy measure* [18] while the latter on the heuristics of *similarity measure*.

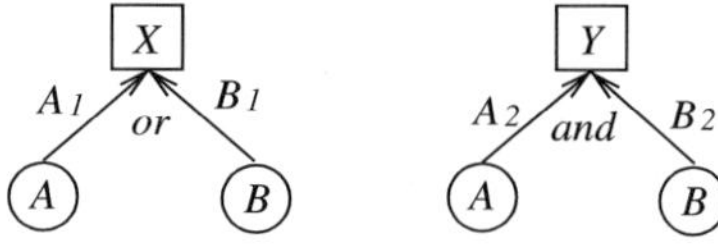

Fig. 9 Alternative pattern trees converted from the decision tree in Fig. 8

4.5 Advantages of Trees

Both decision trees and pattern trees can be converted to conventional fuzzy rules. In terms of the computational complexity, both decision trees and patterns trees have the advantage over conventional rule bases. An example is given below to show how pattern trees can simplify the computation.

Definition 2. Let assume fuzzy variable A, B, C, D, E, and F with each having two fuzzy linguistic terms A_i, B_i, C_i, D_i, E_i and F_i, $i = \{1, 2\}$. One pattern tree associated with the output class X is shown in Fig. 10. The directly extracted fuzzy rule from the pattern tree can be written as

$$IF\ (((A = A_1)\ OR\ (B = B_2))\ AND\ (C = C_1))\ AND$$
$$(((D = D_1)\ AND\ (E = E_2))\ OR\ (F = F_2))$$
$$THEN\ class = X, \tag{14}$$

which is equivalent to the conventional fuzzy rules (without using the OR operator explicitly within a rule):

$$IF\ A = A_1\ AND\ C = C_1\ AND\ D = D_1\ AND\ E = E_2$$
$$THEN\ class = X \tag{15}$$
$$IF\ A = A_1\ AND\ C = C_1\ AND\ F = F_2$$
$$THEN\ class = X \tag{16}$$
$$IF\ B = B_2\ AND\ C = C_1\ AND\ D = D_1\ AND\ E = E_2$$
$$THEN\ class = X \tag{17}$$
$$IF\ B = B_2\ AND\ C = C_1\ AND\ F = F_2$$
$$THEN\ class = X \tag{18}$$

If a data sample needs to be classified with the conventional fuzzy rules, the four rules should fire in turn, leading to the evaluation of $A = A_1$, $B = B_2$, $C = C_1$, $D = D_1$, $E = E_2$, $F = F_2$, the *and* operator, and the *or* operator to be 2, 2, 4, 2, 2, 2, 10 and 3 times respectively. However, for the computation upon the pattern tree directly, they are only computed 1, 1, 1, 1, 1, 1, 3 and 2 times respectively.

From this example, it can be concluded that although the pattern trees (or decision trees) may share the same fuzzy rule base with a conventional rule model, it

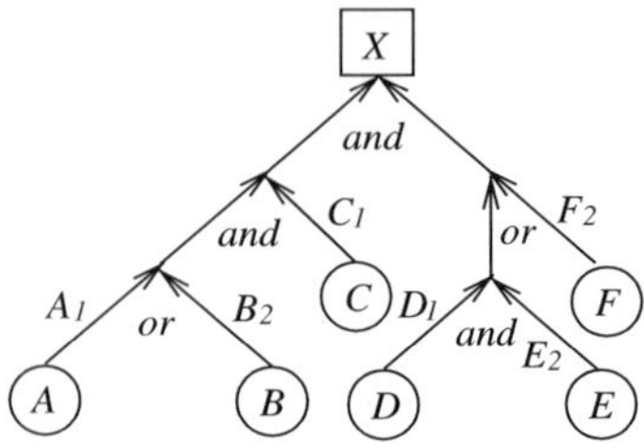

Fig. 10 A example pattern tree

can significantly reduce the computation time in the test stage. This advantage is especially important when quick response is required for fuzzy applications.

4.6 Pattern Trees and the Comparison Results

The aggregations of MIN/MAX, algebraic AND/OR, Šukasiewicz, EINSTEIN, OWA, and WA are considered in building a pattern tree using the proposed method. Two similarity measures as shown in (1) and (3) are used and the best result (using 3) is reported. The threshold tree level for pattern tree induction is set to 3. Both simple pattern tree (with $numCandidateTrees = 1$ and $numLowlevelTrees = 0$) and pattern tree (with $numCandidateTrees = 2$ and $numLowlevelTrees = 3$) configurations are used, resulting in simple pattern trees and pattern trees as shown in Fig. 11 and Fig. 12 respectively. It is worth noting that, if only MIN/MAX are allowed in the simple pattern tree induction, the generated trees using similarity defined in either (1) or (3) are exactly the same as the trees shown in Fig. 9, with *and* being MIN and *or* being MAX.

Now apply this artificial dataset to the rules (or trees) generated by SBM, WSBM, fuzzy decision trees, simple pattern trees, and pattern trees, the results including the number of correctly predicted data samples (No.), the root mean square error (RMSE) of the predictions for class X, Y, and their average (ARMSE) are shown in Table 6.

It is clear that the pattern trees perform the best, in terms of both the correctly predicted number and the mean square error, among all the methods. Simple pattern trees perform slightly worse, but still better than the fuzzy decision tree. This example shows that SBM and WSBM cannot get good results as their fixed pattern tree structures prevent them appropriately representing the structure of the dataset. Fuzzy decision trees can generate different tree structures. However, they lack some candidate search spaces represented by the t-conorm aggregations of fuzzy terms of different variables. For instance, in this example, Fig. 8 considers whether A_1, A_2, $A_1 \vee A_2$, and $A_2 \wedge B_1$ etc. are important or not for the classification, but not $A_1 \vee B_1$ explicitly, thus failing to find such an important rule. In contrast, the proposed pattern tree induction method explicitly considers both t-norm and t-conorm aggregations of fuzzy terms in building trees. Thus, it is more likely to find optimal solutions.

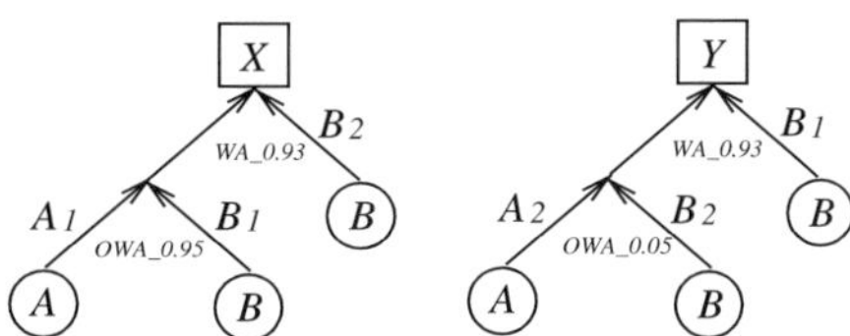

Fig. 11 Pattern trees generated using simple pattern tree configuration

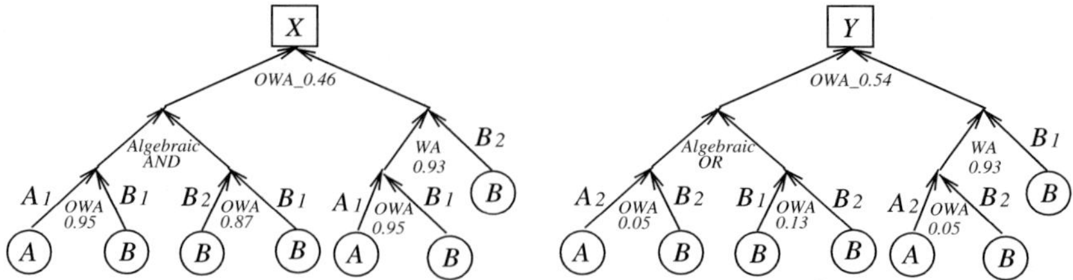

Fig. 12 Pattern trees generated using pattern tree configuration

Table 6 Results of SBM, WSBM, Fuzzy decision tree, and patter trees using the artificial dataset

	No.	RMSE(X)	RMSE(Y)	ARMSE
SBM	4	0.3916	0.2	0.2958
WSBM	4	0.2386	0.3435	0.2911
Fuzzy decision tree	5	0.2	0.2	0.2
Simple pattern trees	6	0.1956	0.1956	0.1956
Pattern trees	6	0.1715	0.1715	0.1715

5 Experimental Results using Saturday Morning Problem Dataset and UCI Datasets

To demonstrate the effectiveness of the proposed pattern tree induction method, Sect. 5.1 presents comparative studies with the fuzzy subsethood based method (SBM) [1], the weighted subsethood based method (WSBM) [12], and the fuzzy decision tree [18] using the Saturday Morning Problem dataset [18, 1, 12]. Section 5.2 presents more comparison to fuzzy decision trees over the datasets obtained from UCI machine learning repository [8], which include Iris-Plant, Wisconsin Breast Cancer, Glass identification, Diabetes, and Wine Recognition. Note that the results of SBM and WSBM over the Iris-Plant dataset, which were reported in [12], are also provided here for comparison.

5.1 Saturday Morning Problem

The Saturday Morning Problem (SMP) dataset has 16 data samples, with each having 4 input variables, *outlook, temperature, humidity* and *wind*. Each variable has fuzzy terms as follows: *outlook = {sunny, cloudy, rain}, temperature ={hot, mild, cool}, humidity ={humid, normal}* and *wind ={windy, not-windy}*. The classification result is one of the plans to be taken: *class ={volleyball, swimming, weight-lifing}*.

Table 7 shows the best performances of SBM, WSBM, fuzzy decision tree, simple pattern trees, and pattern trees respectively. Note that the threshold pattern tree level is set to be 3. No. is the number of correctly predicted data points. RMSE(1), RMSE(2) and RMSE(3) are the root mean square errors for class *volleyball, swimming* and *weight-lifting* respectively and ARMSE is the average of these three.

Table 7 Performance of SBM, WSBM, Fuzzy decision trees, and patter trees using SMP dataset

	No.	RMSE(1)	RMSE(2)	RMSE(3)	ARMSE
SBM	15	0.2046	0.1620	0.1323	0.1663
WSBM	15	0.2818	0.2480	0.4398	0.3232
Fuzzy decision tree	13	0.2681	0.1369	0.1953	0.2001
Simple pattern trees	14	0.3279	0.1750	0.2669	0.2566
Pattern trees	15	0.2078	0.0995	0.2256	0.1776

Among the comparison, SBM obtains the best result, in terms of both the number of correctly predicted data points and the root mean square errors (indicating the differences between the predicted values and the actual values). This is because a default fuzzy rule has been introduced in SBM method to predict class *weight-lifting*, as it cannot generate any "meaningful" rules for this class. This certainly helps in the example, but may not be useful in other datasets – especially when there are more than one class which SBM cannot generate any "meaningful" rules for.

WSBM seems get a good result in terms of the correctly predicted number, it however has the greatest ARMSE among all methods. This reveals that the average difference between the predicted values and the actual values is the highest.

Fuzzy decision tree has a low average RMSE, but only correctly predicts 13 out of 16 data samples. Simple pattern tree predicts 14 correctly, but with a higher average RMSE. Pattern trees perform very well in terms of both correctly predicted number and ARMSE. It is worth noting that, unlike SBM method, three pattern trees are constructed here for three classes respectively. With respect to finding a correct pattern for each class, pattern trees perform nearly the same as SBM in finding the first one, but much better in the second. However, the default rule which SBM uses outperforms pattern trees for the third class.

5.2 UCI Machine Learning Datasets

The datasets of Iris-Plant, Wisconsin Breast Cancer, Glass Identification, Diabetes, and Wine Recognition obtained from UCI machine learning repository [8], which have been widely used as benchmarks in classification applications, are summarized in Table 8.

For all datasets except Iris-Plant, a simple fuzzification method based on six evenly distributed trapezoidal membership functions for each input variable is used to transform the crisp values into fuzzy values. To be comparable to the result reported in [12], Iris-Plant dataset uses three evenly distributed trapezoidal fuzzy sets for each variable.

All datasets (*ds*) are divided into training datasets and test ones. Assume every dataset is labeled for data samples, the training sets (*ds-odd*) contain the odd numbered data samples and the test sets (*ds-even*) contain the even numbered ones. The performances of the SBM and WSBM (only for Iris-Plant dataset),

Table 8 Summary of Iris-Plant, Wisconsin Breast Cancer, Glass Identification, Diabetes, and Wine Recognition Datasets

Name	Number of data	Number of inputs	Number of classes
Iris-Plant	150	4	3
Wisconsin Breast Cancer	699	9	2
Glass identification	214	9	6
Diabetes	768	8	2
Wine recognition	178	13	3

fuzzy decision tree (FDT), simple pattern trees (SimPT), pattern trees with level no more than 5 (PTLevel5) and pattern trees with level no more than 9 (PTLevel9) over different combinations of training-test sets for different datasets are shown in Tables 9, 10, 11, 12 and 13. PTLevel5 trees maintain a good comprehensibility as they have maximal $2^5 = 32$ leaf nodes. However, PTLevel9 trees may be too complex as they have maximal $2^9 = 512$ leaf nodes, although the trees usually have much fewer leaf nodes than this maximum. As the Iris-Plant dataset has only 4 input variables, there is no need to set the threshold tree level to 5. Actually the maximal level the trees reach is no more than 3.

All the results reported for fuzzy decision trees are the best results among the use of different numbers of leaf nodes. The MIN/MAX, algebraic AND/OR, Šukasiewicz, EINSTEIN, OWA, and WA are considered in building pattern trees. Note that $*$ in Table 9 indicates the results are not available in [12].

PTLevel9 trees outperform the other trees over all five datasets. They obtain the highest prediction accuracies in all experiments, except for the training-test sets being *ds* and *ds* in Glass Identification and Diabetes datasets. On the other hand, FDT nearly performs the worst except in Diabetes dataset, in which it outperforms SimPT. SimPT performs roughly the same to PTLevel5 – both of them perform better than FDT, but worse than PTLevel9 trees.

It is worth noting that pattern trees perform in a consistent way for different combinations of training and test datasets, while fuzzy decision trees do not. This can be seen from the Glass Identification and Diabetes datasets. Fuzzy decision trees generate large differences in classification accuracy between the first (or the second) combination of the training-test datasets and the third one, due to the over-fitting problem. The reason is that decision tree induction considers only a portion of the whole training dataset in choosing the branches at low levels of trees. The lack of using the whole training dataset inevitably prevents the method finding better

Table 9 Prediction accuracy of SBM, WSBM, fuzzy decision trees, simple pattern trees, and pattern trees using Iris-Plant dataset

Training	Testing	SBM	WSBM	FDT	SimPT	PTLevel3
ds-odd	ds-even	80%	93.33%	97.33%	97.33%	97.33%
ds-even	ds-odd	78.67%	93.33%	97.33%	98.67%	98.67%
ds	ds	*	*	97.33%	97.33%	97.33%

Table 10 Prediction accuracy of fuzzy decision trees, simple pattern trees, and pattern trees using Wisconsin Breast Cancer dataset

Training	Testing	FDT	SimPT	PTLevel5	PTLevel9
ds-odd	ds-even	93.70%	94.84%	94.84%	95.41%
ds-even	ds-odd	95.71%	96.57%	95.42%	96.57%
ds	ds	96.85%	97.42%	97.13%	98.14%

tree structures for all the dataset. In contrast, pattern trees make use of the whole data in building each level of the tree, which ensures the tree to keep good generality for classifications. Therefore, even complex pattern trees do not suffer from over-fitting.

Simple pattern trees usually have compact structures, and they can be simpler than fuzzy decision trees. For example, Fig. 13 shows the decision tree (with prediction accuracy of 92.13%) generated using Wine Recognition dataset with training set being *ds-odd* and test set being *ds-even*. The ellipses are the input variables and the rectangles are the output classes (0, 1, or 2). Note that the empty rectangles mean no decision class is available. $Fi, i = 0, \ldots, 5$, are the fuzzy terms associated with each input variable.

Figure 14 shows the three simple pattern trees (with prediction accuracy of 93.25%) generated using the same training dataset. In terms of the size of leaf nodes, the three simple pattern trees have $6 \times 3 = 18$ leaf nodes in total as each pattern tree per class has 6 fuzzy terms, while the decision tree has 26. Figure 15 shows the constructed level5PT tree (with prediction accuracy of 94.38%) for class 0, which

Table 11 Prediction accuracy of fuzzy decision trees, simple pattern trees, and pattern trees using Glass Identification dataset

Training	Testing	FDT	SimPT	PTLevel5	PTLevel9
ds-odd	ds-even	55.14%	61.68%	62.61%	62.61%
ds-even	ds-odd	57.94%	55.14%	58.87%	60.74%
ds	ds	87.75%	71.02%	70.09%	72.89%

Table 12 Prediction accuracy of fuzzy decision trees, simple pattern trees, and pattern trees using Diabetes dataset

Training	Testing	FDT	SimPT	PTLevel5	PTLevel9
ds-odd	ds-even	75.26%	72.65%	76.82%	77.60%
ds-even	ds-odd	74.48%	72.13%	74.21%	75.26%
ds	ds	91.15%	75.52%	75.39%	76.30%

Table 13 Prediction accuracy of fuzzy decision trees, simple pattern trees, and pattern trees using Wine Recognition dataset

Training	Testing	FDT	SimPT	PTLevel5	PTLevel9
ds-odd	ds-even	92.13%	93.25%	94.38%	96.62%
ds-even	ds-odd	91.01%	97.75%	97.75%	97.75%
ds	ds	97.75%	98.31%	97.75%	98.31%

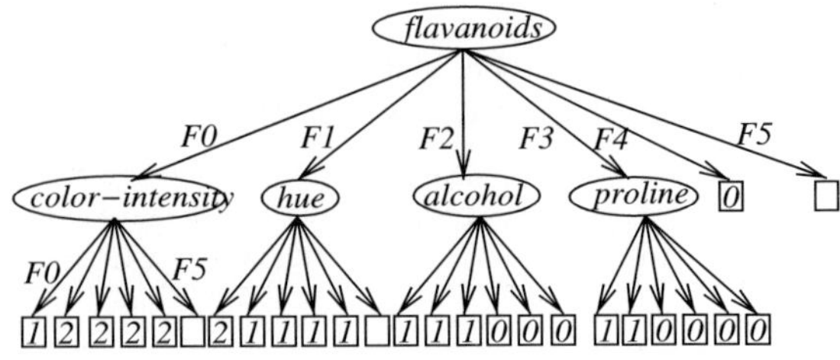

Fig. 13 Decision tree generated using Wine Recognition dataset

has 9 leaf nodes. Its complexity is still acceptable although it is more complex than fuzzy decision trees or simple pattern trees.

The complexity can also be compared in the form of rule representation. For example, if only class 0 is considered, the rules extracted from the decision tree, simple pattern trees, and pattern trees are listed in (19), (20) and (21) respectively. It can be seen that the simple pattern tree has the simplest form.

$$(IF\ flavanoids = F_2\ AND\ alcohol = F_3)\ OR$$
$$(IF\ flavanoids = F_2\ AND\ alcohol = F_4)\ OR$$
$$(IF\ flavanoids = F_2\ AND\ alcohol = F_5)\ OR$$
$$(IF\ flavanoids = F_2\ AND\ proline = F_2)\ OR$$
$$(IF\ flavanoids = F_2\ AND\ proline = F_3)\ OR$$
$$(IF\ flavanoids = F_2\ AND\ proline = F_4)\ OR$$
$$(IF\ flavanoids = F_2\ AND\ proline = F_5)\ OR$$
$$(IF\ flavanoids = F_5)$$
$$THEN\ class = 0 \tag{19}$$

$$(((((IF\ flavanoids = F_3\ OWA\ malic\text{-}acid = F_1)\ OWA$$
$$proline = F_3)\ OWA\ proline = F_4)\ OWA$$
$$proline = F_5)\ WA\ alcalinity = F_1)$$
$$THEN\ class = 0 \tag{20}$$

$$((((((IF\ flavanoids = F_3\ OWA\ malic\text{-}acid = F_1)\ Luc\_OR$$
$$proline = F_3)\ OWA$$

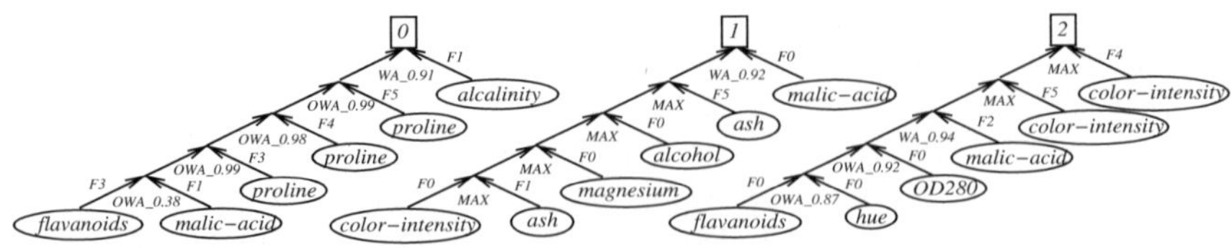

Fig. 14 Simple pattern trees generated using Wine Recognition dataset

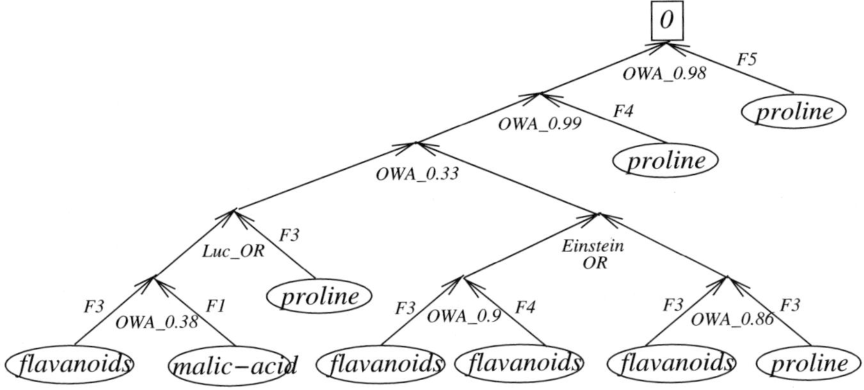

Fig. 15 A pattern tree generated using Wine Recognition dataset for class 0

$$((flavanoids = F_3 \; OWA \, flavanoids = F_4) \; Einstein\_OR$$
$$(flavanoids = F_3 \; OWA \; proline = F_3)))$$
$$OWA \; proline = F_4) \; OWA \; proline = F_5)$$
$$THEN \; class = 0 \qquad (21)$$

With respect to choosing appropriate methods for real world applications, PTLevel9 trees are favored if performance is a critical factor. The drawback is that these trees may compromise the comprehensibility, as the number of leaf nodes tends to be large. If, however, the comprehensibility is critical for the solution to be considered, SimPT trees are a good choice.

6 Weighted Pattern Trees

The classification using pattern trees discussed in Sect. 2.3 is based on the assumption that all pattern trees each have the same confidence on predicting a particular class, though it is not always the case in real world applications. *Weighted trees* are introduced to resolve this problem. For each tree, the similarity of such tree to the output class is served as a degree of confidence, to reflect how confident to use this tree to predict such a class. For example, if the two trees in Fig. 2 have similarities of 0.1 and 0.8 respectively, they can be called weighted pattern trees with weights of 0.1 and 0.8. The prediction using weighted pattern trees is the same as pattern trees, except that the final truth values are multiplied by the weights of trees. As an example, let's revise the classification problem in Sect. 2.3; consider classifying the fuzzy data $A_1 = 0.8$, $A_2 = 0.2$, $B_1 = 0$, and $B_2 = 1$ over pattern trees (with weights of 0.1 and 0.8) in Fig. 2, its truth values over pattern trees for class X and Y change to 0.08 and 0.16 respectively, and Y (rather than X) is therefore chosen as the output class. This reflects the fact that, if a tree has a low weight, even an input data has a high firing strength over such pattern tree, the prediction is not confident.

Note that this example is merely used to show how weighted pattern trees work. In practice, a pattern tree with weight 0.1 may not be trusted to predict a class.

The concept of weighted pattern trees is important. It offers an option to trade off the complexity and performance of pattern trees. The pattern tree building process can stop at very compact trees, if it detects that the similarities (weights) of such trees are already high enough. In addition, it enhances the comprehensibility of pattern trees. For example consider the construction of the pattern tree for class Y in Fig. 2, assume that the tree growing from the primitive tree $B_2 \Rightarrow Y$ to $B_2 \wedge A_2 \Rightarrow Y$ leads to the weight increase from 0.6 to 0.8, this gradual change can be interpreted in a comprehensible way:

$$IF\ B = B_2\ THEN\ it\ is\ possible\ that\ class = Y, \quad (22)$$
$$IF\ B = B_2\ AND\ A = A_2\ THEN\ it\ is\ very\ possible\ that\ class = Y, \quad (23)$$

if users pre-define semantic ranges of weights, say *less possible*: $[0, 0.3)$, *possible*: $[0.3, 0.7)$, and *very possible*: $[0.7, 1]$. Thus, the graduate change of confidence of pattern trees can be monitored from the pattern tree induction process. This provides a very transparent way for fuzzy modeling.

7 Experimental Results using BT Dataset

In this section, different variants of pattern trees, namely simple pattern trees, weighted simple pattern trees, pattern trees, and weighted pattern trees, are applied to a sample customer satisfaction dataset from BT. This dataset has a total of 26 input parameters representing ease of contact, problem understanding, service quality, repair time, and overall event handling. Among the input parameters, 6 are numerical parameters and the rest 20 are category ones, with the number of possible values being from 2 up to 17. The output parameter consists of 7 classes reflecting varying degrees of customer satisfaction.

The BT customer satisfaction dataset has 16698 data points in total. Let ds, *ds-odd* and *ds-even* be the datasets which contains the whole, the odd numbered, and the even numbered data points respectively. The number of data per class for these three datasets are shown in Table 14, with ci, $i = 0, \ldots, 6$ standing for class i. As can be seen, this dataset is not well balanced as the number of data per class varies significantly. The experiments of (weighted) pattern trees are carried out in three combinations of training-test datasets, namely, *odd-even*, *even-odd*, and *ds-ds*. In all experiments, a simple fuzzification method based on three evenly distributed trapezoidal membership functions for each numerical input parameter is used to transform the crisp values into fuzzy values. All aggregations as listed in Table 2 are allowed in pattern trees. The similarity measure as shown in (3) is used.

Table 14 Number of data per class for ds, ds-odd and ds-even datasets

	c0	c1	c2	c3	c4	c5	c6
ds	1895	7289	4027	382	1361	853	891
ds-odd	949	3659	1990	197	660	448	446
ds-even	946	3630	2037	185	701	405	445

7.1 Prediction accuracy and overfitting

The prediction accuracy and rule number of the fuzzy decision trees (FDT) with respect to the minimal number of data per leaf node (used as criteria to terminate the training), over different combinations of training-test sets are shown in Fig. 16. The prediction accuracy of pattern trees (PT) and weighted pattern trees (WPT) with respect to different tree levels, over different combinations of training-test sets is shown in Fig. 17.

The experiments show that weighted pattern trees and pattern trees perform roughly the same. In fact, the former slightly outperform the latter. Table 15 shows the highest prediction accuracy of fuzzy decision trees, (weighted) simple pattern trees and (weighted) pattern trees over different combinations of training-test sets. Both weighted and unweighted pattern trees can obtain higher prediction accuracy than fuzzy decision trees in *odd-even* and *even-odd* combinations. However, if considering *ds-ds* combination, fuzzy decision trees perform much better. This just reflects the overfitting of fuzzy decision trees, since fuzzy decision trees generate large differences in classification accuracy between the *odd-even*, *even-odd* combinations and *ds-ds* one. The reason is that decision tree induction considers only a portion of the whole training dataset in choosing the branches at low levels of trees. The lack of using the whole training dataset inevitably prevents the method finding generalized tree structures for all the dataset. In contrast, pattern trees make use of the whole data in building each level of the tree, which ensures the tree to keep good generality for classifications. Therefore, even complex pattern trees do not suffer from over-fitting.

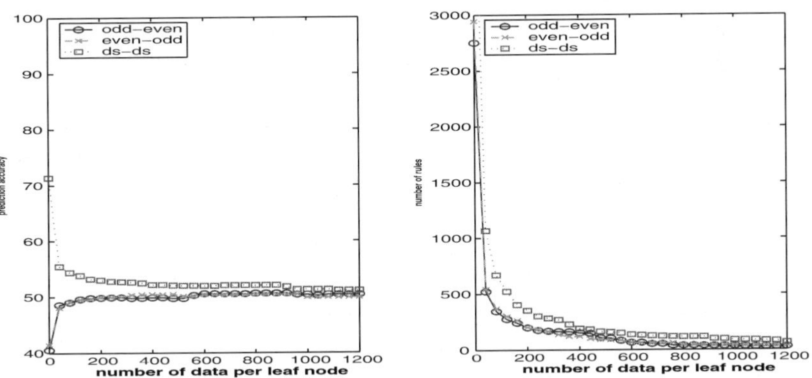

Fig. 16 Prediction accuracy and rule number of fuzzy decision trees with different number of data per leaf node

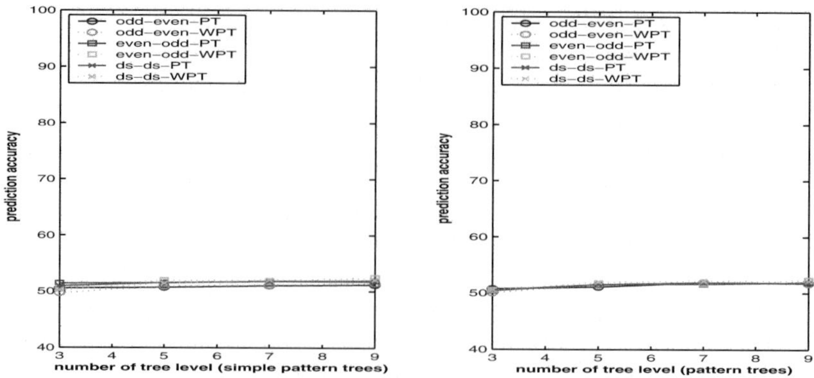

Fig. 17 Prediction accuracy of pattern trees and weighted pattern trees with different tree levels

In addition, the experiments show that (weighted) pattern trees tend to converge to a accuracy rate when the number of tree level becomes large. It has no trend of overfitting. This property is essential to ensure a stable, compact and effective fuzzy model for the problem at hand. In fact, (weighted) pattern trees with two or three level perform very well for all conducted experiments. That means, pattern trees which consist of maximal $2^3 = 8$ leaf nodes can perform well, in contrast to tens, or even hundred rules used in fuzzy decision trees. This provides a superb solution to achieve a highly effective as well as compact fuzzy model.

7.2 Approximate accuracy

Section 7.1 presents the prediction accuracy of trees in a very strict way. That is, if and only if a data is predicted exactly as its class, this prediction is counted as a correct one. In other words, there is no distinction between "close" errors and "gross" errors. In BT customer dataset, this distinction is necessary as it reflects how far the prediction is away from the actual class. It is much worse if a data of class 0 is mis-predicted to class 5 rather than to class 1. To resolve this problem, three accuracy estimations, namely *accuracy 1*, *accuracy 2*, and *accuracy 3* are employed to estimate prediction accuracy which has no tolerance (the same as the one used in Sect. 7.1), tolerance of adjacent mis-prediction, and tolerance of mis-prediction

Table 15 Highest prediction accuracy of fuzzy decision tree, pattern trees, and weighted pattern trees

	FDT	SimPT		PT	
		no weight	weight	no weight	weight
odd-even	50.62%	51.19%	51.45%	51.89%	51.92%
even-odd	50.47%	51.92%	52.47%	51.93%	52.37%
ds-ds	71.36%	51.82%	52.09%	51.88%	52.18%

within two closest neighbor classes in either direction, respectively. For example in the BT dataset, the mis-prediction of a class 0 data to class 2 is still counted as a correct prediction in the estimation of *accuracy 3*, although it is not counted in *accuracy 1* and *accuracy 2*.

Table 16 shows the highest prediction accuracy of fuzzy decision trees, (weighted) simple pattern trees and (weighted) pattern trees over odd-even combination of training-test sets. Both weighted and unweighted pattern trees can obtain higher prediction accuracy than fuzzy decision trees in estimation of accuracy 1 and 2. In estimation of accuracy 3, weighted pattern trees perform the best, and fuzzy decision trees outperform unweighted pattern trees. Generally, accuracy 2 and 3 are very consistent with accuracy 1. Pattern trees with a high value of accuracy 1 usually have high values of accuracy 2 and 3. This table also shows that both fuzzy decision trees and pattern trees can obtain over 80% prediction accuracy if the closest error can be tolerated.

7.3 Interpretation of pattern trees

Each pattern tree can be interpreted as a general rule. Considering building level 5 simple pattern trees using odd dataset, 7 simple pattern trees can be obtained, with each representing one output class. Fig. 18 shows the tree for class 0. The ellipses are the input parameters and the rectangle is the output class 0. Over each branch, i and Fi, $i = 0, \ldots$, are category values and fuzzy terms associated with each input parameter. All aggregators as shown in Table 2 are allowed to be used in pattern trees. For example, A_AND is algebraic AND, and $WA\_0.84$ is weighted average with weight vector $w = (0.84, 0.16)$.

Fig. 18 roughly indicates that one example combination yielding highly satisfied customers are: no call re-routing, fast fault reporting time, high technician competence, being well-informed through the repair process, and high satisfaction with company/product in general. Here, we say roughly, as we use different aggregations such as weighted average (WA), ordered weighted average (OWA), algebraic and (A_AND) etc. rather than simple AND.

These 7 pattern trees obtains an accuracy of 51.46%. In particular, the confusion table is shown in Table 17, where SA and SP are number of data for actual and predicted classes respectively.

Table 16 Highest prediction accuracy of fuzzy decision trees, pattern trees, and weighted pattern trees over odd-even training-test combination

	FDT	SimPT		PT	
		no weight	weight	no weight	weight
accuracy 1	50.62%	51.19%	51.45%	51.89%	51.92%
accuracy 2	84.02%	84.08%	84.68%	84.44%	84.82%
accuracy 3	92.13%	91.74%	92.70%	91.85%	92.29%

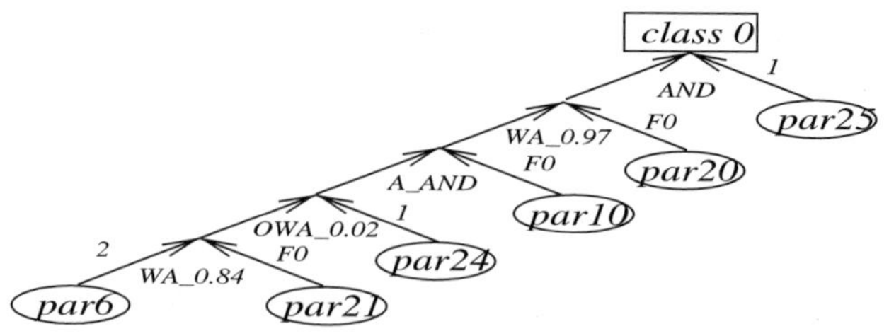

Fig. 18 Pattern tree for class 0 using *odd* dataset

7.4 Limitation

It is a little strange that no prediction is made to $c0$ for all test data in Table 17. It can be seen that nearly all data (884 out of 946 in fact) with class 0 are mis-classified to class 1. A first intuition is to raise the weight of pattern tree for class 0 . However, this does not work; the raise does not only lead to the data of class 0 to be classified correctly, but also lead to the majority of data of class 1 to be classified as class 0. Considering that there are 3630 data of class 1 and only 946 data of class 0 in even dataset, the raise of weight for class 0 tree would therefore cause more mis-classifications. This can be seen in Fig. 19, where the fired values of first 50 data points per class in *even* dataset over pattern trees constructed from *odd* dataset are shown. The real class line indicates the real classes of the data; for example, data numbered from 0 to 49 have class 0, and those from 50 to 99 have class 1.

The phenomena of no prediction on particular classes also occurs in fuzzy decision trees. Considering the highest accuracy of 50.62% which fuzzy decision trees can obtain over odd-even combination, the confusion table in this case is shown in Table 18, where no data is predicted to $c3$, $c4$ or $c5$.

An interesting experiment is carried out trying to improve the prediction accuracy for class 0 in Table 17. The data of classes 0 and 1 in odd dataset are selected as a new training dataset, and the data which are of classes 0 and 1 in even dataset and are classified as class 1 in Table 17 are selected as a new test dataset. Both fuzzy decision trees and pattern trees are applied to the new training data and tested over the new test data. Surprisingly, they obtain the same highest accuracy of 78.76%. Table 19 shows the confusion table, which only has one data

Table 17 Confusion table for pattern tree prediction using odd-even combination

		Prediction							
		c0	c1	c2	c3	c4	c5	c6	SA
	c0	0	884	51	0	3	2	6	946
	c1	0	3283	309	0	10	15	13	3630
	c2	0	1122	751	1	53	72	38	2037
Actual	c3	0	59	94	0	6	20	6	185
	c4	0	122	395	1	30	104	49	701
	c5	0	50	142	0	23	120	70	405
	c6	0	35	129	0	37	131	113	445
	SP	0	5555	1871	2	162	464	295	8349

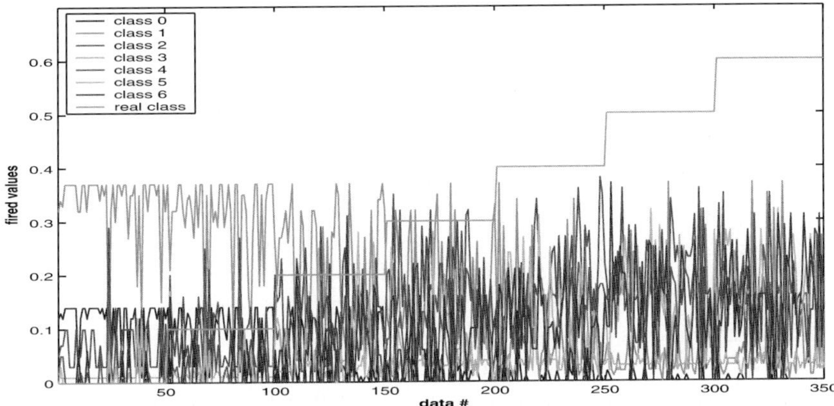

Fig. 19 Fired values of first 50 data points per class in *even* dataset over pattern trees constructed from *odd* dataset

Table 18 Confusion table for fuzzy decision tree prediction using odd-even combination

		Prediction							
		c0	c1	c2	c3	c4	c5	c6	SA
	c0	17	777	151	0	0	0	1	946
	c1	33	2980	605	0	0	0	12	3630
	c2	9	942	1035	0	0	0	51	2037
Actual	c3	0	49	121	0	0	0	15	185
	c4	1	84	509	0	0	0	107	701
	c5	1	26	226	0	0	0	152	405
	c6	0	31	220	0	0	0	194	445
	SP	61	4889	2867	0	0	0	532	8349

Table 19 Confusion table for prediction of both fuzzy decision trees and pattern trees using new training and test datasets

		Prediction		
		c0	c1	SA
	c0	0	884	884
Actual	c1	1	3282	3283
	SP	1	4166	4167

predicted as class 0 (and it is wrong actually). It can be concluded that the data of class 0 and class 1 can not be separated properly by either fuzzy decision trees or pattern trees.

8 Conclusions

This paper has proposed a type of tree termed *pattern trees* which make use of different aggregations. Like decision trees, pattern trees are an effective machine learning approach for classification applications. A novel pattern tree induction method has been proposed. The comparison to other classification methods including SBM, WSBM and fuzzy decision trees using UCI datasets shows that pattern trees can obtain higher accuracy rates in classifications. It also shows that pattern trees perform more consistently than fuzzy decision trees. They are capable of finding classifiers with good generality, while fuzzy decision trees can easily fall into the trap of over-fitting.

According to two different configurations, simple pattern trees and pattern trees have been distinguished. The former not only produce high prediction accuracy, but also preserve compact tree structures, while the latter can produce even better accuracy, but as a compromise produce more complex tree structures. Subject to the particular demands (comprehensibility or performance), simple pattern trees or pattern trees provide an effective methodology for real world applications.

Weighted pattern trees have been proposed to reflect the nature that different trees may have different confidences. The experiments on British Telecom (BT) customer satisfaction dataset show that weighted pattern trees can slightly outperform pattern trees, and both are slightly better than fuzzy decision trees in terms of prediction accuracy. Finally, a limitation of pattern trees as revealed via BT dataset analysis is discussed.

Although the proposed pattern tree induction method shows very promising results, it does not mean that it cannot be improved. Other searching techniques can be used to find alternative pattern trees. For instance, Tabu search[2], which is notable for the capability of escaping from local optima, can be implemented. In addition, the underlying relation between decision trees and pattern trees needs more research work. The conversion between decision trees and pattern trees is worth investigating. Finally, research on assignment of weights to pattern trees is necessary. The current version simply makes use of similarity measures as weights. More sophisticated assignment may be more suitable and can therefore lead to higher accuracy.

Acknowledgments The authors would like to thank Marcus Thint for his helpful discussions on the work presented here.

References

1. Chen, S. M., Lee, S. H. and Lee, C. H., "A new method for generating fuzzy rules from numerical data for handling classification problems," *Applied Artificial Intelligence*, Vol. 15, pp. 645–664, 2001.
2. Glover, Fred and Laguna, Manuel, "Tabu search," *Boston : Kluwer Academic Publishers*, 1997.
3. Huang, Z. H. and Gedeon, T. D., "Pattern trees," *IEEE International Conference on Fuzzy Systems*, pp. 1784–1791, 2006.
4. Huang, Z. H., Gedeon, T. D., and Nikravesh, M.,"Pattern trees," submitted to *Transaction on Fuzzy Systems*, 2006.
5. Huang, Z. H., Nikravesh, M., Azvine, B., and Gedeon, T. D., "Weighted pattern trees: a case study with customer satisfaction dataset," to appear in *World Congress of the International Fuzzy Systems Association (IFSA)*, 2007.

6. Kóczy, L. T., Vámos, T. and Biró, G., "Fuzzy signatures," *EUROFUSE-SIC*, pp. 210–217, 1999.
7. Mendis, B. S. U., Gedeon T. D., and Kóczy, L. T., "Investigation of aggregation in fuzzy signatures," *3rd International Conference on Computational Intelligence, Robotics and Autonomous Systems*, 2005.
8. Newman, D. J., Hettich, S., Blake, C. L. and Merz, C.J., UCI Repository of machine learning databases [http://www.ics.uci.edu/ mlearn/MLRepository.html], 1998.
9. Nikravesh, M., "Soft computing for perception-based decision processing and analysis: web-based BISC-DSS," *Studies in Fuzziness and Soft Computing*, Vol. 164, pp. 93–188, Springer Berlin/Heidelberg, 2005.
10. Quinlan, J. R., "Decision trees and decision making," *IEEE Transactions on Systems, Man, and Cybernetics*, Vol. 20, No. 2, pp. 339–346, 1994.
11. Raju, G. V. S. and Kisner, R. A., "Hierarchical fuzzy control," *International Journal Control*, Vol. 54, No. 5, pp. 1201–1216, 1991.
12. Rasmani, K. A. and Shen, Q., "Weighted linguistic modelling based on fuzzy subsethood values," *IEEE International Conference on Fuzzy Systems*, Vol. 1, pp. 714–719, 2003.
13. Schweizer, B. and Sklar, A., "Associative functions and abstract semigroups," *Publ. Math. Debrecen*, Vol. 10, pp. 69–81, 1963.
14. Wang, L. X. and Mendel, J. M., "Generating fuzzy rules by learning from examples," *IEEE Transactions on Systems, Man, and Cybernetics*, Vol. 22, No. 6, pp. 1414–1427, 1992.
15. Wang, L. X. and Mendel, J. M., "Fuzzy Basis functions, universal approximation, and orthogonal least-squares learning," *IEEE Transactions on Neural Networks*, Vol. 3, No. 5, pp. 807–814, 1992.
16. Wong, K. W, Gedeon, T. D. and Kóczy L. T., "Construction of fuzzy signature from data: an example of SARS pre-clinical diagnosis system," *IEEE International Conference on Fuzzy Systems*, Vol. 3, pp. 1649–1654.
17. Yager R. R., "On ordered weighted averaging aggregation operators in multicritera decison making," *IEEE Transactions on Systems, Man and Cybernetics*, Vol. 18, pp. 183–190, 1988.
18. Yuan, Y. and M. J. Shaw, "Induction of fuzzy decision trees," *Fuzzy Sets and Systems*, Vol. 69, No. 2, pp. 125–139, 1995.
19. Zadeh, L. A., "Fuzzy sets," *Information and Control*, Vol. 8, pp. 338–353, 1965.
20. Zadeh, L. A., "A fuzzy-algorithmic approach to the definition of complex or imprecise concepts," *International Journal of Man-Machine Studies*, Vol. 8, pp. 249–291, 1976.

Printing: Krips bv, Meppel, The Netherlands
Binding: Stürtz, Würzburg, Germany